A FRIENDLY APPROACH TO
FUNCTIONAL ANALYSIS

Essential Textbooks In Mathematics

ISSN: 2059-7657

Published:

A Sequential Introduction to Real Analysis
   by J M Speight (University of Leeds, UK)

A Friendly Approach to Functional Analysis
   by Amol Sasane (London School of Economics, UK)

Essential Textbooks in Mathematics

A FRIENDLY APPROACH TO FUNCTIONAL ANALYSIS

Amol Sasane
London School of Economics, UK

World Scientific

NEW JERSEY · LONDON · SINGAPORE · BEIJING · SHANGHAI · HONG KONG · TAIPEI · CHENNAI · TOKYO

Published by

World Scientific Publishing Europe Ltd.
57 Shelton Street, Covent Garden, London WC2H 9HE
Head office: 5 Toh Tuck Link, Singapore 596224
USA office: 27 Warren Street, Suite 401-402, Hackensack, NJ 07601

Library of Congress Cataloging-in-Publication Data
Names: Sasane, A. (Amol), 1976–
Title: A friendly approach to functional analysis / by Amol Sasane
   (London School of Economics, UK).
Description: New Jersey : World Scientific, 2017. | Series: Essential textbooks in mathematics |
   Includes bibliographical references and index.
Identifiers: LCCN 2017000443 | ISBN 9781786343338 (hardcover : alk. paper) |
   ISBN 9781786343345 (softcover : alk. paper)
Subjects: LCSH: Functional analysis--Textbooks.
Classification: LCC QA320 .S235 2017 | DDC 515/.7--dc23
LC record available at https://lccn.loc.gov/2017000443

British Library Cataloguing-in-Publication Data
A catalogue record for this book is available from the British Library.

Copyright © 2017 by World Scientific Publishing Europe Ltd.

All rights reserved. This book, or parts thereof, may not be reproduced in any form or by any means, electronic or mechanical, including photocopying, recording or any information storage and retrieval system now known or to be invented, without written permission from the Publisher.

For photocopying of material in this volume, please pay a copying fee through the Copyright Clearance Center, Inc., 222 Rosewood Drive, Danvers, MA 01923, USA. In this case permission to photocopy is not required from the publisher.

Printed in Singapore

To Sara

Preface

What is Functional Analysis?

Functional Analysis is Calculus in the setting of (typically infinite dimensional) vector spaces, just like

$\quad\quad\quad$ *Real* Analysis $\;=\;$ Calculus in $\mathbb{R}$, $\mathbb{R}^d$,
$\quad\quad\quad\quad\quad\quad\quad\quad\quad\quad\;$ real (finite dimensional) vector spaces,
$\quad\quad$ *Complex* Analysis $=$ Calculus in $\mathbb{C}$.

Why the adjective "functional"?

There is a historical reason behind this: the subject arose with considerations of problems in vector spaces of *functions*: for example in $C[a,b]$, the vector space of continuous functions on the interval $[a,b] \subset \mathbb{R}$, with pointwise operations.

What do we mean by Calculus?

It is the study of concepts involving "limiting processes", such as convergence of sequences, continuity of functions, differentiation, integration, etc.

Why study Functional Analysis?

Functional analysis plays an important role in the applied sciences as well as in mathematics itself. The impetus came from applications: problems related to ordinary and partial differential equations, calculus of variations, approximation theory, numerical analysis, integral equations, and so on.

$\quad$ In ordinary calculus, one dealt with limiting processes in finite-dimensional vector spaces ($\mathbb{R}$ or $\mathbb{R}^d$), but problems arising in the above applications required a calculus in spaces of functions (which are infinite-dimensional vector spaces). For instance, we mention the following optimisation problem as a motivating example.

Example 0.1. Imagine a copper mining company which is mining in a mountain, that has an estimated amount of Q tonnes of copper, over a period of T years. Suppose that $\mathbf{x}(t)$ denotes the total amount of copper removed up to time $t \in [0, T]$. Since the operation is over a large time period, we may assume that this $\mathbf{x}$ is a function living on the "continuous-time" interval $[0, T]$. The company has the freedom to choose its mining operation: $\mathbf{x}$ can be any nondecreasing function on $[0, T]$ such that

$\mathbf{x}(0) = 0$ (no copper removed initially) and

$\mathbf{x}(T) = Q$ (all copper removed at the end of the mining regime).

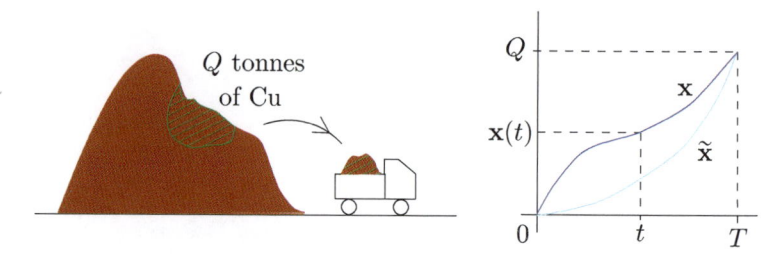

The cost of extracting copper per unit tonne at time t is
$$\mathbf{c}(t) = a\mathbf{x}(t) + b\mathbf{x}'(t), \quad t \in [0, T].$$
Here a, b are given positive constants. The expression is reasonable, since the term $a \cdot \mathbf{x}(t)$ accounts for the fact that when more and more copper is taken out, it becomes more and more difficult to find the leftover copper, while the term $b \cdot \mathbf{x}'(t)$ accounts for the fact that if the rate of removal of copper is high, then the costs increase (for example due to machine replacement costs). We don't need to follow the exact reasoning behind this formula; this is just a model that the optimiser has been given. If the company decides on a particular mining operation $\mathbf{x} : [0, T] \to \mathbb{R}$, then the total cost $f(\mathbf{x}) \in \mathbb{R}$ over the whole mining period $[0, T]$ is given by
$$f(\mathbf{x}) = \int_0^T \big(a\mathbf{x}(t) + b\mathbf{x}'(t)\big)\mathbf{x}'(t)dt.$$
Indeed, $\mathbf{x}'(t)dt$ is the incremental amount of copper removed at time t, and if we multiply this by $\mathbf{c}(t)$, we get the incremental cost at time t. The total cost should be the sum of all these incremental costs over the interval $[0, T]$, and so we obtain the integral expression for $f(\mathbf{x})$ given above.

Hence the mining company is faced with the following natural problem: Which mining operation $\mathbf{x}$ incurs the least cost? In other words, minimise $f : S \to \mathbb{R}$, where S denotes the set of all (continuously differentiable) functions $\mathbf{x} : [0, T] \to \mathbb{R}$ such that $\mathbf{x}(0) = 0$ and $\mathbf{x}(T) = Q$. ◇

Exercise 0.1.
In Example 0.1, find $f(\mathbf{x}_1)$, $f(\mathbf{x}_2)$ where $\mathbf{x}_1(t) = Q\dfrac{t}{T}$, $\mathbf{x}_2(t) = Q\left(\dfrac{t}{T}\right)^2$, $t \in [0, T]$.
Which is smaller among $f(\mathbf{x}_1)$ and $f(\mathbf{x}_2)$?
Which mining operation among $\mathbf{x}_1$ and $\mathbf{x}_2$ will be preferred?

How do we solve optimisation problems in function spaces?

Suppose that instead of an optimisation problem in a function space, we consider a much simpler problem:

Minimise $f : \mathbb{R} \to \mathbb{R}$, where $f(x) := x^2 - 2x + 5$, $x \in \mathbb{R}$.

Then we know how to solve this. Indeed, from ordinary calculus, we know the following two facts.

Fact 1. If $x_* \in \mathbb{R}$ is a minimiser of f, then $f'(x_*) = 0$.
Fact 2. If $f''(x) \geq 0$ for all $x \in \mathbb{R}$ and $f'(x_*) = 0$,
then x_* is a minimiser of f.

Fact 1 allows us to narrow our choice of possible minimisers, since we can calculate $f'(x) = 2x - 2$, and note that $2x - 2 = 0$ if and only if $x = 1$. So if at all there is a minimiser, then it must be $x_* = 1$. On the other hand, $f''(x) = 2 > 0$ for all $x \in \mathbb{R}$, and so Fact 2 confirms that $x_* = 1$ is a minimiser. Thus using these two facts, we have completely solved the optimisation problem: we know that $x_* = 1$ is the only minimiser of f.

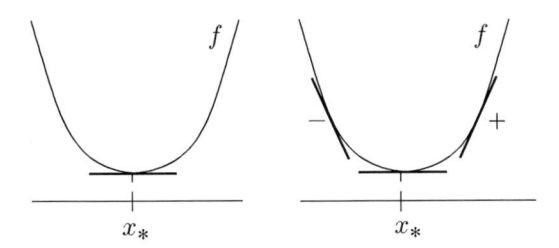

Fig. 0.1 Fact 1 says that at a minimiser x_*, the tangent to the graph of f must be horizontal. Fact 2 says that if f' is increasing ($f'' \geq 0$), and if $f'(x_*) = 0$, then for points x to the left of x_*, f' must be nonpositive, and so f must be decreasing there, and similarly f must be increasing to the right of x_*. This has the consequence that x_* is a minimiser of f.

On the other hand our optimisation problem from Example 0.1 does not fit into the usual framework of calculus, since there we have a real-valued function that lives on a subset of an *infinite* dimensional function space, namely continuously differentiable functions on the interval $[0, T]$.

Thus the need arises for developing calculus in such infinite dimensional vector spaces. Although we have only considered one example, problems requiring calculus in infinite dimensional vector spaces arise from many applications, and from various disciplines such as economics, engineering, physics, and so on. Mathematicians observed that different problems from varied fields often have related features and properties. This fact was used for an effective unifying approach towards such problems, the unification being obtained by the omission of unessential details. Hence the advantage of an *abstract* approach is that it concentrates on the essential facts, so that these facts become clearly visible and one's attention is not disturbed by unimportant details. Moreover, by developing a box of tools in the abstract framework, one is equipped to solve many different problems (that are really the same problem in disguise!) in one go, all at once.

In the abstract approach, one usually starts from a set of elements satisfying certain axioms. The theory then consists of logical consequences which result from the axioms and are derived as theorems once and for all. These general theorems can then later be applied to various concrete special sets satisfying the axioms. The reader is presumably already familiar with such a programme for example in the context of elementary undergraduate linear algebra.

What will we learn in this book?

This aim of this book is to familiarise the reader with the basic concepts, principles and methods of functional analysis and its applications.

We will develop such an abstract scheme for doing calculus in function spaces and other infinite-dimensional spaces, and this is what this book is about. Having done this, we will be equipped with a box of tools for solving many problems, and in particular, we will return to the optimal mining operation problem again and solve it.

The book contains many exercises, which form an integral part of the text, as some results relegated to the exercises are used in proving theorems. The solutions to all the exercises appear at the end of the book.

In this book we have described a few topics from functional analysis which find widespread use, and by no means is the choice of topics "complete". However, equipped with this basic knowledge of the elementary facts in functional analysis, the student can undertake a serious study of a more advanced treatise on the subject, and the bibliography lists a few textbooks which might be suitable for further reading.

Who is the book for?

The book is meant for final year undergraduate students. No prerequisites beyond knowledge of linear algebra and ordinary calculus (with ϵ-δ arguments) are needed to read this book. Knowledge of the Lebesgue integration theory or topology is not assumed. Nevertheless, we will include illustrative examples that involve Lebesgue integrals, with the naive understanding that the Lebesgue integral is simply a generalisation of the usual Riemann integral, which is rather more amenable to limiting processes. We will give a short introduction to the Lebesgue integral, entirely sufficient for treating our subsequent discussion, when we encounter the first such example. Parts of the book marked with an asterisk (∗) may be more challenging as compared to the average level of difficulty of the book, and may be skipped, skim-read or studied, depending on the goals of the student. Thus the book should be accessible to a wide spectrum of students, and may also serve to bridge the gap between linear algebra and advanced functional analysis.

Acknowledgements

I am indebted to the late Prof. Erik Thomas (University of Groningen), for many useful comments and suggestions on the lecture notes used at the London School of Economics, from which this book has grown. I would like to thank Dr. Sara Maad Sasane for going through parts of the manuscript, pointing out typos and mistakes, and offering insightful suggestions. Thanks are also due to Prof. Raymond Mortini (University of Metz) and to MSc student Edvard Johansson (Lund University) for several useful comments. Finally, it is a pleasure to thank the staff at World Scientific; in particular, Laurent Chaminade for his help, and Eng Huay Chionh for overseeing the production of the book, and for valuable copy editorial comments which improved the quality of the book. The book relies on many sources, which are listed in the bibliography. This also applies to the exercises.

Amol Sasane
Lund, 2016

Contents

Preface   vii

1. Normed and Banach spaces   1
   - 1.1 Vector spaces 3
   - 1.2 Normed spaces 7
   - 1.3 Topology of normed spaces 17
   - 1.4 Sequences in a normed space; Banach spaces 24
   - 1.5 Compact sets 44

2. Continuous and linear maps   53
   - 2.1 Linear transformations 54
   - 2.2 Continuous maps 58
   - 2.3 The normed space $CL(X,Y)$ 67
   - 2.4 Composition of continuous linear transformations 82
   - 2.5 (∗) Open Mapping Theorem 92
   - 2.6 Spectral Theory 97
   - 2.7 (∗) Dual space and the Hahn-Banach Theorem 104

3. Differentiation   117
   - 3.1 Definition of the derivative 118
   - 3.2 Fundamental theorems of optimisation 125
   - 3.3 Euler-Lagrange equation 134
   - 3.4 An excursion in Classical Mechanics 145

4. Geometry of inner product spaces   155
   - 4.1 Inner product spaces 156

4.2	Orthogonality	165
4.3	Best approximation	174
4.4	Generalised Fourier series	183
4.5	Riesz Representation Theorem	189
4.6	Adjoints of bounded operators	190
4.7	An excursion in Quantum Mechanics	200

5. Compact operators 209

5.1	Compact operators	210
5.2	The set $K(X, Y)$ of all compact operators	211
5.3	Approximation of compact operators	217
5.4	(∗) Spectral Theorem for Compact Operators	222

6. A glimpse of distribution theory 227

6.1	Test functions, distributions, and examples	230
6.2	Derivatives in the distributional sense	237
6.3	Weak solutions	243
6.4	Multiplication by C^∞ functions	248
6.5	Fourier transform of (tempered) distributions	253

Solutions   259

The Lebesgue integral   359

Bibliography   373

Index   375

A FRIENDLY APPROACH TO
FUNCTIONAL ANALYSIS

Chapter 1

Normed and Banach spaces

As we had discussed in the introduction, we wish to do calculus in vector spaces (such as $C[a,b]$, whose elements are functions). In order to talk about the concepts from calculus such as differentiability, we need a notion of closeness between points of a vector space.

Recall for example, that a real sequence $(a_n)_{n \in \mathbb{N}}$ is said to *converge with limit* $L \in \mathbb{R}$ if for every $\epsilon > 0$, there exists an $N \in \mathbb{N}$ such that whenever $n > N$, $|a_n - L| < \epsilon$. In other words, the sequence converges to L if no matter what *distance* $\epsilon > 0$ is given, one can guarantee that all the terms of the sequence beyond a certain index N are at a *distance* of at most ϵ away from L (this is the inequality $|a_n - L| < \epsilon$). So we notice that in this notion of "convergence of a sequence", indeed the notion of distance played a crucial role. After all, we want to say that the terms of the sequence get "close" to the limit, and to measure closeness, we use the distance between points of $\mathbb{R}$. A similar thing happens with continuity and differentiability. Recall that a function $f : \mathbb{R} \to \mathbb{R}$ is said to be *continuous at* $c \in \mathbb{R}$ if for every $\epsilon > 0$, there exists a $\delta > 0$ such that whenever $|x - c| < \delta$, $|f(x) - f(c)| < \epsilon$. Roughly, given any distance ϵ, I can find a distance δ such that whenever I choose an x not farther than a distance δ from c, I am guaranteed that $f(x)$ is not farther than a distance of ϵ from $f(c)$. Again notice the key role played by the distance in this definition. The distance between points $x, y \in \mathbb{R}$ is taken as $|x - y|$, where $|\cdot| : \mathbb{R} \to [0, \infty)$ is the absolute value function, given by

$$|x| = \begin{cases} x \text{ if } x \geqslant 0, \\ -x \text{ if } x < 0. \end{cases}$$

If we imagine the real numbers depicted on a "number line", then $|x - y|$ is the length of line segment joining x, y visualised on the number line. See the following picture.

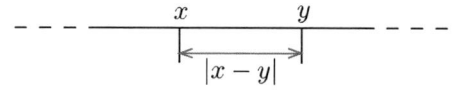

$$\begin{array}{c} x y \\ \text{---}\!\!\mid\!\!\text{---------}\!\!\mid\!\!\text{---} \\ \underset{|x-y|}{\longleftrightarrow} \end{array}$$

But now if one wants to also do calculus in a vector space X (for example $C[a,b]$), there is so far no ready-made available notion of distance between vectors. One way of creating a distance in a vector space is to equip it with a "norm" $\|\cdot\|$, which is the analogue of absolute value $|\cdot|$ in the vector space $\mathbb{R}$. The distance function is then created by taking the norm $\|x-y\|$ of the difference between vectors $x, y \in X$, just like in $\mathbb{R}$ the Euclidean distance between $x, y \in \mathbb{R}$ was taken as $|x-y|$.

Having done this, we have the familiar setting of calculus, and we can talk about notions like the derivative of a function living on a normed space. (Later on, in Chapter 3, we will then also have analogues of the two facts from ordinary calculus relevant to optimisation, namely the vanishing of the derivative for minimisers, and the sufficiency of this condition for minimisation when the function is convex.) Thus the outline of this chapter is as follows.

First of all, we will learn the notion of a "normed space", that is a vector space equipped with a "norm", enabling one to measure distances between vectors in the vector space. This makes it possible to talk about concepts from calculus, and in particular the notion of differentiability of functions between normed spaces, as we shall see later on. Next, we will see lots of examples of normed spaces: we will see that[1]

$$(\mathbb{R}, |\cdot|),$$
$$(\mathbb{R}^d, \|\cdot\|_2),$$
$$(\ell^p, \|\cdot\|_p),$$
$$(C[a,b], \|\cdot\|_\infty),$$
$$(L^2[a,b], \|\cdot\|_2),$$
$$\cdots \text{ and many more } \cdots$$

are all normed spaces, enabling us to do Calculus in each case.

Finally, we will introduce *Banach spaces*, which are special types of normed spaces, namely ones in which "Cauchy sequences converge". We will also motivate this, and see why Banach spaces are nicer than having merely a normed space.

We begin by recalling the notion of vector space.

---

[1] All of this notation will be explained in the course of this chapter.

1.1 Vector spaces

Roughly speaking it is a set of elements, called "vectors". Any two vectors can be "added", resulting in a new vector, and any vector can be multiplied by an element from $\mathbb{R}$ (or $\mathbb{C}$, depending on whether we consider a *real* or *complex* vector space), so as to give a new vector. The precise definition is given below.

Definition 1.1. (Vector space) Let $\mathbb{K} = \mathbb{R}$ or $\mathbb{C}$ (or more generally[2] a *field*). A *vector space* over $\mathbb{K}$, is a set X together with two functions, $+: X \times X \to X$, called *vector addition*, and $\cdot : \mathbb{K} \times X \to X$, called *scalar multiplication* that satisfy the following:

(V1) For all $x_1, x_2, x_3 \in X$, $x_1 + (x_2 + x_3) = (x_1 + x_2) + x_3$.
(V2) There exists an element, denoted by $\mathbf{0}$ (called the *zero vector*) such that for all $x \in X$, $x + \mathbf{0} = x = \mathbf{0} + x$.
(V3) For every $x \in X$, there exists an element, denoted by $-x$, such that $x + (-x) = (-x) + x = \mathbf{0}$.
(V4) For all x_1, x_2 in X, $x_1 + x_2 = x_2 + x_1$.
(V5) For all $x \in X$, $1 \cdot x = x$.
(V6) For all $x \in X$ and all $\alpha, \beta \in \mathbb{K}$, $\alpha \cdot (\beta \cdot x) = (\alpha \beta) \cdot x$.
(V7) For all $x \in X$ and all $\alpha, \beta \in \mathbb{K}$, $(\alpha + \beta) \cdot x = \alpha \cdot x + \beta \cdot x$.
(V8) For all $x_1, x_2 \in X$ and all $\alpha \in \mathbb{K}$, $\alpha \cdot (x_1 + x_2) = \alpha \cdot x_1 + \alpha \cdot x_2$.

Example 1.1. ($\mathbb{R}$). $\mathbb{R}$ is a vector space over $\mathbb{R}$, with vector addition being the usual addition of real numbers, and scalar multiplication being the usual multiplication of real numbers. ($\mathbb{R}$ is also a vector space over the field $\mathbb{Q}$ of rational numbers, but we will always consider the *real* vector space $\mathbb{R}$ unless stated otherwise.) ◇

Example 1.2. ($\mathbb{R}^d$). $\mathbb{R}^d = \mathbb{R} \times \cdots \mathbb{R}$ (d times) is the set of all ordered d-tuples $(x_1, \cdots, x_d)$ of real numbers $x_1, \cdots, x_d$. Then $\mathbb{R}^d$ is a vector space over $\mathbb{R}$, with addition and scalar multiplication defined "component-wise":
$$(x_1, \cdots, x_d) + (y_1, \cdots, y_d) := (x_1 + y_1, \cdots, x_d + y_d),$$
$$\alpha \cdot (x_1, \cdots, x_d) := (\alpha x_1, \cdots, \alpha x_d),$$
for $(x_1, \cdots, x_d), (y_1, \cdots, y_d) \in \mathbb{R}^d$ and $\alpha \in \mathbb{R}$. ◇

Example 1.3. ($C[a,b]$). Let $a, b \in \mathbb{R}$ and $a < b$. Consider the vector space consisting of all continuous functions $\mathbf{x} : [a,b] \to \mathbb{K}$, with addition and scalar multiplication defined in a "pointwise" manner as follows.

---
[2] Unless stated otherwise, the underlying field is always assumed to be $\mathbb{R}$ or $\mathbb{C}$.

If $\mathbf{x}_1, \mathbf{x}_2 \in C[a,b]$, then $\mathbf{x}_1 + \mathbf{x}_2 \in C[a,b]$ is the function given by
$$(\mathbf{x}_1 + \mathbf{x}_2)(t) := \mathbf{x}_1(t) + \mathbf{x}_2(t), \quad t \in [a,b]. \tag{1.1}$$
If $\alpha \in \mathbb{K}$ and $\mathbf{x} \in C[a,b]$, then $\alpha \cdot \mathbf{x} \in C[a,b]$ is the function given by
$$(\alpha \cdot \mathbf{x})(t) := \alpha \mathbf{x}(t), \quad t \in [a,b]. \tag{1.2}$$
It can be checked that the vector space axioms (V1)-(V8) are satisfied. $C[a,b]$ is referred to as a 'function space', since each vector in $C[a,b]$ is a function (from $[a,b]$ to $\mathbb{K}$). The zero vector in $\mathbb{C}[a,b]$ is the zero function $\mathbf{0}$, given by $\mathbf{0}(t) = 0$ for all $t \in [a,b]$. $\diamond$

Example 1.4. ($C^1[a,b]$). Let $C^1[a,b]$ denote the space of continuously differentiable functions on $[a,b]$:
$$C^1[a,b] = \{\mathbf{x} : [a,b] \to \mathbb{R} \mid \mathbf{x} \text{ is continuously differentiable on } [a,b]\}.$$
(Recall that a function $\mathbf{x} : [a,b] \to \mathbb{R}$ is *continuously differentiable* if for all $t \in [a,b]$, the derivative of $\mathbf{x}$ at t, namely $\mathbf{x}'(t)$, exists, and the map $t \mapsto \mathbf{x}'(t) : [a,b] \to \mathbb{R}$ is a continuous function.) We note that
$$C^1[a,b] \subset C[a,b],$$
because whenever a function $\mathbf{x} : [a,b] \to \mathbb{R}$ is differentiable at a point t in $[a,b]$, then $\mathbf{x}$ is continuous at t. In fact, $C^1[a,b]$ is a *subspace* of $C[a,b]$ because it is closed under addition and scalar multiplication, and is nonempty:

(S1) For all $\mathbf{x}_1, \mathbf{x}_2 \in C^1[a,b]$, $\mathbf{x}_1 + \mathbf{x}_2 \in C^1[a,b]$.
(S2) For all $\alpha \in \mathbb{R}$, $\mathbf{x} \in C^1[a,b]$, $\alpha \cdot \mathbf{x} \in C^1[a,b]$.
(S3) $\mathbf{0} \in C^1[a,b]$.

Thus $C^1[a,b]$ is a vector space with the induced operations from $C[a,b]$, namely the same pointwise operations as defined in (1.1) and (1.2). $\diamond$

Example 1.5. (Sequence spaces). For any real p such that $1 \leq p < \infty$,
$$\ell^p := \Big\{ (a_n)_{n \in \mathbb{N}} : \sum_{n=1}^{\infty} |a_n|^p < \infty \Big\}.$$
(Here we take the sequences $(a_n)_{n \in \mathbb{N}}$ with values in $\mathbb{K}$.) We define vector addition and scalar multiplication termwise:
$$(a_n)_{n \in \mathbb{N}} + (b_n)_{n \in \mathbb{N}} := (a_n + b_n)_{n \in \mathbb{N}},$$
$$\alpha \cdot (a_n)_{n \in \mathbb{N}} := (\alpha a_n)_{n \in \mathbb{N}},$$

for elements $(a_n)_{n\in\mathbb{N}}, (b_n)_{n\in\mathbb{N}} \in \ell^p$ and $\alpha \in \mathbb{K}$. It is not yet clear whether the sum of two elements in ℓ^p, defined in the manner above, delivers an element in ℓ^p. That this is indeed true is shown by the elementary chain of inequalities below:

$$\begin{aligned} |a+b|^p &\leq (|a|+|b|)^p \\ &\leq \left(\max\{|a|,|b|\} + \max\{|a|,|b|\}\right)^p \\ &= 2^p \left(\max\{|a|,|b|\}\right)^p \\ &\leq 2^p \left(|a|^p + |b|^p\right). \end{aligned}$$

We can use these inequalities termwise, and the Comparison Test for convergence of real series, to conclude that $(a_n)_{n\in\mathbb{N}} + (b_n)_{n\in\mathbb{N}} \in \ell^p$ whenever $(a_n)_{n\in\mathbb{N}}, (b_n)_{n\in\mathbb{N}} \in \ell^p$.

By ℓ^∞ we denote the vector space of all bounded sequences with values in $\mathbb{K}$, once again with termwise operations. It is easy to see that the sum of two elements from ℓ^∞ is again an element of ℓ^∞.

It is clear that $\ell^p \subset \ell^\infty$ for all $p \in [1, \infty)$: if $(a_n)_{n\in\mathbb{N}} \in \ell^p$, then

$$\sum_{n=1}^\infty |a_n|^p < \infty,$$

and so $\lim_{n\to\infty} |a_n|^p = 0$. In particular, $(a_n)_{n\in\mathbb{N}}$ is a bounded sequence.

So all the ℓ^p spaces with a finite p are subspaces of ℓ^∞. Some other important subspaces are:

$$\begin{array}{ccccccc} c_{00} & \subset & c_0 & \subset & c & \subset & \ell^\infty \\ \| & & \| & & \| & & \\ \left\{\begin{array}{c}\text{all sequences}\\ \text{that are}\\ \text{eventually 0}\end{array}\right\} & & \left\{\begin{array}{c}\text{all sequences}\\ \text{that converge}\\ \text{to 0}\end{array}\right\} & & \left\{\begin{array}{c}\text{all sequences}\\ \text{that are}\\ \text{convergent}\end{array}\right\} & & \end{array}$$

$\diamond$

Example 1.6. ($L^p[a,b]$). For $p \in [1, \infty)$, define

$$L^p[a,b] \text{ `` }= \text{''} \left\{ \mathbf{x} : [a,b] \to \mathbb{R} : \underbrace{\int_a^b |\mathbf{x}(t)|^p dt}_{\text{Lebesgue}} < \infty \right\},$$

where the integral is the "Lebesgue integral" rather than the Riemann integral.

What do we need to know about Lebesgue integrals? Firstly, every Riemann integrable function $\mathbf{x}$ on an interval $[a,b]$ is also Lebesgue integrable on $[a,b]$, and moreover, the Lebesgue integral then coincides with the usual

Riemann integral. However, the class of Lebesgue integrable functions is much larger than the class of continuous functions. For instance, it can be shown that the function

$$\mathbf{x}(t) = \begin{cases} 0 & \text{if } t \in [0,1]\setminus\mathbb{Q} \\ 1 & \text{if } t \in [0,1] \cap \mathbb{Q} \end{cases} \quad (1.3)$$

is Lebesgue integrable, but not Riemann integrable on $[0,1]$. For computation aspects, one can get away without having to go into technical details about Lebesgue integration. (In an appendix called "The Lebesgue Integral" on page 359, we have outlined the key definitions and a few relevant results on the Lebesgue Integral, which the reader might wish to read if so desired, in order to get a better feeling for the L^p spaces.)

We also note that in the above definition of $L^p[a,b]$, we have put quotes around the equality sign. What is that supposed to mean? Strictly speaking, each element of $L^p[a,b]$ is not a function $\mathbf{x}$, but rather an equivalence class $[\mathbf{x}]$ of functions, where

$$[\mathbf{x}] = \left\{ \mathbf{y} : [a,b] \to \mathbb{R} \mid \underset{\text{Lebesgue}}{\int_a^b} |\mathbf{x}(t) - \mathbf{y}(t)|^p dt = 0 \right\}.$$

(The reason for wanting $L^p[a,b]$ to be this set of equivalence classes $[\mathbf{x}]$, rather than the functions $\mathbf{x}$ itself, will become clear when we discuss "norms". It is tied to demanding that the only vector in $L^p[a,b]$ with 0 norm must be the zero vector.) Of course, if $\mathbf{x}, \mathbf{y} \in C[a,b]$, then

$$\underset{\text{Lebesgue}}{\int_a^b} |\mathbf{x}(t) - \mathbf{y}(t)|^p dt = 0 \;\Rightarrow\; \underset{\text{Riemann}}{\int_a^b} |\mathbf{x}(t) - \mathbf{y}(t)|^p dt = 0,$$

and thanks to the continuity of $\mathbf{x}, \mathbf{y}$, we could then conclude that

$$\mathbf{x}(t) = \mathbf{y}(t) \text{ for all } t \in [a,b].$$

But it may happen for functions $\mathbf{x}, \mathbf{y} \in L^p[a,b]$ that they are not equal as functions, but nevertheless

$$\underset{\text{Lebesgue}}{\int_a^b} |\mathbf{x}(t) - \mathbf{y}(t)|^p dt = 0.$$

In fact if $\mathbf{x}$ is the function given by (1.3), and $\mathbf{y} = \mathbf{0}$, then it turns out that

$$\underset{\text{Lebesgue}}{\int_0^1} |\mathbf{x}(t) - \mathbf{y}(t)|^p dt = 0,$$

but clearly $\mathbf{x} \neq \mathbf{y}$! Note however that in this example, "almost everywhere", that is, for "almost all" $t \in [0,1]$, (only the rational t are excluded!) we do have $\mathbf{x}(t) = \mathbf{y}(t)$. These phrases can be made precise using the theory of Lebesgue integral, in that it turns out that if

$$\int_a^b |\mathbf{x}(t) - \mathbf{y}(t)|^p dt = 0$$
_{Lebesgue}

then $\mathbf{x}(t) = \mathbf{y}(t)$ for all $t \in [a,b] \setminus N$, where N has "Lebesgue measure" 0. We won't go into this, but we'll simply bear in mind that for

$$\mathbf{x} \stackrel{L^p}{=} \mathbf{y} \quad \Leftrightarrow \quad \int_a^b |\mathbf{x}(t) - \mathbf{y}(t)|^p dt = 0.$$

So we view elements of $L^p[a,b]$ through "fuzzy glasses", and treat two functions as being identical whenever the integral above is 0.

Analogous to the space ℓ^∞, one can also introduce the space $L^\infty[a,b]$:

$$L^\infty[a,b] = \left\{ \mathbf{x} : [a,b] \to \mathbb{R} \;\middle|\; \begin{array}{l} \mathbf{x} \text{ is Lebesgue measurable and } \exists M > 0 \text{ such} \\ \text{that for almost all } t \in [a,b], |\mathbf{x}(t)| \leq M \end{array} \right\}.$$

Since this example relies on the notion of Lebesgue measure, we won't discuss this any further now. ◇

Exercise 1.1. True or false? The set $V = (0, \infty)$ (positive reals) is a vector space with addition and scalar multiplication given by $x + y = xy$ and $\alpha \cdot x = x^\alpha$ for all positive x, y, and for all $\alpha \in \mathbb{R}$.

Exercise 1.2. ($C[0,1]$ is not finite dimensional.) Show that $C[0,1]$ with the usual pointwise operations is not a finite dimensional vector space.
Hint: One can prove this by contradiction. Let $C[0,1]$ be a finite dimensional vector space with dimension d, say. First show that the set $B = \{t, t^2, \cdots, t^d\}$ is linearly independent. Then B is a basis for $C[0,1]$, and so the constant function **1** should be a linear combination of the functions from B. Derive a contradiction.

Exercise 1.3. Let $S := \{\mathbf{x} \in C^1[a,b] : \mathbf{x}(a) = y_a \text{ and } \mathbf{x}(b) = y_b\}$, where $y_a, y_b \in \mathbb{R}$. Prove that S is a subspace of $C^1[a,b]$ if and only if $y_a = y_b = 0$. (So we see that S is a vector space with pointwise operations if and only if $y_a = y_b = 0$.)

1.2 Normed spaces

We would like to develop calculus in the setting of vector spaces (for example, in function spaces like $C[a,b]$). Underlying all the fundamental concepts in ordinary calculus, is the notion of closeness between points. So in order to generalise the notions from ordinary calculus (where we work with real numbers, and where the absolute value is used to measure distances), to the situation of vector spaces, we need a notion of distance between elements of the vector space. This is done by introducing an additional structure on a vector space, namely, a "norm", which is a real-valued function $\|\cdot\|$ defined on the vector space, and the norm plays a role analogous to the one played by the absolute value in $\mathbb{R}$. Once we have a norm on a vector space X (in other words a "normed space"), then the distance between $x, y \in X$ will be taken as $\|x - y\|$.

Definition 1.2. (Norm; normed space). Let X be a vector space over $\mathbb{K}$ ($\mathbb{R}$ or $\mathbb{C}$). A *norm* on X is a function $\|\cdot\| : X \to [0, +\infty)$ such that:

(N1) (Positive definiteness).
   For all $x \in X$, $\|x\| \geq 0$. If $x \in X$ and $\|x\| = 0$, then $x = \mathbf{0}$.
(N2) For all $\alpha \in \mathbb{K}$ ($\mathbb{R}$ or $\mathbb{C}$) and for all $x \in X$, $\|\alpha x\| = |\alpha| \|x\|$.
(N3) (Triangle inequality) For all $x, y \in X$, $\|x + y\| \leq \|x\| + \|y\|$.

A *normed space* is a vector space X equipped with a norm.

Distance in a normed space. Just like in $\mathbb{R}$, with the absolute value, and where the distance between $x, y \in \mathbb{R}$ is $|x - y|$, now in a normed space $(X, \|\cdot\|)$, we have for $x, y \in X$, that the number $\|x - y\|$ is taken as the distance between $x, y \in X$. Thus $\|x\| = \|x - \mathbf{0}\|$ is the distance of x from the zero vector $\mathbf{0}$ in X.

Remark 1.1. (Metric spaces). A *metric space* is a set X together with a function $d : X \times X \to \mathbb{R}$ satisfying the following properties:

(D1) (Positive definiteness)
   For all $x, y \in X$, $d(x, y) \geq 0$. For all $x \in X$, $d(x, x) = 0$.
   If $x, y \in X$ are such that $d(x, y) = 0$, then $x = y$.
(D2) (Symmetry) For all $x, y \in X$, $d(x, y) = d(y, x)$.
(D3) (Triangle inequality) For all $x, y, z \in X$, $d(x, y) + d(y, z) \geq d(x, z)$.

The reader familiar with "metric spaces" may notice that in a normed space $(X, \|\cdot\|)$, if we define $d : X \times X \to \mathbb{R}$ by $d(x, y) = \|x - y\|$ for $x, y \in X$, then it is easily seen that d satisfies (D1)-(D3), and so (X, d) is a metric space with the metric/distance function d. This distance d is referred to as the *induced distance* in the normed space $(X, \|\cdot\|)$. Then $\|x\| = \|x - \mathbf{0}\| = d(x, \mathbf{0})$, and so the norm of a vector x in the normed space $(X, \|\cdot\|)$ is the induced distance of x to the zero vector.

We now give a few examples of normed spaces, by reconsidering the vector space examples from the previous section, and equipping each of them with norms.

Example 1.7. $(\mathbb{R}, |\cdot|)$. $\mathbb{R}$ is a vector space over $\mathbb{R}$. Define $\|\cdot\| : \mathbb{R} \to \mathbb{R}$ by $\|x\| = |x|$, for $x \in \mathbb{R}$. Then $(\mathbb{R}, |\cdot|)$ is a normed space. (No surprise, since wanting to generalise the situation from ordinary calculus in $\mathbb{R}$ to the case of vector spaces, $|\cdot|$ is what motivated the definition of the norm $\|\cdot\|$!) $\diamond$

Example 1.8. $(\mathbb{R}^d, \|\cdot\|_p)$. $\mathbb{R}^d$ is a vector space over $\mathbb{R}$. Let us define the *Euclidean norm* $\|\cdot\|_2$ by

$$\|\mathbf{x}\|_2 = \sqrt{x_1^2 + \cdots + x_d^2}, \quad \mathbf{x} = (x_1, \cdots, x_d) \in \mathbb{R}^d.$$

Then $\mathbb{R}^d$ is a normed space (see Exercise 1.8.(1) on page 16).

(The motivation behind calling (N3) the triangle inequality is now evident. Indeed, for triangles in Euclidean Geometry of the plane, we know that the sum of the lengths of two sides of a triangle is at least as much as the length of the third side. If we now imagine the points $\mathbf{0}, -\mathbf{x}, \mathbf{y} \in \mathbb{R}^2$ as the three vertices of a triangle, then this is what (N3) says for the $\|\cdot\|_2$ norm; see the following picture.)

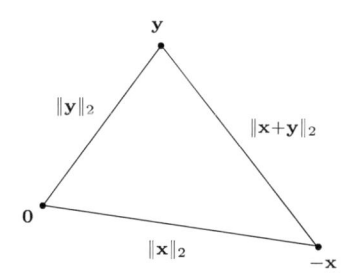

$\|\cdot\|_2$ is not the only norm[3] that can be defined on $\mathbb{R}^d$. For example,

$$\|\mathbf{x}\|_1 = |x_1| + \cdots + |x_d|, \quad \text{and} \quad \|\mathbf{x}\|_\infty = \max\{|x_1|, \cdots, |x_d|\},$$

are also examples[4]

Note that $(\mathbb{R}^d, \|\cdot\|_2)$, $(\mathbb{R}^d, \|\cdot\|_1)$ and $(\mathbb{R}^d, \|\cdot\|_\infty)$ are all *different* normed spaces. This illustrates the important fact that from a given vector space, we can obtain various normed spaces by choosing different norms. What norm is considered depends on the particular application at hand. We illustrate this in the next paragraph.

---
[3] However, unless otherwise stated, we will always use the Euclidean norm on $\mathbb{R}^d$.
[4] Just like the $\|\cdot\|_1$ and $\|\cdot\|_2$-norms, more generally, for any $p \in [1, \infty)$, one can define the $\|\cdot\|_p$-norm on $\mathbb{R}^d$, given by $\|\mathbf{x}\|_p := \left(|x_1|^p + \cdots + |x_d|^p\right)^{1/p}$, $x \in \mathbb{R}^d$. of norms (see Exercise 1.8.(1) on page 16).

Imagine a city (like New York) in which there are streets and avenues with blocks in between, forming a square grid as shown in the picture below. Then if we take a taxi/cab to go from point A to point B in the city, it is clear that it isn't the Euclidean norm in $\mathbb{R}^2$ which is relevant, but rather the $\|\cdot\|_1$-norm in $\mathbb{R}^2$. (It is for this reason that the $\|\cdot\|_1$-norm is sometimes called the *taxicab norm*.)

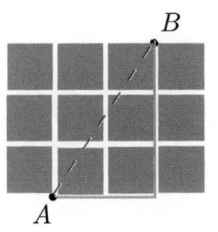

So what norm one uses depends on the situation at hand, and is something that the modeller decides. It is not something that falls out of the sky! ◇

Example 1.9. ($C[a,b]$ as a normed space). Consider the vector space $C[a,b]$ defined earlier. Define

$$\|\mathbf{x}\|_\infty = \sup_{t\in[a,b]} |\mathbf{x}(t)| = \max_{t\in[a,b]} |\mathbf{x}(t)|, \quad \mathbf{x} \in C[a,b]. \tag{1.4}$$

Then $\|\cdot\|_\infty$ is a norm on $C[a,b]$, and is referred to as the "supremum norm." The second equality above, guaranteeing that the supremum is attained, that is, that there is a $c \in [a,b]$ such that

$$|\mathbf{x}(c)| = \sup_{t\in[a,b]} |\mathbf{x}(t)|,$$

follows from the Extreme Value Theorem[5] for continuous functions.

Exercise 1.4. In $C[0,1]$ equipped with the $\|\cdot\|_\infty$-norm, calculate the norms of t, $-t$, t^n and $\sin(2\pi n t)$, where $n \in \mathbb{N}$.

Let us check that $\|\cdot\|_\infty$ on $C[a,b]$ does satisfy (N1), (N2), (N3).

(N1) For $\mathbf{x} \in C[a,b]$, $|\mathbf{x}(t)| \geq 0$ for all $t \in [a,b]$. So $\|\mathbf{x}\|_\infty = \max_{t\in[a,b]} |\mathbf{x}(t)| \geq 0$.

Also, if $\mathbf{x} \in C[a,b]$ is such that $\|\mathbf{x}\|_\infty = 0$, then for each $t \in [a,b]$,

$$0 \leq |\mathbf{x}(t)| \leq \max_{t\in[a,b]} |\mathbf{x}(t)| = \|\mathbf{x}\|_\infty = 0.$$

So for all $t \in [a,b]$, $|\mathbf{x}(t)| = 0$, and so $\mathbf{x}(t) = 0$.

In other words, $\mathbf{x} = \mathbf{0}$, the zero function in $C[a,b]$.

---
[5]See for example [Sasane (2015), §3.4].

(N2) If $\alpha \in \mathbb{R}$ and $\mathbf{x} \in C[a,b]$, then $|(\alpha \cdot \mathbf{x})(t)| = |\alpha \mathbf{x}(t)| = |\alpha||\mathbf{x}(t)|$, for $t \in [a,b]$, and so $\|\alpha \cdot \mathbf{x}\|_\infty = \max_{t \in [a,b]} |\alpha||\mathbf{x}(t)| = |\alpha| \max_{t \in [a,b]} |\mathbf{x}(t)| = |\alpha|\|\mathbf{x}\|_\infty$.

(N3) Let $\mathbf{x}_1, \mathbf{x}_2 \in C[a,b]$. If $t \in [a,b]$, then
$$|(\mathbf{x}_1 + \mathbf{x}_2)(t)| = |\mathbf{x}_1(t) + \mathbf{x}_2(t)|$$
$$\leq |\mathbf{x}_1(t)| + |\mathbf{x}_2(t)|$$
$$\leq \max_{s \in [a,b]} |\mathbf{x}_1(s)| + \max_{s \in [a,b]} |\mathbf{x}_2(s)| = \|\mathbf{x}_1\|_\infty + \|\mathbf{x}_2\|_\infty.$$

As this holds for *all* $t \in [a,b]$, $\max_{t \in [a,b]} |(\mathbf{x}_1 + \mathbf{x}_2)(t)| \leq \|\mathbf{x}_1\|_\infty + \|\mathbf{x}_2\|_\infty$.

Thus $\|\mathbf{x}_1 + \mathbf{x}_2\|_\infty \leq \|\mathbf{x}_1\|_\infty + \|\mathbf{x}_2\|_\infty$.

So $C[a,b]$ with the supremum norm $\|\cdot\|_\infty$ is a normed space. Thus we can use $\|\mathbf{x}_1 - \mathbf{x}_2\|_\infty$ as the distance between $\mathbf{x}_1, \mathbf{x}_2 \in C[a,b]$.

Geometric meaning of the distance in $C[a,b]$ equipped with the supremum norm. We ask the question: what does it mean geometrically when we say that $\mathbf{x}$ is close to $\mathbf{x}_0$? In other words, what does the set of points $\mathbf{x}$ that are close to (say within a distance of ϵ from) $\mathbf{x}_0$ look like?

In $(\mathbb{R}, |\cdot|)$, we know that the set of points x whose distance to x_0 is less than ϵ is an interval:
$$|x - x_0| < \epsilon \Leftrightarrow [x - x_0 < \epsilon \text{ and } -(x - x_0) < \epsilon] \Leftrightarrow x_0 - \epsilon < x < x_0 + \epsilon,$$
and so $\{x \in \mathbb{R} : |x - x_0| < \epsilon\} = (x_0 - \epsilon, x_0 + \epsilon)$.

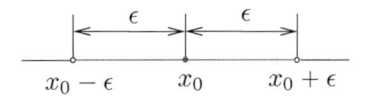

Now we ask: can we visualise the set $\{\mathbf{x} \in C[a,b] : \|\mathbf{x} - \mathbf{x}_0\|_\infty < \epsilon\}$? We have that
$$\|\mathbf{x} - \mathbf{x}_0\|_\infty < \epsilon \Leftrightarrow \max_{t \in [a,b]} |\mathbf{x}(t) - \mathbf{x}_0(t)| < \epsilon$$
$$\Leftrightarrow \text{for all } t \in [a,b], \; |\mathbf{x}(t) - \mathbf{x}_0(t)| < \epsilon$$
$$\Leftrightarrow \text{for all } t \in [a,b], \; \mathbf{x}(t) \in (\mathbf{x}_0(t) - \epsilon, \mathbf{x}_0(t) + \epsilon).$$

We can imagine translating the graph of $\mathbf{x}_0$ upward by a distance of ϵ, and downward through a distance of ϵ, so as to obtain the shaded strip depicted in the following picture. Then the graph of $\mathbf{x}$ has to lie in this shaded strip, because at each t, $\mathbf{x}_0(t) - \epsilon < \mathbf{x}(t) < \mathbf{x}_0(t) + \epsilon$. So for example at the particular t indicated in the following picture, $\mathbf{x}(t)$ has to lie on the line segment AB. Since this has to happen at each $t \in [a,b]$, we see that the graph of $\mathbf{x}$ lies in the shaded strip.

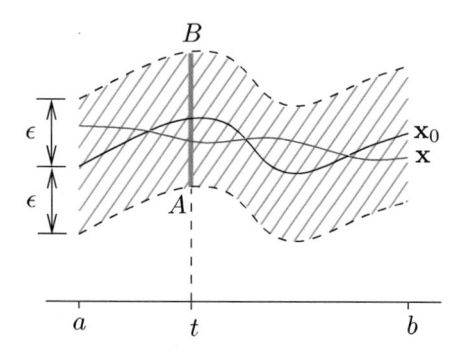

Fig. 1.1 The set of all continuous functions $\mathbf{x}$ whose graph lies between the two dashed curves is the "ball" $B(\mathbf{x}_0, \epsilon) = \{\mathbf{x} \in C[a,b] : \|\mathbf{x} - \mathbf{x}_0\|_\infty < \epsilon\}$.

Here are examples of some other frequently used norms in $C[a,b]$:

$$\|\mathbf{x}\|_1 := \int_a^b |\mathbf{x}(t)| dt, \tag{1.5}$$

$$\|\mathbf{x}\|_2 := \sqrt{\int_a^b |\mathbf{x}(t)|^2 dt}, \tag{1.6}$$

for $\mathbf{x} \in C[a,b]$. The $\|\cdot\|_1$-norm can be thought of as a continuous analogue of the taxicab norm, while the $\|\cdot\|_2$ norm is the continuous analogue of the Euclidean norm. The verification that $\|\cdot\|_1$ is indeed a norm on $C[a,b]$ will be done in Exercise 1.10. We'll postpone checking that $\|\cdot\|_2$ is also a norm on $C[a,b]$ until Chapter 4, where we will first check that $C[a,b]$ can be endowed with an "inner product"

$$\langle \mathbf{x}, \mathbf{y} \rangle := \int_a^b \mathbf{x}(t)\big(\mathbf{y}(t)\big)^* dt, \quad \mathbf{x}, \mathbf{y} \in C[a,b],$$

and then

$$\|\mathbf{x}\|_2 = \sqrt{\langle \mathbf{x}, \mathbf{x} \rangle} = \sqrt{\int_a^b |\mathbf{x}(t)|^2 dt},$$

will automatically become a norm! Right now we'll just accept the fact that $\|\cdot\|_2$ is a norm on $C[a,b]$.

We will see later on that $(C[a,b], \|\cdot\|_\infty)$ is "complete", that is, {Cauchy sequences} = {convergent sequences}, while $(C[a,b], \|\cdot\|_2)$ is not complete. On the other hand, $(C[a,b], \|\cdot\|_2)$ has a "nicer geometry", allowing one to talk about orthogonality[6]. What is the remedy? This motivates the consideration of $(L^2[a,b], \|\cdot\|_2)$, which besides allowing the nice geometry, also turns out to be complete. We will introduce this normed space in Example 1.12 below. ◇

Example 1.10. $C^1[a,b]$. Recall our optimal mining problem from Example 0.1, where the function to be minimised was defined on a subset of the subspace $C^1[a,b]$. So we see that the space $C^1[a,b]$ also arises naturally in applications. What norm do we use in $C^1[a,b]$? In general, if X is a normed space and Y is a subspace of the vector space X, then we can make Y into a normed space by simply using the restriction of the norm in X to Y. This is called the induced norm in Y, and in Exercise 1.7, we will see that this does give a norm on Y. So surely $C^1[a,b]$, being a subspace of $C[a,b]$ (which is a normed space with the supremum norm), is also a normed space with the supremum norm $\|\cdot\|_\infty$. However, it turns out that in applications, this is not a good choice, essentially because the differentiation map

$$\frac{d}{dt} : C^1[a,b] \to C[a,b], \ \mathbf{x} \mapsto \mathbf{x}' \ (\mathbf{x} \in C^1[a,b])$$

is not "continuous" (we will see this later on). There is a different norm on $C^1[a,b]$, denoted by $\|\cdot\|_{1,\infty}$, given below, which we shall use:

$$\|\mathbf{x}\|_{1,\infty} := \|\mathbf{x}\|_\infty + \|\mathbf{x}'\|_\infty = \max_{t \in [a,b]} |\mathbf{x}(t)| + \max_{t \in [a,b]} |\mathbf{x}'(t)|, \quad \mathbf{x} \in C^1[a,b].$$

In $C^1[0,1]$, for example

$$\|t\|_{1,\infty} = \|t\|_\infty + \|\mathbf{1}\|_\infty = 1 + 1 = 2,$$
$$\|t^n\|_{1,\infty} = \|t^n\|_\infty + \|nt^{n-1}\|_\infty = 1 + n, \quad n \in \mathbb{N}.$$

Roughly, two functions in $(C^1[a,b], \|\cdot\|_{1,\infty})$ are regarded as close together if both the functions themselves *and* their first derivatives are close together. Indeed, $\|\mathbf{x}_1 - \mathbf{x}_2\|_{1,\infty} < \epsilon$ implies that

$$|\mathbf{x}_1(t) - \mathbf{x}_2(t)| < \epsilon \text{ and } |\mathbf{x}_1'(t) - \mathbf{x}_2'(t)| < \epsilon \text{ for all } t \in [a,b], \qquad (1.7)$$

and conversely, (1.7) implies that $\|\mathbf{x}_1 - \mathbf{x}_2\|_{1,\infty} < 2\epsilon$. We will see later (when discussing continuity of maps between normed spaces), that the differentiation mapping from $C^1[a,b]$ to $C[a,b]$ is continuous if $C^1[a,b]$ is equipped with the $\|\cdot\|_{1,\infty}$-norm and $C[a,b]$ is equipped with the $\|\cdot\|_\infty$-norm. ◇

---
[6]This is in turn useful in applications, for example to solve shortest distance problems via projections.

Example 1.11. (Sequence spaces). For $1 \leq p < \infty$, ℓ^p is a normed space with the $\|\cdot\|_p$ norm, given by

$$\|(a_n)_{n\in\mathbb{N}}\|_p := \Big(\sum_{n=1}^{\infty} |a_n|^p\Big)^{1/p}, \quad (a_n)_{n\in\mathbb{N}} \in \ell^p.$$

Checking that the triangle inequality holds can be done using an inequality called *Hölder's Inequality*; see Exercise 1.8.

When $p = \infty$, that is, for the sequence space ℓ^∞, we define

$$\|(a_n)_{n\in\mathbb{N}}\|_\infty = \sup_{n\in\mathbb{N}} |a_n|, \quad (a_n)_{n\in\mathbb{N}} \in \ell^\infty.$$

Then it is easy to check that $\|\cdot\|_\infty$ is a norm, and so $(\ell^\infty, \|\cdot\|_\infty)$ is a normed space. ◇

Example 1.12. For $1 \leq p < \infty$, $L^p[a,b]$ is the normed space with the $\|\cdot\|_p$ norm, given by

$$\|\mathbf{x}\|_p := \Big(\int_a^b |\mathbf{x}(t)| dt\Big)^{1/p}, \quad \mathbf{x} \in L^p[a,b].$$

We won't check the validity of (N3) here. Also, the space $L^\infty[a,b]$ is a normed space with the norm given by

$$\|\mathbf{x}\|_\infty := \operatorname{esssup}|\mathbf{x}(t)|$$
$$:= \inf\{M : |\mathbf{x}(t)| \leq M \text{ for almost all } t \in [a,b]\}.$$

Again, we won't try to make "almost all" precise, as it relies of the notion of Lebesgue measure. The $\|\cdot\|_\infty$-norm here is referred to as the "essential supremum norm".

We remark that even more generally, if $(\Omega, \mathcal{M}, \mu)$ is any "measure space", where μ is a positive measure, then $L^p(\Omega)$ denotes the collection of all real-valued measurable functions $\mathbf{x}$ on Ω with

$$\|\mathbf{x}\|_p := \Big(\int_\Omega |\mathbf{x}(\omega)|^p d\mu(\omega)\Big)^{1/p}.$$

It turns out that $(L^p(\Omega), \|\cdot\|_p)$ is a normed space, which is moreover, complete. This normed space arises in applications, for example when $(\Omega, \mathcal{M}, \mu)$ is a probability space, where the $\mathbf{x}$ are "random variables", and $\|\mathbf{x}\|_p^p$ then has the interpretation of being the expected value $\mathbb{E}(|\mathbf{x}|^p)$. ◇

Exercise 1.5. (Triangle Inequality). Let $(X, \|\cdot\|)$ be a normed space. Prove that for all $x, y \in X$, $\big|\|x\| - \|y\|\big| \leq \|x - y\|$.

Exercise 1.6. If $x \in \mathbb{R}$, then let $\|x\| = |x|^2$. Is $\|\cdot\|$ a norm on $\mathbb{R}$?

Exercise 1.7. Let X be a normed space with norm $\|\cdot\|_X$, and Y be a subspace of X. Prove that Y is also a normed space with the norm $\|\cdot\|_Y$ defined simply as the restriction of the norm $\|\cdot\|_X$ to Y. This norm on Y is called the *induced norm*.

Exercise 1.8. Let $1 < p < \infty$ and q be defined by $\dfrac{1}{p} + \dfrac{1}{q} = 1$.
Then *Hölder's inequality* says that if $x_1, \cdots, x_d, y_1, \cdots, y_d \in \mathbb{C}$, then
$$\sum_{n=1}^{d} |x_n y_n| \leqslant \Big(\sum_{n=1}^{d} |x_n|^p \Big)^{1/p} \Big(\sum_{n=1}^{d} |y_n|^q \Big)^{1/q}.$$
Let's quickly establish this inequality. Suppose that $a, b \in \mathbb{R}$ and $a, b \geqslant 0$. We begin by showing that
$$\frac{a}{p} + \frac{b}{q} \geqslant a^{1/p} b^{1/q}. \tag{1.8}$$
If $a = 0$ or $b = 0$, then the conclusion is clear, and so we assume that both a and b are positive. We will use the following result.

Claim: If $\alpha \in (0,1)$, then for all $x \in [1, \infty)$, $\alpha(x-1) + 1 \geqslant x^\alpha$.
Given $\alpha \in (0,1)$, define $f_\alpha : [1, \infty) \to \mathbb{R}$ by $f_\alpha(x) = \alpha(x-1) - x^\alpha + 1$, for $x \geqslant 1$. Note that $f_\alpha(1) = \alpha \cdot 0 - 1^\alpha + 1 = 0$, and for all $x \geqslant 1$,
$$f'_\alpha(x) = \alpha - \alpha \cdot x^{\alpha - 1} = \alpha \Big(1 - \frac{1}{x^{1-\alpha}} \Big) \geqslant 0.$$
By the Fundamental Theorem of Calculus, for any $x > 1$,
$$f_\alpha(x) - f_\alpha(1) = \int_0^x f'_\alpha(y) dy \geqslant 0,$$
and so we obtain $f_\alpha(x) \geqslant 0$ for all $x \in [1, \infty)$, completing the proof of the claim.
As $p \in (1, \infty)$, it follows that $1/p \in (0,1)$. Applying the above with $\alpha = 1/p$ and
$$x := \begin{cases} \dfrac{a}{b} & \text{if } a \geqslant b, \\ \dfrac{b}{a} & \text{if } a \leqslant b, \end{cases}$$
we obtain inequality (1.8).
Hölder's inequality is obvious if $\sum_{n=1}^{d} |x_n|^p = 0$ or $\sum_{n=1}^{d} |y_n|^q = 0$.
So we assume that neither is 0, and proceed as follows.
Define $a_m = |x_m|^p \Big/ \sum_{n=1}^{d} |x_n|^p$ and $b_m = |y_m|^q \Big/ \sum_{n=1}^{d} |y_n|^q$, $1 \leqslant m \leqslant d$.
Applying the inequality (1.8) to a_m, b_m, we obtain for each m that:
$$\frac{|x_m y_m|}{\Big(\sum_{n=1}^{d} |x_n|^p \Big)^{1/p} \Big(\sum_{n=1}^{d} |y_n|^q \Big)^{1/q}} \leqslant \frac{|x_m|^p}{p \sum_{n=1}^{d} |x_n|^p} + \frac{|y_m|^p}{q \sum_{n=1}^{d} |y_n|^q}.$$
Adding these d inequalities, we obtain Hölder's inequality.

If $1 \leq p \leq \infty$, and $d \in \mathbb{N}$, then for $\mathbf{x} = (x_1, \cdots, x_d) \in \mathbb{R}^d$, define

$$\|\mathbf{x}\|_p = \Big(\sum_{n=1}^d |x_n|^p \Big)^{1/p} \text{ if } 1 \leq p < \infty, \text{ and } \|\mathbf{x}\|_\infty = \max\{|x_1|, \cdots, |x_d|\}. \qquad (1.9)$$

(1) Show that the function $\mathbf{x} \mapsto \|\mathbf{x}\|_p$ is a norm on $\mathbb{R}^d$.
    Hint: In the case when $1 < p < \infty$, use Hölder's inequality to obtain
    $$\sum_{n=1}^d |x_n||x_n+y_n|^{p-1} \leq \|\mathbf{x}\|_p \|\mathbf{x}+\mathbf{y}\|_p^{p/q}, \quad \sum_{n=1}^d |y_n||x_n+y_n|^{p-1} \leq \|\mathbf{y}\|_p \|\mathbf{x}+\mathbf{y}\|_p^{p/q},$$
    and use $\|\mathbf{x}+\mathbf{y}\|_p^p = \sum_{n=1}^d |x_n+y_n||x_n+y_n|^{p-1}$
    $$\leq \sum_{n=1}^d |x_n||x_n+y_n|^{p-1} + \sum_{n=1}^d |y_n||x_n+y_n|^{p-1}.$$

(2) Let $d = 2$. For $1 \leq p \leq \infty$, the "open unit ball" $B_p(\mathbf{0}, 1)$, is defined by $B_p(\mathbf{0}, 1) := \{\mathbf{x} \in \mathbb{R}^2 : \|\mathbf{x}\|_p < 1\}$. Sketch $B_p(\mathbf{0}, 1)$ for $p = 1, 2, \infty$.

(3) (Explanation of the notation for the maximum norm "$\|\cdot\|_\infty$".) Let $\mathbf{x} \in \mathbb{R}^2$. Prove that $(\|\mathbf{x}\|_p)_{p \in \mathbb{N}}$ is a convergent sequence in $\mathbb{R}$, and $\lim_{p \to \infty} \|\mathbf{x}\|_p = \|\mathbf{x}\|_\infty$. Describe qualitatively what happens to the sets $B_p(\mathbf{0}, 1)$ as p tends to ∞.

Exercise 1.9. A subset C of a vector space is said to be *convex* if for all $x, y \in C$, and all $\alpha \in (0, 1)$, $(1 - \alpha)x + \alpha y \in C$; see the following picture for examples of convex and nonconvex sets in $\mathbb{R}^2$.

convex | not convex

(1) In any normed space $(X, \|\cdot\|)$, show that the "closed unit ball" $\overline{B(\mathbf{0}, 1)}$, defined by $\overline{B(\mathbf{0}, 1)} := \{x \in X : \|x\| \leq 1\}$ is convex.
(2) Depict the set $\{(x_1, x_2) \in \mathbb{R}^2 : \sqrt{|x_1|} + \sqrt{|x_2|} = 1\}$ in the plane.
(3) (Explanation of why we've been taking p in $[1, \infty)$ rather than just all $p > 0$). Prove that $\|\mathbf{x}\|_{1/2} := (\sqrt{|x_1|} + \sqrt{|x_2|})^2$, $\mathbf{x} = (x_1, x_2) \in \mathbb{R}^2$, does not define a norm on $\mathbb{R}^2$.

Exercise 1.10. Show that (1.5) on page 12 defines a norm on $C[a, b]$.

Exercise 1.11. Let $C^n[a,b]$ be the set of n times continuously differentiable functions on $[a,b]$: $C^n[a,b] = \{\mathbf{x} : [a,b] \to \mathbb{R} \text{ such that } \mathbf{x}', \mathbf{x}'', \cdots, \mathbf{x}^{(n)} \in C[a,b]\}$, equipped with pointwise operations, and the norm

$$\|\mathbf{x}\|_{n,\infty} = \|\mathbf{x}\|_\infty + \|\mathbf{x}'\|_\infty + \cdots + \|\mathbf{x}^{(n)}\|_\infty, \quad \mathbf{x} \in C^n[a,b]. \qquad (1.10)$$

Show that $\|\cdot\|_{n,\infty}$ is a norm on $C^n[a,b]$.

Exercise 1.12. ("p-adic norm"). Consider the vector space of the rational numbers $\mathbb{Q}$ over the field $\mathbb{Q}$. Let p be a prime number. If the integer q divides the integer n, we write $q \mid n$, and if not, then we write $q \nmid n$.
Define the p-adic norm $|\cdot|_p$ on $\mathbb{Q}$ as follows:

$|0|_p := 0$, and if $r \in \mathbb{Q}\setminus\{0\}$, then $|r|_p := \dfrac{1}{p^k}$ where $r = p^k \dfrac{m}{n}$, $k, m, n \in \mathbb{Z}$, $p \nmid m, n$.

So in this context, a rational number is close to 0 precisely when it is "highly divisible" by p.

(1) Show that $|\cdot|_p$ is well-defined on $\mathbb{Q}$.
(2) If $r \in \mathbb{Q}$, then prove that $|r|_p \geq 0$, and that if $|r|_p = 0$ then $r = 0$.
(3) For all $r_1, r_2 \in \mathbb{Q}$, show that $|r_1 r_2|_p = |r_1|_p |r_2|_p$.
(4) For all $r_1, r_2 \in \mathbb{Q}$, prove that $|r_1 + r_2|_p \leq \max\{|r_1|_p, |r_2|_p\}$.
In particular, for all $r_1, r_2 \in \mathbb{Q}$, $|r_1 + r_2|_p \leq |r_1|_p + |r_2|_p$.

Exercise 1.13. Consider the vector space $\mathbb{R}^{m \times n}$ of matrices with m rows and n columns of real numbers, with the usual entrywise addition and scalar multiplication. Let the entry in the ith row and jth column of $\mathbf{M}$ be denoted by m_{ij}. For $\mathbf{M} \in \mathbb{R}^{m \times n}$, define $\|\mathbf{M}\|_\infty := \max_{\substack{1 \leq i \leq m \\ 1 \leq j \leq n}} |m_{ij}|$. Show that $\|\cdot\|_\infty$ is a norm on $\mathbb{R}^{m \times n}$.

1.3 Topology of normed spaces

In a normed space, we can describe "neighbourhoods" of points by considering sets which include all points whose distance to the given point is not too large.

Definition 1.3. (Open ball).
Let $(X, \|\cdot\|)$ be a normed space, $x \in X$, and $r > 0$.
The *open ball* $B(x,r)$ *with centre* x *and radius* r is defined by

$$B(x,r) = \{y \in X : \|x - y\| < r\}.$$

Thus $B(x,r)$ is the set of all points in X whose distance to the centre x is strictly less than r.

We'll keep the following picture in mind.

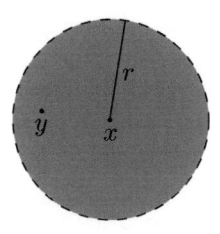

In the sequel, for example in our study of continuous functions, *open sets* will play an important role. Here is the definition.

Definition 1.4. (Open set). Let $(X, \|\cdot\|)$ be a normed space. A set $U \subset X$ is said to be *open* if for every $x \in U$, there exists an $r > 0$ such that $B(x, r) \subset U$.

Note that the radius r may depend on the choice of the point x. See the following picture. Roughly speaking, no matter which point you take in an open set, there is always some "room" around it consisting only of points of the open set.

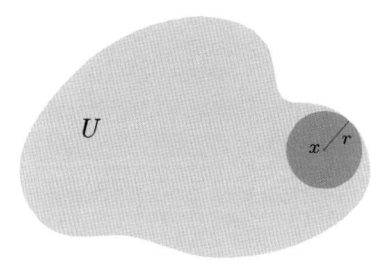

Example 1.13. Let us show that the "open interval" (a, b) is open in $\mathbb{R}$. Given any $x \in (a, b)$, we have $a < x < b$. Motivated by the following picture, let us take $r = \min\{x - a, b - x\}$. Then $r > 0$, and if $|y - x| < r$, then $-r < y - x < r$. So $a = x - (x - a) \leqslant x - r < y < x + r \leqslant x + (b - x) = b$, that is, $y \in (a, b)$. Hence $B(x, r) \subset (a, b)$. Consequently, (a, b) is open.

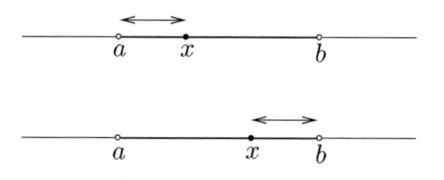

On the other hand, the interval $[a,b]$ is not open: with $x := a \in [a,b]$, we have that no matter how small an $r > 0$ we take, the set
$$B(a,r) = \{y \in \mathbb{R} : |y - a| < r\} = (a - r, a + r)$$
contains points that do not belong to $[a,b]$: for example,
$$a - \frac{r}{2} \in B(a,r) \text{ but } a - \frac{r}{2} \notin [a,b].$$

The picture above illustrates this. ◇

Example 1.14. The set X is open, since given an $x \in X$, we can take any $r > 0$, and notice that $B(x,r) \subset X$ trivially.

The empty set $\emptyset$ is also open ("vacuously"). Indeed, the reasoning is as follows: can one show an x for which there is no $r > 0$ such that $B(x,r) \subset \emptyset$? And the answer is no, because there is no x in the empty set (let alone an x which has the extra property that there is no $r > 0$ such that $B(x,r) \subset \emptyset$!). ◇

Exercise 1.14. Let $(X, \|\cdot\|)$ be a normed space, $x \in X$ and $r > 0$. Show that the open ball $B(x,r)$ is an open set.

Exercise 1.15. We know that the segment $(0,1)$ is open in $\mathbb{R}$. Show that the segment $(0,1)$ considered as a subset of the plane, that is, the set
$$I = \{(x,y) \in \mathbb{R}^2 : 0 < x < 1, \ y = 0\}$$
is not open in $(\mathbb{R}^2, \|\cdot\|_2)$.

Exercise 1.16. (Euclidean, taxicab, and maximum norm topologies coincide). Recall the three norms $\|\cdot\|_2$ (Euclidean), $\|\cdot\|_1$ (taxicab) and $\|\cdot\|_\infty$ (maximum) on $\mathbb{R}^2$ from Example 1.8 on page 9. Give a pictorial "proof without words" to show that a set U is open in $\mathbb{R}^2$ in the Euclidean metric if and only if it is open when $\mathbb{R}^2$ is equipped with the metric d_1 or the metric d_∞. *Hint:* Inside every square you can draw a circle, and inside every circle, you can draw a square!

Lemma 1.1. *Any finite intersection of open sets is open.*

Proof. It is enough to consider two open sets, as the general case follows immediately by induction on the number of sets.

Let U_1, U_2 be two open sets. Let $x \in U_1 \cap U_2$. Then there exist $r_1 > 0$, $r_2 > 0$ such that $B(x, r_1) \subset U_1$ and $B(x, r_2) \subset U_2$. Take $r = \min\{r_1, r_2\}$. Then $r > 0$, and we claim that $B(x,r) \subset U_1 \cap U_2$. To see this, let y be an element of $B(x,r)$. Then $\|x - y\| < r = \min\{r_1, r_2\}$, and so $\|x - y\| < r_1$ and $\|x - y\| < r_2$. So $y \in B(x, r_1) \cap B(x, r_1) \subset U_1 \cap U_2$. □

Example 1.15. The finiteness condition in the above lemma cannot be dropped: In $\mathbb{R}$, consider the open sets $U_n := (-1/n, 1/n)$, $n \in \mathbb{N}$. Then we have $\bigcap_{n \in \mathbb{N}} U_n = \{0\}$, which is not open in $\mathbb{R}$. ◇

Lemma 1.2. *Any union of open sets is open.*

Proof. Let U_i, $i \in I$, be a family of open sets indexed[7] by the set I. If $x \in \bigcup_{i \in I} U_i$, then we have that $x \in U_{i_*}$ for some $i_* \in I$. But as U_{i_*} is open, there exists a $r > 0$ such that $B(x,r) \subset U_{i_*}$. Thus $B(x,r) \subset U_{i_*} \subset \bigcup_{i \in I} U_i$. So the union $\bigcup_{i \in I} U_i$ is open. □

Definition 1.5. (Closed set). Let $(X, \|\cdot\|)$ be a normed space. A set F is *closed* if its complement $X \backslash F$ is open.

Example 1.16. The "closed interval" $[a,b]$ is closed in $\mathbb{R}$. Indeed, its complement $\mathbb{R} \backslash [a,b]$ is the union of the two open sets $(-\infty, a)$ and (b, ∞). Hence $\mathbb{R} \backslash [a,b]$ is open, and so $[a,b]$ is closed.

The set $(-\infty, b]$ is closed in $\mathbb{R}$. (Why?)

The sets $(a,b]$, $[a,b)$ are neither open nor closed in $\mathbb{R}$. (Why?) ◇

Example 1.17. X, $\emptyset$ are closed. ◇

Exercise 1.17. Show that arbitrary intersections of closed sets are closed. Prove that a finite union of closed sets is closed. Can the finiteness condition be dropped in the previous claim?

Exercise 1.18. Let $(X, \|\cdot\|)$ be a normed space, $x \in X$ and $r > 0$. Show that the "closed ball" $\overline{B(x,r)} := \{y \in X : \|y - x\| \leq r\}$ is a closed set.

Exercise 1.19. Determine if the following statements are true or false.

(1) If a set is not open, then it is closed.
(2) If a set is open, then it is not closed.
(3) There are sets which are both open and closed.
(4) There are sets which are neither open nor closed.
(5) $\mathbb{Q}$ is open in $\mathbb{R}$.
(6) (∗) $\mathbb{Q}$ is closed in $\mathbb{R}$.
(7) $\mathbb{Z}$ is closed in $\mathbb{R}$.

---

[7]This means that we have a set I, and for each $i \in I$, there is a set U_i. The set I is referred to as the *index set*, and any particular $i \in I$ as the *index* of U_i.

Exercise 1.20. Let $(X, \|\cdot\|)$ be a normed space.
Show that the unit sphere $\mathbb{S} := \{\mathbf{x} \in X : \|x\| = 1\}$ is closed.

Exercise 1.21. Let $(X, \|\cdot\|)$ be a normed space.
Show that a singleton (a subset of X having exactly one element) is always closed. Conclude that every finite subset F of X is closed.

Exercise 1.22. (∗) A subset D of a normed space $(X, \|\cdot\|)$ is said to be *dense* in X if for all $x \in X$ and all $\epsilon > 0$, there exists a $y \in D$ such that $\|x - y\| < \epsilon$.

That is, if we take any $x \in X$ and consider any ball $B(x, \epsilon)$ centred at x, it contains a point from D. In everyday language, we may say for example that "These woods have a dense growth of birch trees", and the picture we then have in mind is that in any small area of the woods, we find a birch tree. A similar thing is conveyed by the above: no matter what "patch" (described by $B(x,\epsilon)$) we take in X (thought of as the woods), we can find an element of D (analogous to birch trees) in that patch.

Show that $\mathbb{Q}$ is dense in $\mathbb{R}$ by proceeding as follows.

If $x, y \in \mathbb{R}$ and $x < y$, then show that there is a $q \in \mathbb{Q}$ such that $x < q < y$. (By the Archimedean Property[8] of $\mathbb{R}$, there is a positive integer n such that $n(y-x) > 1$. Next there are positive integers m_1, m_2 such that $m_1 > nx$ and $m_2 > -nx$ so that $-m_2 < nx < m_1$. Hence there is an integer m such that $m - 1 \leqslant nx < m$. Consequently $nx < m \leqslant 1 + nx < ny$, which gives the desired result.)

Exercise 1.23. Is the set $\mathbb{R}\backslash\mathbb{Q}$ of irrational numbers dense in $\mathbb{R}$? *Hint:* Take any $x \in \mathbb{R}$. If x is irrational itself, then we may just take y to be x and we are done; whereas if x is rational, then take $y = x + \sqrt{2}/n$ with a sufficiently large n.

Exercise 1.24. Show that c_{00} is dense in ℓ^2.

Exercise 1.25. (Separable spaces.) A normed space X is called *separable* if it has a countable dense set, that is, there exists a set $D := \{x_1, x_2, x_3, \cdots\}$ in X such that for every $r > 0$ and every $x \in X$, there exists an $x_n \in D$ such that $\|x_n - x\| < r$. For example $\mathbb{R}$ is separable, since we can simply take $D = \mathbb{Q}$.

Show that ℓ^1 is separable. (Analogously it can be shown that ℓ^p is separable for all $1 \leqslant p < \infty$.)

On the other hand, ℓ^∞ is not separable. Suppose that $D = \{\mathbf{x}_1, \mathbf{x}_2, \mathbf{x}_3, \cdots\}$ is a dense subset of ℓ^∞. Consider the set A of all sequences with all terms equal to either 0 or 1. If $(a_n)_{n \in \mathbb{N}}$, $(b_n)_{n \in \mathbb{N}}$ are distinct elements of A, then their mutual distance is 1, since $a_n \neq b_n$ for at least one n. Now by the density of D in ℓ^∞, it follows that for each $\mathbf{a} \in A$, we can choose an element $\mathbf{x}_{n(\mathbf{a})} \in B(\mathbf{a}, 1/3)$. As the balls $B(\mathbf{a}, 1/3)$, $\mathbf{a} \in A$, are all mutually disjoint, it follows that we get an injective map $A \ni \mathbf{a} \mapsto n(\mathbf{a}) \in \mathbb{N}$, a contradiction, since A is uncountable (as it is in one-to-one correspondence with all real numbers between 0 and 1 via binary expansion).

---

[8]The Archimedean Property of $\mathbb{R}$ says that if $x, y \in \mathbb{R}$ and $x > 0$, then there exists an $n \in \mathbb{N}$ such that $y < nx$. See for example [Sasane (2015), page 18].

Separability is a sort of a topological limitation on size. It plays a role in constructive mathematics, since many theorems have constructive proofs only for separable spaces even though the theorem is true for nonseparable ones. Such constructive proofs can sometimes be turned into algorithms for use in numerical analysis.

Exercise 1.26. (Weierstrass's Approximation Theorem).
The aim of this exercise is to show that polynomials are dense in $(C[a,b], \|\cdot\|_\infty)$. By considering the map $\mathbf{x} \mapsto \mathbf{x}(a + \cdot(b-a)) : C[a,b] \to C[0,1]$, we see that there is no loss of generality in assuming that $a = 0$ and $b = 1$. For $\mathbf{x} \in C[0,1]$ and $n \in \mathbb{N}$, let $B_n\mathbf{x}$ be the polynomial[9] given by

$$(B_n\mathbf{x})(t) := \sum_{k=0}^{n} \mathbf{x}\left(\frac{k}{n}\right)\binom{n}{k}t^k(1-t)^{n-k}, \quad t \in [0,1].$$

Let us introduce the auxiliary polynomials

$$\mathbf{p}_{n,k}(t) := \binom{n}{k}t^k(1-t)^{n-k}, \quad t \in [0,1], \ 0 \leq k \leq n, \ n \in \mathbb{N}.$$

Show that:

$$\sum_{k=0}^{n} \mathbf{p}_{k,n}(t) = 1,$$

$$\sum_{k=0}^{n} k\,\mathbf{p}_{k,n}(t) = nt,$$

$$\sum_{k=0}^{n} (k-nt)^2 \mathbf{p}_{k,n}(t) = nt(1-t).$$

The proof of Weierstrass's Approximation Theorem can now be completed as follows. For $\delta > 0$, we have

$$\sum_{k:\,|\frac{k}{n}-t|\geq\delta} \mathbf{p}_{n,k}(t) \leq \sum_{k:\,|\frac{k}{n}-t|\geq\delta} \mathbf{p}_{n,k}(t) \cdot \underbrace{\frac{(k-nt)^2}{\delta^2 n^2}}_{\geq 1}$$

$$\leq \frac{1}{n^2\delta^2}\sum_{k=0}^{n}(k-nt)^2\mathbf{p}_{k,n}(t) = \frac{t(1-t)}{n\delta^2}$$

$$\leq \frac{1}{4n\delta^2},$$

where we used the observation

$$0 \leq (\sqrt{t}-\sqrt{1-t})^2 = 1 - 2\sqrt{t(1-t)} \quad \text{for all } t \in [0,1],$$

in order to obtain the last inequality.
Now for $\delta > 0$, set $\omega_\delta(\mathbf{x}) := \sup_{|t-s|\leq\delta} |\mathbf{x}(t) - \mathbf{x}(s)|$.

---

[9] The notation B is after Sergei Bernstein, 1880–1968, who did fundamental work in *constructive function theory*, where smoothness properties of a function are related to its approximability by polynomials.

Then we have
$$|(B_n\mathbf{x})(t) - \mathbf{x}(t)|$$
$$= |(B_n\mathbf{x})(t) - \mathbf{x}(t) \cdot 1| = \Big|(B_n\mathbf{x})(t) - \mathbf{x}(t) \cdot \sum_{k=0}^{n} \mathbf{p}_{n,k}(t)\Big|$$
$$= \Big|\sum_{k=0}^{n} \mathbf{x}\Big(\frac{k}{n}\Big)\mathbf{p}_{n,k}(t) - \mathbf{x}(t)\sum_{k=0}^{n} \mathbf{p}_{n,k}(t)\Big|$$
$$\leqslant \sum_{k=0}^{n} \Big|\mathbf{x}\Big(\frac{k}{n}\Big) - \mathbf{x}(t)\Big|\mathbf{p}_{n,k}(t)$$
$$= \sum_{k:|\frac{k}{n}-t|<\delta} \Big|\mathbf{x}\Big(\frac{k}{n}\Big) - \mathbf{x}(t)\Big|\mathbf{p}_{n,k}(t) + \sum_{k:|\frac{k}{n}-t|\geqslant\delta} \Big|\mathbf{x}\Big(\frac{k}{n}\Big) - \mathbf{x}(t)\Big|\mathbf{p}_{n,k}(t)$$
$$\leqslant \omega_\delta(\mathbf{x}) \sum_{k:|\frac{k}{n}-t|<\delta} \mathbf{p}_{n,k}(t) + 2\|\mathbf{x}\|_\infty \frac{1}{4n\delta^2} \leqslant \omega_\delta(\mathbf{x}) \cdot 1 + \frac{\|\mathbf{x}\|_\infty}{2n\delta^2}.$$

Let $\epsilon > 0$. Since $\mathbf{x}$ is uniformly continuous[10], we can choose $\delta > 0$ such that $\omega_\delta(\mathbf{x}) < \epsilon/2$. Next choose $n > \|\mathbf{x}\|_\infty/(\epsilon\delta^2)$. Then it follows from the above that $\|B_n\mathbf{x} - \mathbf{x}\|_\infty < \epsilon$, completing the proof of the Weierstrass Approximation Theorem.

Remark 1.2. (Topology). If we look at the collection $\mathcal{O}$ of all open sets in a normed space $(X, \|\cdot\|)$, we notice that it has the following three properties:

(T1) $\varnothing, X \in \mathcal{O}$.

(T2) If $U_i \in \mathcal{O}$ for all $i \in I$, then $\bigcup_{i \in I} U_i \in \mathcal{O}$.

(T3) If $U_1, \cdots, U_n$ is a finite collection of sets from $\mathcal{O}$, then $\bigcap_{i=1}^{n} U_i \in \mathcal{O}$.

More generally, if X is any set (not necessarily a normed space), then any collection $\mathcal{O}$ of subsets of X that satisfy properties (T1), (T2), (T3) is called a *topology on* X and $(X, \mathcal{O})$ is called a *topological space*. Elements of $\mathcal{O}$ are called *open sets* in $(X, \mathcal{O})$. So for a normed space X, if we take $\mathcal{O}$ to be the family of open sets in $(X, \|\cdot\|)$, then we obtain a topological space. The following picture displays the hierarchy of structures[11].

---

[10] Recall that every continuous function on a *compact* interval is *uniformly* continuous there; see for example [Sasane (2015), Proposition 3.11, page 113].
[11] We remark here that every vector space can be made into a normed space. For the details, see Remark 4.3 on page 162.

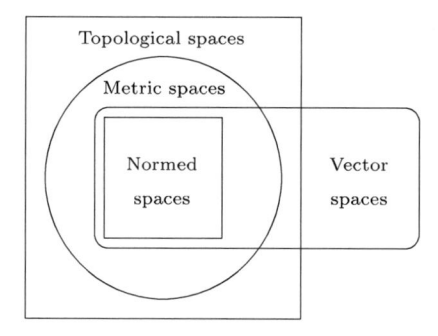

It turns out that one can in fact extend some of the notions from Calculus (such as convergence of sequences and continuity of maps) in the even more general set-up of topological spaces, devoid of any metric or norm, where the notion of closeness is specified by considering arbitrary open neighbourhoods provided by elements of $\mathcal{O}$. In some applications this is exactly the right thing needed, but we will not go into such abstractions here. In fact, this is a very broad subdiscipline of mathematics called *Topology*.

1.4 Sequences in a normed space; Banach spaces

In a normed space, we have a notion of "distance" between vectors, and we can say when two vectors are close by, and when they are far away. So we can talk about convergent sequences. In the same way as in $\mathbb{R}$ or $\mathbb{C}$, we can define convergent sequences and Cauchy sequences in a normed space:

Definition 1.6. (Convergent sequence). Let $(x_n)_{n\in\mathbb{N}}$ be a sequence in X and let $L \in X$. The sequence $(x_n)_{n\in\mathbb{N}}$ is said to be *convergent* (in X) *with limit L* if

$\forall \epsilon > 0$	$\exists N \in \mathbb{N}$	such that $\forall n > N$,	$\|x_n - L\| < \epsilon$
for every distance ϵ	there is an index	such that all terms beyond that index	have distance to L less than ϵ

In the above, we have used the symbol "$\forall$", which is read "for every". Also the symbol "$\exists$" means "there exists a/an".

Note that the definition says that the convergence of $(x_n)_{n\in\mathbb{N}}$ to L is the same as the real sequence $(\|x_n - L\|)_{n\in\mathbb{N}}$ converging to 0:

$$\lim_{n\to\infty} \|x_n - L\| = 0,$$

that is the distance of the vector x_n to the limit L tends to zero, and this matches our geometric intuition. One can show in the same way as with $\mathbb{R}$, that the limit is unique: a convergent sequence has only one limit. We write $\lim\limits_{n\to\infty} x_n = L$.

Theorem 1.1. *A convergent sequence has a unique limit.*

Proof. Let $(x_n)_{n\in\mathbb{N}}$ be convergent with limits L_1 and L_2, with $L_1 \neq L_2$. Let $\epsilon := \|L_1 - L_2\|/3 > 0$, where the positivity of the ϵ follows from the fact that $L_1 \neq L_2$. Since L_1 is a limit of the sequence $(x_n)_{n\in\mathbb{N}}$, there exists an $N_1 \in \mathbb{N}$ such that for all $n > N_1$, $\|x_n - L_1\| < \epsilon$. Since L_2 is a limit of the sequence $(x_n)_{n\in\mathbb{N}}$, there exists an $N_2 \in \mathbb{N}$ such that for all $n > N_2$, $\|x_n - L_2\| < \epsilon$. So for $n > N_1 + N_2$, we have $n > N_1$ and $n > N_2$, and
$$\|L_1 - L_2\| = \|L_1 - x_n + x_n - L_2\| \leq \|L_1 - x_n\| + \|x_n - L_2\| < \epsilon + \epsilon = \frac{2}{3}\|L_1 - L_2\|.$$
So we arrive at the contradiction that $1 < 2/3$. Hence our assumption was incorrect, and so a convergent sequence must have a unique limit. $\square$

Example 1.18. Consider the sequence $(\mathbf{x}_n)_{n\in\mathbb{N}}$ in the normed space $(C[0,1], \|\cdot\|_\infty)$, where $\mathbf{x}_n = \dfrac{\sin(2\pi n t)}{n}$, $t \in [0,1]$.

The first few terms of the sequence are shown in the following picture.

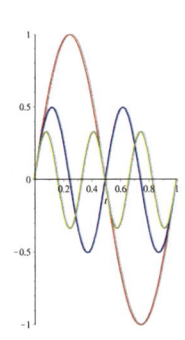

From the figure, we see that the terms seem to converge to the zero function. Indeed we have
$$\|\mathbf{x}_n - \mathbf{0}\|_\infty = \max_{t\in[0,1]} |\mathbf{x}_n(t) - \mathbf{0}(t)| = \max_{t\in[0,1]} \left|\frac{\sin(2\pi n t)}{n} - 0\right|$$
$$= \max_{t\in[0,1]} \frac{1}{n}|\sin(2\pi n t)| = \frac{1}{n} \max_{t\in[0,1]} |\sin(2\pi n t)|$$
$$= \frac{1}{n} \cdot 1 = \frac{1}{n}.$$

Given $\epsilon > 0$, let $N \in \mathbb{N}$ be such that $N > 1/\epsilon$. Then for all $n > N$,
$$\|\mathbf{x}_n - \mathbf{0}\|_\infty = \frac{1}{n} < \frac{1}{N} < \epsilon.$$
So $(\mathbf{x}_n)_{n \in \mathbb{N}}$ is convergent in the normed space $(C[0,1], \|\cdot\|_\infty)$ to $\mathbf{0}$. $\diamond$

Definition 1.7. (Cauchy sequence). A sequence $(x_n)_{n \in \mathbb{N}}$ in a normed space $(X, \|\cdot\|)$ is called a *Cauchy sequence* if for every $\epsilon > 0$, there exists an $N \in \mathbb{N}$ such that for all $m, n \in \mathbb{N}$ satisfying $m, n > N$, $\|x_m - x_n\| < \epsilon$.

Roughly speaking, we can make the terms of the sequence arbitrarily close to each other provided we go far enough in the sequence.

Proposition 1.1. *Every convergent sequence is Cauchy.*

Proof. Let $(x_n)_{n \in \mathbb{N}}$ be a sequence in $(X, \|\cdot\|)$ that converges to $L \in X$. Let $\epsilon > 0$. (We want to find N which guarantees for $n, m > N$ that $\|x_n - x_m\| < \epsilon$. But we *do* know that the terms x_n, x_m can both be made close to L if n, m are large enough. So we introduce L artificially: $\|x_n - x_m\| = \|x_n - L + L - x_m\|$ and use the triangle inequality to complete the argument. The details are given below.)

Then there exists an $N \in \mathbb{N}$ such that for $n > N$, we have $\|x_n - L\| < \dfrac{\epsilon}{2}$. Thus for $n, m > N$, we have
$$\|x_n - x_m\| = \|x_n - L + L - x_m\| \leq \|x_n - L\| + \|x_m - L\| < \frac{\epsilon}{2} + \frac{\epsilon}{2} = \epsilon.$$
So the sequence $(x_n)_{n \in \mathbb{N}}$ is a Cauchy sequence. $\square$

Convergent sequences

Showing membership here needs knowledge of limit (Harder!)

Cauchy sequences

Showing membership here needs no knowledge of limit, but only an investigation of the mutual behaviour of the terms of the sequence (Easier!)

We recall from ordinary calculus that in $\mathbb{R}$,

$$\{ \text{ convergent sequences } \} = \{ \text{ Cauchy sequences } \}.$$

(We will recall the proof of this fact below, in Theorem 1.4 on page 31.) This raises the tempting question of whether this equality is true in general normed spaces too:

$$\{ \text{ convergent sequences } \} \overset{?}{\underset{\checkmark}{\supset}} \{ \text{ Cauchy sequences } \}.$$

If the two sets coincide, then one can conclude that a sequence is convergent by just checking Cauchyness. This is the basis of many *existence* results in Analysis: for example, the convergence tests in Calculus, the existence results for differential equations, the Riesz Representation Theorem[12], etc. Once existence is known, (and after showing uniqueness, if valid), one can justify and use numerical approximations. So this prompts the question:

Q. Is it true in *all* normed spaces that

$$\{ \text{ convergent sequences } \} = \{ \text{ Cauchy sequences } \}?$$

Answer: No. It is true in *some* normed spaces, for example

$$(\mathbb{R}, |\cdot|),$$
$$(\mathbb{C}, |\cdot|),$$
$$(\ell^p, \|\cdot\|_p),$$
$$(C[a,b], \|\cdot\|_\infty),$$
$$(L^2[a,b], \|\cdot\|_2),$$
$$\cdots,$$

but *not true* in others, for example

$$(\mathbb{Q}, |\cdot|),$$
$$(C[a,b], \|\cdot\|_2),$$
$$\cdots.$$

(We will soon justify these claims.)

In light of the above answer, it makes sense to give normed spaces in which

$$\{ \text{ convergent sequences } \} = \{ \text{ Cauchy sequences } \}$$

a special name. These are called *Banach spaces*, after the Polish mathematician Stefan Banach (1892–1945), who laid the foundations of the study of such spaces in his doctoral dissertation from 1920.

Definition 1.8. (Banach space). A normed space in which the set of Cauchy sequences is equal to the set of convergent sequences is called a *Banach space*. Sometimes, we also call it a *complete normed space*.

---
[12] To be studied in Chapter 5.

Thus in a complete normed space, or Banach space, the Cauchy condition is sufficient for convergence: the sequence $(x_n)_{n\in\mathbb{N}}$ converges if and only if it is a Cauchy sequence. So we can determine convergence a priori without the knowledge of the limit. Just as it was possible to introduce new numbers in $\mathbb{R}$ as the limits of Cauchy sequences, now in a Banach space, it is possible to show the existence of elements with some property of interest, by making use of the Cauchyness. In this manner, one can sometimes show that certain equations possess a solution. In many cases, one cannot write the solution explicitly. But after existence and uniqueness of the solution is demonstrated, one can do numerical approximations.

$(\mathbb{R}, |\cdot|)$ is a Banach space

The completeness of $\mathbb{R}$ will be used fundamentally in checking all of our other examples of Banach spaces. While the fact that real Cauchy sequences are always convergent may be familiar to the reader, we reprove this here for the sake of completeness. We will first establish the following elementary lemma, which is valid in *all* normed space, not just in $\mathbb{R}$.

Lemma 1.3. *Every Cauchy sequence in a normed space is bounded*[13].

Proof. Suppose that $(x_n)_{n\in\mathbb{N}}$ is a Cauchy sequence in the normed space $(X, \|\cdot\|)$. Choose any positive ϵ, say $\epsilon = 1$. Then there exists an $N \in \mathbb{N}$ such that for all $n, m > N$, $\|x_n - x_m\| < \epsilon$. In particular, with $m = N+1 > N$, and $n > N$, $\|x_n - x_{N+1}\| < \epsilon$. By the Triangle Inequality, for all $n > N$, $\|x_n\| = \|x_n - x_{N+1} + x_{N+1}\| \leq \|x_n - x_{N+1}\| + \|x_{N+1}\| < 1 + \|x_{N+1}\|$. On the other hand, for $n \leq N$, $\|x_n\| \leq \max\{\|x_1\|, \cdots, \|x_N\|, 1 + \|x_{N+1}\|\} =: M$. So $\|x_n\| \leq M$ for all $n \in \mathbb{N}$, that is, the sequence $(x_n)_{n\in\mathbb{N}}$ is bounded. $\square$

Next we'll show that:

Theorem 1.2. *Every real sequence has a monotone*[14] *subsequence.*

Before giving the formal proof, we give an illustration of the idea behind this proof[15]. If $(x_n)_{n\in\mathbb{N}}$ is the given sequence, then imagine that there is an infinite chain of hotels along a line, where the nth hotel has height x_n, and at the horizon, there is a sea. A hotel is said to have the *seaview property* if it is higher than all hotels following it (so that from the roof of the hotel, one can view the sea). There are only two possibilities:

---

[13]That is, the norms of the terms of the sequence form a bounded real sequence.
[14]That is, either the terms are increasing, or the terms are decreasing.
[15]This illustrative analogy stems from [Bryant (1990)].

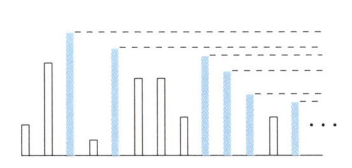

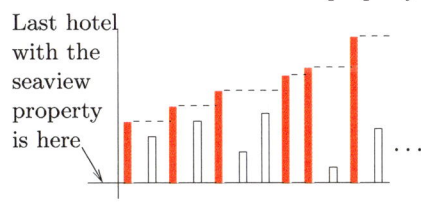

1° Infinitely many hotels have the seaview property	2° Finitely many hotels have the seaview property

Last hotel with the seaview property is here

1° There are infinitely many hotels with the seaview property.	2° There are finitely many hotels with the seaview property.
Then by taking successively the heights of the hotels with the seaview property we get a *decreasing* subsequence.	Then after the last hotel with the seaview property, one can start with any hotel and then always find one that is at least as high, which is taken as the next hotel, and then finding yet another that is at least as high as that one, and so on. The heights of these hotels form an *increasing* subsequence.

Proof. Let $(x_n)_{n\in\mathbb{N}}$ be a real sequence, and let

$$S = \{m \in \mathbb{N} : \text{for all } n > m,\ x_n < x_m\}.$$

(This is the collection of indices of hotels with the seaview property.) Then we have the following two cases.

<u>1°</u> S is infinite.
Arrange the elements of S in increasing order: $n_1 < n_2 < n_3 < \ldots$. Then $(x_{n_k})_{k\in\mathbb{N}}$ is a *decreasing* subsequence of $(x_n)_{n\in\mathbb{N}}$.

<u>2°</u> S is finite.
If S is empty, then define $n_1 = 1$, and otherwise let $n_1 = \max S + 1$. Define inductively $n_{k+1} = \min\{m \in \mathbb{N} : m > n_k \text{ and } x_m \geqslant x_{n_k}\}$.
(n_{k+1} is the index of the first hotel blocking the view from the top of the n_kth hotel.) The minimum exists as $\{m \in \mathbb{N} : m > n_k \text{ and } x_m \geqslant x_{n_k}\}$ is a nonempty subset of $\mathbb{N}$. (Otherwise if it were empty, then $n_k \in S$, and this is not possible if S was empty, and also impossible if S was not empty, since $n_k > \max S$.) Then $(x_{n_k})_{k\in\mathbb{N}}$ is an *increasing* subsequence of $(x_n)_{n\in\mathbb{N}}$. □

Theorem 1.3.
If a real sequence is monotone and bounded, then it is convergent.

Proof.
$\underline{1}°$ We will first consider the case of *increasing* sequences which are bounded. Let $(x_n)_{n \in \mathbb{N}}$ be an increasing and bounded sequence. We want to show that $(x_n)_{n \in \mathbb{N}}$ is convergent. But with what limit?

The picture above suggests that the limit should be the smallest number bigger than each of the terms of this sequence, that is, the supremum of the set $\{x_n : n \in \mathbb{N}\}$. Since $(x_n)_{n \in \mathbb{N}}$ is bounded, it follows that the set $S := \{x_n : n \in \mathbb{N}\}$ has an upper bound and so $\sup S$ exists. We show that in fact $(x_n)_{n \in \mathbb{N}}$ converges to $\sup S$. Let $\epsilon > 0$. Since $\sup S - \epsilon < \sup S$, it follows that $\sup S - \epsilon$ is *not* an upper bound for S, and so there exists an $x_N \in S$ such that $\sup S - \epsilon < x_N$, that is $\sup S - x_N < \epsilon$. Since $(x_n)_{n \in \mathbb{N}}$ is an increasing sequence, for $n > N$, we have $x_N \leqslant x_n$. Since $\sup S$ is an upper bound for S, $x_n \leqslant \sup S$ and so $|x_n - \sup S| = \sup S - x_n$, Thus for $n > N$ we obtain $|x_n - \sup S| = \sup S - x_n \leqslant \sup S - x_N < \epsilon$.

$\underline{2}°$ If $(x_n)_{n \in \mathbb{N}}$ is a *decreasing* and bounded sequence, then clearly $(-x_n)_{n \in \mathbb{N}}$ is an increasing sequence. Furthermore if $(x_n)_{n \in \mathbb{N}}$ is bounded, then $(-x_n)_{n \in \mathbb{N}}$ is bounded as well ($|-x_n| = |x_n| \leqslant M$). Hence by the case considered above, it follows that $(-x_n)_{n \in \mathbb{N}}$ is a convergent sequence with limit $\sup\{-x_n : n \in \mathbb{N}\} = -\inf\{x_n : n \in \mathbb{N}\} = -\inf S$, where $S = \{x_n : n \in \mathbb{N}\}$. So given $\epsilon > 0$, there exists an $N \in \mathbb{N}$ such that for all $n > N$, $|-x_n - (-\inf S)| < \epsilon$, that is, $|x_n - \inf S| < \epsilon$. Thus $(x_n)_{n \in \mathbb{N}}$ is convergent with limit $\inf S$. $\square$

Corollary 1.1. (Bolzano-Weierstrass Theorem).
Every bounded real sequence has a convergent subsequence.

Proof. Let $(x_n)_{n \in \mathbb{N}}$ be a bounded real sequence. The sequence $(x_n)_{n \in \mathbb{N}}$ has a monotone subsequence, say $(x_{n_k})_{k \in \mathbb{N}}$. Then $(x_{n_k})_{k \in \mathbb{N}}$ is bounded too. We have that $(x_{n_k})_{k \in \mathbb{N}}$ is monotone and bounded, and hence it is convergent in $\mathbb{R}$. $\square$

We are now ready to prove that $(\mathbb{R}, |\cdot|)$ is a Banach space.

Theorem 1.4. *Every real Cauchy sequence in $\mathbb{R}$ is convergent.*

Proof. Let $(x_n)_{n \in \mathbb{N}}$ be Cauchy in $\mathbb{R}$. Then $(x_n)_{n \in \mathbb{N}}$ is bounded. By the Bolzano-Weierstrass Theorem, $(x_n)_{n \in \mathbb{N}}$ has a convergent subsequence, say $(x_{n_k})_{k \in \mathbb{N}}$, with limit, say $L \in \mathbb{R}$. We will now show that $(x_n)_{n \in \mathbb{N}}$ is also convergent with limit L. Let $\epsilon > 0$. Then there exists an $N \in \mathbb{N}$ such that for all $n, m > N$,
$$|x_n - x_m| < \frac{\epsilon}{2}.$$
Also, since $(x_{n_k})_{k \in \mathbb{N}}$ converges to L, we can find an $n_K > N$ such that
$$|x_{n_K} - L| < \frac{\epsilon}{2}.$$
Thus we have for all $n > N$ that
$$|x_n - L| = |x_n - x_{n_K} + x_{n_K} - L| \leq |x_n - x_{n_K}| + |x_{n_K} - L| < \frac{\epsilon}{2} + \frac{\epsilon}{2} = \epsilon.$$
Thus $(x_n)_{n \in \mathbb{N}}$ is also convergent with limit L. □

Example 1.19. ($\mathbb{Q}$ is not complete). Consider the sequence $(x_n)_{n \in \mathbb{N}}$ in $\mathbb{Q}$ defined by $x_1 = 3/2$, and for $n > 1$, recursively by
$$x_{n+1} = \frac{4 + 3x_n}{3 + 2x_n}.$$
Then it can be shown by induction that $(x_n)_{n \in \mathbb{N}}$ is bounded below by $\sqrt{2}$, and that $(x_n)_{n \in \mathbb{N}}$ is monotone decreasing.

(A) $x_n \geq \sqrt{2}$ for all n.

If $n = 1$, then $x_1 = \frac{3}{2} \geq \sqrt{2}$ (as $\frac{9}{4} \geq 2$). If $x_n \geq \sqrt{2}$ for some n, then
$$x_{n+1}^2 - 2 = \frac{(4 + 3x_n)^2}{(3 + 2x_n)^2} - 2$$
$$= \frac{16 + 24x_n + 9x_n^2 - 18 - 24x_n - 8x_n^2}{(3 + 2x_n)^2} = \frac{x_n^2 - 2}{(3 + 2x_n)^2} \geq 0.$$
So this gives, since $x_{n+1} \geq 0$, that $x_{n+1} \geq \sqrt{2}$, and the claim follows.

(B) $x_n \geq x_{n+1}$ for all n.

We have
$$x_n - x_{n+1} = x_n - \frac{4 + 3x_n}{3 + 2x_n} = \frac{3x_n + 2x_n^2 - 4 - 3x_n}{3 + 2x_n} = \frac{2(x_n^2 - 2)}{3 + 2x_n} \geq 0,$$
where the last inequality follows from part (A).

So this sequence is convergent in $\mathbb{R}$. Hence it is also Cauchy in $\mathbb{R}$. But as each term x_n is a rational number for all $n \in \mathbb{N}$, it follows that $(x_n)_{n \in \mathbb{N}}$ is also Cauchy in $\mathbb{Q}$. However, we now show that $(x_n)_{n \in \mathbb{N}}$ is not convergent in $\mathbb{Q}$. Suppose, on the contrary, that $(x_n)_{n \in \mathbb{N}}$ converges to $L \in \mathbb{Q}$. Then from the recurrence relation, we obtain using the Algebra of Limits that

$$L = \frac{4 + 3L}{3 + 2L},$$

and so $L^2 = 2$. As L must be positive (the sequence is bounded below by $\sqrt{2}$), it follows that $L = \sqrt{2}$. But this is a contradiction, since we know that there is no rational number whose square is 2.

(Alternately, consider the real number c with the decimal expansion

$$c = 0.101001000100001\cdots.$$

This number c is irrational because it has a nonterminating and nonrepeating decimal expansion. If we consider the sequence of rational numbers $0.1, 0.101, 0.101001, 0.1010010001, 0.101001000100001, \cdots$, obtained by truncation, then this sequence converges with limit c.) $\Diamond$

Example 1.20.
$\frac{1}{1^1}, \frac{1}{1^1} + \frac{1}{2^2}, \frac{1}{1^1} + \frac{1}{2^2} + \frac{1}{3^3}, \cdots$ converges in $\mathbb{R}$, as it is Cauchy: for $n > m$,

$$\left|\left(\frac{1}{1^1} + \cdots + \frac{1}{n^n}\right) - \left(\frac{1}{1^1} + \cdots + \frac{1}{m^m}\right)\right| = \frac{1}{(m+1)^{m+1}} + \cdots + \frac{1}{n^n}$$

$$\leq \frac{1}{2^{m+1}} + \cdots + \frac{1}{2^n} + \cdots$$

$$= \frac{\frac{1}{2^{m+1}}}{1 - \frac{1}{2}} = \frac{1}{2^m},$$

which can be made as small as we please by taking m large enough. We remark that it is not yet known if the limit is rational or irrational! $\Diamond$

The completeness of $\mathbb{R}$ is the basis for the completeness of other normed spaces, and we'll see this now.

Finite-dimensional normed spaces are Banach

Theorem 1.5. $(\mathbb{R}^d, \|\cdot\|_2)$ *is a Banach space.*

Proof.
(Essentially, this is because $\mathbb{R}$ is complete, and one has d copies of $\mathbb{R}$ in $\mathbb{R}^d$.)
Suppose that $(\mathbf{x}_n)_{n \in \mathbb{N}}$ is a Cauchy sequence in $\mathbb{R}^d$; $\mathbf{x}_n = (x_n^{(1)}, \cdots, x_n^{(d)})$.

We have $|x_n^{(k)} - x_m^{(k)}| \leq \|\mathbf{x}_n - \mathbf{x}_m\|_2$ ($n, m \in \mathbb{N}$, $k = 1, \cdots, d$), from which it follows that each of the real sequences $(x_n^{(k)})_{n\in\mathbb{N}}$, $k = 1, \cdots, d$, is Cauchy in $\mathbb{R}$, and hence convergent, with respective limits, say $L^{(1)}, \cdots, L^{(d)} \in \mathbb{R}$. So given $\epsilon > 0$, there exists a large enough N such that whenever $n > N$, we have $|x_n^{(k)} - L^{(k)}| < \dfrac{\epsilon}{\sqrt{d}}$ ($k = 1, \cdots, d$).

Set $\mathbf{L} = (L^{(1)}, \cdots, L^{(d)}) \in \mathbb{R}^d$.

Then for $n > N$, $\|\mathbf{x}_n - \mathbf{L}\|_2 = \Big(\sum_{k=1}^d |x_n^{(k)} - x^{(k)}|^2\Big)^{1/2} < \Big(\sum_{k=1}^d \dfrac{\epsilon^2}{d}\Big)^{1/2} = \epsilon$.

Consequently, the sequence $(\mathbf{x}_n)_{n\in\mathbb{N}}$ converges to $\mathbf{L}$. $\square$

Corollary 1.2. $(\mathbb{C}, |\cdot|)$ *is a Banach space.*

Proof. This follows from the fact that $(\mathbb{R}^2, \|\cdot\|_2)$ is a Banach space. $\square$

Exercise 1.27. $(*)$[16] (Equivalent norms).
Let X be a vector space, and let $\|\cdot\|_a$, $\|\cdot\|_b$ be norms on X. $\|\cdot\|_a$ is said to be *equivalent to* $\|\cdot\|_b$, denoted by $\|\cdot\|_a \sim \|\cdot\|_b$, if there exist positive constants m and M such that $m\|x\|_b \leq \|x\|_a \leq M\|x\|_b$.

(1) Show that $\sim$ defines an equivalence relation on the set of all norms on X.

(2) Prove for equivalent norms on X, their respective collections of open sets, convergent sequences, and Cauchy sequences coincide.

One can show that all norms are equivalent on $\mathbb{R}^d$ as follows. (It follows from here that all finite dimensional normed spaces are Banach since $\mathbb{R}^d$ is complete!) In view of the fact that $\sim$ is an equivalence relation, it is enough to show that any norm $\|\cdot\| \sim \|\cdot\|_2$, the Euclidean norm. We do this in three steps:

Step 1. First we will show that there is a positive M such that $\|\mathbf{x}\| \leq M\|\mathbf{x}\|_2$ for all $\mathbf{x} \in \mathbb{R}^d$. Let $\mathbf{e}_1, \cdots, \mathbf{e}_d$ be the standard basis in $\mathbb{R}^d$. Then every $\mathbf{x} \in \mathbb{R}^d$ can be decomposed uniquely as $\mathbf{x} = x_1 \mathbf{e}_1 + \cdots + x_d \mathbf{e}_d$, where $x_1, \cdots, x_d$ are scalars. So $\|\mathbf{x}\| = \|x_1\mathbf{e}_1 + \cdots + x_d\mathbf{e}_d\| \leq |x_1|\|\mathbf{e}_1\| + \cdots + |x_d|\|\mathbf{e}_d\|$ (using (N2) and (N3))
$$\leq \underbrace{\sqrt{\|\mathbf{e}_1\|^2 + \cdots + \|\mathbf{e}_d\|^2}}_{=:M} \|\mathbf{x}\|_2 \text{ (Cauchy-Schwarz)}.$$

Step 2. Let $K := \{\mathbf{y} \in \mathbb{R}^d : \|\mathbf{y}\|_2 = 1\}$. Then K is a compact set in the $\|\cdot\|_2$ norm topology since it is closed and bounded. The map $\|\cdot\| : K \to \mathbb{R}$ is continuous from $(K, \|\cdot\|_2)$ to $(\mathbb{R}, |\cdot|)$: $\forall \mathbf{y}_1, \mathbf{y}_2 \in K$, $\big|\|\mathbf{y}_1\| - \|\mathbf{y}_2\|\big| \leq \|\mathbf{y}_1 - \mathbf{y}_2\| \leq M\|\mathbf{y}_1 - \mathbf{y}_2\|_2$. By Weierstrass's Theorem, $\|\cdot\| : K \to \mathbb{R}$ attains a minimum value m on K. But this m can't be zero, since if $\|\mathbf{y}\| = 0$, then $\mathbf{y} = \mathbf{0} \notin K$. So this m ought to be positive. Conclusion: $\|\mathbf{y}\| \geq m$ for all $\mathbf{y}$'s with $\|\mathbf{y}\|_2 = 1$.

---

[16] This exercise assumes familiarity with the notion of continuity of real-valued maps on compact sets in $(\mathbb{R}^d, \|\cdot\|_2)$ and Weierstrass's Theorem saying that such a map assumes a minimum value. We will prove Weierstrass's Theorem in Chapter 2; see page 66.

Step 3. Now we will show that $m\|\mathbf{x}\|_2 \leq \|\mathbf{x}\|$ for all $\mathbf{x} \in \mathbb{R}^n$. This is obvious if $\mathbf{x} = \mathbf{0}$, since both sides of the inequality are zero in this case.
If $\mathbf{x} \neq \mathbf{0}$, then $\mathbf{y} := \mathbf{x}/\|\mathbf{x}\|_2$ satisfies $\|\mathbf{y}\|_2 = 1$, so that $\mathbf{y} \in K$.
Thus $m \leq \|\mathbf{y}\| = \|\mathbf{x}/\|\mathbf{x}\|_2\| = \|\mathbf{x}\|/\|\mathbf{x}\|_2$. Rearranging, we obtain $m\|\mathbf{x}\|_2 \leq \|\mathbf{x}\|$.
So we've shown that for all $\mathbf{x} \in \mathbb{R}^n$, $m\|\mathbf{x}\|_2 \leq \|\mathbf{x}\| \leq M\|\mathbf{x}\|_2$, that is, $\|\cdot\| \sim \|\cdot\|_2$.

$(C[a,b], \|\cdot\|_\infty)$ is a Banach space

The following theorem is an important result, and lies at the core of several results, for example the result on the existence of solutions for Ordinary Differential Equations (ODEs).

Theorem 1.6. $(C[a,b], \|\cdot\|_\infty)$ is a Banach space.

Proof. The idea behind the proof is similar to the proof of the completeness of $\mathbb{R}^d$. If $(\mathbf{x}_n)_{n\in\mathbb{N}}$ is a Cauchy sequence, then we think of the $\mathbf{x}_n(t)$ as being the "components" of $\mathbf{x}_n$ indexed by $t \in [a,b]$. We first freeze a $t \in [a,b]$, and show that $(\mathbf{x}_n(t))_{n\in\mathbb{N}}$ is a Cauchy sequence in $\mathbb{R}$, and hence convergent to a number (which depends on t), and which we denote by $\mathbf{x}(t)$. Next we show that the function $t \mapsto \mathbf{x}(t)$ is continuous, and finally that $(\mathbf{x}_n)_{n\in\mathbb{N}}$ does converge to $\mathbf{x}$ in the supremum norm.

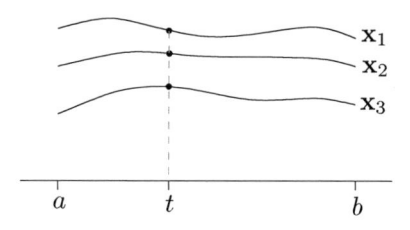

Let $(\mathbf{x}_n)_{n\in\mathbb{N}}$ be a Cauchy sequence. Let $t \in [a,b]$. We claim that $(\mathbf{x}_n(t))_{n\in\mathbb{N}}$ is a Cauchy sequence in $\mathbb{R}$. Let $\epsilon > 0$. Then there exists an $N \in \mathbb{N}$ such that for all $n, m > N$, $\|\mathbf{x}_n - \mathbf{x}_m\|_\infty < \epsilon$. But

$$|\mathbf{x}_n(t) - \mathbf{x}_m(t)| \leq \max_{\tau \in [a,b]} |\mathbf{x}_n(\tau) - \mathbf{x}_m(\tau)| = \|\mathbf{x}_n - \mathbf{x}_m\|_\infty < \epsilon,$$

for $n, m > N$. This shows that indeed $(\mathbf{x}_n(t))_{n\in\mathbb{N}}$ is a Cauchy sequence in $\mathbb{R}$. But $\mathbb{R}$ is complete, and so the Cauchy sequence $(\mathbf{x}_n(t))_{n\in\mathbb{N}}$ is in fact convergent, with a limit which depends on which $t \in [a,b]$ we had frozen at the outset. To highlight this dependence on t, we denote the limit of $(\mathbf{x}_n(t))_{n\in\mathbb{N}}$ by $\mathbf{x}(t)$. (Thus for example $\mathbf{x}(a)$ is the number which is the limit of the convergent sequence $(\mathbf{x}_n(a))_{n\in\mathbb{N}}$, $\mathbf{x}(b)$ is the number which is the limit of the convergent sequence $(\mathbf{x}_n(b))_{n \in \mathbb{N}}$, and so on.) So we have a function

$$t \mapsto \boxed{\text{the limit of the convergent sequence } (\mathbf{x}_n(t))_{n \in \mathbb{N}}} : [a,b] \to \mathbb{R}.$$

We call this function $\mathbf{x}$. This will serve as the limit of the sequence $(\mathbf{x}_n)_{n \in \mathbb{N}}$. But first we have to see if it belongs to $C[a,b]$, that is, we need to check that this $\mathbf{x}$ is continuous on $[a,b]$.

Let $t \in [a,b]$. We will show that $\mathbf{x}$ is continuous at t. Recall that in order to do this, we have to show that for each $\epsilon > 0$, there exists a $\delta > 0$ such that whenever $|\tau - t| < \delta$, we have $|\mathbf{x}(\tau) - \mathbf{x}(t)| < \epsilon$. Let $\epsilon > 0$. Choose N large enough so that for all $n, m > N$, $\|\mathbf{x}_n - \mathbf{x}_m\|_\infty < \epsilon/3$. Let $\tau \in [a,b]$. Then for $n > N$, $|\mathbf{x}_n(\tau) - \mathbf{x}_{N+1}(\tau)| \leq \|\mathbf{x}_n - \mathbf{x}_{N+1}\|_\infty < \epsilon/3$. Now let $n \to \infty$: $|\mathbf{x}(\tau) - \mathbf{x}_{N+1}(\tau)| = \lim_{n \to \infty} |\mathbf{x}_n(\tau) - \mathbf{x}_{N+1}(\tau)| \leq \epsilon/3$.

The choice of $\tau \in [a,b]$ was arbitrary, and so for all $\tau \in [a,b]$

$$|\mathbf{x}(\tau) - \mathbf{x}_{N+1}(\tau)| \leq \epsilon/3.$$

Now $\mathbf{x}_{N+1} \in C[a,b]$. So there exists a $\delta > 0$ such that whenever $|\tau - t| < \delta$,

$$|\mathbf{x}_{N+1}(\tau) - \mathbf{x}_{N+1}(t)| < \epsilon/3.$$

Thus whenever $|\tau - t| < \delta$, we have

$$\begin{aligned}|\mathbf{x}(\tau) - \mathbf{x}(t)| &= |\mathbf{x}(\tau) - \mathbf{x}_{N+1}(\tau) + \mathbf{x}_{N+1}(\tau) - \mathbf{x}_{N+1}(t) + \mathbf{x}_{N+1}(t) - \mathbf{x}(t)| \\ &\leq |\mathbf{x}(\tau) - \mathbf{x}_{N+1}(\tau)| + |\mathbf{x}_{N+1}(\tau) - \mathbf{x}_{N+1}(t)| + |\mathbf{x}_{N+1}(t) - \mathbf{x}(t)| \\ &\leq \epsilon/3 + \epsilon/3 + \epsilon/3 = \epsilon.\end{aligned}$$

This shows that $\mathbf{x}$ is continuous at t. As the choice of $t \in [a,b]$ was arbitrary, $\mathbf{x}$ is continuous on $[a,b]$.

Finally, we show that $(\mathbf{x}_n)_{n \in \mathbb{N}}$ does converge to $\mathbf{x}$. Let $\epsilon > 0$. Choose N large enough so that for all $n, m > N$, $\|\mathbf{x}_n - \mathbf{x}_m\|_\infty < \epsilon$. Fix $n > N$. Let $t \in [a,b]$. Then for all $m > N$, $|\mathbf{x}_n(t) - \mathbf{x}_m(t)| \leq \|\mathbf{x}_n - \mathbf{x}_m\|_\infty < \epsilon$. Thus

$$|\mathbf{x}_n(t) - \mathbf{x}(t)| = \lim_{m \to \infty} |\mathbf{x}_n(t) - \mathbf{x}_m(t)| \leq \epsilon.$$

But $t \in [a,b]$ was arbitrary. Hence $\|\mathbf{x}_n - \mathbf{x}\|_\infty = \max_{t \in [a,b]} |\mathbf{x}_n(t) - \mathbf{x}(t)| \leq \epsilon$.

But we could have fixed any $n > N$ at the outset and obtained the same result. So for all $n > N$, $\|\mathbf{x}_n - \mathbf{x}\|_\infty \leq \epsilon$. Thus $\lim_{n \to \infty} \mathbf{x}_n = \mathbf{x}$ in $(C[a,b], \|\cdot\|_\infty)$.

This completes the proof. $\square$

Example 1.21. ($C[a,b]$ is not a a Banach space with the $\|\cdot\|_2$-norm.) We will work with $[a,b] = [0,2]$ for computational ease. Consider the sequence $(\mathbf{x}_n)_{n \in \mathbb{N}}$ in $C[0,2]$, where $\mathbf{x}_n$ has a graph as shown below.

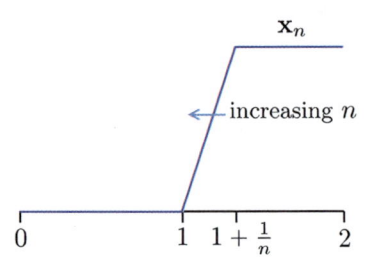

$(\mathbf{x}_n)_{n \in \mathbb{N}}$ is a Cauchy sequence in $(C[0,2], \|\cdot\|_2)$: indeed, for $n > m$,

$$\|\mathbf{x}_n - \mathbf{x}_m\|_2^2 = \int_0^2 \big(\mathbf{x}_n(t) - \mathbf{x}_m(t)\big)^2 dt = \int_1^2 \big(\mathbf{x}_n(t) - \mathbf{x}_m(t)\big)^2 dt$$

$$= \int_1^{1+\frac{1}{m}} \big(\mathbf{x}_n(t) - \mathbf{x}_m(t)\big)^2 dt \quad (\text{since } 1 < 1 + \frac{1}{n} < 1 + \frac{1}{m})$$

$$\leq \int_1^{1+\frac{1}{m}} 1 \, dt = \frac{1}{m} \xrightarrow{m \to \infty} 0.$$

Suppose that $(\mathbf{x}_n)_{n \in \mathbb{N}}$ converges to $\mathbf{x} \in C[0,2]$ in $(C[0,2], \|\cdot\|_2)$. Then:

$$\int_0^1 \big(\mathbf{x}(t)\big)^2 dt = \int_0^1 \big(0 - \mathbf{x}(t)\big)^2 dt = \int_0^1 \big(\mathbf{x}_n(t) - \mathbf{x}(t)\big)^2 dt$$

$$\leq \int_0^2 \big(\mathbf{x}_n(t) - \mathbf{x}(t)\big)^2 dt = \|\mathbf{x}_n - \mathbf{x}\|_2^2 \xrightarrow{n \to \infty} 0.$$

As $\mathbf{x} \in C[0,1]$, $\int_0^1 \big(\mathbf{x}(t)\big)^2 dt = 0$ implies that $\mathbf{x}(t) = 0$ for $t \in [0,1]$.

Let $N \in \mathbb{N}$. Then for all $n > N$,

$$\int_{1+\frac{1}{N}}^2 \big(1 - \mathbf{x}(t)\big)^2 dt = \int_{1+\frac{1}{N}}^2 \big(\mathbf{x}_n(t) - \mathbf{x}(t)\big)^2 dt$$

$$\leq \int_0^2 \big(\mathbf{x}_n(t) - \mathbf{x}(t)\big)^2 dt = \|\mathbf{x}_n - \mathbf{x}\|_2^2 \xrightarrow{n \to \infty} 0,$$

and so $\int_{1+\frac{1}{N}}^{2}\left(1-\mathbf{x}(t)\right)^{2}dt = 0$. As $\mathbf{x} \in C[1+\frac{1}{N},2]$, this implies

$$\mathbf{x}(t) = 1 \text{ for all } 1+\frac{1}{N} \leq t \leq 2.$$

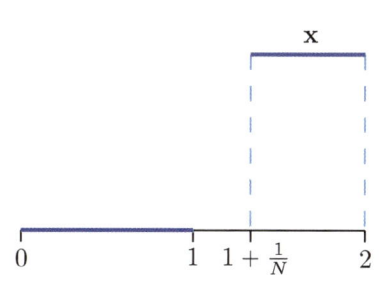

Since $N \in \mathbb{N}$ was arbitrary, it follows that $\mathbf{x}(t) = 1$ for all $t \in (0,1]$. Conclusion:

$$\mathbf{x}(t) = \begin{cases} 0 & \text{if } t \in [0,1], \\ 1 & \text{if } t \in (1,2]. \end{cases}$$

But then $\mathbf{x} \notin C[0,2]$ (as it has a discontinuity at $t=1$), a contradiction. $\diamond$

The following is an instance where one uses "Cauchyness $\Rightarrow$ convergence" in Banach spaces.

Theorem 1.7. *In a Banach space, absolutely convergent series converge, that is:*

If $(x_n)_{n \in \mathbb{N}}$ is a sequence in a Banach space $(X, \|\cdot\|)$ such that $\sum_{n=1}^{\infty} \|x_n\| < \infty$, then $\sum_{n=1}^{\infty} x_n$ converges in X. Moreover, $\left\|\sum_{n=1}^{\infty} x_n\right\| \leq \sum_{n=1}^{\infty} \|x_n\|.$

Proof. Let $s_n = x_1 + \cdots + x_n$, $n \in \mathbb{N}$. We want to show that $\sum_{n=1}^{\infty} x_n$ converges, that is, the sequence $(s_n)_{n \in \mathbb{N}}$ of partial sums converges in X. As X is a Banach space, it is enough to show that $(s_n)_{n \in \mathbb{N}}$ is Cauchy. We are given that the real series $\sum_{n=1}^{\infty} \|x_n\|$ converges, that is its sequence $(\sigma_n)_{n \in \mathbb{N}}$ of partial sums converges, where $\sigma_n = \|x_1\| + \cdots + \|x_n\|$, $n \in \mathbb{N}$.

In particular, $(\sigma_n)_{n\in\mathbb{N}}$ is Cauchy. For $n > m$,
$$0 \leq \|s_n - s_m\| = \|x_{m+1} + \cdots + x_n\|$$
$$\leq \|x_{m+1}\| + \cdots + \|x_n\| = \sigma_n - \sigma_m = |\sigma_n - \sigma_m|,$$
and this can be made as small as we please for all $n > m > N$ with a large enough N. (The rightmost equality above follows from the leftmost inequality.) Thus $(s_n)_{n\in\mathbb{N}}$ is Cauchy in X, and hence convergent in X (as X is a Banach space), to, say, $L \in X$. Let $\epsilon > 0$. Then there exists an n such that $\|s_n - L\| < \epsilon$. Thus
$$\|L\| \leq \|L - s_n\| + \|s_n\| < \epsilon + \sigma_n < \epsilon + \sum_{n=1}^{\infty} \|x_n\|.$$
As the choice of $\epsilon > 0$ was arbitrary, $\left\|\sum_{n=1}^{\infty} x_n\right\| = \|L\| \leq \sum_{n=1}^{\infty} \|x_n\|.$ □

Example 1.22. $\sum_{n=1}^{\infty} \dfrac{\sin(n\cdot)}{n^2}$ converges in $(C[0, 2\pi], \|\cdot\|_\infty)$.
(Here $\sin(n\cdot)$ means the function $t \mapsto \sin(nt) : [0, 2\pi] \to \mathbb{R}$.)
Indeed, we have $\left\|\dfrac{\sin(n\cdot)}{n^2}\right\|_\infty = \dfrac{1}{n^2}$, and $\sum_{n=1}^{\infty} \dfrac{1}{n^2} < \infty$.
So $\mathbf{x} := \sum_{n=1}^{\infty} \dfrac{\sin(n\cdot)}{n^2}$ defines a continuous function on $[0, 2\pi]$.
We can get a good idea of the limit by computing the first N terms (with a large enough N) and plotting the resulting function; the error can then be bounded as follows:
$$\left\|\sum_{n=N+1}^{\infty} \dfrac{\sin(n\cdot)}{n^2}\right\|_\infty \leq \sum_{n=N+1}^{\infty} \left\|\dfrac{\sin(nt)}{n^2}\right\|_\infty \leq \sum_{n=N+1}^{\infty} \dfrac{1}{n^2}.$$
For example, if $N = 100$, then the error is bounded above by
$$\sum_{n=101}^{\infty} \dfrac{1}{n^2} \leq \int_{100}^{\infty} \dfrac{1}{x^2} dx = \dfrac{1}{100} = 0.01.$$
Using Maple, we have plotted the partial sum of $\mathbf{x}$ with $N = 100$.

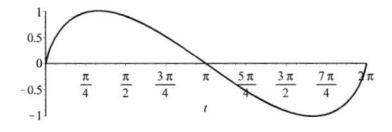

Thus the sum converges to a continuous function that lies in the strip of width 0.01 around the graph shown in the figure. ◊

Later on, we will use this theorem to show that e^A converges, where A belongs to $CL(X)$. Here $CL(X)$ denotes a certain Banach space, namely the space of all "continuous linear transformations" from X to itself, with the "operator norm". For example, when $X = \mathbb{R}^d$, $CL(X)$ turns out to be the space of all square $d \times d$ real matrices. Why fuss over e^A? The answer is that it plays a role in differential equations: the *initial value problem*

$$\begin{cases} \dfrac{d\mathbf{x}}{dt}(t) = A\mathbf{x}(t), & t \in \mathbb{R}, \\ \mathbf{x}(0) = x_0 \in X, \end{cases}$$

has the unique solution $\mathbf{x}(t) = e^{tA}x_0$, $t \in \mathbb{R}$.

Also, using the fact that $(C[a,b], \|\cdot\|_\infty)$ is a Banach space, one can show the Fundamental Theorem of Ordinary Differential Equations (ODEs):

Theorem 1.8. (Existence and Uniqueness of ODEs).
If there exists an $r > 0$ and an $L > 0$ such that $f : \mathbb{R} \times \mathbb{R} \to \mathbb{R}$ satisfies

$$\forall t \geqslant 0, \ \forall x, y \in \mathbb{R}, \ |f(x,t) - f(y,t)| \leqslant L|x - y|, \tag{L}$$

then for all $x_0 \in \mathbb{R}$, there exists a $T > 0$ and there exists an $\mathbf{x} \in C^1[0,T]$ solving the Initial Value Problem

$$\text{(IVP)} : \begin{cases} \dfrac{d\mathbf{x}}{dt}(t) = f(\mathbf{x}(t), t), & t \geqslant 0, \\ \mathbf{x}(0) = x_0, \end{cases}$$

on $[0,T]$, and moreover (IVP) has a unique solution.

Condition (L) on f is expressed as: f is "Lipschitz in x, uniformly in t".

Proof. (Uniqueness) Let $\mathbf{x}_1, \mathbf{x}_2$ be two solutions to (IVP) on $[0,T]$ for some $T > 0$. Let $t_* := \max\{t \in [0,T] : \mathbf{x}_1(\tau) = \mathbf{x}_2(\tau), \forall \tau \leqslant t\}$.

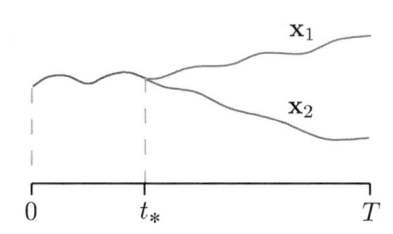

Then

$$\mathbf{x}_1(t) - \mathbf{x}_1(t_*) = \int_{t_*}^t \mathbf{x}_1'(\tau)d\tau = \int_{t_*}^t f(\mathbf{x}_1(\tau), \tau)d\tau,$$

$$\mathbf{x}_2(t) - \mathbf{x}_2(t_*) = \int_{t_*}^t \mathbf{x}_2'(\tau)d\tau = \int_{t_*}^t f(\mathbf{x}_2(\tau), \tau)d\tau.$$

So $\mathbf{x}_1(t) - \mathbf{x}_2(t) = \int_{t_*}^{t} \big(f(\mathbf{x}_1(\tau), \tau) - f(\mathbf{x}_2(\tau), \tau)\big) d\tau.$

Let[17] $N > \max\left\{1, \dfrac{1}{L}, \dfrac{1}{L(T-t_*)}\right\}$, and $M := \max\limits_{t \in [t_*, t_* + \frac{1}{LN}]} |\mathbf{x}_1(t) - \mathbf{x}_2(t)|.$

Note that $t_* + \dfrac{1}{LN} < T$ as $\dfrac{1}{LN} < T - t_*$. Then for all $t \in \left[t_*, t_* + \dfrac{1}{LN}\right]$,

$$|\mathbf{x}_1(t) - \mathbf{x}_2(t)| = \left|\int_{t_*}^{t} \big(f(\mathbf{x}_1(\tau), \tau) - f(\mathbf{x}_2(\tau), \tau)\big) d\tau\right|$$

$$\leq \int_{t_*}^{t} |f(\mathbf{x}_1(\tau), \tau) - f(\mathbf{x}_2(\tau), \tau)| d\tau$$

$$\leq \int_{t_*}^{t} L|\mathbf{x}_1(\tau) - \mathbf{x}_2(\tau)| d\tau \leq \int_{t_*}^{t} L \cdot M\, d\tau = LM(t - t_*)$$

$$\leq LM\left(t_* + \dfrac{1}{LN} - t_*\right) = \dfrac{LM}{LN} = \dfrac{M}{N}.$$

Thus
$$\forall\, t \in \left[t_*, t_* + \dfrac{1}{LN}\right],\ |\mathbf{x}_1(t) - \mathbf{x}_2(t)| \leq \dfrac{M}{N},$$

and so $M \leq \dfrac{M}{N}$, that is, $N \leq 1$, a contradiction. This shows the uniqueness.

(Existence) We will write down a sequence of recursively defined functions, which are not solutions, but serve as "good approximations":

$$\mathbf{x}_0(t) = x_0,$$

$$\mathbf{x}_1(t) = x_0 + \int_0^t f(\mathbf{x}_0(\tau), \tau) d\tau,$$

$$\mathbf{x}_2(t) = x_0 + \int_0^t f(\mathbf{x}_1(\tau), \tau) d\tau,$$

$$\cdots$$

$$\mathbf{x}_{n+1}(t) = x_0 + \int_0^t f(\mathbf{x}_n(\tau), \tau) d\tau,$$

$$\cdots$$

We will show that $(\mathbf{x}_n)_{n \geq 0}$ converges to $\mathbf{x}$ in $\left(C\left[0, \dfrac{1}{2L}\right], \|\cdot\|_\infty\right)$, and this $\mathbf{x}$ solves (IVP)! (So, in particular, we'll take $T = \dfrac{1}{2L} > 0$.)

We note that $\mathbf{x}_{n+1} = \mathbf{x}_0 + \sum\limits_{k=0}^{n} (\mathbf{x}_{n+1} - \mathbf{x}_n).$

---

[17] We will see the rationale behind these seemingly strange choice of N soon enough.

Also, for $0 \leqslant t \leqslant \dfrac{1}{2L}$,

$$|\mathbf{x}_{n+1}(t) - \mathbf{x}_n(t)| = \left|\int_0^t \big(f(\mathbf{x}_n(\tau), \tau) - f(\mathbf{x}_{n-1}(\tau), \tau)\big) d\tau\right|$$

$$\leqslant \int_0^t |f(\mathbf{x}_n(\tau), \tau) - f(\mathbf{x}_{n-1}(\tau), \tau)| d\tau$$

$$\leqslant \int_0^t L|\mathbf{x}_n(\tau) - \mathbf{x}_{n-1}(\tau)| d\tau$$

$$\leqslant \int_0^t L\|\mathbf{x}_n - \mathbf{x}_{n-1}\|_\infty d\tau = L\|\mathbf{x}_n - \mathbf{x}_{n-1}\|_\infty t$$

$$\leqslant L\|\mathbf{x}_n - \mathbf{x}_{n-1}\|_\infty \cdot \dfrac{1}{2L} = \dfrac{1}{2}\|\mathbf{x}_n - \mathbf{x}_{n-1}\|_\infty.$$

Thus $\|\mathbf{x}_{n+1} - \mathbf{x}_n\|_\infty \leqslant \dfrac{1}{2}\|\mathbf{x}_n - \mathbf{x}_{n-1}\|_\infty$, and so $\|\mathbf{x}_{n+1} - \mathbf{x}_n\|_\infty \leqslant \dfrac{1}{2^n}\|\mathbf{x}_1 - \mathbf{x}_0\|_\infty$.

Hence $\displaystyle\sum_{n=0}^\infty \|\mathbf{x}_{n+1} - \mathbf{x}_n\|_\infty \leqslant \|\mathbf{x}_1 - \mathbf{x}_0\|_\infty \sum_{n=0}^\infty \dfrac{1}{2^n} < \infty.$

So $\mathbf{x}_0 + \displaystyle\sum_{n=0}^\infty (\mathbf{x}_{n+1} - \mathbf{x}_n)$ converges in $(C[0,T], \|\cdot\|_\infty)$, to, say, $\mathbf{x} \in C[0,T]$.

We know that $\mathbf{x}_{n+1}(t) = x_0 + \displaystyle\int_0^t f(\mathbf{x}_n(\tau), \tau) d\tau,\ t \in [0,T]$.

Passing the limit as $n \to \infty$, we have (see the explanation below):

$$\mathbf{x}(t) = x_0 + \int_0^t f(\mathbf{x}(\tau), \tau) d\tau, \quad t \in [0, T]. \tag{1.11}$$

(Here's the justification. Define the continuous $\mathbf{g}_n$ ($n = 0, 1, 2, 3, \cdots$) by:

$$\mathbf{g}_n(t) = f(\mathbf{x}_n(t), t).$$

Then the sequence $\mathbf{g}_0, \mathbf{g}_1, \mathbf{g}_2, \cdots$ is the sequence of partial sums of the series

$$\mathbf{g}_0 + \sum_{k=0}^n (\mathbf{g}_{k+1} - \mathbf{g}_k). \tag{1.12}$$

We have

$$|\mathbf{g}_{k+1}(t) - \mathbf{g}_k(t)| = |f(\mathbf{x}_{k+1}(t), t) - f(\mathbf{x}_k(t), t)| \leqslant L|\mathbf{x}_{k+1}(t) - \mathbf{x}_k(t)|$$

$$\leqslant L\|\mathbf{x}_{k+1} - \mathbf{x}_k\|_\infty \leqslant \dfrac{1}{2^k}\|\mathbf{x}_1 - \mathbf{x}_0\|_\infty.$$

So (1.12) converges absolutely to some $\mathbf{g}$ in $(C[0,T], \|\cdot\|_\infty)$. We have

$$\mathbf{g}(t) = \lim_{n\to\infty} \mathbf{g}_n(t) = \lim_{n\to\infty} f(\mathbf{x}_n(t), t) = f(\mathbf{x}(t), t).$$

We'll now use the fact that if $\sum_{k=1}^{\infty} \mathbf{f}_k$ converges to $\mathbf{f}$ in $(C[a,b], \|\cdot\|_\infty)$, then

$$\sum_{k=1}^{\infty} \int_a^b \mathbf{f}_k(t)dt = \int_a^b \mathbf{f}(t)dt,$$

and this is precisely the content of Exercise 2.14 on page 73, which will be dealt with after discussing continuity of linear transformations. Using this,

$$\lim_{n\to\infty} \int_0^T f(\mathbf{x}_n(t),t)dt = \lim_{n\to\infty} \int_0^T \left(\mathbf{g}_0(t) + \sum_{k=0}^{n-1}(\mathbf{g}_{k+1}(t) - \mathbf{g}_k(t))\right)dt$$
$$= \int_0^T f(\mathbf{x}(t),t)dt,$$

that is, we have proved (1.11).)

Thus $\mathbf{x}(0) = x_0 + 0 = x_0$, and by the Fundamental Theorem of Calculus, $\mathbf{x}'(t) = 0 + f(\mathbf{x}(t),t)$ for all $t \in [0,T]$. $\square$

Exercise 1.28. (Nonuniqueness when non-Lipschitz).
(1) Let $f(x) := \sqrt{|x|}$, $x \in \mathbb{R}$. Show that f is not *Lipschitz*, that is, there is no constant $L > 0$ such that for all $x, y \in \mathbb{R}$, $|f(x) - f(y)| \leq L|x-y|$.
(2) Check that $\mathbf{x}_1 \equiv 0$ and $\mathbf{x}_2(t) = t^2/4$ are solutions to the Initial Value Problem
$$\begin{cases} \dfrac{d\mathbf{x}}{dt}(t) = \sqrt{|\mathbf{x}(t)|}, \ t \geq 0, \\ \mathbf{x}(0) = 0. \end{cases}$$

$(\ell^p, \|\cdot\|_p)$ are Banach spaces

Theorem 1.9. *Let* $1 \leq p \leq +\infty$. *Then* $(\ell^p, \|\cdot\|_p)$ *is a Banach space.*

Proof. We had already seen that ℓ^p is a vector space, and the fact that $\|\cdot\|_p$ defines a norm will be established in Exercise 1.35 (page 44). We must now show that ℓ^p is complete. Let $(\mathbf{x}_n)_{n\in\mathbb{N}}$ be a Cauchy sequence in ℓ^p. Denote the kth term of $\mathbf{x}_n$ by $x_n^{(k)}$. The proof will be carried out in 3 steps.

Step 1. We have $|x_n^{(k)} - x_m^{(k)}| \leq \|\mathbf{x}_n - \mathbf{x}_m\|_p$ and so $(x_n^{(k)})_{n\in\mathbb{N}}$ is a Cauchy sequence in $\mathbb{K}$ ($= \mathbb{R}$ or $\mathbb{C}$), and consequently, it is convergent, with limit, say, $x^{(k)}$. Set $\mathbf{x} = (x^{(k)})_{k\in\mathbb{N}}$.

Step 2. We show that $\mathbf{x}$ belongs to ℓ^p. Let $\epsilon > 0$. Then there exists an $N \in \mathbb{N}$ such that for all $n, m > N$, $\|\mathbf{x}_n - \mathbf{x}_m\|_p < \epsilon$. Fix any $n > N$. If $K \in \mathbb{N}$, then for $p < \infty$, $\sum_{k=1}^K |x_n^{(k)} - x_m^{(k)}|^p \leq \epsilon^p$.

Passing the limit as m goes to ∞ yields $\sum_{k=1}^{K} |x_n^{(k)} - x^{(k)}|^p \leq \epsilon^p$.
As the choice of K was arbitrary,
$$\sum_{k=1}^{\infty} |x_n^{(k)} - x^{(k)}|^p \leq \epsilon^p. \tag{1.13}$$
So $\mathbf{x}_n - \mathbf{x}$ belongs to ℓ^p. But $\mathbf{x}_n \in \ell^p$. Hence $(\mathbf{x} - \mathbf{x}_n) + \mathbf{x}_n = \mathbf{x} \in \ell^p$ too.
The $p = \infty$ case can be seen as follows. Fix $n > N$ and $k \in \mathbb{N}$. Then for all $m > N$, $|x_n^{(k)} - x_m^{(k)}| < \epsilon$. Passing the limit as m goes to ∞ yields $|x_n^{(k)} - x^{(k)}| \leq \epsilon$. As k was arbitrary,
$$\sup_{k \in \mathbb{N}} |x_n^{(k)} - x^{(k)}| \leq \epsilon, \tag{1.14}$$
that is, $\mathbf{x}_n - \mathbf{x}$ belongs to ℓ^∞. As $\mathbf{x}_n \in \ell^\infty$, it now follows that $\mathbf{x} \in \ell^\infty$ too.

Step 3. Finally, we'll show that $(\mathbf{x}_n)_{n \in \mathbb{N}}$ converges to $\mathbf{x}$. In the case when $p < \infty$, proceeding as in Step 2, (1.13) gives for all $n > N$, $\|\mathbf{x}_n - \mathbf{x}\|_p \leq \epsilon$. When $p = \infty$, (1.14) gives $\|\mathbf{x}_n - \mathbf{x}\|_\infty \leq \epsilon$ for all $n > N$. $\square$

Exercise 1.29. (Characterisation of closed sets).
Let X be a normed space and F be a subset of X. Show that the following two statements are equivalent:
(1) F is closed.
(2) For every sequence $(x_n)_{n \in \mathbb{N}}$ in F ($\forall n \in \mathbb{N}$, $x_n \in F$), which is convergent in X with limit $x \in X$, we have that $x \in F$.

Exercise 1.30. Show that c_{00}, the set of all sequences with compact support (that is sequences which have all terms equal to zero eventually), is a subspace of ℓ^2 which is not closed.

Exercise 1.31. (Closure of a set).
Let X be a normed space and S be a subset of X. A point $L \in X$ is a *limit point of S* if there exists a sequence $(x_n)_{n \in \mathbb{N}}$ in $S \setminus \{L\}$ with limit L. The set consisting of all points and limit points of S is denoted by $\overline{S}$, and is called the *closure of S*.
(1) Prove that $\overline{S}$ is the smallest closed set which contains S.
(2) Show that if Y is a subspace of X, then $\overline{Y}$ is also a subspace of X.
(3) Prove that if C is a convex subset of X, then $\overline{C}$ is also convex.
(4) Show that a subset D of X is dense if and only if $\overline{D} = X$.

Exercise 1.32. Show that $\ell^1 \subsetneq \ell^2$.
Is ℓ^1 a Banach space with the topology induced from ℓ^2?

Exercise 1.33. Let c_0 be the set if all sequences convergent with limit 0. Then c_0 is a subspace of the normed space ℓ^∞. Prove that c_0 is a Banach space.

Exercise 1.34. Let $(X, \|\cdot\|)$ be a normed space, and let $(x_n)_{n\in\mathbb{N}}$ be a convergent sequence in X with limit x. Prove that $(\|x_n\|)_{n\in\mathbb{N}}$ is a convergent sequence in $\mathbb{R}$ and that $\lim\limits_{n\to\infty} \|x_n\| = \|x\|$.

Exercise 1.35. Show that if $1 \leq p \leq \infty$, then ℓ^p is a normed space.

Exercise 1.36. Show that $(C^1[a,b], \|\cdot\|_{1,\infty})$ is a Banach space.

Exercise 1.37. (∗) We have seen that if X is a Banach space, then every absolutely convergent series is convergent. The aim of this exercise is to show the converse. That is, prove that if X is a normed space with the property that every absolutely convergent series converges, then X is a Banach space. *Hint:* Construct a subsequence $(x_{n_k})_{k\in\mathbb{N}}$ of a given Cauchy sequence $(x_n)_{n\in\mathbb{N}}$ possessing the property that if $n > n_k$, then $\|x_n - x_{n_k}\| < 1/2^k$. Define $u_1 = x_{n_1}$, $u_{k+1} = x_{n_{k+1}} - x_{n_k}$, $k \in \mathbb{N}$, and consider the series with terms u_k.

Exercise 1.38. (Finite product of normed spaces).
If X, Y are normed spaces, then $X \times Y$ is a vector space with component-wise operations. Show that $\|(x,y)\| := \max\{\|x\|, \|y\|\}$, $(x,y) \in X \times Y$, defines a norm on $X \times Y$. Prove that if X, Y are Banach, then so is $X \times Y$.

1.5 Compact sets

In this section, we study an important class of subsets of a normed space, called *compact sets*. Before we learn the definition, let us give some motivation for this concept.

Of the different types of intervals in $\mathbb{R}$, perhaps the most important are those of the form $[a,b]$, where a, b are finite real numbers. Why are such intervals so important? We know of an important result, the Extreme Value Theorem[18], where such intervals play a vital role. Recall that the Extreme Value Theorem asserts that any continuous function $f : [a,b] \to \mathbb{R}$ attains a maximum and a minimum value on $[a,b]$. This result does not hold in general for continuous functions $f : I \to \mathbb{R}$ with $I = (a,b)$ or $I = [a,b)$ or $I = (a,\infty)$, and so on. Besides its theoretical importance in Analysis, the Extreme Value Theorem is also a fundamental result in Optimisation Theory. It turns out that when we want to generalise this result, the notion of "compact sets" is pertinent, and later on, we will learn the following analogue of the Extreme Value Theorem: If K is a compact subset of a normed space X and $f : K \to \mathbb{R}$ is continuous, then f assumes a maximum and a minimum on K. Here is the definition of a compact set.

---
[18]See for example [Sasane (2015), §3.4].

Definition 1.9. (Compact set). Let $(X, \|\cdot\|)$ be a normed space. A subset K of X is said to be *compact* if every sequence in K has a convergent subsequence with limit in K, that is, if $(x_n)_{n\in\mathbb{N}}$ is a sequence such that $x_n \in K$ for each $n \in \mathbb{N}$, then there exists a subsequence $(x_{n_k})_{k\in\mathbb{N}}$ which converges to some $L \in K$.

Example 1.23. (Compact intervals in $\mathbb{R}$). The interval $[a,b]$ is a compact subset of $\mathbb{R}$. Indeed, every sequence $(a_n)_{n\in\mathbb{N}}$ contained in $[a,b]$ is bounded, and thus by the Bolzano-Weierstrass Theorem, possesses a convergent subsequence, say $(a_{n_k})_{k\in\mathbb{N}}$, with limit L. But since $a \leq a_{n_k} \leq b$, for all k's, by letting $k \to \infty$, we obtain $a \leq L \leq b$, that is, $L \in [a,b]$. Hence $[a,b]$ is compact.

On the other hand, (a,b) is not compact, since the sequence
$$\left(a + \frac{b-a}{2n}\right)_{n\in\mathbb{N}}$$
is contained in (a,b), but it has no convergent subsequence whose limit belongs to (a,b). Indeed this is because the sequence is convergent, with limit a, and so every subsequence of this sequence is also convergent with limit a, which doesn't belong to (a,b).

$\mathbb{R}$ is not compact since the sequence $(n)_{n\in\mathbb{N}}$ cannot have a convergent subsequence. Indeed, if such a convergent subsequence existed, it would also be Cauchy, but the distance between any two distinct terms, being distinct integers, is at least 1, contradicting the Cauchyness. $\diamond$

In the above list of nonexamples, note that $\mathbb{R}$ is not bounded, and that (a,b) is not closed. On the other hand, in the example $[a,b]$, we see that $[a,b]$ is both bounded and closed. It turns out that **in $\mathbb{R}^d$**, having the property "closed and bounded" is a characterisation of compact sets, and we will show this below.

Theorem 1.10.
A subset K of $\mathbb{R}^d$ is compact if and only if K is closed and bounded.

Before showing this, we prove a technical result, which besides being interesting on its own, will also somewhat simplify the proof of the above theorem.

Lemma 1.4. *Every bounded sequence in $\mathbb{R}^d$ has convergent subsequence.*

Proof. As all norms on $\mathbb{R}^d$ are equivalent, it suffices to work with the $\|\cdot\|_2$ norm. We prove this using induction on d. Let us consider the case when $d = 1$. Then the statement is precisely the Bolzano-Weierstrass Theorem!

Suppose that the result has been proved in $\mathbb{R}^d$ for a $d \geq 1$. We'll show that it holds in $\mathbb{R}^{d+1}$. Let $(\mathbf{x}_n)_{n \in \mathbb{N}}$ be a bounded sequence. We split each $\mathbf{x}_n$ into its first d components and its last component in $\mathbb{R}$, and write $\mathbf{x}_n = (\boldsymbol{\alpha}_n, \beta_n)$, where $\boldsymbol{\alpha}_n \in \mathbb{R}^d$ and $\beta_n \in \mathbb{R}$. Since $\|\boldsymbol{\alpha}_n\|_2 \leq \sqrt{\|\boldsymbol{\alpha}_n\|_2^2 + \beta_n^2} = \|\mathbf{x}_n\|_2$, we see that $(\boldsymbol{\alpha}_n)_{n \in \mathbb{N}}$ is a bounded sequence in $\mathbb{R}^d$. By the induction hypothesis, it has a convergent subsequence, say $(\boldsymbol{\alpha}_{n_k})_{k \in \mathbb{N}}$ which converges to, say $\boldsymbol{\alpha} \in \mathbb{R}^d$. Now consider the sequence $(\beta_{n_k})_{k \in \mathbb{N}}$ in $\mathbb{R}$. Then $(\beta_{n_k})_{k \in \mathbb{N}}$ is bounded, and so by the Bolzano-Weierstrass Theorem, it has a convergent subsequence $(\beta_{n_{k_\ell}})_{\ell \in \mathbb{N}}$, with limit, say $\beta \in \mathbb{R}$. Then we have
$$\mathbf{x}_{n_{k_\ell}} = (\boldsymbol{\alpha}_{n_{k_\ell}}, \beta_{n_{k_\ell}}) \overset{\ell \to \infty}{\longrightarrow} (\boldsymbol{\alpha}, \beta) =: L \in \mathbb{R}^{d+1}.$$
So the bounded sequence $(\mathbf{x}_n)_{n \in \mathbb{N}}$ has $(\mathbf{x}_{n_{k_\ell}})_{\ell \in \mathbb{N}}$ as a convergent subsequence. $\square$

Also, we note that the "only if" part of Theorem 1.10 holds in *all* normed spaces.

Proposition 1.2. *Let $(X, \|\cdot\|)$ be a normed space, and $K \subset X$ be compact. Then K must be closed and bounded.*

Proof.
(K is closed:) Let $(\mathbf{x}_n)_{n \in \mathbb{N}}$ be a sequence in K that converges to $\mathbf{L}$. Then there is a convergent subsequence, say $(\mathbf{x}_{n_k})_{k \in \mathbb{N}}$ that is convergent to a limit $\mathbf{L}' \in K$. But as $(\mathbf{x}_{n_k})_{k \in \mathbb{N}}$ is a subsequence of a convergent sequence with limit $\mathbf{L}$, it is also convergent to $\mathbf{L}$. By the uniqueness of limits, $\mathbf{L} = \mathbf{L}' \in K$. Thus K is closed.

(K is bounded:) Suppose that K is not bounded. Then given any $n \in \mathbb{N}$, we can find an $\mathbf{x}_n \in K$ such that $\|\mathbf{x}_n\| > n$. But this implies that no subsequence of $(\mathbf{x}_n)_{n \in \mathbb{N}}$ is bounded. So no subsequence of $(\mathbf{x}_n)_{n \in \mathbb{N}}$ can be convergent either. This contradicts the compactness of K. Thus our assumption was incorrect, that is, K is bounded. $\square$

Now we return to the task of proving of Theorem 1.10.

Proof. It remains to just prove the "if" part. Let K be closed and bounded. Let $(\mathbf{x}_n)_{n \in \mathbb{N}}$ be a sequence in K. Then $(\mathbf{x}_n)_{n \in \mathbb{N}}$ is bounded, and so it has a convergent subsequence, with limit $\mathbf{L}$. But since K is closed and since each term of the sequence belongs to K, it follows that $\mathbf{L} \in K$. Consequently, K is compact. $\square$

Example 1.24. If $a, b \in \mathbb{R}$ and $a < b$, then the intervals $(a, b]$, $[a, b)$ are not compact in $\mathbb{R}$, since although they are bounded, they are not closed.

The intervals $(-\infty, b]$, $[a, \infty)$ are not compact, since although they are closed, they are not bounded. ◇

Let us consider an interesting compact subset of the real line, called the *Cantor set*.

Example 1.25. (Cantor set). The Cantor set is constructed as follows. Let $F_1 := [0, 1]$, and delete from F_1 the open interval $(\frac{1}{3}, \frac{2}{3})$ which is its middle third, and denote the remaining set by F_2. Thus we have that $F_2 = [0, \frac{1}{3}] \cup [\frac{2}{3}, 1]$. Next, delete from F_2 the middle thirds of its two pieces, namely the open intervals $(\frac{1}{9}, \frac{2}{9})$ and $(\frac{7}{9}, \frac{8}{9})$, and denote the remaining set by F_3. It can be checked that $F_3 = [0, \frac{1}{9}] \cup [\frac{2}{9}, \frac{1}{3}] \cup [\frac{2}{3}, \frac{7}{9}] \cup [\frac{8}{9}, 1]$. Continuing this process, that is, at each stage deleting the open middle third of each interval remaining from the previous stage, we obtain a sequence of sets F_n, each of which contains all of its successors.

The *Cantor set* is defined by $C := \bigcap_{n=1}^{\infty} F_n$.

C is contained in $[0, 1]$, and consists of those points in the interval $[0, 1]$ which "ultimately remain" after the removal of all the open intervals $(\frac{1}{3}, \frac{2}{3})$, $(\frac{1}{9}, \frac{2}{9})$, $(\frac{7}{9}, \frac{8}{9})$, $\cdots$. What points do remain? C clearly contains the endpoints of the intervals which make up each set F_n:

$$0, 1, \frac{1}{3}, \frac{2}{3}, \frac{1}{9}, \frac{2}{9}, \frac{7}{9}, \frac{8}{9}, \cdots.$$

Does C contain any other points? Actually, C contains many more points than the above list of end points. After all, the above list of endpoints is countable, but it can be shown that C is uncountable, see Example 7.1 on page 362.

As C is an intersection of closed sets, it is closed. Moreover it is contained in $[0, 1]$ and so it is also bounded. Consequently it is compact. (It turns out that the Cantor set is a very intricate mathematical object, and is often a source of interesting examples/counterexamples in Analysis. For example, it can be shown that the Lebesgue measure of C is 0, and so C is an example of an uncountable set with Lebesgue measure 0; see Example 7.1 on page 362.) ◇

Remark 1.3. Since all norms on a finite dimensional normed space are equivalent, we have the following consequence of Theorem 1.10.

Corollary 1.3. *Let X be a finite dimensional normed space. A subset $K \subset X$ is compact if and only if K is closed and bounded.*

However, in *infinite* dimensional normed spaces, although compact sets continue to be necessarily closed and bounded, it turns out that closed and bounded sets may fail to be compact. We give two examples below, the closed unit ball in ℓ^2 (Example 1.26) and in $C[0,1]$ (Example 1.28).

Example 1.26. (The closed unit ball in ℓ^2 is *not* compact.)
Consider the closed unit ball with centre 0 in the normed space ℓ^2:
$$K = \{\mathbf{x} \in \ell^2 : \|\mathbf{x}\|_2 \leqslant 1\}.$$
Then K is bounded, it is closed (since its complement is easily seen to be open), but K is not compact, as shown below.

For example, take the sequence $(\mathbf{e}_n)_{n \in \mathbb{N}}$, where $\mathbf{e}_n$ is the sequence with only the nth term equal to 1, and all other terms are equal to 0:
$$\mathbf{e}_n := (0, \cdots, 0, 1, 0, \cdots) \in K \subset \ell^2.$$
Then this sequence $(\mathbf{e}_n)_{n \in \mathbb{N}}$ in $K \subset \ell^2$ can have no convergent subsequence. Indeed, whenever $n \neq m$, $\|\mathbf{e}_n - \mathbf{e}_m\|_2 = \sqrt{2}$, and so any subsequence of $(\mathbf{e}_n)_{n \in \mathbb{N}}$ must be non-Cauchy, and hence also not convergent! $\diamond$

Example 1.27. (∗)(The Hilbert cube in ℓ^2 is compact.)
Let C denote the set of all real sequences $(x_n)_{n \in \mathbb{N}}$, whose nth term satisfies $0 \leqslant x_n \leqslant 1/n$ for all $n \in \mathbb{N}$. Then it is clear that C is a subset of the real vector space ℓ^2. It can be shown that C is a compact subset of ℓ^2, and we include a proof below, even though it is somewhat technical. The proof relies on creating subsequences of subsequences, and eventually using a "diagonal" sequence created in this process. We will also use a similar process in the proof of Theorem 5.4 (page 213) in Chapter 5. If the reader so wishes, he/she can skip the proof below for now, and move on to Example 1.28.

Let $(\mathbf{x}_m)_{m \in \mathbb{N}}$ be a sequence in C. The task is to produce a subsequence of this sequence which converges in ℓ^2 to an $\mathbf{x}$ in C. Let $\mathbf{x}_m = (x_m^{(n)})_{n \in \mathbb{N}}$. Then for all n and m, $0 \leqslant x_m^{(n)} \leqslant 1/n$.

In particular, $0 \leqslant x_1^{(1)}, x_2^{(1)}, x_3^{(1)}, \cdots \leqslant 1$, and as $[0,1]$ is compact, there is a subsequence $m_1(1), m_1(2), m_1(3), \cdots$ of $1, 2, 3, \cdots$ such that $(x_{m_1(j)}^{(1)})_{j \in \mathbb{N}}$ is convergent, with limit, say, $x^{(1)} \in [0,1]$.

Now $0 \leqslant x^{(2)}_{m_1(2)}, x^{(2)}_{m_1(3)}, \cdots \leqslant 1/2$, and as $[0, 1/2]$ is compact, there is a subsequence $(m_2(j))_{j \geqslant 2}$ of $(m_1(j))_{j \geqslant 2}$ such that $(x^{(2)}_{m_2(j)})_{j \geqslant 2}$ is convergent, with limit, say, $x^{(2)} \in [0, 1/2]$.

Proceeding in this manner, we get for all ℓ that there is a subsequence $(m_\ell(j))_{j \geqslant \ell}$ of $(m_{\ell-1}(j))_{j \geqslant \ell}$ such that $(x^{(2)}_{m_\ell(j)})_{j \geqslant \ell}$ is convergent, with limit $x^{(\ell)} \in [0, 1/\ell]$. We claim that $(\mathbf{x}_{m_j(j)})_{j \in \mathbb{N}}$ converges in ℓ^2 to the sequence $\mathbf{x} := (x^{(n)})_{n \in \mathbb{N}}$. See the schematic diagram below.

$$
\begin{array}{cccc}
1 & 2 & 3 & \cdots \\
m_1(1) & m_1(2) & m_1(3) & \cdots \\
& m_2(2) & m_2(3) & \cdots \\
& & m_3(3) & \cdots \\
& & & \cdots
\end{array}
\quad
\begin{array}{l}
[0,1] \ni x^{(1)}_{m_1(1)}, x^{(1)}_{m_1(2)}, x^{(1)}_{m_1(3)} \cdots \longrightarrow x^{(1)} \\
[0, \tfrac{1}{2}] \ni \quad\quad\quad x^{(2)}_{m_2(2)}, x^{(2)}_{m_2(3)}, \cdots \longrightarrow x^{(2)} \\
[0, \tfrac{1}{3}] \ni \quad\quad\quad\quad\quad\quad x^{(3)}_{m_3(3)}, \cdots \longrightarrow x^{(3)} \\
\quad\quad\quad\quad\quad\quad\quad\quad\quad\quad \cdots
\end{array}
$$

First, let us note that $\mathbf{x} \in C$ because for all $n \in \mathbb{N}$, $0 \leqslant x^{(n)} \leqslant 1/n$.

Now the plan is to show that $\|\mathbf{x}_{m_j(j)} - \mathbf{x}\|_2$ is small for all js sufficiently large. To do this, we will split this quantity into two parts, and estimate them separately:

$$\|\mathbf{x}_{m_j(j)} - \mathbf{x}\|_2^2 = \sum_{n=1}^{N} \left(x^{(n)}_{m_j(j)} - x^{(n)}\right)^2 + \sum_{n=N+1}^{\infty} \left(x^{(n)}_{m_j(j)} - x^{(n)}\right)^2. \qquad (1.15)$$

Let us see how to handle the second summand on the right-hand side above. Let $\epsilon > 0$. Let N be such that $\sum_{n=N+1}^{\infty} \dfrac{1}{n^2} < \epsilon$. Then we have

$$\sum_{n=N+1}^{\infty} \left(x^{(n)}_{m_j(j)} - x^{(n)}\right)^2 = \sum_{n=N+1}^{\infty} \left(x^{(n)}_{m_j(j)}\right)^2 + \sum_{n=N+1}^{\infty} \left(x^{(n)}\right)^2 - 2 \sum_{n=N+1}^{\infty} x^{(n)}_{m_j(j)} x^{(n)}$$

$$\leqslant \epsilon + \epsilon + 2 \Big(\sum_{n=N+1}^{\infty} \left(x^{(n)}_{m_j(j)}\right)^2 \Big)^{\frac{1}{2}} \Big(\sum_{n=N+1}^{\infty} \left(x^{(n)}\right)^2 \Big)^{\frac{1}{2}}$$

$$\leqslant \epsilon + \epsilon + 2\sqrt{\epsilon}\sqrt{\epsilon} = 4\epsilon.$$

Having accomplished this, let us now estimate the first summand in (1.15). For all $n \leqslant N$, $x^{(n)}_{m_N(j)} \xrightarrow{j \to \infty} x^{(n)}$. As $(m_j(j))_{j \geqslant N}$ is a subsequence of $(m_N(j))_{j \geqslant N}$, and hence also of each $(m_n(j))_{j \geqslant n}$ for $n \leqslant N$, it follows that for all $n \leqslant N$, $x^{(n)}_{m_j(j)} \xrightarrow{j \to \infty} x^{(n)}$.

So we can find a J such that for all $j > J$, $\sum_{n=1}^{N}\left(x_{m_j(j)}^{(n)} - x^{(n)}\right)^2 \leq \epsilon$. Thus

$$\|\mathbf{x}_{m_j(j)} - \mathbf{x}\|_2^2 = \sum_{n=1}^{N}\left(x_{m_j(j)}^{(n)} - x^{(n)}\right)^2 + \sum_{n=N+1}^{\infty}\left(x_{m_j(j)}^{(n)} - x^{(n)}\right)^2 \leq \epsilon + 4\epsilon = 5\epsilon.$$

As $\epsilon > 0$ was arbitrary, we have indeed shown that $(\mathbf{x}_{m_j(j)})_{j\in\mathbb{N}}$ converges in ℓ^2 to $\mathbf{x} \in C$. This completes the proof of the compactness of the Hilbert cube C. $\diamond$

Example 1.28. (The closed unit ball in $(C[0,1], \|\cdot\|_\infty)$ is *not* compact.) Consider the closed unit ball with centre $\mathbf{0}$ in $(C[0,1], \|\cdot\|_\infty)$:

$$K = \{\mathbf{x} \in C[0,1] : \|\mathbf{x}\|_\infty \leq 1\}.$$

Then K is bounded, and also it is closed (since its complement is open). But K is not compact, and this can be demonstrated by considering the sequence $(\mathbf{x}_n)_{n\in\mathbb{N}}$, where the graphs of the $\mathbf{x}_n$s have "narrowing" tents of height 1, with the supports of the tents moving to the right, on half of each remaining interval, as shown in the following picture:

Then this sequence does not have a convergent subsequence, since if it did, then the convergent subsequence would be Cauchy, but whenever $n \neq m$, $\|\mathbf{x}_n - \mathbf{x}_m\|_\infty = 1$, a contradiction to the Cauchyness. $\diamond$

Exercise 1.39. Let K be a compact subset of $\mathbb{R}^d$ (with the Euclidean norm $\|\cdot\|_2$), and $F \subset \mathbb{R}^d$ be a closed subset. Show that $F \cap K$ is compact in $\mathbb{R}^d$.

Exercise 1.40. Show that the unit sphere $\mathbb{S}^{d-1} := \{\mathbf{x} \in \mathbb{R}^d : \|\mathbf{x}\|_2 = 1\}$ with centre $\mathbf{0}$ in $\mathbb{R}^d$, is compact in $\mathbb{R}^d$.

Exercise 1.41. (∗)
Consider the normed space $(\mathbb{R}^{2\times 2}, \|\cdot\|_\infty)$ from Exercise 1.13, page 17.
(1) Show that the set $O(2) := \{\mathbf{R} \in \mathbb{R}^{2\times 2} : \mathbf{R}^\top \mathbf{R} = \mathbf{I}\}$ of orthogonal matrices is a compact set.
(2) Is the indefinite orthogonal group $O(1,1) := \{\mathbf{R} \in \mathbb{R}^{2\times 2} : \mathbf{R}^\top J \mathbf{R} = J\}$, where $J := \begin{bmatrix} 1 & 0 \\ 0 & -1 \end{bmatrix}$ also compact?

Hint: Consider "hyperbolic rotations" $\mathbf{R} := \begin{bmatrix} \cosh t & \sinh t \\ \sinh t & \cosh t \end{bmatrix}$.

Exercise 1.42. Show that $\left\{1, \frac{1}{2}, \frac{1}{3}, \cdots\right\} \bigcup \{0\}$ is compact in $\mathbb{R}$.

Remark 1.4. (Definition of compactness.) The notion of a compact set that we have defined is really *sequential* compactness. In the context of the more general topological spaces, one defines the notion of compactness as follows.

Definition 1.10. Let X be a topological space with the topology given by the family of open sets $\mathcal{O}$. Let $K \subset X$. A collection $\mathcal{C} = \{U_i : i \in I\}$ of open sets is said to be an *open cover of* K if $K \subset \bigcup_{i \in I} U_i$.
$K \subset X$ is said to be a *compact set* if every open cover of K has a finite subcover, that is, given any open cover $\mathcal{C} = \{U_i : i \in I\}$ of K, there exist finitely many indices $i_1, \cdots, i_n \in I$ such that $K \subset U_{i_1} \cup \cdots \cup U_{i_n}$.

In the case of *normed spaces*, it can be shown that the set of compact sets *coincides* with the set of sequentially compact sets. But in general topological spaces, these two notions may not be the same; see for instance [Steen and Seebach (1995)].

Chapter 2

Continuous and linear maps

A normed space has two structures: a linear one (the underlying vector space), and a topological one (the norm). So when we study *maps* between normed spaces, it is natural to focus on maps which are well-behaved with these structures, and we'll do this now. In particular, we'll study:

(1) linear transformations
(well-behaved with respect to the linear structure),
(2) continuous maps
(well-behaved with respect to the topological structure),
(3) continuous linear transformations
(well-behaved with respect to both structures).

In the context of normed spaces, continuous linear transformations are most important, and these are sometimes also called *bounded linear operators*.

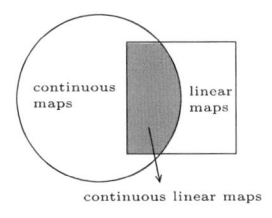

continuous maps | linear maps
continuous linear maps

The reason for this terminology will become clear in Theorem 2.6 (page 67). We'll see that the set of all bounded linear operators is itself a vector space, with obvious pointwise operations of addition and scalar multiplication, and it also has a natural notion of a norm, called the *operator norm*. Equipped with the operator norm, the vector space of bounded linear operators is a Banach space, provided that the co-domain is a Banach space. This is a useful result, which we will use in order to prove the existence of solutions to integral and differential equations.

2.1 Linear transformations

Linear transformations are maps that respect vector space operations.

Definition 2.1. (Linear transformation).
Let X and Y be vector spaces over $\mathbb{K}$ ($\mathbb{R}$ or $\mathbb{C}$).
A map $T : X \to Y$ is called a *linear transformation* if:
 (L1) For all $x_1, x_2 \in X$, $T(x_1 + x_2) = T(x_1) + T(x_2)$.
 (L2) For all $x \in X$ and all $\alpha \in \mathbb{K}$, $T(\alpha \cdot x) = \alpha \cdot T(x)$.

Example 2.1. (Linear galore!)
(1) $D : C^1[a,b] \to C[a,b]$ given by $D\mathbf{x} = \mathbf{x}'$, $\mathbf{x} \in C^1[a,b]$ is a linear transformation, since
 (L1) $D(\mathbf{x}+\mathbf{y}) = (\mathbf{x}+\mathbf{y})' = \mathbf{x}' + \mathbf{y}' = D\mathbf{x} + D\mathbf{y}$ for all $\mathbf{x}, \mathbf{y} \in C^1[a,b]$;
 (L2) $D(\alpha\mathbf{x}) = (\alpha\mathbf{x})' = \alpha \cdot \mathbf{x}'$ for all $\alpha \in \mathbb{R}$ and $\mathbf{x} \in C^1[a,b]$.

(2) Let $m, n \in \mathbb{N}$ and $X = \mathbb{R}^n$ and $Y = \mathbb{R}^m$.

If $A = \begin{bmatrix} a_{11} & \cdots & a_{1n} \\ \vdots & & \vdots \\ a_{m1} & \cdots & a_{mn} \end{bmatrix} \in \mathbb{R}^{m \times n}$, then $T_A : \mathbb{R}^n \to \mathbb{R}^m$ defined by

$$T_A \begin{bmatrix} x_1 \\ \vdots \\ x_n \end{bmatrix} = \begin{bmatrix} a_{11}x_1 + \cdots + a_{1n}x_n \\ \vdots \\ a_{m1}x_1 + \cdots + a_{mn}x_n \end{bmatrix} \text{ for all } \begin{bmatrix} x_1 \\ \vdots \\ x_n \end{bmatrix} \in \mathbb{R}^n, \qquad (2.1)$$

is a linear transformation from $\mathbb{R}^n$ to $\mathbb{R}^m$. Indeed,

$$T_A \left(\begin{bmatrix} x_1 \\ \vdots \\ x_n \end{bmatrix} + \begin{bmatrix} y_1 \\ \vdots \\ y_n \end{bmatrix} \right) = T_A \begin{bmatrix} x_1 \\ \vdots \\ x_n \end{bmatrix} + T_A \begin{bmatrix} y_1 \\ \vdots \\ y_n \end{bmatrix}, \quad \begin{bmatrix} x_1 \\ \vdots \\ x_n \end{bmatrix}, \begin{bmatrix} y_1 \\ \vdots \\ y_n \end{bmatrix} \in \mathbb{R}^n,$$

and so (L1) holds. Moreover,

$$T_A \left(\alpha \cdot \begin{bmatrix} x_1 \\ \vdots \\ x_n \end{bmatrix} \right) = \alpha \cdot T_A \begin{bmatrix} x_1 \\ \vdots \\ x_n \end{bmatrix} \text{ for all } \alpha \in \mathbb{R} \text{ and all } \begin{bmatrix} x_1 \\ \vdots \\ x_n \end{bmatrix} \in \mathbb{R}^n,$$

and so (L2) holds as well. Hence T_A is a linear transformation.

(3) Let $X = Y = \ell^2$. Define the *left/right shift operators* L, R as follows:
if $\mathbf{x} = (x_n)_{n \in \mathbb{N}} \in \ell^2$, then
$$L\big((x_1, x_2, x_3, \cdots)\big) = (x_2, x_3, x_4, \cdots) \text{ and}$$
$$R\big((x_1, x_2, x_3, \cdots)\big) = (0, x_1, x_2, x_3, \cdots).$$
Then it is easy to see that R and L are linear transformations.

(4) Let $X := c \ (\subset \ell^\infty)$, the space of all real valued convergent sequences, and $Y = \mathbb{R}$. The map $L : c \to \mathbb{R}$, $L\big((a_n)_{n\in\mathbb{N}}\big) := \lim_{n\to\infty} a_n$, for $(a_n)_{n\in\mathbb{N}}$, is a linear transformation (using the algebra of limits).

Recall that given a linear transformation $T : X \to Y$, we can associate with T two natural subspaces, of X and Y, respectively,

the kernel of T, $\ker T := \{x \in X : Tx = \mathbf{0}_Y\} \subset X$, and

the range of T, $\operatorname{ran} T := \{y \in Y : \exists x \in X \text{ such that } y = Tx\} \subset Y$.

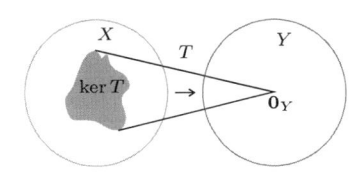

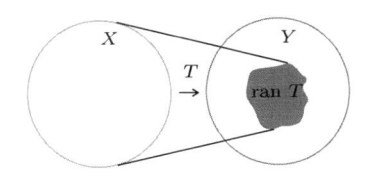

In the above example of the linear transformation L, we have

$\ker L = c_0$ (set of sequences convergent with limit 0),

$\operatorname{ran} L = \mathbb{R}$ (since for every $r \in \mathbb{R}$, the constant sequence $(r)_{n\in\mathbb{N}}$ converges to r).

(5) The map $I : C[a,b] \to \mathbb{R}$, given by $I(\mathbf{x}) = \int_a^b x(t) dt$ for all $\mathbf{x} \in C[a,b]$, is a linear transformation.

(6) Let $S := \{\mathbf{h} \in C^1[a,b] : \mathbf{h}(a) = \mathbf{h}(b) = 0\}$. From Exercise 1.3 (page 7), we see that S is a subspace of $C^1[a,b]$. Let $\mathbf{A}, \mathbf{B} \in C[a,b]$ be fixed functions. Let $L : S \to \mathbb{R}$ be given by $L(\mathbf{h}) = \int_a^b \big(\mathbf{A}(t)\mathbf{h}(t) + \mathbf{B}(t)\mathbf{h}'(t)\big) dt$, $\mathbf{h} \in S$.

Let us check that L is a linear transformation. We have:

(L1) For all $\mathbf{h}_1, \mathbf{h}_2 \in S$,

$L(\mathbf{h}_1 + \mathbf{h}_2)$

$= \int_a^b \big(\mathbf{A}(t)(\mathbf{h}_1(t) + \mathbf{h}_2(t)) + \mathbf{B}(t)(\mathbf{h}_1'(t) + \mathbf{h}_2'(t))\big) dt$

$= \int_a^b \big(\mathbf{A}(t)\mathbf{h}_1(t) + \mathbf{B}(t)\mathbf{h}_1'(t) + \mathbf{A}(t)\mathbf{h}_2(t) + \mathbf{B}(t)\mathbf{h}_2'(t)\big) dt$

$= \int_a^b \big(\mathbf{A}(t)\mathbf{h}_1(t) + \mathbf{B}(t)\mathbf{h}_1'(t)\big) dt + \int_a^b \big(\mathbf{A}(t)\mathbf{h}_2(t) + \mathbf{B}(t)\mathbf{h}_2'(t)\big) dt$

$= L(\mathbf{h}_1) + L(\mathbf{h}_2)$.

(L2) For all $\mathbf{h} \in C^1[0,1]$ and all $\alpha \in \mathbb{R}$,
$$L(\alpha \cdot \mathbf{h}) = \int_a^b \big(\mathbf{A}(t)\alpha\mathbf{h}(t) + \mathbf{B}(t)\alpha\mathbf{h}'(t)\big)dt$$
$$= \alpha \int_a^b \big(\mathbf{A}(t)\mathbf{h}(t) + \mathbf{B}(t)\mathbf{h}'(t)\big)dt = \alpha L(\mathbf{h}).$$
Thus L is a linear transformation.

(7) Let $C^1(\mathbb{R})$, $C^2(\mathbb{R})$ denote the vector spaces of once, respectively twice continuously differentiable real-valued functions in $\mathbb{R}$ with pointwise operations. For $\mathbf{f} \in C^2(\mathbb{R})$ and $\mathbf{g} \in C^1(\mathbb{R})$, consider the initial value problem for the one (spatial) dimensional wave equation:

$$\text{(IVP)}: \begin{cases} \dfrac{\partial^2 u}{\partial t^2} - c^2 \dfrac{\partial^2 u}{\partial x^2} = 0 & (t \geqslant 0,\ x \in \mathbb{R}), \\ u(x,0) = f(x) & (x \in \mathbb{R}), \\ \dfrac{\partial u}{\partial t}(x,0) = g(x) & (x \in \mathbb{R}). \end{cases}$$

Let $C^2(\mathbb{R} \times [0,\infty))$ denote the vector space of all twice continuously differentiable functions $(x,t) \overset{u}{\mapsto} u(x,t) : \mathbb{R} \times [0,\infty) \longrightarrow \mathbb{R}$, again with pointwise operations. Then it can be shown that the unique solution $u_{\mathbf{f},\mathbf{g}}$ in $C^2(\mathbb{R} \times [0,\infty))$ to (IVP) is given by *d'Alembert's Formula*,

$$u_{\mathbf{f},\mathbf{g}}(x,t) = \frac{\mathbf{f}(x+ct) + \mathbf{f}(x-ct)}{2} + \frac{1}{2c}\int_{x-ct}^{x+ct} g(\xi)d\xi, \quad x \in \mathbb{R},\ t \geqslant 0.$$

Then the map $(\mathbf{f},\mathbf{g}) \mapsto u_{\mathbf{f},\mathbf{g}} : C^2(\mathbb{R}) \times C^1(\mathbb{R}) \to C^2(\mathbb{R} \times [0,\infty))$ is a linear transformation. $\diamond$

Here are some non examples.

Example 2.2. (Not quite linear!)

(1) If $\cdot^*$ denotes complex conjugation, then the complex conjugation map $z \mapsto z^* : \mathbb{C} \to \mathbb{C}$ is not a linear transformation, since although (L1) *is* satisfied: $(z+w)^* = z^* + w^*$ $(z, w \in \mathbb{C})$, we see that (L2) *isn't*: indeed, $(i \cdot 1)^* = i^* = -i \neq i \cdot 1^*$.

(2) Consider the map $T : \mathbb{R}^2 \to \mathbb{R}^2$ defined by

$$T\begin{bmatrix} x \\ y \end{bmatrix} = \begin{cases} \begin{bmatrix} x^2/y \\ y \end{bmatrix} & \text{if } xy \neq 0 \\ \begin{bmatrix} x \\ y \end{bmatrix} & \text{if } xy = 0 \end{cases}, \quad \text{for } \begin{bmatrix} x \\ y \end{bmatrix} \in \mathbb{R}^2.$$

Then T is not a linear transformation since (L1) is *not* satisfied.
Indeed, $T\left(\begin{bmatrix} -1 \\ 0 \end{bmatrix} + \begin{bmatrix} 0 \\ 1 \end{bmatrix}\right) = T\left(\begin{bmatrix} -1 \\ 1 \end{bmatrix}\right) = \begin{bmatrix} (-1)^2/1 \\ 1 \end{bmatrix} = \begin{bmatrix} 1 \\ 1 \end{bmatrix},$
while $T\begin{bmatrix} -1 \\ 0 \end{bmatrix} + T\begin{bmatrix} 0 \\ 1 \end{bmatrix} = \begin{bmatrix} -1 \\ 0 \end{bmatrix} + \begin{bmatrix} 0 \\ 1 \end{bmatrix} = \begin{bmatrix} -1 \\ 1 \end{bmatrix}.$

If $\alpha \in \mathbb{R} \setminus \{0\}$ and $\begin{bmatrix} x \\ y \end{bmatrix} \in \mathbb{R}^2$, then we have

$$T\left(\alpha \cdot \begin{bmatrix} x \\ y \end{bmatrix}\right) = T\begin{bmatrix} \alpha x \\ \alpha y \end{bmatrix} = \begin{cases} \begin{bmatrix} (\alpha^2 x^2)/(\alpha y) \\ \alpha y \end{bmatrix} & \text{if } xy \neq 0 \\ \begin{bmatrix} \alpha x \\ \alpha y \end{bmatrix} & \text{if } xy = 0 \end{cases}$$

$$= \begin{cases} \alpha \cdot \begin{bmatrix} x^2/y \\ y \end{bmatrix} & \text{if } xy \neq 0 \\ \alpha \cdot \begin{bmatrix} x \\ y \end{bmatrix} & \text{if } xy = 0 \end{cases} = \alpha \cdot \left(T\begin{bmatrix} x \\ y \end{bmatrix}\right).$$

If $\alpha = 0$ and $\begin{bmatrix} x \\ y \end{bmatrix} \in \mathbb{R}^2$, $T\left(\alpha \cdot \begin{bmatrix} x \\ y \end{bmatrix}\right) = T\begin{bmatrix} 0 \\ 0 \end{bmatrix} = \begin{bmatrix} 0 \\ 0 \end{bmatrix} = 0 \cdot \left(T\begin{bmatrix} x \\ y \end{bmatrix}\right).$

So for all $\alpha \in \mathbb{R}$, $\begin{bmatrix} x \\ y \end{bmatrix} \in \mathbb{R}^2$, $T\left(\alpha \cdot \begin{bmatrix} x \\ y \end{bmatrix}\right) = \alpha \cdot \left(T\begin{bmatrix} x \\ y \end{bmatrix}\right).$ Thus (L2) holds. $\diamond$

Notation 2.1. We will denote the set of all linear transformations from the vector space X to the vector space Y by $L(X, Y)$. Recall from elementary linear algebra that $L(X, Y)$ is itself a vector space (over the common field $\mathbb{K}$ for X, Y) with pointwise operations: if $T, S \in L(X, Y)$, then we define $T + S \in L(X, Y)$ by $(T + S)(x) = Tx + Sx$, for all $x \in X$, and if $\alpha \in \mathbb{K}$ and $T \in L(X, Y)$, then we define $\alpha \cdot T \in L(X, Y)$ by $(\alpha \cdot T)(x) = \alpha \cdot (Tx)$, for all $x \in X$. What is the zero vector in this vector space $L(X, Y)$? It is the "zero linear transformation" $\mathbf{0} : X \to Y$, given by $\mathbf{0}x = \mathbf{0}_Y$, for all $x \in X$, where $\mathbf{0}_Y$ denotes the zero vector in Y.

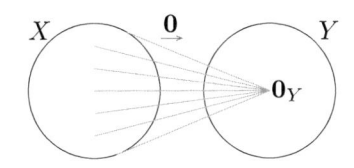

If $X = Y$, then we write $L(X)$ instead of $L(X, X)$.

Exercise 2.1. Consider the two maps $S_1, S_2 : C[0,1] \to \mathbb{R}$ given by
$$S_1(\mathbf{x}) = \int_0^1 (\mathbf{x}(t))^2 dt \text{ and } S_2(\mathbf{x}) = \int_0^1 \mathbf{x}(t^2) dt \text{ for } \mathbf{x} \in C[0,1].$$
Show that S_1 is not a linear transformation, while S_2 is.

Exercise 2.2. Let a, b be nonzero real numbers, and consider the two real-valued functions $\mathbf{f}_1, \mathbf{f}_2$ defined on $\mathbb{R}$ by $\mathbf{f}_1(t) = e^{at} \cos(bt)$ and $\mathbf{f}_2(t) = e^{at} \sin(bt)$, $t \in \mathbb{R}$. $\mathbf{f}_1$ and $\mathbf{f}_2$ are vectors belonging to the infinite dimensional vector space $C^1(\mathbb{R})$, consisting of all continuously differentiable functions from $\mathbb{R}$ to $\mathbb{R}$. Denote by $S_{\mathbf{f}_1,\mathbf{f}_2}$ the span of the two functions $\mathbf{f}_1$ and $\mathbf{f}_2$.

 (1) Prove that $\mathbf{f}_1$ and $\mathbf{f}_2$ are linearly independent in $C^1(\mathbb{R})$.
 (2) Show that the differentiation map, $\mathbf{f} \stackrel{D}{\mapsto} \mathbf{f}' : S_{\mathbf{f}_1,\mathbf{f}_2} \longrightarrow S_{\mathbf{f}_1,\mathbf{f}_2}$, is a linear transformation.
 (3) What is the matrix $[D]_B$ of D with respect to the (ordered) basis $B = (\mathbf{f}_1, \mathbf{f}_2)$?
 (4) Prove that D is invertible, and write down the matrix corresponding to the inverse of D.
 (5) Compute the indefinite integrals $\int e^{at} \cos(bt) dt$ and $\int e^{at} \sin(bt) dt$.

2.2 Continuous maps

Let X and Y be normed spaces. As there is a notion of distance between pairs of vectors in either space (provided by the norm of the difference of the pair of vectors in each respective space), one can talk about continuity of maps. Within the huge collection of all maps, the class of continuous maps form an important subset. Continuous maps play a prominent role in functional analysis since they possess some useful properties.

Before discussing the case of a function between normed spaces, let us first of all recall the notion of continuity of a function $f : \mathbb{R} \to \mathbb{R}$.

Continuity of functions from $\mathbb{R}$ to $\mathbb{R}$

In everyday speech, a 'continuous' process is one that proceeds without gaps of interruptions or sudden changes. What does it mean for a function $f : \mathbb{R} \to \mathbb{R}$ to be continuous? The common informal definition of this concept states that a function f is continuous if one can sketch its graph without lifting the pencil. In other words, the graph of f has no breaks in it. If a break does occur in the graph, then this break will occur at some point. Thus (based on this visual view of continuity), we first give the formal definition of the continuity of a function *at a point* below. Next, if a function is continuous at *each* point, then it will be called continuous. If a function has a break at a point, say x_0, then even if points x are close to x_0, the points $f(x)$ do not get close to $f(x_0)$.

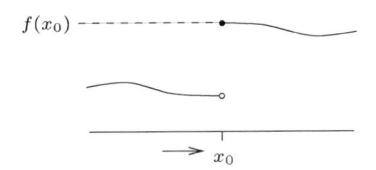

This motivates the definition of continuity in calculus, which guarantees that if a function is continuous at a point x_0, then we can make $f(x)$ as close as we like to $f(x_0)$, by choosing x sufficiently close to x_0.

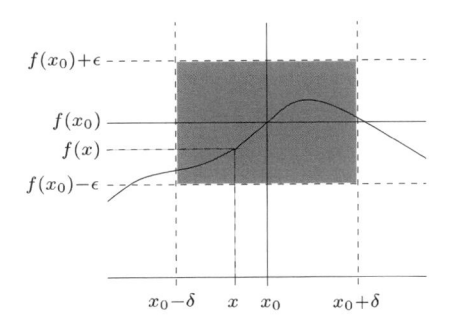

Definition 2.2. A function $f : \mathbb{R} \to \mathbb{R}$ is *continuous at* x_0 if for every $\epsilon > 0$, there exists a $\delta > 0$ such that for all $x \in \mathbb{R}$ satisfying $|x - x_0| < \delta$, we have that $|f(x) - f(x_0)| < \epsilon$.

$f : \mathbb{R} \to \mathbb{R}$ is *continuous* if for every $x_0 \in \mathbb{R}$, f is continuous at x_0.

Continuity of functions between normed spaces

We now define the set of continuous maps from a normed space X to a normed space Y.

We observe that in the definition of continuity in ordinary calculus, if x, y are real numbers, then $|x-y|$ is a measure of the distance between them, and that the absolute value $|\cdot|$ is a norm in the finite (one) dimensional normed space $\mathbb{R}$. So it is natural to define continuity in arbitrary normed spaces by simply replacing the absolute values by the corresponding norms, since the norm provides a notion of distance between vectors.

Definition 2.3. (Continuity of maps between normed spaces).
Let X and Y be normed spaces over $\mathbb{K}$ ($\mathbb{R}$ or $\mathbb{C}$). Let $x_0 \in X$. A map $f : X \to Y$ is *continuous at* x_0 if for every $\epsilon > 0$, there exists a $\delta > 0$ such that for all $x \in X$ satisfying $\|x - x_0\| < \delta$, we have $\|f(x) - f(x_0)\| < \epsilon$.
$f : X \to Y$ is *continuous* if for all $x_0 \in X$, f is continuous at x_0.

We will soon study when linear transformations are continuous, but first let us consider some examples of nonlinear maps.

Example 2.3. Consider the map $S : C[0,1] \to \mathbb{R}$, given by
$$S(\mathbf{x}) = \int_0^1 \big(\mathbf{x}(t)\big)^2 dt, \quad \mathbf{x} \in C[0,1].$$
We'll show that S is continuous. (As usual, $C[0,1]$ is endowed with the supremum norm.) Suppose that $\mathbf{x}_0 \in C[0,1]$. Let $\epsilon > 0$. As we would like to make $|S(\mathbf{x}) - S(\mathbf{x}_0)|$ small, let us first consider this expression. We have
$$|S(\mathbf{x}) - S(\mathbf{x}_0)| = \left| \int_0^1 \big(\mathbf{x}(t)\big)^2 dt - \int_0^1 \big(\mathbf{x}_0(t)\big)^2 dt \right|$$
$$= \left| \int_0^1 \big(\mathbf{x}(t) - \mathbf{x}_0(t)\big)\big(\mathbf{x}(t) - \mathbf{x}_0(t)\big) dt \right|$$
$$\leq \int_0^1 |(\mathbf{x} - \mathbf{x}_0)(t)| |(\mathbf{x} - \mathbf{x}_0)(t) + 2\mathbf{x}_0(t)| dt$$
$$\leq \int_0^1 \|\mathbf{x} - \mathbf{x}_0\|_\infty \big(\|\mathbf{x} - \mathbf{x}_0\|_\infty + 2\|\mathbf{x}_0\|_\infty\big) dt$$
$$< \int_0^1 \delta(\delta + 2\|\mathbf{x}_0\|_\infty) dt = \delta(\delta + 2\|\mathbf{x}_0\|_\infty),$$
if $\|\mathbf{x} - \mathbf{x}_0\|_\infty < \delta$, where $\delta > 0$ is some number. We ought to choose $\delta > 0$ suitably so as to make the right-hand side above smaller than ϵ. There is no unique way to do this, and anything one can justify works. We set
$$\delta := \min\left\{1, \frac{\epsilon}{1 + 2\|\mathbf{x}_0\|_\infty}\right\} > 0.$$
Whenever $\|\mathbf{x} - \mathbf{x}_0\|_\infty < \delta$, in light of the above computation, we have
$$|S(\mathbf{x}) - S(\mathbf{x}_0)| < \delta(\delta + 2\|\mathbf{x}_0\|_\infty) \leq \delta(1 + 2\|\mathbf{x}_0\|_\infty)$$
$$\leq \frac{\epsilon}{1 + 2\|\mathbf{x}_0\|_\infty}(1 + 2\|\mathbf{x}_0\|_\infty) = \epsilon.$$
Thus S is continuous at $\mathbf{x}_0$. As the choice of $\mathbf{x}_0$ was arbitrary, it follows that S is continuous (on $C[0,1]$). ◇

Example 2.4. c_{00} is the subspace of ℓ^∞ of all finitely supported sequences. c_{00} is a normed space with the supremum norm inherited from ℓ^∞.

Consider the map $s : c_{00} \to \mathbb{R}$ given by $s(\mathbf{a}) = \sum_{n=1}^{\infty} a_n^2$, $\mathbf{a} = (a_n)_{n \in \mathbb{N}} \in c_{00}$.

We'll show that s is not continuous at $\mathbf{0}$. Suppose on the contrary that it is. With $\epsilon = 1/4 > 0$, there exists a $\delta > 0$ such that if $\|\mathbf{a}\|_\infty = \|\mathbf{a} - \mathbf{0}\|_\infty < \delta$, then we are guaranteed that $|s(\mathbf{a}) - s(\mathbf{0})| = |s(\mathbf{a}) - 0| = |s(\mathbf{a})| < \epsilon = 1/4$.

If $\mathbf{a}_m = \left(\frac{1}{\sqrt{m+1}}, \cdots, \frac{1}{\sqrt{2m}}, 0, \cdots \right) \in c_{00}$, $m \in \mathbb{N}$, then $\|\mathbf{a}_m\|_\infty = \frac{1}{\sqrt{m+1}} \xrightarrow{m \to \infty} 0$.

So for all m sufficiently large, we must have $\|\mathbf{a}_m\|_\infty < \delta$, giving in turn that $|s(\mathbf{a}_m)| < 1/4$. But for all m we have

$$|s(\mathbf{a}_m)| = \left(\frac{1}{\sqrt{m+1}}\right)^2 + \cdots + \left(\frac{1}{\sqrt{2m}}\right)^2 + 0^2 + \cdots = \frac{1}{m+1} + \cdots + \frac{1}{2m}$$
$$> \underbrace{\frac{1}{2m} + \cdots + \frac{1}{2m}}_{m \text{ times}} = m \cdot \frac{1}{2m} = \frac{1}{2} > \frac{1}{4},$$

a contradiction. Hence s is not continuous at $\mathbf{0}$. ◇

Exercise 2.3. (Rationale for the $C^1[a,b]$ norm.)
This exercise concerns the norm on $C^1[a,b]$ we have chosen to use. Since we want to be able to use ordinary analytic operations such as passage to the limit, then, given a function $f : C^1[0,1] \to \mathbb{R}$, it is reasonable to choose a norm such that f is continuous. As our f, let us take the *arc length* function given by

$$f(\mathbf{x}) = \int_0^1 \sqrt{1 + (\mathbf{x}'(t))^2}\, dt, \quad \mathbf{x} \in C^1[0,1].$$

We show in the following sequence of exercises that f is not continuous if we equip $C^1[0,1]$ with the supremum norm $\|\cdot\|_\infty$ induced from $C[0,1]$.

(1) Calculate $f(\mathbf{0})$. (The arc length of the graph of the constant function taking value 0 everywhere on $[0,1]$ is obviously 1, and check that the above formula delivers this.)

(2) Now consider $\mathbf{x}_n := \frac{\cos(2\pi n t)}{\sqrt{n}}$, $t \in [0,1]$, $n \in \mathbb{N}$. Using $\sin(2\pi n t) \geq \frac{1}{\sqrt{2}}$ for all $t \in [\frac{1}{8n}, \frac{3}{8n}]$, and the periodicity of $\sin(2\pi n t)$ (the graph of $\sin(2\pi n t)$ on $[0, \frac{1}{n}]$ is repeated n times in $[0,1]$), conclude that

$$f(\mathbf{x}_n) \geq n \cdot \int_{\frac{1}{8n}}^{\frac{3}{8n}} \sqrt{1 + 4\pi^2 n (\sin(2\pi n t))^2}\, dt \geq \frac{\sqrt{1 + 2\pi^2 n}}{4}.$$

(3) Show that f is *not* continuous at $\mathbf{0}$. (Prove this by contradiction. Note that by taking larger and larger n, $\|\mathbf{x}_n - \mathbf{0}\|_\infty$ can be made as small as we please, but $f(\mathbf{x}_n)$ doesn't stay close to $f(\mathbf{0})$.)

Show that the arc length function f is continuous if we equip $C^1[0,1]$ with the norm $\|\cdot\|_{1,\infty}$. It may be useful to note that by using the triangle inequality in $(\mathbb{R}^2, \|\cdot\|_2)$, we have for $a, b \in \mathbb{R}$ that

$$\big|\, \|(1,a)\|_2 - \|(1,b)\|_2 \,\big| = \big|\sqrt{1^2 + a^2} - \sqrt{1^2 + b^2}\big|$$
$$\leqslant \|(1,a) - (1,b)\|_2 = \sqrt{(1-1)^2 + (a-b)^2} = |a-b|.$$

Exercise 2.4. Let $(X, \|\cdot\|)$ be a normed space. Show that the norm $\|\cdot\| : X \to \mathbb{R}$ is a continuous map.

Continuity and open sets

We'll now learn an important property of continuous maps:

"inverse images" of open sets under a continuous map are open.

In fact, we shall see that this property is a characterisation of continuity. First let's some notation. Let $f : X \to Y$ be a map between the normed spaces X and Y, and let $V \subset Y$. We set $f^{-1}(V) := \{x \in X : f(x) \in V\}$, and call it the *inverse image of V under f*. Clearly $f^{-1}(Y) = X$ and $f^{-1}(\emptyset) = \emptyset$.

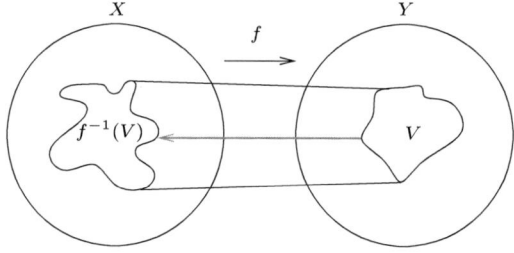

Exercise 2.5. Let $f : \mathbb{R} \to \mathbb{R}$ be given by $f(x) = \cos x$ $(x \in \mathbb{R})$.
Find $f^{-1}(V)$, where $V = \{-1, 1\}$, $V = \{1\}$, $V = [-1, 1]$, $V = \mathbb{R}$, $V = (-\tfrac{1}{2}, \tfrac{1}{2})$.

On the other hand if $U \subset X$, then we set $f(U) := \{f(x) \in Y : x \in U\}$, and call it the *image of U under f*.

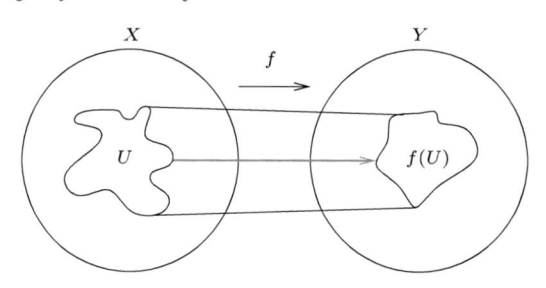

Exercise 2.6. Let $f : \mathbb{R} \to \mathbb{R}$ be given by $f(x) = \cos x$ $(x \in \mathbb{R})$.
Find $f(U)$, where $U = \mathbb{R}$, $U = [0, 2\pi]$, $U = [\delta, \delta + 2\pi]$ where $\delta > 0$.

Theorem 2.1. *Let X, Y be normed spaces and $f : X \to Y$ be a map. Then $\boxed{f \text{ is continuous on } X}$ if and only if $\boxed{\text{for every } V \text{ open in } Y, \ f^{-1}(V) \text{ is open in } X}$.*

Proof.
(**If**) Let $c \in X$, and let $\epsilon > 0$. Consider the open ball $B(f(c), \epsilon)$ with center $f(c)$ and radius ϵ in Y. We know that this open ball $V := B(f(c), \epsilon)$ is an open set in Y. Thus we also know that $f^{-1}(V) = f^{-1}(B(f(c), \epsilon))$ is an open set in X. But the point $c \in f^{-1}(B(f(c), \epsilon))$, because $f(c) \in B(f(c), \epsilon)$ ($\|f(c), f(c)\| = 0 < \epsilon$!). So by the definition of an open set, there is a $\delta > 0$ such that $B(c, \delta) \subset f^{-1}(B(f(c), \epsilon))$. In other words, whenever $x \in X$ satisfies $\|x - c\| < \delta$, we have $x \in f^{-1}(B(f(c), \epsilon))$, that is, $f(x) \in B(f(c), \epsilon)$, which implies $\|f(x) - f(c)\| < \epsilon$. Hence f is continuous at c. But the choice of $c \in X$ was arbitrary. Consequently f is continuous on X. See the picture on the left.

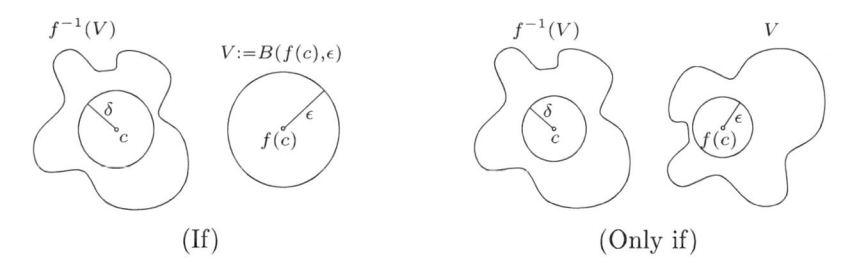

(If)   (Only if)

(**Only if**) Now let f be continuous, and let V be an open subset of Y. We would like to show that $f^{-1}(V)$ is open. So let $c \in f^{-1}(V)$. Then $f(c) \in V$. As V is open, there is a small open ball $B(f(c), \epsilon)$ with center $f(c)$ and radius ϵ that is contained in V. By the continuity of f at c, there is a $\delta > 0$ such that whenever $\|x - c\| < \delta$, we have $\|f(x) - f(c)\| < \epsilon$, that is, $f(x) \in V$. But this means that $B(c, \delta) \subset f^{-1}(V)$. Indeed, if $x \in B(c, \delta)$, then $\|x - c\| < \delta$ and so by the above, $f(x) \in V$, that is, $x \in f^{-1}(V)$. Consequently, $f^{-1}(V)$ is open in X. See the picture on the right above. □

Note that the theorem does *not* claim that for every U open in X, $f(U)$ is open in Y. Consider for example $X = Y = \mathbb{R}$ equipped with the Euclidean norm, and the constant function $f(x) = c$ ($x \in \mathbb{R}$), which is clearly continuous. But note that direct images of open sets are not always open under f: indeed $X = \mathbb{R}$ *is* open in $X = \mathbb{R}$, but $f(X) = \{c\}$ is *not* open in $Y = \mathbb{R}$.

Corollary 2.1. *Let X, Y be normed spaces and $f : X \to Y$ be a map. Then $\boxed{f \text{ is continuous on } X}$ if and only if $\boxed{\text{for every } F \text{ closed in } Y, f^{-1}(F) \text{ is closed in } X}$.*

Proof. If $F \subset Y$, then $f^{-1}(Y \backslash F) = X \backslash (f^{-1}(F))$. □

Exercise 2.7. Fill in the details of the proof of Corollary 2.1.

Theorem 2.2. *Let X, Y, Z be normed spaces, and $f : X \to Y$, $g : Y \to Z$ be continuous maps. Then the composition map $g \circ f : X \to Z$, defined by $(g \circ f)(x) := g(f(x))$ $(x \in X)$, is continuous.*

Proof. Let W be open in Z. Then since g is continuous, $g^{-1}(W)$ is open in Y. Also, since f is continuous, $f^{-1}(g^{-1}(W))$ is open in X. Finally, we note that $(g \circ f)^{-1}(W) = f^{-1}(g^{-1}(W))$. So $g \circ f$ is continuous. □

Exercise 2.8. In the proof of Theorem 2.2, we used $(g \circ f)^{-1}(W) = f^{-1}(g^{-1}(W))$. Check this.

Exercise 2.9. Let X be a normed space and $f : X \to \mathbb{R}$ be a continuous map. Determine if the following statements are true or false.
  (1) $\{x \in X : f(x) < 1\}$ is an open set.
  (2) $\{x \in X : f(x) > 1\}$ is an open set.
  (3) $\{x \in X : f(x) = 1\}$ is an open set.
  (4) $\{x \in X : f(x) \leq 1\}$ is a closed set.
  (5) $\{x \in X : f(x) = 1\}$ is a closed set.
  (6) $\{x \in X : f(x) = 1 \text{ or } f(x) = 2\}$ is a closed set.
  (7) $\{x \in X : f(x) = 1\}$ is a compact set.

Continuity and convergence

We have the following characterisation of continuous maps in terms of convergence of sequences: "Continuous maps preserve convergent sequences".

Theorem 2.3. *Let X, Y be normed spaces, $c \in X$, and let $f : X \to Y$. Then the following two statements are equivalent:*

(1) *f is continuous at c.*
(2) *For every sequence $(x_n)_{n \in \mathbb{N}}$ in X such that $(x_n)_{n \in \mathbb{N}}$ converges to c, $(f(x_n))_{n \in \mathbb{N}}$ converges to $f(c)$.*

Proof.
(1) $\Rightarrow$ (2): Suppose that f is continuous at c. Let $(x_n)_{n \in \mathbb{N}}$ be a sequence in X such that $(x_n)_{n \in \mathbb{N}}$ converges to c. Let $\epsilon > 0$. Then there exists a $\delta > 0$

such that for all $x \in X$ satisfying $\|x - c\| < \delta$, we have $\|f(x) - f(c)\| < \epsilon$. As the sequence $(x_n)_{n \in \mathbb{N}}$ converges to c, for this $\delta > 0$, there exists an $N \in \mathbb{N}$ such that whenever $n > N$, $\|x_n - c\| < \delta$. But then by the above, $\|f(x_n) - f(c)\| < \epsilon$. So we have shown that for every $\epsilon > 0$, there is an $N \in \mathbb{N}$ such that for all $n > N$, $\|f(x_n) - f(c)\| < \epsilon$. In other words, the sequence $(f(x_n))_{n \in \mathbb{N}}$ converges to $f(c)$.

(2) $\Rightarrow$ (1): Suppose that f is not continuous at c. Thus there is an $\epsilon > 0$ such that for every $\delta > 0$, there is an $x \in X$ such that $\|x - c\| < \delta$, but $\|f(x) - f(c)\| > \epsilon$. We will use this statement to construct a sequence $(x_n)_{n \in \mathbb{N}}$ for which the conclusion in (2) does not hold. Let $\delta = 1/n$, for $n \in \mathbb{N}$, and denote a corresponding x as x_n: thus, $\|x_n - c\| < \delta = 1/n$, but $\|f(x_n) - f(c)\| > \epsilon$. Clearly the sequence $(x_n)_{n \in \mathbb{N}}$ is convergent with limit c, but $(f(x_n))_{n \in \mathbb{N}}$ does not converge to $f(c)$ since $\|f(x_n) - f(c)\| > \epsilon$ for all $n \in \mathbb{N}$. Consequently if (1) does not hold, then (2) does not hold. In other words, we have shown that (2) $\Rightarrow$ (1). □

Exercise 2.10. Let X, Y be normed spaces. Find all continuous maps $f : X \to Y$ such that for all $x \in X$, $f(x) + f(2x) = \mathbf{0}$. *Hint:* $f(x) = -f(\frac{x}{2}) = f(\frac{x}{4}) = \cdots$.

Exercise 2.11. (∗)(Continuity of the determinant; {invertible matrices} is open). Show that the determinant $\mathbf{M} \mapsto \det \mathbf{M} : (\mathbb{R}^{n \times n}, \|\cdot\|_\infty) \to (\mathbb{R}, |\cdot|)$ is continuous. Prove that the set of invertible matrices is open in $(\mathbb{R}^{n \times n}, \|\cdot\|_\infty)$. *Hint:* $\det^{-1}\{0\}$.

Continuity and compactness

In this section we will learn about a very useful result in Optimisation Theory, on the existence of global minimisers of real-valued continuous functions on compact sets.

Theorem 2.4.
If (1) *K is a compact subset of a normed space X,*
(2) *Y is a normed space, and*
(3) *$f : X \to Y$ is function that is continuous at each $x \in K$,*
then $f(K)$ is a compact subset of Y.

Proof. Suppose that $(y_n)_{n \in \mathbb{N}}$ is a sequence contained in $f(K)$. Then for each $n \in \mathbb{N}$, there exists an $x_n \in K$ such that $y_n = f(x_n)$. Thus we obtain a sequence $(x_n)_{n \in \mathbb{N}}$ in the set K. As K is compact, there exists a convergent subsequence, say $(x_{n_k})_{k \in \mathbb{N}}$, with limit $L \in K$. As f is continuous, it preserves convergent sequences. So $(f(x_{n_k}))_{k \in \mathbb{N}} = (y_{n_k})_{k \in \mathbb{N}}$ is convergent with limit $f(L) \in f(K)$. Consequently, $f(K)$ is compact. □

Now we prove the aforementioned result which turns out to be very useful in Optimisation Theory, namely that a real-valued continuous function on a compact set attains its maximum/minimum on the compact set. This is a generalisation of the Extreme Value Theorem we had learnt earlier, where the compact set in question was just the interval $[a, b]$.

Theorem 2.5. (Weierstrass).
If (1) K *is a nonempty compact subset of a normed space* X, *and*
(2) $f : X \to \mathbb{R}$ *is a function that is continuous at each* $x \in K$,
then there exists a $c \in K$ *such that* $f(c) = \sup\{f(x) : x \in K\}$.

We note that since $c \in K$, $f(c) \in \{f(x) : x \in K\}$, and so the supremum above is actually a maximum:
$$f(c) = \sup\{f(x) : x \in K\} = \max\{f(x) : x \in K\}.$$
Also, under the same hypothesis of the above result, there exists a minimiser in K, that is, there exists a $d \in K$ such that
$$f(d) = \inf\{f(x) : x \in K\} = \min\{f(x) : x \in K\}.$$
This follows from the above result by just looking at $-f$, that is by applying the above result to the function $g : X \to \mathbb{R}$ given by $g(x) = -f(x)$ ($x \in X$).

Proof. (Of Theorem 2.5.) We know that the image of K under f, namely the set $f(K)$ is compact and hence bounded. So $\{f(x) : x \in K\}$ is bounded. It is also nonempty since K is nonempty. But by the least upper bound property of $\mathbb{R}$, a nonempty bounded subset of $\mathbb{R}$ has a least upper bound. Thus $M := \sup\{f(x) : x \in K\} \in \mathbb{R}$. Now consider $M - 1/n$ ($n \in \mathbb{N}$). This number cannot be an upper bound for $\{f(x) : x \in K\}$. So there must be an $x_n \in K$ such that $f(x_n) > M - 1/n$. In this manner we get a sequence $(x_n)_{n \in \mathbb{N}}$ in K. As K is compact, $(x_n)_{n \in \mathbb{N}}$ has a convergent subsequence $(x_{n_k})_{k \in \mathbb{N}}$ with limit, say c, belonging to K. As f is continuous, $(f(x_{n_k}))_{k \in \mathbb{N}}$ is convergent as well with limit $f(c)$. But from the inequalities $f(x_n) > M - 1/n$ ($n \in \mathbb{N}$), it follows that $f(c) \geq M$. On the other hand, from the definition of M, we also have that $f(c) \leq M$. So $f(c) = M$. □

Example 2.5. Since the set $K = \{x \in \mathbb{R}^3 : x_1^2 + x_2^2 + x_3^2 = 1\}$ is compact in $\mathbb{R}^3$ and since the function $x \mapsto x_1 + x_2 + x_3$ is continuous on $\mathbb{R}^3$, it follows that the optimisation problem
$$\left\{\begin{array}{l} \text{minimise} \quad x_1 + x_2 + x_3 \\ \text{subject to} \quad x_1^2 + x_2^2 + x_3^2 = 1 \end{array}\right\},$$
has a minimiser. ◇

Remark 2.1. (∗) In Optimisation Theory, one often meets *necessary* conditions for a minimiser, that is, results of the following form:

If $\hat{x}$ is a minimiser to the optimisation problem
$$\left\{ \begin{array}{l} \text{minimise } f(x) \\ \text{subject to } x \in \mathcal{F} \ (\subset \mathbb{R}^n) \end{array} \right\},$$
then $\hat{x}$ satisfies $\boxed{***}$.

(Here $\boxed{***}$ are certain mathematical conditions, such as the Lagrange multiplier equations.) Now such a result has limited use as such since even if we find all $\hat{x}$ which satisfy $\boxed{***}$, we can't conclude that there is one that is a minimiser. But now suppose that we know that f is continuous on $\mathcal{F}$ and that $\mathcal{F}$ is compact. Then we know that a minimiser exists, and so we know that among the $\hat{x}$ that satisfy $\boxed{***}$, there is at least one which is a minimiser.

Notation 2.2. We will denote the set of all continuous maps from the normed space X to the normed space Y by $C(X, Y)$.

2.3 The normed space $CL(X, Y)$

In this section we study those linear transformations from a normed space X to a normed space Y that are also continuous.

Notation 2.3. We denote the set of all continuous linear transformations from the normed space X to the normed space Y by $CL(X, Y)$, that is, $CL(X, Y) := C(X, Y) \bigcap L(X, Y)$. If $X = Y$, then we denote $CL(X, X)$ simply by $CL(X)$.

We begin by giving a characterisation of continuous linear transformations.

When is a linear transformation continuous?

Theorem 2.6.
Let X and Y be normed spaces, and $T : X \to Y$ be a linear transformation. Then the following properties of T are equivalent:
  (1) T is continuous.
  (2) T is continuous at $\mathbf{0}$.
  (3) There exists an $M > 0$ such that for all $x \in X$, $\|Tx\|_Y \leqslant M\|x\|_X$.

We'll see the proof below. But let us first remark that the useful part is the equivalence of (1) and (3), since by just showing the existence/lack of of the bound M, we can conclude the continuity/lack of continuity of the given linear transformation. So we don't have to go through the rigmarole of verifying the ϵ-δ definition: rather, a simple estimate, as stipulated in (3), suffices. Note also that it seems miraculous that continuity at just one point (at $\mathbf{0}$) delivers continuity everywhere on X! This miracle happens because the map T is not any old map, but rather a *linear transformation*. Here is an elementary example.

Example 2.6. (The left shift and right shift operators).
The left shift operator, $L : \ell^2 \to \ell^2$, given by

$$L(a_1, a_2, a_3, \cdots) := (a_2, a_3, \cdots), \quad (a_n)_{n \in \mathbb{N}} \in \ell^2,$$

is a linear transformation. We have for all $(a_n)_{n \in \mathbb{N}} \in \ell^2$ that

$$\|L(a_n)_{n \in \mathbb{N}}\|_2 = \sqrt{|a_2|^2 + |a_3|^2 + \cdots}$$
$$\leq \sqrt{|a_1|^2 + |a_2|^2 + |a_3|^2 + \cdots} = 1 \cdot \|(a_n)_{n \in \mathbb{N}}\|_2,$$

and so $L \in C(\ell^2, \ell^2)$. The right shift operator $R : \ell^2 \to \ell^2$, given by $R(a_1, a_2, a_3, \cdots) := (0, a_1, a_2, \cdots)$, $(a_n)_{n \in \mathbb{N}} \in \ell^2$, is also a linear transformation which is continuous, thanks to the equality

$$\|R(a_n)_{n \in \mathbb{N}}\|_2 = \sqrt{0^2 + |a_1|^2 + |a_2|^2 + \cdots} = \|(a_n)_{n \in \mathbb{N}}\|_2,$$

for all $(a_n)_{n \in \mathbb{N}} \in \ell^2$. ◇

Proof. (Of Theorem 2.6.) We will show the three implications (1)⇒(2), (2)⇒(3), and (3)⇒(1), which are enough to get all the three equivalences (and six implications) given in the statement of the theorem.

(1)⇒(2). This is just the definition of continuity on X. Indeed, T has to be continuous at each point in X, and in particular at $\mathbf{0} \in X$.

(2)⇒(3). Take $\epsilon := 1 > 0$. Then there exists a $\delta > 0$ such that whenever $\|x - \mathbf{0}\| = \|x\| < \delta$, we have that $\|Tx - T\mathbf{0}\| = \|Tx - \mathbf{0}\| = \|Tx\| < 1$. Let's check that this yields:

$$\|Tx\| \leq \frac{2}{\delta}\|x\| \text{ for all } x \in X. \tag{2.2}$$

First consider $x = \mathbf{0}$. Then

$$\text{the left-hand side} = \|Tx\| = \|T\mathbf{0}\| = \|\mathbf{0}\| = 0, \text{ while}$$
$$\text{the right-hand side} = \frac{2}{\delta}\|x\| = \frac{2}{\delta}\|\mathbf{0}\| = 0.$$

And so the claim in (2.2) holds because we have in fact an equality.
On the other hand, now suppose that $x \neq \mathbf{0}$. Set $y := \dfrac{\delta}{2\|x\|} \cdot x$. Then
$$\|y\| = \left\|\dfrac{\delta}{2\|x\|} \cdot x\right\| = \dfrac{\delta}{2\|x\|}\|x\| = \dfrac{\delta}{2} < \delta,$$
and so $\|Ty\| < 1$, that is $\left\|T\left(\dfrac{\delta}{2\|x\|} \cdot x\right)\right\| = \left\|\dfrac{\delta}{2\|x\|} \cdot Tx\right\| = \dfrac{\delta}{2\|x\|}\|Tx\| < 1.$

Upon rearranging, we obtain (2.2). So the claim in (3) holds with $M = \dfrac{2}{\delta}$.

(3)$\Rightarrow$(1). Let $M > 0$ be such that for all $x \in X$, $\|Tx\| \leq M\|x\|$. Let $x_0 \in X$, and $\epsilon > 0$. Set $\delta := \epsilon/M > 0$. Then whenever $\|x - x_0\| < \delta$, we have
$$\|Tx - Tx_0\| = \|T(x - x_0)\| \leq M\|x - x_0\| < M \cdot \delta = M \cdot \dfrac{\epsilon}{M} = \epsilon.$$
So T is continuous at x_0. But as $x_0 \in X$ was arbitrary, T is continuous. □

Example 2.7. (Norm on $C^1[a,b]$ revisited). Consider the differentiation mapping $D : C^1[0,1] \to C[0,1]$ defined by $(D\mathbf{x})(t) = \mathbf{x}'(t)$, $t \in [0,1]$, $\mathbf{x} \in C^1[0,1]$. We had seen that D is a linear transformation. Let's now investigate if D is also continuous.

(1) We will show that D is *not* continuous if both $C^1[0,1]$ and $C[0,1]$ are equipped with the $\|\cdot\|_\infty$ norm. Suppose on the contrary that the map D is continuous. Because D is a linear transformation, it follows from Theorem 2.6 that there exists an $M > 0$ such that for all $\mathbf{x} \in C^1[0,1]$,
$$\|D\mathbf{x}\|_\infty = \|\mathbf{x}'\|_\infty \leq M\|\mathbf{x}\|_\infty.$$
But if we take $\mathbf{x} = t^n$ ($n \in \mathbb{N}$), then we have
$$\|\mathbf{x}\|_\infty = \|t^n\|_\infty = 1, \text{ and}$$
$$\|\mathbf{x}'\|_\infty = \|nt^{n-1}\|_\infty = n\|t^{n-1}\|_\infty = n,$$
and so $\|D\mathbf{x}\|_\infty = \|\mathbf{x}'\|_\infty = n \leq M\|\mathbf{x}\|_\infty = M \cdot 1$, that is, $n \leq M$ for all $n \in \mathbb{N}$, which is clearly not true. So D is not continuous.

(2) However, D *is* continuous if $C^1[0,1]$ is equipped with the $\|\cdot\|_{1,\infty}$ norm:
$$\|\mathbf{x}\|_{1,\infty} := \|\mathbf{x}\|_\infty + \|\mathbf{x}'\|_\infty, \quad \mathbf{x} \in C^1[0,1],$$
while $C[0,1]$ has the usual supremum norm $\|\cdot\|_\infty$. Indeed, we have for all $\mathbf{x} \in C^1[0,1]$, that $\|D\mathbf{x}\|_\infty = \|\mathbf{x}'\|_\infty \leq \|\mathbf{x}\|_\infty + \|\mathbf{x}'\|_\infty = \|\mathbf{x}\|_{1,\infty}$. ◇

Example 2.8. (If X, Y are finite dimensional, then $L(X,Y) = CL(X,Y)$.) Let $X = (\mathbb{R}^n, \|\cdot\|_2)$, $Y = (\mathbb{R}^m, \|\cdot\|_2)$ and let $A \in \mathbb{R}^{m \times n}$ be given by
$$A = \begin{bmatrix} a_{11} & \cdots & a_{1n} \\ \vdots & & \vdots \\ a_{m1} & \cdots & a_{mn} \end{bmatrix} \in \mathbb{R}^{m \times n}.$$

Let $T_A : \mathbb{R}^n \to \mathbb{R}^m$ be the linear transformation given by $T_A\mathbf{x} := A\mathbf{x}$ for all $\mathbf{x} \in \mathbb{R}^n$. Then for all $\mathbf{x} \in \mathbb{R}^n$,

$$\|T_A\mathbf{x}\|_2^2 = \|A\mathbf{x}\|_2^2 = \sum_{i=1}^m \Big(\sum_{j=1}^n a_{ij}x_j\Big)^2$$
$$\leq \sum_{i=1}^m \Big(\sum_{j=1}^n a_{ij}^2\Big)\Big(\sum_{j=1}^n x_j^2\Big) \quad \text{(Cauchy-Schwarz)}$$
$$= \|\mathbf{x}\|_2^2 \Big(\sum_{i=1}^m \sum_{j=1}^n a_{ij}^2\Big),$$

and so $\|T_A\mathbf{x}\|_2 \leq \sqrt{\sum_{i=1}^m \sum_{j=1}^n a_{ij}^2} \cdot \|\mathbf{x}\|_2$. Hence T_A is continuous. $\diamond$

Remark 2.2. We know that *every* linear transformation on finite dimensional vector spaces X, Y can be represented by T_A once bases for X, Y have been chosen. Also we know that *all* norms on finite-dimensional normed spaces are equivalent to each other. It follows from these two facts that *every* linear transformation between finite dimensional normed spaces *is* continuous.

Example 2.9. Let $A = \begin{bmatrix} a_{11} & a_{12} & a_{13} & \cdots \\ a_{21} & a_{22} & a_{23} & \cdots \\ a_{31} & a_{32} & a_{33} & \cdots \\ \vdots & \vdots & \vdots & \ddots \end{bmatrix}$, and let $\sum_{i=1}^\infty \sum_{j=1}^\infty a_{ij}^2 < \infty$.

For $\mathbf{x} = (x_j)_{j \in \mathbb{N}} \in \ell^2$, set $T_A\mathbf{x} := A\mathbf{x} := \Big(\sum_{j=1}^\infty a_{ij}x_j\Big)_{i \in \mathbb{N}}$.

We claim that $T_A : \ell^2 \to \ell^2$ is a continuous linear transformation on ℓ^2. Firstly, $T_A\mathbf{x} \in \ell^2$, since

$$\|T_A\mathbf{x}\|_2^2 = \sum_{i=1}^\infty \Big(\sum_{j=1}^\infty a_{ij}x_j\Big)^2 \leq \sum_{i=1}^\infty \Big(\sum_{j=1}^\infty a_{ij}^2\Big)\Big(\sum_{j=1}^\infty x_j^2\Big) = \|\mathbf{x}\|_2^2 \Big(\sum_{i=1}^\infty \sum_{j=1}^\infty a_{ij}^2\Big) < \infty.$$

Moreover, it is easily seen that $T_A \in L(\ell^2)$. Moreover, by Theorem 2.6, the computation above shows that $T_A \in CL(\ell^2)$. $\diamond$

Example 2.10. (Integral operators).

Suppose that $A : [0,1] \times [0,1] \to \mathbb{R}$ be such that $\int_0^1 \int_0^1 \big(A(t,\tau)\big)^2 d\tau dt < \infty$.

We think of A as a "non-discrete/continuous" analogue of a square matrix: the indices i, j are replaced by the "non-discrete/continuous" indices t, τ.

Then the map $T_A : L^2[0,1] \to L^2[0,1]$, defined by

$$(T_A\mathbf{x})(t) = \int_0^1 A(t,\tau)\mathbf{x}(\tau)d\tau, \quad \mathbf{x} \in L^2[0,1],$$

is a continuous linear transformation. The following picture illustrates the action of T_A on $\mathbf{x}$ schematically, highlighting the analogy with matrix multiplication.

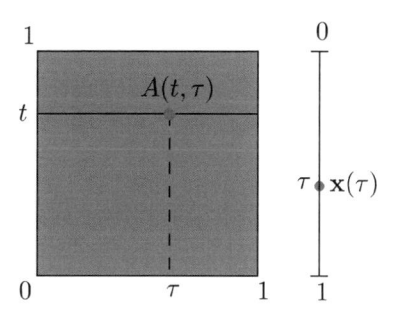

We note that for $\mathbf{x} \in L^2[0,1]$,

$$\|T_A\mathbf{x}\|_2^2 = \int_0^1 \left(\int_0^1 A(t,\tau)\mathbf{x}(\tau)d\tau \right)^2 dt$$

$$\leqslant \int_0^1 \left(\int_0^1 (A(t,\tau))^2 d\tau \right) \left(\int_0^1 (\mathbf{x}(\tau))^2 d\tau \right) dt$$

$$= \left(\int_0^1 (\mathbf{x}(\tau))^2 d\tau \right) \cdot \int_0^1 \left(\int_0^1 (A(t,\tau))^2 d\tau \right) dt$$

$$= \left(\int_0^1 \int_0^1 (A(t,\tau))^2 d\tau dt \right) \cdot \|\mathbf{x}\|_2^2 < \infty.$$

(The inequality in the second line above, is the Cauchy-Schwarz inequality in $L^2[0,1]$, and it follows from the general Cauchy-Schwarz inequality in inner product spaces, which will be shown in Theorem 4.1, page 157; see also Example 4.3, page 159. We'll accept this for now.) So $T_A\mathbf{x} \in L^2[0,1]$, and $T_A \in CL(L^2[0,1])$.

Operators T_A are called *integral operators*. It used to be common to call the function A that plays the role of the matrix, as the "kernel"[1] of the integral operator. Many variations of the integral operator are possible. ◇

---
[1] This has nothing to do with the null space: $\ker T_A := \{\mathbf{x} \in L^2[0,1] : T_A\mathbf{x} = \mathbf{0}\}$, which is also called the kernel of the integral operator.

Example 2.11. We had seen on page 55 that with

$$S := \{\mathbf{h} \in C^1[a,b] : \mathbf{h}(a) = \mathbf{h}(b) = 0\},$$

and $\mathbf{A}, \mathbf{B} \in C[a,b]$, the map $L : S \to \mathbb{R}$ given by

$$L(\mathbf{h}) = \int_a^b \big(\mathbf{A}(t)\mathbf{h}(t) + \mathbf{B}(t)\mathbf{h}'(t)\big)dt, \quad \mathbf{h} \in S$$

is a linear transformation. Now we ask: is L continuous? Here we equip $S \subset C^1[0,1]$ with the norm $\|\cdot\|_{1,\infty}$. For $\mathbf{h} \in S$,

$$|L(\mathbf{h})| = \left|\int_a^b \big(\mathbf{A}(t)\mathbf{h}(t) + \mathbf{B}(t)\mathbf{h}'(t)\big)dt\right|$$

$$\leq \int_a^b |\mathbf{A}(t)\mathbf{h}(t) + \mathbf{B}(t)\mathbf{h}'(t)|dt$$

$$\leq \int_a^b \big(|\mathbf{A}(t)||\mathbf{h}(t)| + |\mathbf{B}(t)||\mathbf{h}'(t)|\big)dt$$

$$\leq \int_a^b \big(|\mathbf{A}(t)|\|\mathbf{h}\|_\infty + |\mathbf{B}(t)|\|\mathbf{h}'\|_\infty\big)dt$$

$$\leq \int_a^b \big(|\mathbf{A}(t)|\|\mathbf{h}\|_{1,\infty} + |\mathbf{B}(t)|\|\mathbf{h}\|_{1,\infty}\big)dt$$

$$\leq \left(\int_a^b \big(|\mathbf{A}(t)| + |\mathbf{B}(t)|\big)dt\right)\|\mathbf{h}\|_{1,\infty} = M\|\mathbf{h}\|_{1,\infty},$$

where $M := \int_a^b \big(|\mathbf{A}(t)| + |\mathbf{B}(t)|\big)dt$. In the above, we have used

$$\|\mathbf{h}\|_{1,\infty} = \|\mathbf{h}\|_\infty + \|\mathbf{h}'\|_\infty \geq \|\mathbf{h}\|_\infty \text{ and}$$
$$\|\mathbf{h}\|_{1,\infty} = \|\mathbf{h}\|_\infty + \|\mathbf{h}'\|_\infty \geq \|\mathbf{h}'\|_\infty.$$

Hence L is continuous. ◇

Example 2.12. (∗)(Fourier transform).
Let $L^1(\mathbb{R})$ be the space of all complex valued Lebesgue integrable functions on $\mathbb{R}$, with the usual L^1-norm:

$$\|\mathbf{f}\|_1 := \int_{-\infty}^\infty |\mathbf{f}(x)|dx < \infty, \quad \mathbf{f} \in L^1(\mathbb{R}).$$

Its *Fourier transform* is the function $\widehat{\mathbf{f}} : \mathbb{R} \to \mathbb{C}$ defined by

$$\widehat{\mathbf{f}}(\xi) := \int_{-\infty}^\infty \mathbf{f}(x)e^{-2\pi i \xi x}dx, \quad \xi \in \mathbb{R}.$$

Then $\hat{\mathbf{f}}$ is a continuous function on $\mathbb{R}$, and it is also bounded because

$$\left|\int_{-\infty}^{\infty} \mathbf{f}(x)e^{-2\pi i \xi x} dx\right| \leqslant \int_{-\infty}^{\infty} |\mathbf{f}(x)||e^{-2\pi i \xi x}| dx = \int_{-\infty}^{\infty} |\mathbf{f}(x)| \cdot 1 dx = \|\mathbf{f}\|_1.$$

The vector space $C_b(\mathbb{R})$ of all complex-valued continuous functions on $\mathbb{R}$ that are bounded, is a normed space with the supremum norm:

$$\|\mathbf{g}\|_\infty := \sup_{\xi \in \mathbb{R}} |\mathbf{g}(\xi)|, \quad \mathbf{g} \in C_b(\mathbb{R}).$$

(We won't check this; the proof is analogous to Example 1.9, page 10.) Thus from the above, we have $\hat{\mathbf{f}} \in C_b(\mathbb{R})$. It is also easy to check that $\hat{\cdot} : L^1(\mathbb{R}) \to C_b(\mathbb{R})$ is a linear transformation, and it is continuous, thanks to the estimate above, giving $\|\hat{\mathbf{f}}\|_\infty \leqslant \|\mathbf{f}\|_1$. ◇

Remark 2.3. Owing to the characterisation of continuous linear transformations by the existence of a bound as in item (3) of Theorem 2.6 above, they are sometimes called *bounded* linear operators.

Exercise 2.12. Show that if $A \in \mathbb{R}^{m \times n}$, then $\ker A = \{\mathbf{x} \in \mathbb{R}^n : A\mathbf{x} = \mathbf{0}\}$ is a closed subspace of $\mathbb{R}^n$.

Exercise 2.13. (∗) Prove that *every* subspace of $\mathbb{R}^n$ is closed.
Hint: Construct a linear transformation whose kernel is the given subspace.

Exercise 2.14. Let $C[a,b]$ be endowed with the $\|\cdot\|_\infty$-norm.
(1) Show that $\mathbf{x} \xrightarrow{T} \int_a^b \mathbf{x}(t) dt : C[a,b] \to \mathbb{R}$ is a continuous linear transformation.
(2) Prove that if $\sum_{k=1}^\infty \mathbf{f}_k$ converges to $\mathbf{f}$ in $C[a,b]$, then $\sum_{k=1}^\infty \int_a^b \mathbf{f}_k(t) dt = \int_a^b \mathbf{f}(t) dt$.

Exercise 2.15. (Convolution operator).
If $\mathbf{f} \in L^1(\mathbb{R})$, then the corresponding *convolution operator* $\mathbf{f}* : L^\infty(\mathbb{R}) \to L^\infty(\mathbb{R})$ is given by

$$(\mathbf{f} * \mathbf{g})(t) = \int_{-\infty}^{\infty} \mathbf{f}(t-\tau)\mathbf{g}(\tau) d\tau, \quad t \in \mathbb{R}, \ \mathbf{g} \in L^\infty(\mathbb{R}).$$

Show that $\mathbf{f}*$ is well-defined and that $\mathbf{f}* \in CL(L^\infty(\mathbb{R}))$.

Exercise 2.16. Let $Y = \{\mathbf{f} \in L^2(\mathbb{R}) : \mathbf{f} = \check{\mathbf{f}} := \mathbf{f}(-\cdot)\}$ be the set of all even functions in $L^2(\mathbb{R})$. Show that Y is a closed subspace of $L^2(\mathbb{R})$.
Hint: View Y as the kernel of a suitable map in $CL(L^2(\mathbb{R}))$.

Operator norm and the normed space $CL(X,Y)$

Consider the set $CL(X,Y)$ of all continuous linear transformations from a normed space X to a normed space Y. We will show that $CL(X,Y)$ is a normed space, with pointwise operations (inherited from $L(X,Y)$), and the "operator norm" $\|\cdot\| : CL(X,Y) \to \mathbb{R}$ given by

$$\|T\| := \sup\{\|Tx\| : x \in X,\ \|x\| \leq 1\}, \quad T \in CL(X,Y).$$

Let us first show that $CL(X,Y)$ is a subspace of $L(X,Y)$, making it a vector space in its own right.

Proposition 2.1. *$CL(X,Y)$ is a subspace of $L(X,Y)$.*

Proof. We have:

(S1) Let $S, T \in CL(X,Y)$. Then there exist $M_S, M_T > 0$ such that for all $x \in X$, $\|Sx\| \leq M_S \|x\|$ and $\|Tx\| \leq M_T \|x\|$. So
$$\|(S+T)x\| = \|Sx + Tx\| \leq \|Sx\| + \|Tx\|$$
$$\leq M_S \|x\| + M_T \|x\| = (M_S + M_T)\|x\|.$$
Thus $S + T \in CL(X,Y)$ too.

(S2) Let $\alpha \in \mathbb{R}$ and $T \in CL(X,Y)$. There exists an $M > 0$ such that for all $x \in X$, $\|Tx\| \leq M\|x\|$. So $\|(\alpha T)x\| = \|\alpha(Tx)\| = |\alpha|\|Tx\| \leq |\alpha|M\|x\|$. Hence $\alpha T \in CL(X,Y)$.

(S3) The zero linear transformation $\mathbf{0} \in CL(X,Y)$ because for all $x \in X$, $\|\mathbf{0}x\| = \|\mathbf{0}_Y\| = 0 \leq 1 \cdot \|x\|$.

Consequently, $CL(X,Y)$ is a subspace of $L(X,Y)$. □

Next we show that the operator norm $\|\cdot\| : CL(X,Y) \to \mathbb{R}$ given by

$$\|T\| := \sup\{\|Tx\| : x \in X,\ \|x\| \leq 1\}, \quad T \in CL(X,Y)$$

is indeed a norm on $CL(X,Y)$. First let us check that is a well-defined number. If we set $S := \{\|Tx\| : x \in X,\ \|x\| \leq 1\}$, then we note that this is a subset of the real numbers. Let us observe that this is a nonempty bounded set:

(1) $S \neq \emptyset$ because if we take $x = \mathbf{0}_X \in X$, then $\|x\| = \|\mathbf{0}_X\| = 0 \leq 1$, and so $\|Tx\| = \|T\mathbf{0}_X\| = \|\mathbf{0}_Y\| = 0 \in S$.
(2) S is bounded above. As $T \in CL(X,Y)$, there is an $M > 0$ such that for all $x \in X$, $\|Tx\| \leq M\|x\|$. We claim that M is an upper bound of S. Indeed, if $x \in X$ and $\|x\| \leq 1$, then $\|Tx\| \leq M\|x\| \leq M \cdot 1 = M$.

Since S is a nonempty subset of $\mathbb{R}$ which is bounded above, it follows from the Least Upper Bound Property of $\mathbb{R}$ that the supremum of S exists: so for all $T \in CL(X,Y)$, $\|T\| := \sup\{\|Tx\| : x \in X, \|x\| \leq 1\} < \infty$. In order to do our verification that this operator norm $\|\cdot\|$ is a norm on $CL(X,Y)$, the following two results will be useful.

Lemma A. *Let $T \in CL(X,Y)$.*
If $M > 0$ is such that $\boxed{\text{for all } x \in X, \|Tx\| \leq M\|x\|}$, *then $\|T\| \leq M$.*

Proof. If $x \in X$ and $\|x\| \leq 1$, then $\|Tx\| \leq M\|x\| \leq M \cdot 1 = M$. So M is an upper bound of $S = \{\|Tx\| : x \in X, \|x\| \leq 1\}$. Thus $\sup S \leq M$, that is, $\|T\| \leq M$. □

Lemma B. *Let $T \in CL(X,Y)$. Then for all $x \in X$, $\|Tx\| \leq \|T\|\|x\|$.*

Proof.
1° $x = \mathbf{0}$. Then $\|Tx\| = \|T\mathbf{0}\| = \|\mathbf{0}\| = 0 = \|T\|0 = \|T\|\|\mathbf{0}\| = \|T\|\|x\|$.
2° Suppose that $\mathbf{0}_X \neq x \in X$. Let $y := \dfrac{1}{\|x\|}x \in X$. Then $\|y\| = \dfrac{1}{\|x\|}\|x\| = 1$.
Thus $\|Ty\| \in S$, and so $\|Ty\| \leq \sup S = \|T\|$, that is,
$$\frac{\|Tx\|}{\|x\|} = \left\|\frac{1}{\|x\|}Tx\right\| = \left\|T\left(\frac{1}{\|x\|}x\right)\right\| = \|Ty\| \leq \|T\|.$$
Rearranging, we get $\|Tx\| \leq \|T\|\|x\|$. □

Lemmas A and B together tell us that for a $T \in CL(X,Y)$, $\|T\|$ is allowed as an "M" in
$$\boxed{\text{for all } x \in X, \|Tx\| \leq M\|x\|},$$
and moreover it is the smallest possible such number M, in the sense that any other allowed M has got to be at least as large as $\|T\|$.

Theorem 2.7. *The operator norm, $\|\cdot\| : CL(X,Y) \to \mathbb{R}$, given by*
$$\|T\| := \sup\{\|Tx\| : x \in X, \|x\| \leq 1\}, \quad T \in CL(X,Y),$$
is a norm on $CL(X,Y)$.

Proof. We have:
(N1) For $T \in CL(X,Y)$, $\|T\| = \sup\limits_{\substack{x \in X \\ \|x\| \leq 1}} \|Tx\| \geq 0$ since $\|Tx\| \geq 0$ for all x.

If $T \in CL(X,Y)$ and $\|T\| = 0$, then $\|Tx\| \leq \|T\|\|x\| = 0\|x\| = 0$, and so $\|Tx\| = 0$, that is, $Tx = \mathbf{0}_Y$ for all $x \in X$. So $T = \mathbf{0}$, the zero linear transformation.

(N2) For $\alpha \in \mathbb{K}$ and $T \in CL(X,Y)$,

$$\|\alpha \cdot T\| = \sup_{\substack{x \in X \\ \|x\| \leq 1}} \|(\alpha \cdot T)x\| = \sup_{\substack{x \in X \\ \|x\| \leq 1}} \|\alpha \cdot (Tx)\|$$

$$= \sup_{\substack{x \in X \\ \|x\| \leq 1}} |\alpha| \|Tx\| = |\alpha| \sup_{\substack{x \in X \\ \|x\| \leq 1}} \|Tx\| = |\alpha| \|T\|.$$

(N3) Let $T, S \in CL(X,Y)$. Then for all $x \in X$,

$$\|(T+S)x\| = \|Tx + Sx\| \leq \|Tx\| + \|Sx\| \leq (\|T\| + \|S\|)\|x\|,$$

from which it follows (Lemma A) that $\|T + S\| \leq \|T\| + \|S\|$. $\square$

Example 2.13. Recall Example 2.8, page 69.
Let $\mathbb{R}^n$ and $\mathbb{R}^m$ be equipped with the Euclidean $\|\cdot\|_2$-norm.
Let $A = [A_{ij}] \in \mathbb{R}^{m \times n}$, and $T_A \in CL(\mathbb{R}^n, \mathbb{R}^m)$ be the continuous linear transformation given by $T_A \mathbf{x} = A\mathbf{x}$, $\mathbf{x} \in \mathbb{R}^n$.
Then we'd seen that for all $\mathbf{x} \in \mathbb{R}^n$, $\|T_A \mathbf{x}\|_2^2 \leq \left(\sum_{i=1}^m \sum_{j=1}^n a_{ij}^2\right) \cdot \|\mathbf{x}\|_2^2$. So

$$\|T_A\| \leq \left(\sum_{i=1}^m \sum_{j=1}^n a_{ij}^2\right)^{\frac{1}{2}}.$$

So we have an estimate for $\|T_A\|$ in terms of the matrix coefficients a_{ij}. But there does not exist a general "formula" for $\|T_A\|$ in terms of the matrix coefficients except in the special cases $n = 1$ or $m = 1$, when $\|T_A\| = |a_{11}|$. It can be seen that the map

$$A \mapsto \|A\|_2 := \left(\sum_{i=1}^m \sum_{j=1}^n a_{ij}^2\right)^{\frac{1}{2}}, \quad A = [a_{ij}] \in \mathbb{R}^{m \times n},$$

is also a norm on $\mathbb{R}^{m \times n}$, and is called the *Hilbert-Schmidt norm* of A. $\diamond$

Exercise 2.17. (Diagonal operator norm; operator norm needn't be attained.) Let $(\lambda_n)_{n \in \mathbb{N}}$ be a bounded sequence in $\mathbb{K}$, and let $\Lambda \in CL(\ell^2)$ be given by $\Lambda(a_1, a_2, a_3, \cdots) = (\lambda_1 a_1, \lambda_2 a_2, \lambda_3 a_3, \cdots)$ for all $(a_1, a_2, a_3, \cdots) \in \ell^2$.
Show that $\Lambda \in CL(\ell^2)$ and $\|\Lambda\| = \sup_{n \in \mathbb{N}} |\lambda_n|$.
Now let $\lambda_n = 1 - \frac{1}{n}$, $n \in \mathbb{N}$. Show that there is no $\mathbf{x} \in \ell^2$ such that $\|\mathbf{x}\|_2 \leq 1$ and $\|\Lambda \mathbf{x}\|_2 = \|\Lambda\|$. This gives an example showing that the operator norm need not be attained.

Exercise 2.18. (Schauder basis). Let X be a Banach space. A sequence of vectors $(e_n)_{n \in \mathbb{N}}$ in X is a *Schauder basis* for X if for every $x \in X$, there exists a unique sequence of numbers $(\xi_n)_{n \in \mathbb{N}}$ such that $x = \sum_{n=1}^{\infty} \xi_n e_n$.

Let $1 \leqslant p < \infty$, and $\mathbf{e}_n = (0, \cdots, 0, 1, 0, \cdots)$ be the sequence in ℓ^p with nth term equal to 1 and all others 0. Show that $\{\mathbf{e}_n : n \in \mathbb{N}\}$ is a Schauder basis for ℓ^p.
Hint: For uniqueness use the continuity of the "coordinate map" $\varphi_n : \mathbf{x} \mapsto x_n$, selecting the nth term of the sequence $\mathbf{x}$.

Remark. A Banach space X that has a Schauder basis is separable, that is, there exists a countable dense subset in X (for example the linear combinations of the e_n with rational coefficients). The converse of the above, namely if every separable Banach space had a Schauder basis, was an open problem for a long time. In 1973, the Swedish mathematician Per Enflo finally constructed an example of a separable Banach space that does not have a Schauder basis.

Exercise 2.19. (Invariant subspace, and the Invariant Subspace Problem)
(1) Prove that the *averaging operator*[2] $A : \ell^\infty \to \ell^\infty$, defined by
$$A(x_1, x_2, x_3, \cdots) = \left(x_1, \frac{x_1 + x_2}{2}, \frac{x_1 + x_2 + x_3}{3}, \cdots\right), \quad (x_n)_{n \in \mathbb{N}} \in \ell^\infty,$$
is a continuous linear transformation. What is the operator norm of A?

(2)(∗) A subspace Y of a normed space X is said to be an *invariant subspace* with respect to a linear transformation $T : X \to X$ if $TY \subset Y$. Let $A \in CL(\ell^\infty)$ be the averaging operator from part (1). Show that the subspace c of ℓ^∞, consisting of all convergent sequences, is an invariant subspace of the averaging operator A. *Hint:* Show that if $\mathbf{x} \in c$ has limit L, then $A\mathbf{x}$ has limit L.

Remark. Invariant subspaces are useful since they are helpful in studying complicated operators by breaking them down into smaller operators acting on invariant subspaces. This is already familiar to the student from the diagonalisation procedure in linear algebra, where one decomposes the vector space into eigenspaces, and in these eigenspaces the linear transformation acts trivially. One of the open problems in functional analysis is the *invariant subspace problem*:

> Does every $T \in CL(H)$ on a separable complex Hilbert space H have a non-trivial invariant subspace?

Hilbert spaces are just special types of Banach spaces, in which the norm is induced by an inner product, and we will learn about Hilbert spaces in Chapter 4. Non-trivial means that the invariant subspace must be different from $\{\mathbf{0}\}$ or H. In the case of Banach spaces, the answer to the above question is "no": during the annual meeting of the American Mathematical Society in Toronto in 1976, Per Enflo (again!) announced the existence of a Banach space and a bounded linear operator on it without any non-trivial invariant subspace.

Now that we know $CL(X, Y)$ is a normed space with the operator norm, it is natural to ask if $CL(X, Y)$ is complete, that is, if $CL(X, Y)$ is a Banach space. It turns out that $CL(X, Y)$ is a Banach space if and only if Y is a Banach space, and we'll show this in the next section.

---
[2] In Fourier/Harmonic Analysis, this is sometimes called the *Cesáro summation operator*.

When is $CL(X,Y)$ complete?

We'll see that $CL(X,Y)$ is a Banach space if and only if Y is a Banach space. In this section the "if" part will be shown, and the "only if" part will be done in Remark 2.9, page 109.

Theorem 2.8. *If Y is a Banach space, and X is any normed space, then $CL(X,Y)$ is a Banach space.*

Proof. Let $(T_n)_{n \in \mathbb{N}}$ be a Cauchy sequence in $CL(X,Y)$. Let $x \in X$.
Claim: $(T_n x)_{n \in \mathbb{N}}$ is Cauchy in Y.
  Indeed, for all n, m, $\|T_n x - T_m x\| \leq \|T_n - T_m\| \|x\|$.
As Y is Banach, $(T_n x)_{n \in \mathbb{N}}$ converges in Y, with limit, say $Tx \in Y$.
So we get a map $x \mapsto Tx : X \mapsto Y$.
Questions: (a) Is $T \in CL(X,Y)$?
  (b) Does $T_n \stackrel{n \to \infty}{\longrightarrow} T$ in $CL(X,Y)$?

(a) Is T a linear transformation?
If $x_1, x_2 \in X$, then $(T_n x_1)_{n \in \mathbb{N}}$ converges to $T x_1$ in Y, and $(T_n x_2)_{n \in \mathbb{N}}$ converges to $T x_2$ in Y. Thus $(T_n x_1 + T_n x_2)_{n \in \mathbb{N}} = (T_n(x_1 + x_2))_{n \in \mathbb{N}}$ converges to $T x_1 + T x_2$ in Y. But we know that $(T_n(x_1 + x_2))_{n \in \mathbb{N}}$ converges to $T(x_1 + x_2)$ in Y. By the uniqueness of limits, $T(x_1 + x_2) = T x_1 + T x_2$.
Let $\alpha \in \mathbb{K}$ and $x \in X$. Then $(T_n x)_{n \in \mathbb{N}}$ converges to Tx in Y. So we have $(\alpha \cdot (T_n x))_{n \in \mathbb{N}} = (T_n(\alpha \cdot x))_{n \in \mathbb{N}}$ converges to $\alpha \cdot (Tx)$ in Y. But $(T_n(\alpha \cdot x))_{n \in \mathbb{N}}$ converges to $T(\alpha \cdot x)$ in Y. So $\alpha \cdot T(x) = T(\alpha \cdot x)$.
Is T continuous? Let $\epsilon = 1$. Then there exists an $N \in \mathbb{N}$ such that for all $n, m > N$, $\|T_n - T_m\| \leq \epsilon = 1$. So for all $n > N$, $\|T_n - T_{N+1}\| \leq 1$. Thus for $n > N$ and $x \in X$, $\|T_n x - T_{N+1} x\| \leq \|T_n - T_{N+1}\| \|x\| \leq 1 \cdot \|x\|$. Passing the limit $n \to \infty$, we obtain $\|Tx - T_{N+1} x\| \leq \|x\|$ for all $x \in X$. So for all $x \in X$, $\|Tx\| \leq \|Tx - T_{N+1} x\| + \|T_{N+1} x\| \leq (1 + \|T_{N+1}\|) \|x\|$.
Conclusion: $T \in CL(X,Y)$.

(b) Is it true that $\lim_{n \to \infty} T_n = T$ in $CL(X,Y)$?

Let $\epsilon > 0$. Then there exists an $N \in \mathbb{N}$ such that for all $n, m > N$, we have $\|T_n - T_m\| \leq \epsilon$. So for all $n, m > N$ and all $x \in X$, we obtain that $\|T_n x - T_m x\| \leq \|T_n - T_m\| \cdot \|x\| \leq \epsilon \|x\|$. Passing to the limit as $m \to \infty$, we get that for all $n > N$ and $x \in X$, $\|T_n x - Tx\| \leq \epsilon \|x\|$. Hence for all $n > N$, $\|T_n - T\| \leq \epsilon$. □

Corollary 2.2. *If X is a normed space over $\mathbb{K}$, then the dual space of X, $X' := CL(X, \mathbb{K})$, is a Banach space with the operator norm.*

Corollary 2.3. *If X is a Banach space, then $CL(X) := CL(X, X)$ is a Banach space with the operator norm.*

Remark 2.4. ("Hilbert" versus Banach spaces). In Chapter 4, we'll meet Hilbert spaces: a Hilbert space is a special type of a Banach space in which the norm is induced by an "inner product". If instead of Banach spaces, we are interested only in Hilbert spaces, then the notion of a Banach space is still indispensable, since for a Hilbert space H, the normed space $CL(H)$ is typically only a Banach space, and not a Hilbert space in general.

(∗) Strong and weak operator topologies on $CL(X, Y)$

Many claims in this section won't be proved, but are included to provide the reader with a "road map". The main content of the section are the definitions of the three operator topologies and the illustrative examples. One who wants to know more could embark on a deeper study, as offered for example in [Pedersen (1989)] or [Rudin (1976)].

Let a set $\mathcal{X}$ be equipped with two topologies, and let $\mathcal{X}_1$ (respectively $\mathcal{X}_2$) denote the set $\mathcal{X}$ equipped with the first (respectively second) topology. If the identity map $x \mapsto x : \mathcal{X}_1 \to \mathcal{X}_2$ is continuous, namely if every set open in $\mathcal{X}_2$ is open in $\mathcal{X}_1$, one says that first topology is *stronger* than the second, or that the second topology is *weaker/coarser/smaller* than the first. Of all the topologies on the set $\mathcal{X}$, there is a strongest one (*discrete topology*), namely the one for which *all* subsets of $\mathcal{X}$ are open, and there is a weakest one (*trivial topology*), namely the one for which only $\mathcal{X}$, $\varnothing$ are open.

Now suppose we have a set $\mathcal{X}$, and a family $\mathcal{F} = \{f_i : \mathcal{X} \to \mathbb{R} \mid i \in I\}$ of maps. Then of course there exists at least one topology on $\mathcal{X}$ with respect to which all the maps f_i are continuous, namely the discrete topology on $\mathcal{X}$. However, there is also a "less wasteful/more efficient/weakest" topology on $\mathcal{X}$ that makes all the maps f_i, $i \in I$, continuous, characterized by the following: U is *open* in this topology on $\mathcal{X}$ if for every $x \in U$, there exist a finite number of indices $i_1, \cdots, i_n \in I$ and intervals $(a_1, b_1), \cdots, (a_n, b_n)$ such that $x \in \{y \in \mathcal{X} : f_{i_k}(y) \in (a_k, b_k), \ k = 1, \cdots, n\} \subset U$. It can be shown that this gives a topology $\mathcal{T}$ on $\mathcal{X}$, and for any other topology $\mathcal{T}'$ on $\mathcal{X}$ that makes the maps f_i, $i \in I$, continuous, we have $\mathcal{T} \subset \mathcal{T}'$.

We had seen that $CL(X, Y)$ is a normed space with the operator norm $\|T\| := \sup\{\|Tx\| : x \in X, \ \|x\| \leq 1\}$, $T \in CL(X, Y)$. We call the resulting

topology the *uniform operator topology* on $CL(X,Y)$, and is the weakest topology making each map in the family
$$\mathcal{F} := \{S \mapsto \|S - T\| : CL(X,Y) \to \mathbb{R} \mid T \in CL(X,Y)\}$$
continuous. A subset $U \subset CL(X,Y)$ is *open* in the uniform operator topology on $CL(X,Y)$ if for each $T \in U$, there exists an $\epsilon > 0$ such that $\{S \in CL(X,Y) : \|S - T\| < \epsilon\} \subset U$. A sequence $(T_n)_{n \in \mathbb{N}}$ *converges* to $T \in CL(X,Y)$ in the uniform operator topology if $\lim_{n \to \infty} \|T_n - T\| = 0$.

We remark that besides the uniform operator topology on $CL(X,Y)$, there are weaker topologies (with fewer open sets), on $CL(X,Y)$, called the Strong Topology and the Weak Topology. Here are the definitions, although in this basic introduction, we won't use these useful alternative topologies much.

Definition 2.4. (Strong Operator Topology)
Let X, Y be normed spaces. Then the weakest topology on $CL(X,Y)$ which makes each map in the family
$$\mathcal{F} := \{S \xmapsto{p_x} \|Sx - Tx\| : CL(X,Y) \to \mathbb{R} \mid x \in X, \, T \in CL(X,Y)\},$$
continuous, is called the *strong operator topology on $CL(X,Y)$*. A subset $U \subset CL(X,Y)$ is *open* in the strong operator topology on $CL(X,Y)$ if for each $T \in U$, there exists an $\epsilon > 0$ and finitely many $x_1, \cdots, x_n \in X$ such that $\{S \in CL(X,Y) : \|Sx_k - Tx_k\| < \epsilon, \, k = 1, \cdots, n\} \subset U$. A sequence $(T_n)_{n \in \mathbb{N}}$ *converges* to $T \in CL(X,Y)$ in the strong operator topology if for all $x \in X$, $\lim_{n \to \infty} \|T_n x - Tx\| = 0$.

Example 2.14. (Strong but not uniform convergence).
For $n \in \mathbb{N}$, let $P_n \in CL(\ell^2)$ be the "projection operator" given by
$$\ell^2 \ni \mathbf{a} = (a_1, a_2, a_3, \cdots) \xmapsto{P_n} (a_1, \cdots, a_n, 0, \cdots) \in \ell^2.$$
We claim that $(P_n)_{n \in \mathbb{N}}$ converges to the identity operator $I \in CL(\ell^2)$ in the strong operator topology.

Indeed, $\|I\mathbf{a} - P_n \mathbf{a}\|_2^2 = \|(0, \cdots, 0, a_{n+1}, a_{n+2}, \cdots)\|_2^2 = \sum_{k=n+1}^{\infty} |a_k|^2 \xrightarrow{n \to \infty} 0$.

But the sequence $(P_n)_{n \in \mathbb{N}}$ does *not* converge to the identity $I \in CL(\ell^2)$ in the uniform operator topology. Let's show this by contradiction.

Suppose it does converge to I with respect to the operator norm. With $\epsilon := 1/2 > 0$, there exists an $N \in \mathbb{N}$ such that $\|P_N - I\| < 1/2$. So if $\mathbf{e}_{N+1} \in \ell^2$ is the sequence with the $(N+1)$st term 1 and all others 0, then we have

$$1 = \|\mathbf{e}_{N+1}\|_2 = \|(I - P_N)\mathbf{e}_{N+1}\|_2 \leqslant \|I - P_N\| \|\mathbf{e}_{N+1}\|_2 < \frac{1}{2} \cdot 1 = \frac{1}{2},$$

a contradiction! ◇

A yet weaker topology than the strong operator topology is the weak operator topology, defined below.

Definition 2.5. (Weak Operator Topology)
Let X, Y be normed spaces. Let $Y' := CL(Y, \mathbb{K})$. Then the weakest topology on $CL(X, Y)$ which makes each map in the family

$$\mathcal{F} := \{S \overset{p_{x,\varphi}}{\mapsto} |\varphi(Sx - Tx)| : CL(X, Y) \to \mathbb{R} \mid x \in X, \ \varphi \in Y', \ T \in CL(X, Y)\}$$

continuous, is called the *weak operator topology on $CL(X,Y)$*. A subset $U \subset CL(X, Y)$ is *open* in the weak operator topology on $CL(X, Y)$ if for all $T \in U$, there exists an $\epsilon > 0$, finitely many $x_1, \cdots, x_n \in X$, and $\varphi_1, \cdots, \varphi_n \in Y'$ such that

$$\{S \in CL(X, Y) : |\varphi_k(Sx_k - Tx_k)| < \epsilon, \ k = 1, \cdots, n\} \subset U.$$

A sequence $(T_n)_{n \in \mathbb{N}}$ *converges* to $T \in CL(X, Y)$ in the weak operator topology if for all $\varphi \in Y'$ and for all $x \in X$, $\lim_{n \to \infty} |\varphi(T_n x - Tx)| = 0$.

The following table summarises this:

Uniform	Strong	Weak
\multicolumn{3}{c}{$(T_n)_{n\in\mathbb{N}}$ converges to T if}
$\|T_n - T\| \overset{n\to\infty}{\longrightarrow} 0$	$\forall x \in X,$ $\|T_n x - Tx\| \overset{n\to\infty}{\longrightarrow} 0$	$\forall \varphi \in Y',$ $\forall x \in X,$ $|\varphi(T_n x - Tx)| \overset{n\to\infty}{\longrightarrow} 0$

more open sets ←——————————————

weaker/coarser/smaller topology ←——————————————→

Example 2.15. (Weak but not strong convergence).
Let $R \in CL(\ell^2)$ be given by $\ell^2 \ni (a_1, a_2, a_3, \cdots) \stackrel{R}{\mapsto} (0, a_1, a_2, a_3, \cdots)$, the right shift operator. We claim that $(R^n)_{n\in\mathbb{N}}$ converges to $\mathbf{0} \in CL(\ell^2)$ in the weak operator topology. We'll use a result which will be proved later on in Theorem 2.14, page 104 (and also in Chapter 4, Theorem 4.10, page 189): For each $\varphi \in CL(\ell^2, \mathbb{C}) =: (\ell^2)'$, there is an $\mathbf{x}_\varphi = (\mathbf{x}_\varphi(k))_{k\in\mathbb{N}} \in \ell^2$, such that
$$\varphi(\mathbf{a}) = \sum_{k=1}^{\infty} \mathbf{a}(k)(\mathbf{x}_\varphi(k))^*, \text{ for all } \mathbf{a} = (\mathbf{a}(k))_{k\in\mathbb{N}} \in \ell^2.$$
(Here $\cdot^*$ denotes complex conjugation.)
Using the Cauchy-Schwarz inequality (page 159), for all $\mathbf{a} \in \ell^2$, $\varphi \in (\ell^2)'$,
$$|\varphi(R^n \mathbf{a})|^2 = \left| \sum_{k=1}^{\infty} \mathbf{a}(k)(\mathbf{x}_\varphi(n+k))^* \right|^2 \stackrel{\text{Cauchy-Schwarz}}{\leqslant} \|\mathbf{a}\|_2^2 \sum_{k=1}^{\infty} |\mathbf{x}_\varphi(n+k)|^2 \stackrel{n\to\infty}{\longrightarrow} 0.$$
Thus $(R^n)_{n\in\mathbb{N}}$ converges to $\mathbf{0} \in CL(\ell^2)$ in the weak operator topology.

If $\mathbf{e}_1 := (1, 0, 0, \cdots) \in \ell^2$, then $R^n \mathbf{e}_1 = (0, \cdots, 0, 1, 0, \cdots)$, the sequence with $(n+1)$st term 1 and all others 0. So $\|R^n \mathbf{e}_1\|_2 = 1$, $n \in \mathbb{N}$. Thus it is *not* the case that $\lim_{n\to\infty} R^n \mathbf{e}_1 = \mathbf{0} = 0\mathbf{e}_1$.

So $(R^n)_{n\in\mathbb{N}}$ does not converge to $\mathbf{0}$ in the strong operator topology. $\diamond$

2.4 Composition of continuous linear transformations

If $T \in CL(X, Y)$, $S \in CL(Y, Z)$, then the *composition* $ST : X \to Z$ of T, S is defined by $(ST)(x) = S(T(x))$, $x \in X$.

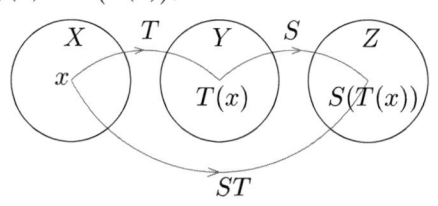

It is easily checked that ST is linear. Moreover, it is continuous too, since for all $x \in X$, we have $\|(ST)(x)\| = \|S(T(x))\| \leqslant \|S\|\|Tx\| \leqslant \|S\|\|T\|\|x\|$. Moreover, the above inequality shows that $\|ST\| \leqslant \|S\|\|T\|$.

In particular, if X is a normed space, then $CL(X)$, besides possessing a natural addition and scalar multiplication (both defined pointwise), also possesses a natural multiplication of elements of $CL(X)$, namely composition $(S, T) \mapsto ST : CL(X) \times CL(X) \to CL(X)$. So $CL(X)$ is an "algebra". Loosely speaking, an *algebra* is a vector space in which there is also available a nice way of multiplying vectors and producing new vectors.

Definition 2.6. (Algebra). An *algebra* is a vector space V in which an associative and distributive multiplication is defined, that is,
$$u(vw) = (uv)w, \quad (u+v)w = uw + vw, \quad u(v+w) = uv + uw$$
for all $u, v, w \in V$, and which is related to scalar multiplication so that
$$\alpha(uv) = u(\alpha v) = (\alpha u)v$$
for all $u, v \in V$ and all $\alpha \in \mathbb{K}$. We call $e \in V$ a *multiplicative identity element* if for all $v \in V$, one has $ev = v = ve$.

The algebra $V := CL(X)$ has a multiplicative identity element, namely the identity operator I. The *identity operator* is the map $I : X \mapsto X$, given by $Ix = x$, $x \in X$. The operator I clearly belongs to $CL(X)$ (with $\|I\| = 1$), and I serves as the multiplicative identity element of the algebra $CL(X)$: $IT = T = TI$ for all $T \in CL(X)$.

Definition 2.7. (Normed and Banach algebras).
A *normed algebra* is an algebra V equipped with a norm $\|\cdot\|$ that satisfies:
$$\|uv\| \leq \|u\|\|v\| \text{ for all } u, v \in V.$$
A *Banach algebra* is a normed algebra which is complete.

We note that $V := CL(X)$ is a normed algebra. We'd seen earlier that $CL(X)$ is a Banach space if X is a Banach space. So $CL(X)$ is a Banach algebra if X is a Banach algebra.

Let us note that as opposed to vector *addition* in $CL(X)$, vector multiplication (that is, composition) in $CL(X)$ is in general *not* commutative. Here is an example. Take $X = \mathbb{R}^2$. Let T be clockwise rotation by $\pi/2$, and S be reflection in the x-axis, that is,
$$T = \begin{bmatrix} 0 & 1 \\ -1 & 0 \end{bmatrix}, \quad S = \begin{bmatrix} 1 & 0 \\ 0 & -1 \end{bmatrix}.$$
Then one can check that $TS \neq ST$. This can also be observed visually by observing the distinct fates of the point $(1, 0)$ under TS and under ST:

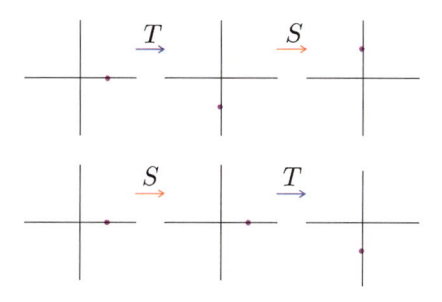

The *commutator* of $A, B \in CL(X)$ is defined by $[A, B] = AB - BA$, and "measures" the lack of commutativity of A and B. The above example shows that the commutator may not be necessarily $\mathbf{0}$. In Exercise 2.22, page 86, we will investigate the "largeness" of the commutator in finite and infinite dimensional spaces X. This plays a role in Quantum Mechanics. We'll show in Chapter 4 (page 204) that for "observables" A, B, the Heisenberg Uncertainty Relation holds:
$$(\Delta_\psi A)(\Delta_\psi B) \geq \frac{1}{2}|\langle [A, B]\rangle_\psi|.$$
We won't explain this[3] right now, but we simply notice that the commutator makes an appearance on the right-hand side.

If $\dim X = d < \infty$, and $T, S \in CL(X)$ are such that $TS = I$, then $ST = I$ too. So $TS = I \Rightarrow TS = ST = I$. (Let us show this. First of all, if $TS = I$, then $\ker S = \{\mathbf{0}\}$. Indeed, if $Sx = \mathbf{0}$, then
$$x = Ix = (TS)x = T(Sx) = T(\mathbf{0}) = \mathbf{0}.$$
Next observe that if $\{v_1, \cdots, v_d\}$ is a basis for X, then $\{Sv_1, \cdots, Sv_d\}$ are linearly independent: if α_ks are scalars such that $\alpha_1 Sv_1 + \cdots + \alpha_d Sv_d = \mathbf{0}$, then $S(\alpha_1 v_1 + \cdots + \alpha_d v_d) = \mathbf{0}$, and so $\alpha_1 v_1 + \cdots + \alpha_d v_d = \mathbf{0}$, making all α_ks zeros. Hence $\{Sv_1, \cdots, Sv_d\}$ must be a basis for X. For $x \in X$, there exist β_ks in $\mathbb{K}$ such that $x = \beta_1 Sv_1 + \cdots + \beta_d Sv_d = S(\beta_1 v_1 + \cdots + \beta_d v_d)$; and so $STx = STS(\beta_1 v_1 + \cdots + \beta_d v_d) = SI(\beta_1 v_1 + \cdots + \beta_d v_d) = x$.)

However, if $\dim X = \infty$, then it can happen that $TS = I$, but $ST \neq I$. Consider for example the left/right shift operators on ℓ^2. We have $LR = I$ as $LR(a_1, a_2, a_3, \cdots) = L(0, a_1, a_2, \cdots) = (a_1, a_2, \cdots) = I(a_1, a_2, a_3, \cdots)$, for all $(a_1, a_2, a_3, \cdots) \in \ell^2$. But $RL \neq I$ since
$$RL(1, 0, 0, \cdots) = R(0, 0, 0, \cdots) = (0, 0, 0, \cdots) \neq I(1, 0, 0, \cdots).$$
This prompts the following definition.

Definition 2.8. (Invertible operator) Let X be a normed space. An element $A \in CL(X)$ is said to be *invertible* if there exists a $B \in CL(X)$ such that $AB = I = BA$.

Inverses are unique. This follows from the associativity of composition.

Proposition 2.2. *If $A \in CL(X)$ is invertible, then there exists a unique $B \in CL(X)$ such that $AB = I = BA$.*

The unique inverse of an invertible $A \in CL(X)$ is denoted by $A^{-1} \in CL(X)$.

---

[3]That is, what Δ_ψ, $\langle \cdot \rangle_\psi$, etc. mean.

Proof. If $B_1, B_2 \in CL(X)$ satisfy $AB_1 = I = B_1 A$ and $AB_2 = I = B_2 A$, then $B_1 = IB_1 = (B_2 A)B_1 = B_2(AB_1) = B_2 I = B_2$. $\square$

Proposition 2.3. *If $A \in CL(X)$ is invertible, then A is bijective.*

Proof. If $x, y \in X$ are such that $Ax = Ay$, then $A^{-1}(Ax) = A^{-1}(Ay)$, that is, $Ix = Iy$, and so $x = y$. Thus A is injective/one-to-one.

If $y \in X$, then $x := A^{-1}y \in X$, and so $Ax = A(A^{-1}y) = Iy = y$. Hence A is surjective/onto too. $\square$

If $A \in CL(X)$ is bijective, then the inverse map is automatically a linear transformation. In the case when $\dim X < \infty$, we have $L(X) = CL(X)$. So in this case the inverse is automatically continuous too. So if $\dim X < \infty$, then $A \in CL(X)$ is invertible if and only if A is a bijection.

In the infinite dimensional case, is it still true that if $A \in CL(X)$ is a bijection, then A must be invertible? The answer is "yes" *if* X is a Banach space. The proof is not immediate, and we will show this below, using a deep result called the "Open Mapping Theorem". But first, let us see an example showing that in non-Banach spaces, the inverses of continuous bijections may fail to be continuous.

Example 2.16. (Bijection, but not invertible.)
Recall that c_{00} is the subspace of ℓ^∞ of all finitely supported sequences. Consider the map $A : c_{00} \to c_{00}$ given by
$$A(x_1, x_2, x_3, \cdots) = \left(x_1, \frac{x_2}{2}, \frac{x_3}{3}, \cdots\right), \quad (x_n)_{n \in \mathbb{N}} \in c_{00}.$$
Then A is linear, and continuous (because $\|A\mathbf{x}\|_\infty \leq \|\mathbf{x}\|_\infty$ for all $\mathbf{x} \in c_{00}$). It is also easily seen that A is injective and surjective. So A is a bijection. However, it is not invertible. Indeed, if otherwise, $B \in CL(c_{00})$ is the inverse, then we would have, with $\mathbf{e}_m := (0, \cdots, 0, 1, 0, \cdots)$ (mth term 1, all others 0), that
$$1 = \|\mathbf{e}_m\|_\infty = \|BA\mathbf{e}_m\|_\infty \leq \|B\| \|A\mathbf{e}_m\|_\infty = \|B\| \cdot \frac{1}{m},$$
giving $m \leq \|B\|$ for all $m \in \mathbb{N}$, a contradiction. But we aren't shocked by this example, since c_{00} is *not* complete with the supremum norm, and the equivalence of bijectivity with invertibility is supposed to hold for operators in a *Banach* space. $\diamond$

Exercise 2.20. (When is the diagonal operator invertible?)
Let $(\lambda_n)_{n \in \mathbb{N}}$ be a bounded sequence in $\mathbb{K}$, and consider $\Lambda \in CL(\ell^2)$ given by
$$\Lambda(a_1, a_2, a_3, \cdots) = (\lambda_1 a_1, \lambda_2 a_2, \lambda_3 a_3, \cdots) \text{ for all } (a_1, a_2, a_3, \cdots) \in \ell^2.$$
Show that Λ is invertible in $CL(\ell^2)$ if and only if $\inf_{n \in \mathbb{N}} |\lambda_n| > 0$.

Exercise 2.21. Let X be a normed space, and suppose that $A, B \in CL(X)$. Show that if $I + AB$ is invertible, then $I + BA$ is also invertible, with the inverse $(I + BA)^{-1}$ given by $I - B(I + AB)^{-1}A$.

Remark. This identity can be used to show that the nonzero spectrum of AB and BA coincide. λ is said to be in the spectrum of an operator T if $\lambda I - T$ is *not* invertible in $CL(X)$.

Exercise 2.22. ($[A, B]$ can't be "large" for $A, B \in CL(X)$.)

(1) The *trace*, $\operatorname{tr}(A)$, of a square matrix $A = [a_{ij}] \in \mathbb{C}^{d \times d}$ is the sum of its diagonal entries: $\operatorname{tr}(A) = a_{11} + \cdots + a_{dd}$. It can be shown that $\operatorname{tr}(A+B) = \operatorname{tr}(A) + \operatorname{tr}(B)$ and that $\operatorname{tr}(AB) = \operatorname{tr}(BA)$. Prove that there cannot exist A, B in $\mathbb{C}^{d \times d}$ such that $AB - BA = I$, where I denotes the $d \times d$ identity matrix.

(2) Let X be a normed space, and A, B be in $CL(X)$. Show that if $AB - BA = I$, then for all $n \in \mathbb{N}$, $AB^n - B^n A = nB^{n-1}$, where we set $B^0 := I$. Taking the operator norm on both sides of $AB^n - B^n A = nB^{n-1}$, conclude that we can never have $AB - BA = I$ with $A, B \in CL(X)$.

(3) Let $C^\infty(\mathbb{R})$ denote the set of all functions $\mathbf{f} : \mathbb{R} \to \mathbb{R}$ such that for all $n \in \mathbb{N}$, $\mathbf{f}^{(n)}$ exists. It is clear that $C^\infty(\mathbb{R})$ is a vector space with pointwise operations. Consider the operators $A, B : C^\infty(\mathbb{R}) \to C^\infty(\mathbb{R})$ given as follows:
$$(A\Psi)(x) = \frac{d\Psi}{dx}(x) \text{ and } (B\Psi)(x) = x\Psi(x), \quad x \in \mathbb{R}, \ \Psi \in C^\infty(\mathbb{R}).$$
(The operators A and B appear as the momentum operator and the position operator in Quantum Mechanics.) Show that $AB - BA = I$, where I denotes the identity on $C^\infty(\mathbb{R})$.

The Neumann Series Theorem.

Theorem 2.9. (Neumann[4] Series Theorem).
Let X be a Banach space, and $A \in CL(X)$ be such that $\|A\| < 1$.
Then (1) $I - A$ is invertible in $CL(X)$,

(2) $(I - A)^{-1} = I + A + \cdots + A^n + \cdots = \sum_{n=0}^{\infty} A^n$,

(3) $\|(I - A)^{-1}\| \leq \dfrac{1}{1 - \|A\|}$.

In particular, $I - A : X \to X$ is bijective: for each $y \in X$, there exists a unique solution $x \in X$ of the equation $x - Ax = y$, and moreover,
$$\|x\| \leq \frac{1}{1 - \|A\|}\|y\|,$$
so that x depends continuously on y.

---

[4]The "geometric" series in (2) is called the *Neumann series*, after the German mathematician Carl Neumann, who used it in connection with the Dirichlet problem.

This plays a role in integral equation theory:

$$\mathbf{x}(t) = \mathbf{y}(t) + \int_a^b k(t,\tau)\mathbf{x}(\tau)d\tau, \quad t \in [a,b],$$

where $\mathbf{y}$, k are given, and $\mathbf{x}$ is the unknown function.
(This is called the *Fredholm equation of the second type*.)

Proof. (Of the Neumann Series Theorem). For all $n \in \mathbb{N}$, $\|A^n\| \leq \|A\|^n$. As $\|A\| < 1$, $\sum_{n=1}^{\infty} \|A\|^n$ converges. By comparison, $\sum_{n=0}^{\infty} \|A^n\|$ converges too. As X is Banach, so is $CL(X)$. Since all absolutely convergent series in the Banach space $CL(X)$ converge, it follows that

$$S := \sum_{n=0}^{\infty} A^n$$

converges in $CL(X)$. Is this S the inverse of $I - A$? For $n \in \mathbb{N}$, define

$$S_n := I + A + A^2 + \cdots + A^n.$$

Then we know that $\lim_{n \to \infty} S_n = S$ in $CL(X)$. We have

$$S_n A = A S_n = A + A^2 + A^3 + \cdots + A^{n+1} = S_{n+1} - I.$$

Since $\|AS_n - AS\| \leq \|A\|\|S_n - S\|$ and $\|S_n A - SA\| \leq \|A\|\|S_n - S\|$, it follows that $SA = AS = S - I$. This gives $(I-A)S = I = S(I-A)$. Hence $I - A$ is invertible in $CL(X)$ and

$$(I-A)^{-1} = S = \sum_{n=0}^{\infty} A^n.$$

Moreover, $\|(I-A)^{-1}\| = \left\| \sum_{n=0}^{\infty} A^n \right\| \leq \sum_{n=0}^{\infty} \|A^n\| \leq \sum_{n=0}^{\infty} \|A\|^n = \frac{1}{1-\|A\|}.$ □

Exercise 2.23. Consider the system

$$\begin{cases} x_1 = \dfrac{1}{2}x_1 + \dfrac{1}{3}x_2 + 1, \\ x_2 = \dfrac{1}{3}x_1 + \dfrac{1}{4}x_2 + 2, \end{cases} \quad (2.3)$$

in the unknown variables $(x_1, x_2) \in \mathbb{R}^2$. If I denotes the 2×2 identity matrix, then this system can be written as $(I - K)x = y$, where

$$K = \begin{bmatrix} 1/2 & 1/3 \\ 1/3 & 1/4 \end{bmatrix}, \quad \mathbf{y} = \begin{bmatrix} 1 \\ 2 \end{bmatrix} \text{ and } \mathbf{x} = \begin{bmatrix} x_1 \\ x_2 \end{bmatrix}.$$

(1) Show that if $\mathbb{R}^2$ is equipped with the norm $\|\cdot\|_2$, then $\|K\| < 1$. Conclude that (2.3) has a unique solution (denoted by $\mathbf{x}$ in the sequel).
(2) Find out the unique solution $\mathbf{x}$ by computing $(I - K)^{-1}$.
(3) Write a computer program to compute $\mathbf{x}_n = (I + K + \cdots + K^n)\mathbf{y}$ and the relative error $\|\mathbf{x} - \mathbf{x}_n\|_2/\|\mathbf{x}\|_2$ for various values of n (say, until the relative error is less than 1%). Note the slow convergence of the Neumann series.

Exercise 2.24. Let X be a Banach space, and let $A \in CL(X)$ be such that $\|A\| < 1$. For $n \in \mathbb{N}$, let $P_n := (I + A)(I + A^2)(I + A^4) \cdots (I + A^{2^n})$.
(1) Using induction, show that $(I - A)P_n = I - A^{2^{n+1}}$ for all $n \in \mathbb{N}$.
(2) Prove that $(P_n)_{n \in \mathbb{N}}$ is convergent in $CL(X)$ to $(I - A)^{-1}$.

Exercise 2.25. (∗)(The set of invertibles is open, and $\cdot^{-1}$ is continuous.) Let X be a Banach space and $GL(X)$ denote the set of all invertible continuous linear transformations on X.
(1) Prove that $GL(X)$ is an open subset of $CL(X)$ in the usual operator norm topology.
(2) Prove that $T \mapsto T^{-1}$ is continuous on $GL(X)$, that is, for all $T_0 \in CL(X)$ and each $\epsilon > 0$, there exists a $\delta > 0$ such that if $T \in CL(X)$ satisfies $\|T - T_0\| < \delta$, then $T \in GL(X)$ and $\|T^{-1} - T_0^{-1}\| < \epsilon$.

The exponential of an operator. Let X be a Banach space and let $A \in CL(X)$. We will now study the *exponential operator* $e^A \in CL(X)$. For $a \in \mathbb{R}$, one defines the exponential $e^a \in \mathbb{R}$ by

$$e^a := 1 + \frac{1}{1!}a + \frac{1}{2!}a^2 + \frac{1}{3!}a^3 + \cdots.$$

The exponential function $e^{\cdot}$ is useful, because it provides a solution to the initial value problem for the most basic differential equation

$$\begin{cases} \dfrac{d\mathbf{x}}{dt}(t) = a\mathbf{x}(t), & t \in \mathbb{R}, \\ \mathbf{x}(0) = x_0. \end{cases}$$

(Here $\mathbf{x}(t) \in \mathbb{R}$ and $x_0 \in \mathbb{R}$.) The unique solution is given by $\mathbf{x}(t) = e^{ta}x_0$, $t \in \mathbb{R}$. This fundamental differential equation arises in all sorts of applications, for example, radioactive decay, Newton's law of cooling, continuous compound interest, population growth, etc.

For $A \in CL(X)$, we will show that an analogous definition,

$$e^A := I + \frac{1}{1!}A + \frac{1}{2!}A^2 + \frac{1}{3!}A^3 + \cdots,$$

(where we have simply replaced the little a by capital A!) works, and the series converges in $CL(X)$. Then the map $t \mapsto e^{tA}x_0$ provides a solution to the analogous initial value problem, but now in the Banach space X, with the initial condition $x_0 \in X$.

Theorem 2.10. *Let X be a Banach space, and $A \in CL(X)$. Then $e^A := \sum_{n=0}^{\infty} \frac{1}{n!} A^n$ converges in $CL(X)$.*

Proof. The real series $\sum_{n=0}^{\infty} \frac{1}{n!} \|A\|^n$ converges (to $e^{\|A\|}$). Since for $n \in \mathbb{N}$ we have $\left\| \frac{1}{n!} A^n \right\| \leq \frac{\|A\|^n}{n!}$, by the Comparison Test, $\sum_{n=0}^{\infty} \frac{1}{n!} A^n$ converges absolutely. So $\sum_{n=0}^{\infty} \frac{1}{n!} A^n$ converges in the Banach space $CL(X)$. $\square$

Remark 2.5. (∗) Recall that when $a \in \mathbb{R}$, we have $\frac{d}{dt} e^{ta} = a e^{ta}$. Similarly, it can be shown that when $A \in CL(X)$, $\frac{d}{dt} e^{tA} = A e^{tA} = e^{tA} A$. The last equality is not superfluous, since commutativity of multiplication in $CL(X)$ is not always guaranteed, but it turns out that A does commute with e^{tA}. Formally, the above result is not surprising, as can be seen by differentiating the series for e^{tA} termwise with respect to t:

$$\frac{d}{dt} e^{tA} = \frac{d}{dt} \left(I + \frac{t}{1!} A + \frac{t^2}{2!} A^2 + \frac{t^3}{3!} A^3 + \cdots \right)$$
$$= 0 + \frac{1}{1!} A + \frac{2t}{2!} A^2 + \frac{3t^2}{3!} A^3 + \cdots$$
$$= A + \frac{1}{1!} A^2 + \frac{1}{2!} A^3 + \cdots = A e^{tA} = e^{tA} A.$$

A rigorous justification can be given using the fact that $e^{(t+s)A} = e^{tA} e^{sA}$ for all $s, t \in \mathbb{R}$. In general, if $A, B \in CL(X)$ commute, that is, $AB = BA$, then $e^{A+B} = e^A e^B$. This shows that e^A is *always* invertible in $CL(X)$. Indeed, since A commutes with $-A$, we have $e^{-A} e^A = e^{A-A} = e^0 = I = e^A e^{-A}$. Now let $x_0 \in X$, $A \in CL(X)$, and consider the initial value problem:

$$\begin{cases} \dfrac{d\mathbf{x}}{dt}(t) = A\mathbf{x}(t), & t \in \mathbb{R}, \\ \mathbf{x}(0) = x_0. \end{cases}$$

Then $\mathbf{x}(t) := e^{tA} x_0$, $t \in \mathbb{R}$, solves the initial value problem because

$$\frac{d}{dt} \mathbf{x}(t) = \frac{d}{dt} e^{tA} x_0 = A e^{tA} x_0 = A\mathbf{x}(t), \quad t \in \mathbb{R},$$

with $\mathbf{x}(0) = e^{0A}x_0 = e^{\mathbf{0}}x_0 = Ix_0 = x_0$.

Moreover, the solution is unique, since if $\tilde{\mathbf{x}}$ is any solution, then

$$\frac{d}{dt}\left(e^{-tA}\tilde{\mathbf{x}}(t)\right) = e^{-tA}(-A)\tilde{\mathbf{x}}(t) + e^{-tA}\frac{d\tilde{\mathbf{x}}}{dt}(t)$$
$$= e^{-tA}(-A)\tilde{\mathbf{x}}(t) + e^{-tA}A\tilde{\mathbf{x}}(t) = \mathbf{0},$$

so that $e^{-tA}\tilde{\mathbf{x}}(t) = e^{-0A}\tilde{\mathbf{x}}(0) = Ix_0 = x_0$ for all t, giving

$$\tilde{\mathbf{x}}(t) = I\tilde{\mathbf{x}}(t) = e^{tA}e^{-tA}\tilde{\mathbf{x}}(t) = e^{tA}x_0$$

for all $t \in \mathbb{R}$. Hence the solution $t \mapsto e^{tA}x_0$, $t \in \mathbb{R}$, is unique.

Initial value problems in Banach spaces of the above type arise from initial boundary value problems for partial differential equations and their discretisations. More generally, the operator A in the initial value problem is then "unbounded", and similar to $t \mapsto e^{tA}$, one can then associate a "C_0-semigroup $(e^{tA})_{t \geq 0}$ generated by the infinitesimal operator A". The solution to the initial value problem is given by $\mathbf{x}(t) = e^{tA}x_0$ for $t \geq 0$. For example, the initial value problem for the diffusion equation with the homogeneous Dirichlet boundary conditions

$$\begin{cases} \dfrac{\partial u}{\partial t}(x,t) = \dfrac{\partial^2 u}{\partial x^2} & (0 \leq x \leq 1,\ t \geq 0), \\ u(0,t) = 0 \text{ and } u(1,t) = 0 & (t \geq 0), \\ u(x,0) = u_0(x) & (0 \leq x \leq 1) \end{cases}$$

gives the initial value problem for the following ordinary differential equation in the Banach space $L^2[0,1]$:

$$\begin{cases} \dfrac{d\mathbf{x}}{dt}(t) = A\mathbf{x}(t), & t \geq 0, \\ \mathbf{x}(0) = u_0(\cdot) \in L^2[0,1], \end{cases}$$

where $\mathbf{x}(t) = u(\cdot, t) \in L^2[0,1]$, and $A : D(A) (\subset L^2[0,1]) \to L^2[0,1]$ is an unbounded operator given by

$$Af = \frac{d^2 f}{dx^2}, \quad f \in D(A),$$

and $D(A) = \left\{ f \in L^2[0,1] : \dfrac{d^2 f}{dx^2} \in L^2[0,1],\ f(0) = f(1) = 0 \right\}$.

This completes our (rather long!) Remark 2.5.

Example 2.17. (Computing e^A for diagonalisable A). Consider the system
$$\begin{cases} \mathbf{x}_1'(t) = \mathbf{x}_1(t) + 2\mathbf{x}_2(t), \\ \mathbf{x}_2'(t) = -2\mathbf{x}_1(t) + \mathbf{x}_2(t). \end{cases} \tag{2.4}$$

With $\mathbf{x} = (\mathbf{x}_1, \mathbf{x}_2)$, this system can be written as $\mathbf{x}'(t) = A\mathbf{x}(t)$, where
$$A = \begin{bmatrix} 1 & 2 \\ -2 & 1 \end{bmatrix}.$$

We know that given the initial condition $x_0 = (\mathbf{x}_1(0), \mathbf{x}_2(0)) \in \mathbb{R}^2$, the unique solution is $\mathbf{x}(t) = e^{tA} x_0$. This raises the question:

$$\boxed{\text{How does one compute } e^{tA}?}$$

There are several ways, but let us consider a method which works for diagonalisable As. First we note that if

$$A := \begin{bmatrix} \lambda_1 & & \\ & \ddots & \\ & & \lambda_d \end{bmatrix} \in \mathbb{C}^{d \times d}, \text{ then } A^n = \begin{bmatrix} \lambda_1^n & & \\ & \ddots & \\ & & \lambda_d^n \end{bmatrix},$$

and so

$$e^A = \begin{bmatrix} 1 & & \\ & \ddots & \\ & & 1 \end{bmatrix} + \frac{1}{1!}\begin{bmatrix} \lambda_1 & & \\ & \ddots & \\ & & \lambda_d \end{bmatrix} + \frac{1}{2!}\begin{bmatrix} \lambda_1^2 & & \\ & \ddots & \\ & & \lambda_d^2 \end{bmatrix} + \cdots = \begin{bmatrix} e^{\lambda_1} & & \\ & \ddots & \\ & & e^{\lambda_d} \end{bmatrix}.$$

Note in particular that $e^0 = I$, and so calculating e^A *cannot* be the same as taking exponentials of the entries of A!

Now suppose that A is diagonalisable, that is, $A = PDP^{-1}$ where D is diagonal and P is invertible. Then $A^n = PD^n P^{-1}$ and so

$$e^A = I + \frac{1}{1!}PDP^{-1} + \frac{1}{2!}PD^2 P^{-1} + \frac{1}{3!}PD^3 P^{-1} + \cdots$$
$$= P\Big(I + D + \frac{1}{2!}D^2 + \frac{1}{3!}D^3 + \cdots\Big)P^{-1} = Pe^D P^{-1}.$$

Let's see this method in action when $A = \begin{bmatrix} a & b \\ -b & a \end{bmatrix}$ where $a, b \in \mathbb{R}$.

By computing the eigenvalues and eigenvectors of A, we can write

$$A = \underbrace{\begin{bmatrix} 1 & 1 \\ i & -i \end{bmatrix}}_{P} \underbrace{\begin{bmatrix} a + ib & \\ & a - ib \end{bmatrix}}_{D} \frac{1}{2}\begin{bmatrix} 1 & -i \\ 1 & i \end{bmatrix},$$

and so

$$e^{tA} = \begin{bmatrix} 1 & 1 \\ i & -i \end{bmatrix} e^{ta} \begin{bmatrix} \cos(bt) + i\sin(bt) & \\ & \cos(bt) - i\sin(bt) \end{bmatrix} \frac{1}{2} \begin{bmatrix} 1 & -i \\ 1 & i \end{bmatrix}$$

$$= e^{ta} \begin{bmatrix} \cos(bt) & \sin(bt) \\ -\sin(bt) & \cos(bt) \end{bmatrix}.$$

In particular, our initial value problem for (2.4) has the solution (putting $a = 1$ and $b = 2$ above)

$$\begin{bmatrix} \mathbf{x}_1(t) \\ \mathbf{x}_2(t) \end{bmatrix} = e^t \begin{bmatrix} \cos(2t) & \sin(2t) \\ -\sin(2t) & \cos(2t) \end{bmatrix} \begin{bmatrix} \mathbf{x}_1(0) \\ \mathbf{x}_2(0) \end{bmatrix}, \quad t \in \mathbb{R}.$$

So we've seen how to compute e^A if the matrix A is diagonalisable. However, not all matrices are diagonalisable. For example, consider the matrix

$$A = \begin{bmatrix} 0 & 1 \\ 0 & 0 \end{bmatrix}.$$

The eigenvalues of this matrix are both 0, and so if it were diagonalisable, say $A = PDP^{-1}$, then the diagonal matrix D must be the zero matrix. But then $A = PDP^{-1} = P\mathbf{0}P^{-1} = \mathbf{0}$, and we have arrived at a contradiction since $A \neq \mathbf{0}$! So this A is *not* diagonalisable.

In general, however, every matrix has what is called a *Jordan canonical form*, that is, there exists an invertible P such that $P^{-1}AP = D + N$, where D is diagonal, N is *nilpotent* (that is, there exists an $n \in \mathbb{N}$ such that $N^n = 0$), and D and N commute. Then the exponential of A is:

$$e^A = Pe^D \left(I + N + \frac{1}{2!}N^2 + \cdots + \frac{1}{n!}N^n \right) P^{-1}.$$

But the computation of a P taking A to its Jordan form requires some sophisticated linear algebra, and we won't treat this here. The interested reader is referred to [Hirsch and Smale (1974), Chapter 6]. ◇

Exercise 2.26. ($e^{A+B} \neq e^A e^B$).
Compute e^A and e^B, where A, B are the nilpotent matrices $\begin{bmatrix} 0 & 1 \\ 0 & 0 \end{bmatrix}, \begin{bmatrix} 0 & 0 \\ -1 & 0 \end{bmatrix}$.
Give an example of matrices $A, B \in \mathbb{R}^{2 \times 2}$ for which $e^{A+B} \neq e^A e^B$.

2.5 (∗) Open Mapping Theorem

In this section, we will show Theorem 2.11, the "Open Mapping Theorem". The proofs in this section are somewhat more technical than the rest of the sections of this chapter.

Definition 2.9. (Open map) Let X, Y be normed spaces. $T \in CL(X, Y)$ is called *open* if for all open sets $U \subset X$, $T(U)$ is open in Y.

Proposition 2.4.
Let X, Y be normed spaces, $T \in CL(X, Y)$, and $B := \{x \in X : \|x\| \leq 1\}$. Then the following are equivalent:

(1) T is open.

(2) There exists a $\delta > 0$ such that $\boxed{B(\mathbf{0}_Y, \delta) \subset T(B)}$.

Proof.
(2)⇒(1): Suppose that there exists a $\delta > 0$ such that $B(\mathbf{0}_Y, \delta) \subset T(B)$. Let U be open in X. If $y_0 \in T(U)$, then $y_0 = Tx_0$ for some $x_0 \in U$. As U is open, there exists a $r > 0$ such that the open ball $B(x_0, r)$ with centre x_0 and radius r is contained in U. We claim that the open ball $B(y_0, \delta r/2)$ is contained in $T(U)$. If $y \in B(y_0, \delta r/2)$, then $\|y - y_0\| < \delta r/2$, that is, $\|(2/r)(y - y_0)\| < \delta$, and so $(2/r)(y - y_0) \in B(\mathbf{0}_Y, \delta) \subset T(B)$. Hence there exists an $x \in B$ such that $(2/r)(y - y_0) = Tx$, that is, we have $y = T((r/2)x + x_0)$. But as $\|((r/2)x + x_0) - x_0\| = (r/2)\|x\| \leq (r/2) \cdot 1 < r$, we see that $(r/2)x + x_0 \in B(x_0, r) \subset U$. Consequently, $y \in T(U)$, as desired.

(1)⇒(2): Suppose that T is open. Then $T(B(\mathbf{0}_X, 1))$, the image of the open set $B(\mathbf{0}_X, 1)$, must be open. But $\mathbf{0}_Y = T\mathbf{0}_X \in T(B(\mathbf{0}_X, 1))$, and so, there must exist a $\delta > 0$ such that the open ball $B(\mathbf{0}_Y, \delta) \subset T(B(\mathbf{0}_X, 1)) \subset T(B)$, as wanted. □

Lemma 2.1. (Baire Lemma)

Let (1) X *be a Banach space, and*

(2) $(F_n)_{n \in \mathbb{N}}$ *be a sequence of closed sets in X such that* $X = \bigcup_{n \in \mathbb{N}} F_n$.

Then there exist an $n \in \mathbb{N}$ and a nonempty open set U such that $U \subset F_n$.

Proof. We assume none of the sets F_n contain a nonempty open subset and construct a Cauchy sequence that converges to a point, which lies in none of the F_n, contradicting the fact that the F_ns cover X.

First let us observe that whenever a closed set F in X does not contain any open set, we have that $\complement F$ is dense in X. (To see this, let $x \in X$, and $r > 0$. We'd like to show that $B(x, r) \cap \complement F \neq \emptyset$. If $x \in \complement F$, then $x \in B(x, r) \cap \complement F$, and we are done. On the other hand, if $x \notin \complement F$, then $x \in F$. But as F doesn't contain any open set, it won't, in particular, contain $B(x, r)$. So there must be an element y in $B(x, r)$ which is not in F. But this means that $y \in \complement F$, and so we've got $y \in B(x,r) \cap \complement F$, as wanted.) By our assumption, it follows that $\complement F_n$ is dense in X for all $n \in \mathbb{N}$.

Let x_1 be any element in the nonempty (dense!) open set $\complement F_1$. Let $r_1 > 0$ be such that $\overline{B(x_1, r_1)} \subset \complement F_1$. As $\complement F_2$ is dense in X, there exists an $x_2 \in B(x_1, r_1) \cap \complement F_2$. As $B(x_1, r_1) \cap \complement F_2$ is open, we can find an $r_2 < r_1/2$ such that $\overline{B(x_2, r_2)} \subset B(x_1, r_1) \cap \complement F_2$. As $\complement F_3$ is dense in X, there exists an $x_3 \in B(x_2, r_2) \cap \complement F_3$. As $B(x_2, r_2) \cap \complement F_3$ is open, we can find an $r_3 < r_1/4$ such that $\overline{B(x_3, r_3)} \subset B(x_2, r_2) \cap \complement F_3$.

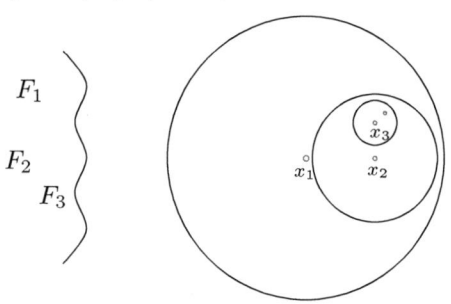

Proceeding in this manner, we obtain a sequence $(x_n)_{n \in \mathbb{N}}$, with the term $x_{n+1} \in B(x_n, r_n)$. If $n > m$, then $B(x_n, r_n) \subset B(x_m, r_m)$, and so we have $\|x_n - x_m\| < r_m < r_1/2^{m-1} \stackrel{m \to \infty}{\longrightarrow} 0$. Thus $(x_n)_{n \in \mathbb{N}}$ is Cauchy, and as X is Banach, also convergent, say, to $x \in X$. With a fixed m, in the inequality above, if we pass the limit as $n \to \infty$, then we obtain $\|x - x_m\| \leq r_m$, that is, $x \in \overline{B(x_m, r_m)} \subset \complement F_m$. As the choice of $m \in \mathbb{N}$ was arbitrary, for all $m \in \mathbb{N}$, $x \notin F_m$. But this contradicts the fact that the F_ms cover X. □

Exercise 2.27.
Show that the Hamel basis[5] of a Banach space can only be finite or uncountable.

Before proving the Open Mapping Theorem, we'll give some notation and a useful technical result. For subsets A, B of a normed space X and a scalar α, we set $\alpha A := \{\alpha a : a \in A\}$, and $A + B := \{a + b : a \in A, b \in B\}$.

Lemma 2.2. *Let X be a normed space, and $A \subset X$ satisfy*

(1) *A is symmetric, that is, $-A = A$,*

(2) *A is mid-point convex, that is, for all $x, y \in A$, $\dfrac{x+y}{2} \in A$, and*

(3) *there is a nonempty open set $U \subset A$.*

Then there exists a $\delta > 0$ such that $B(\mathbf{0}, \delta) \subset A$.

---
[5]For the definition/existence of a Hamel basis, see Exercise 2.39, page 115.

Proof. First note that for a fixed scalar $\alpha \neq 0$, and an $a \in X$, the maps $x \mapsto x + a : X \to X$ and $x \mapsto \alpha x : X \to X$, are both continuous, with the continuous inverses ($x \mapsto x - a$ and $x \mapsto \alpha^{-1} x$).
Hence if U is open in X, then $U + \{-a\}$ is open in X.
So $U + (-A) = \bigcup_{a \in A} (U + \{-a\})$ is open in X. Thus $\dfrac{U + (-A)}{2}$ is open in X.
If $a \in U$, then $\mathbf{0} = \dfrac{a - a}{2} \in \underbrace{\dfrac{U + (-A)}{2}}_{\text{open}} \subset \dfrac{A + (-A)}{2} = \dfrac{A + A}{2} \subset A$.

Thus there exists a $\delta > 0$ such that $B(\mathbf{0}, \delta) \subset \dfrac{U + (-A)}{2} \subset \dfrac{A + A}{2} \subset A$. $\square$

Theorem 2.11. (Open Mapping Theorem).
Let X, Y be Banach spaces, and $T \in CL(X, Y)$ be surjective. Then T is open.

Proof. Let $B := \{x \in X : \|x\| \leq 1\}$. Then $X = \bigcup_{n \in \mathbb{N}} nB$. Thanks to the surjectivity of T, we have $Y = \bigcup_{n \in \mathbb{N}} T(nB)$. Thus certainly $Y = \bigcup_{n \in \mathbb{N}} \overline{T(nB)}$.
It can be checked that $\overline{T(nB)} = n\overline{T(B)}$. By the Baire Lemma, there exists an $n \in \mathbb{N}$ such that $n\overline{T(B)}$ contains a nonempty open set. But since the map $x \mapsto nx : X \to X$ is continuous with a continuous inverse, it follows that $\overline{T(B)}$ contains a nonempty open set too. By Lemma 2.2, there exists a $\delta > 0$ such that $B(\mathbf{0}_Y, \delta) \subset \overline{T(B)}$. We will now show that this implies

$$B(\mathbf{0}_Y, \delta/2) \subset T(B), \tag{2.5}$$

giving the required openness of T by Proposition 2.4. Let y such that $\|y\| < \delta/2$. We must show that there exists a $x \in B$ with $y = Tx$. Using $B(\mathbf{0}_Y, \delta) \subset \overline{T(B)}$, it can be seen that

$$B(\mathbf{0}_Y, \delta/2^n) \subset \overline{T(B/2^n)}. \tag{2.6}$$

From (2.6), with $n = 1$, it follows that we can arbitrarily closely approximate y by elements from $T(B/2)$. Thus there exists an x_1 with $\|x_1\| \leq 1/2$ such that $\|y - Tx_1\| \leq \delta/4$ that is, $y - Tx_1 \in B(\mathbf{0}, \delta/4)$. From (2.6) again it follows (with $n = 2$) that we can arbitrarily closely approximate $y - Tx_1$ by an element Tx_2 with $\|x_2\| \leq 1/4$: $\|y - Tx_1 - Tx_2\| \leq \delta/8$. Proceeding in this manner, we can inductively construct a sequence $(x_n)_{n \in \mathbb{N}}$ such that: $\|x_n\| \leq 1/2^n$ and $\|y - Tx_1 - Tx_2 - \cdots - Tx_{n-1}\| \leq \delta/2^n$.

As $\|x_n\| \leq \dfrac{1}{2^n}$, $\displaystyle\sum_{n=1}^{\infty} x_n$ is absolutely convergent, and $\left\| \displaystyle\sum_{n=1}^{\infty} x_n \right\| \leq \displaystyle\sum_{n=1}^{\infty} \dfrac{1}{2^n} = 1$.

If we denote the sum of the series $\sum_{n=1}^{\infty} x_n$ by x, then

$$y = \lim_{n\to\infty} \sum_{k=1}^n Tx_k = \lim_{n\to\infty} T\Big(\sum_{k=1}^n x_k\Big) = T\Big(\lim_{n\to\infty} \sum_{k=1}^n x_k\Big) = Tx,$$

thanks to the continuity of T. Since $\|x\| \leqslant 1$, this proves the desired inclusion (2.5). □

Corollary 2.4. *If X, Y are Banach spaces, and $T \in CL(X,Y)$ is bijective, then $T^{-1} \in L(Y,X)$ is continuous.*

We then refer to T as a normed space *isomorphism*, and say that X, Y are *isomorphic* (as normed spaces), written $X \simeq Y$.

Proof. T is open, and so if U is open in X, $T(U)$ is open in Y. But $(T^{-1})^{-1}(U) = \{y \in Y : T^{-1}y \in U\} = \{y \in Y : y \in T(U)\} = T(U)$. Thus the inverse images of open sets under T^{-1} are open, showing that T^{-1} is continuous. □

Exercise 2.28. Construct a continuous and surjective, but not open, $f : \mathbb{R} \to \mathbb{R}$.

Exercise 2.29. (Closed Graph Theorem).
The aim of this exercise is to prove the Closed Graph Theorem:
    Let X, Y be Banach spaces and $T : X \to Y$ be a linear transformation.
    Then T is continuous if and only if its graph $\mathcal{G}(T)$ is closed in $X \times Y$.
Here $X \times Y$ has the norm $\|(x,y)\| := \max\{\|x\|, \|y\|\}$, $(x,y) \in X \times Y$, and the set $\mathcal{G}(T) := \{(x, Tx) : x \in X\} \subset X \times Y$ is the *graph of T*.
The "only if" part is easy to see. If $(x_n, Tx_n) \to (x,y)$, then $x_n \to x$, and as T is continuous, $\|Tx_n - Tx\| \leqslant \|T\| \|x_n - x\|$, so that $Tx_n \to Tx$. But $Tx_n \to y$, and so, by the uniqueness of limits, $Tx = y$. Thus $(x_n, Tx_n) \to (x, Tx) \in \mathcal{G}(T)$, showing that $\mathcal{G}(T)$ is closed.
Show the "if" part. *Hint:* Consider $p : \mathcal{G}(T) \to X$, where $p\big((x, Tx)\big) = x$, $x \in X$.

Uniform Boundedness Principle.

We give below another important application of the Baire Lemma.

Theorem 2.12. (Uniform Boundedness Principle).
Suppose that
  (1) X and Y are Banach spaces,
  (2) $T_i \in CL(X,Y)$, $i \in I$, is a "pointwise bounded" family, that is,

$$\text{for all } x \in X, \ \sup_{i\in I} \|T_i x\| < +\infty.$$

Then the family is "uniformly bounded", that is, $\sup_{i\in I} \|T_i\| < +\infty$.

Proof. For $n \in \mathbb{N}$, $F_n := \left\{x \in X : \sup_{i \in I} \|T_i x\| \leq n\right\} = \bigcap_{i \in I} \{x \in X : \|T_i x\| \leq n\}$ is mid-point convex, symmetric, and closed, as F_n is the intersection of the mid-point convex, symmetric, and closed sets $\{x \in X : \|T_i x\| \leq n\}$, $i \in I$. From (2), we have $X = \bigcup_{n \in \mathbb{N}} F_n$, and so by the Baire Lemma, there exists an n such that F_n contains a nonempty open set. By Lemma 2.2, there exists a $\delta > 0$ such that the ball $B(\mathbf{0}, \delta)$ with center $\mathbf{0}$ and radius δ is contained in F_n, that is, if $\|x\| < \delta$, then for all $i \in I$ we have $\|T_i x\| \leq n$. We claim that $\|T_i x\| \leq (2n/\delta)\|x\|$ for all $x \in X$ and all $i \in I$. Clearly this is true if $x = \mathbf{0}$, since then both sides of the inequality are equal to 0. On the other hand, if $x \neq \mathbf{0}$, then $y := \frac{\delta}{2\|x\|} x$ has norm $\|y\| = \delta/2 < \delta$, and so we must have $\|T_i y\| \leq n$, which, using the linearity of T_i and the positive homogeneity of the norm, delivers, upon a rearrangement, the desired inequality. Thus $\|T_i\| \leq 2n/\delta$ for all $i \in I$, and thus $\sup_{i \in I} \|T_i\| \leq 2n/\delta$. □

Corollary 2.5. (Banach-Steinhauss Theorem).
Let (1) X, Y be Banach spaces, and
 (2) $(T_n)_{n \in \mathbb{N}}$ in $CL(X, Y)$ be such that $\lim_{n \to \infty} T_n x$ exists for all $x \in X$.
Then $x \mapsto \lim_{n \to \infty} T_n x : X \to Y$ belongs to $CL(X, Y)$.

Proof. It is clear that the map $x \mapsto \lim_{n \to \infty} T_n x : X \to Y$ is linear. It remains to show that it is continuous too. Set $Tx := \lim_{n \to \infty} T_n x$, $x \in X$. For each $x \in X$, $(T_n x)_{n \in \mathbb{N}}$ is convergent, and in particular, bounded:
$$\text{for all } x \in X, \sup_{n \in \mathbb{N}} \|T_n x\| < +\infty.$$
Hence by the Uniform Boundedness Principle, there exists an M such that for all $n \in \mathbb{N}$, $\|T_n\| \leq M$. This gives, for each fixed $x \in X$, that
$$\text{for all } n \in \mathbb{N}, \|T_n x\| \leq M \|x\|.$$
Passing the limit $n \to \infty$ yields $\|Tx\| \leq M\|x\|$. As the choice of x was arbitrary, this holds for all x, and consequently, the linear transformation T is continuous. □

2.6 Spectral Theory

For a linear transformation $T \in L(X)$ on a *finite* dimensional vector space X over $\mathbb{C}$, the set of eigenvalues of T is known as its *spectrum* $\sigma(T)$, and has cardinality at most $\dim X$. But in infinite dimensional complex vector spaces, strange things may happen, for example linear transformations may have no eigenvalues at all or finitely many or (countably/uncountably) infinitely many! First of all, here is a natural definition of eigenvectors and eigenvalues, extending our prior familiarity with eigenvalues from elementary linear algebra. We remind the reader that the prefix *eigen* is derived from German, meaning "one's own".

Definition 2.10. (Eigenvalues and eigenvectors). Let X be a normed space and $T \in CL(X)$. Then $\lambda \in \mathbb{C}$ is called an *eigenvalue of T* if there exists a nonzero vector $x \in X$ such that $Tx = \lambda x$. Such a nonzero vector x is then called an *eigenvector of T corresponding to the eigenvalue λ*.

Example 2.18. (Uncountably many eigenvalues).
Let $\lambda \in \mathbb{D} := \{z \in \mathbb{C} : |z| < 1\}$. If $\mathbf{x} := (1, \lambda, \lambda^2, \lambda^3, \cdots)$, then as $|\lambda| < 1$,
$$\sum_{n=0}^{\infty} |\lambda^n|^2 = \sum_{n=0}^{\infty} \left(|\lambda|^2\right)^n = \frac{1}{1 - |\lambda|^2} < \infty,$$
and so $\mathbf{x} \in \ell^2$. Clearly $\mathbf{x} \neq \mathbf{0}$ too.
We see that $\mathbf{x}$ is an eigenvector of the left shift operator $L \in CL(\ell^2)$ because
$$L\mathbf{x} = L(1, \lambda, \lambda^2, \lambda^3, \cdots) = (\lambda, \lambda^2, \lambda^3, \cdots) = \lambda \cdot (1, \lambda, \lambda^2, \cdots) = \lambda \cdot \mathbf{x}.$$
Thus each point in the open unit disk[6] is an eigenvalue of L. ◇

Example 2.19. (No eigenvalues). On the other hand, the right shift operator $R \in CL(\ell^2)$ has *no* eigenvalues. Suppose that $\lambda \in \mathbb{C}$ is such that $R\mathbf{x} = \lambda \mathbf{x}$ for some $\mathbf{x} = (x_n)_{n \in \mathbb{N}} \in \ell^2$. Then
$$R\mathbf{x} = R(x_1, x_2, x_3, \cdots) = (0, x_1, x_2, x_3, \cdots) = (\lambda x_1, \lambda x_2, \lambda x_3, \cdots) = \lambda \mathbf{x}.$$
Suppose first that $\lambda \neq 0$. Then from the above, $\lambda x_1 = 0$ gives $x_1 = 0$. Next, $\lambda x_2 = x_1$ now gives $x_2 = 0$. Proceeding in this manner, we obtain $x_1 = x_2 = x_3 = \cdots = 0$, and so $\mathbf{x} = \mathbf{0}$.

---

[6] We remark that if we look at the "matrix" corresponding to L, while thinking of vectors in ℓ^2 as an "infinite columns", then the action of L is described by
$$\begin{bmatrix} 0 & 1 & 0 & 0 & \cdots \\ 0 & 0 & 1 & 0 & \cdots \\ 0 & 0 & 0 & 1 & \cdots \\ \vdots & \vdots & \vdots & \vdots & \ddots \end{bmatrix},$$
a matrix with all diagonal entries equal to 0, and with 1s along an "upper" diagonal. So in this case, our "matricial intuition" would have led us astray, since based on the above matrix, reminiscent of a Jordan block in finite dimensional linear algebra, one would be tempted to hastily guess that L has the only eigenvalue 0!

On the other hand, if $\lambda = 0$, then $(0, x_1, x_2, x_3, \cdots) = (\lambda x_1, \lambda x_2, \lambda x_3, \cdots)$ shows immediately that $x_1 = x_2 = x_3 = \cdots = 0$, and so $\mathbf{x} = \mathbf{0}$.
Consequently, R has no eigenvalues. ◇

Note that when $\dim X < \infty$, and $T \in CL(X)$, then

$\boxed{\lambda \in \mathbb{C} \text{ is an eigenvalue of } T}$ if and only if $\boxed{\lambda I - T \text{ is not invertible.}}$

So the points in the spectrum $\sigma(T)$ are exactly the ones where $\lambda I - T$ fails to be invertible in $\sigma(T)$. This prompts the following natural concept in the general case, that is, when $\dim X \not< \infty$.

Definition 2.11. (Spectrum and resolvent).
Let X be a normed space and $T \in CL(X)$.
We say that $\lambda \in \mathbb{C}$ belongs to the *spectrum* $\sigma(T)$ *of* T if $\lambda I - T$ is not invertible in $CL(X)$. Thus

$$\rho(T) := \mathbb{C} \backslash \sigma(T) = \{\lambda \in \mathbb{C} : \lambda I - T \in CL(X) \text{ is invertible in } CL(X)\}.$$

The set $\rho(T)$ is called the *resolvent set of* T.

The set $\sigma_p(T)$ of all eigenvalues of T is called the *point spectrum of* T.

We have that $\sigma_p(T) \subset \sigma(T)$, since if $\lambda \in \sigma_p(T)$, then there exists a nonzero vector x such that $Tx = \lambda x$, that is, $(\lambda I - T)x = \mathbf{0}$, showing that $\lambda I - T$ is not injective, and hence can't be invertible either!

We'll now show that if X is Banach and $T \in CL(X)$, then $\sigma(T)$ is a compact nonempty subset of $\mathbb{C}$.

Theorem 2.13. *Let X be a Banach space and $T \in CL(X)$. Then*

(1) $\sigma(T) \subset \{\lambda \in \mathbb{C} : |\lambda| \leq \|T\|\}$.
(2) $\rho(T)$ *is an open subset of* $\mathbb{C}$.
(3) $\sigma(T)$ *is a compact subset of* $\mathbb{C}$.
(4) $\sigma(T)$ *is nonempty.*

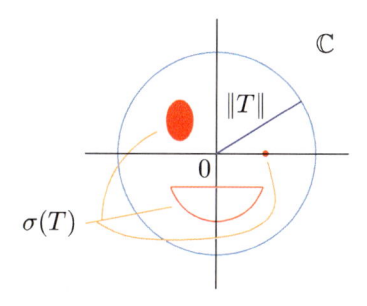

Proof.

(1) Let $|\lambda| > \|T\| \geq 0$. Then $\left\|\frac{1}{\lambda}T\right\| < 1$, and so $I - \frac{1}{\lambda}T$ is invertible in $CL(X)$. Thus, as $\lambda \neq 0$, we have that
$$\lambda I - T = \lambda\left(I - \frac{1}{\lambda}T\right)$$
is invertible in $CL(X)$ too.

(2) Let $\lambda_0 \in \rho(T)$. Then for $\lambda \in \mathbb{C}$,
$$I - (\lambda_0 I - T)^{-1}(\lambda I - T) = (\lambda_0 I - T)^{-1}(\lambda_0 I - T - \lambda I + T)$$
$$= (\lambda - \lambda_0)(\lambda_0 I - T)^{-1}.$$
So $I - \underbrace{(\lambda_0 - \lambda)(\lambda_0 I - T)^{-1}}_{=:A} = (\lambda_0 I - T)^{-1}(\lambda I - T).$

For $\lambda_0 \approx \lambda$, A has small norm, in particular, < 1. Hence it follows that $(\lambda_0 I - T)^{-1}(\lambda I - T) =: S$ is invertible in $CL(X)$. So we conclude that $\lambda I - T = (\lambda_0 I - T)S$ (being the product of two invertible operators in $CL(X)$) is also invertible in $CL(X)$.

(3) $\sigma(T)$ is bounded (as $\sigma(T) \subset B(0, \|T\|) := \{z \in \mathbb{C} : |z| \leq \|T\|\}$), and also it is closed (because its complement $\mathbb{C}\setminus\sigma(T) = \rho(T)$ is open). So $\sigma(T)$ is compact.

(4)(∗)[7] Let $\sigma(T) = \emptyset$. Then $f(z) := (zI - T)^{-1} \in CL(X)$ for all $z \in \mathbb{C}$.
In particular, T^{-1} exists, and is not **0**.
Let $\varphi \in (CL(X))'$ be such that $\varphi(T^{-1}) \neq 0$.
Such a φ exists by the Hahn-Banach Theorem (Exercise 2.38, page 109).
Let $g : \mathbb{R}^2 \to \mathbb{C}$ be given by $g(r, \theta) = \varphi(f(re^{i\theta}))$, for all $(r, \theta) \in \mathbb{R}^2$.
We will show that $g \in C^1(\mathbb{R}^2, \mathbb{C})$ by showing that it has continuous first order partial derivatives (which will in turn be used in the calculations, and also to justify a differentiation under the integral sign).
Using the resolvent identity (Exercise 2.30, page 102), we have
$$\frac{g(r,\theta) - g(r_0, \theta)}{r - r_0} = \frac{\varphi(f(re^{i\theta})) - \varphi(f(r_0 e^{i\theta}))}{r - r_0}$$
$$= \varphi\left(\frac{(re^{i\theta}I - T)^{-1} - (r_0 e^{i\theta}I - T)^{-1}}{r - r_0}\right)$$
$$= \varphi\left(\frac{(r - r_0)e^{i\theta}(re^{i\theta}I - T)^{-1}(r_0 e^{i\theta}I - T)^{-1}}{r - r_0}\right)$$
$$= e^{i\theta}\varphi\left((re^{i\theta}I - T)^{-1}(r_0 e^{i\theta}I - T)^{-1}\right).$$

---

[7]The usual proof of this is by using some tools from complex analysis. We will instead follow the proof from [Singh (2006)] relying on real analysis techniques.

Using continuity of φ and that of the inverse operation (Exercise 2.25, page 88), it follows from the above calculation, that

$$\begin{aligned}\frac{\partial g}{\partial r}(r_0, \theta) &= \lim_{r \to r_0} \frac{g(r, \theta) - g(r_0, \theta)}{r - r_0} \\ &= \lim_{r \to r_0} e^{i\theta} \varphi\big((re^{i\theta}I - T)^{-1}(r_0 e^{i\theta}I - T)^{-1}\big) \\ &= e^{i\theta} \varphi\big((r_0 e^{i\theta}I - T)^{-1}(r_0 e^{i\theta}I - T)^{-1}\big) \\ &= e^{i\theta} \varphi\Big(\big(f(r_0 e^{i\theta})\big)^2\Big).\end{aligned}$$

Similarly, $\frac{\partial g}{\partial r}(r, \theta_0) = r i e^{i\theta_0} \varphi\Big(\big(f(re^{i\theta_0})\big)^2\Big)$. Set $F(r) := \int_0^{2\pi} g(r, \theta) d\theta$.
By differentiating under the integral sign, we obtain

$$F'(r) = \int_0^{2\pi} \frac{\partial g}{\partial r}(r, \theta) d\theta = \int_0^{2\pi} e^{i\theta} \varphi\Big(\big(f(re^{i\theta})\big)^2\Big) d\theta.$$

Consequently,

$$\begin{aligned}irF'(r) &= \int_0^{2\pi} i r e^{i\theta} \varphi\Big(\big(f(re^{i\theta})\big)^2\Big) d\theta = \int_0^{2\pi} \frac{\partial g}{\partial r}(r, \theta) d\theta \\ &= g(r, 2\pi) - g(r, 0) = \varphi\big(f(r \cdot 1)\big) - \varphi\big(f(r \cdot 1)\big) = 0.\end{aligned}$$

Hence F is constant, and we have

$$\begin{aligned}F(r) = F(0) &= \int_0^{2\pi} g(0, \theta) d\theta = \int_0^{2\pi} \varphi\big(f(0)\big) d\theta \\ &= \int_0^{2\pi} \varphi\big(-T^{-1}\big) d\theta = -2\pi \varphi(T^{-1}).\end{aligned}$$

Now

$$\begin{aligned}\varphi\big(f(re^{i\theta})\big) &= \varphi\big((re^{i\theta}I - T)^{-1}\big) \\ &= \varphi\big((I - r^{-1}e^{-i\theta}T)^{-1}\big) \cdot r^{-1} e^{-i\theta} \xrightarrow{r \to \infty} \varphi(I) \cdot 0 = 0.\end{aligned}$$

Fix r such that $|\varphi(f(re^{i\theta}))| < \dfrac{|\varphi(T^{-1})|}{2}$. Then

$$\begin{aligned}2\pi |\varphi(T^{-1})| = |F(r)| &\leq \int_0^{2\pi} |\varphi(f(re^{i\theta}))| d\theta \\ &< \int_0^{2\pi} \frac{|\varphi(T^{-1})|}{2} d\theta = \pi |\varphi(T^{-1})|,\end{aligned}$$

giving $2 < 1$, a contradiction. This completes the proof. □

Example 2.20. (Spectrum of the left shift operator).
Consider the left shift operator $L \in CL(\ell^2)$. Then $\|L\| \leq 1$. So it follows that $\sigma(L) \subset \{z \in \mathbb{C} : |z| \leq 1\}$. As $\{z \in \mathbb{C} : |z| < 1\} \subset \sigma_p(L) \subset \sigma(L)$, and because $\sigma(L)$ is closed, it follows that $\{z \in \mathbb{C} : |z| \leq 1\} \subset \sigma(L)$ too. So $\sigma(L) = \{z \in \mathbb{C} : |z| \leq 1\}$.

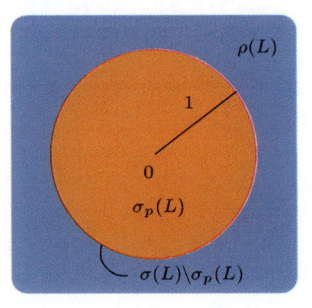

We now claim that $\sigma_p(L) = \{z \in \mathbb{C} : |z| < 1\}$. We had seen earlier that $\{z \in \mathbb{C} : |z| < 1\} \subset \sigma_p(L)$. Now we'll show the reverse inclusion.
To this end, let $\lambda \in \sigma_p(L)$ with eigenvector $\mathbf{x} = (x_n)_{n \in \mathbb{N}}$.
Then $(x_2, x_3, \cdots) = L(x_1, x_2, x_3, \cdots) = \lambda(x_1, x_2, x_3, \cdots)$.
So $\lambda x_n = x_{n+1}$ for all n, giving (by induction) $x_n = \lambda^{n-1} x_1$ for all n.
As $\ell^2 \ni \mathbf{x} \neq \mathbf{0}$, we have

$$0 \neq \|\mathbf{x}\|_2^2 = |x_1|^2 \sum_{n=0}^{\infty} \left(|\lambda|^2\right)^n < \infty,$$

so that $x_1 \neq 0$, and the geometric series with common ratio $|\lambda|^2$ converges. So $|\lambda| < 1$, and we get the reverse inclusion $\sigma_p(T) \subset \{z \in \mathbb{C} : |z| < 1\}$. $\diamond$

We will return to this topic on spectral theory when we deal with operators on a Hilbert space, and also in the context of compact operators.

Exercise 2.30. (Resolvent Identity). Let X be a normed space, $T \in CL(X)$ and $\lambda, \mu \in \rho(T)$. Prove that $(\lambda I - T)^{-1} - (\mu I - T)^{-1} = (\mu - \lambda)(\lambda I - T)^{-1}(\mu I - T)^{-1}$.

Exercise 2.31. (Spectral radius). Let X be a Banach space, and $T \in CL(X)$. Define the *spectral radius of T* by $r_\sigma(T) := \sup_{\lambda \in \sigma(T)} |\lambda|$.

(1) Prove that $r_\sigma(T) \leq \|T\|$.
(2) Show that for $T_A \in CL(\mathbb{R}^2)$, $A := \begin{bmatrix} 1 & 1 \\ 0 & 1 \end{bmatrix}$, then $r_\sigma(T_A) < \|T_A\|$.

Here $\mathbb{R}^2$ has the usual Euclidean $\|\cdot\|_2$-norm.
Remark. In this connection, the *Gelfand-Beurling Formula*[8] says that:
If X is Banach and $T \in CL(X)$, then $r_\sigma(T) = \lim_{n \to \infty} \|T^n\|^{1/n}$.

---
[8] See for example [Taylor and Lay (1980), page 287, Theorem 3.2].

Exercise 2.32. Let X be a Banach space, $T \in CL(X)$, and $\lambda \in \sigma(T)$.
Prove that λ^2 belongs to the spectrum of T^2.
Hint: Use $(\lambda^2 I - T^2) = (\lambda I - T)(\lambda I + T) = (\lambda I + T)(\lambda I - T)$.
Remark. More generally, the *Spectral Mapping Theorem*[9] says that:
If X is a Banach space, $T \in CL(X)$, $p = c_0 + c_1 z + \cdots + c_d z^d \in \mathbb{C}[z]$ (a polynomial
with complex coefficients), and $p(T) := c_0 I + c_1 T + \cdots + c_d T^d$, then we have
$\sigma(p(T)) = p(\sigma(T)) := \{p(\lambda) : \lambda \in \sigma(T)\}$.

Exercise 2.33. (Spectrum of the diagonal operator).
Let $(\lambda_n)_{n \in \mathbb{N}}$ be sequence in $\mathbb{C}$ which is convergent to 0, and consider $\Lambda \in CL(\ell^2)$
given by $\Lambda(a_1, a_2, a_3, \cdots) = (\lambda_1 a_1, \lambda_2 a_2, \lambda_3 a_3, \cdots)$ for all $(a_1, a_2, a_3, \cdots) \in \ell^2$.
Show that $\{\lambda_n : n \in \mathbb{N}\} \subset \sigma_p(\Lambda) \subset \{\lambda_n : n \in \mathbb{N}\} \bigcup \{0\} = \sigma(\Lambda)$.
Remark. (Spectral Theorem for Compact Operators).
In Chapter 5, we will learn that this Λ is an example of a "compact operator";
see Example 5.3 on page 214. More generally, one can show the *Spectral Theorem
for Compact Operators*, which says that for a compact operator K on an infinite
dimensional Hilbert space H,
  (1) $\sigma(K) \setminus \{0\} = \sigma_p(K) \setminus \{0\}$, and $\sigma(K)$ is countable,
  (2) 0 is the only accumulation point of $\sigma(K)$,
  (3) For all $\lambda \in \sigma_p(K) \setminus \{0\}$, $\dim \ker(\lambda I - K) = \dim \ker(\lambda^* I - K^*) < \infty$.

Exercise 2.34. (Approximate spectrum).
(1) Let X be a Banach space, and $T \in CL(X)$. A number $\lambda \in \mathbb{C}$ is said to belong
to the *approximate spectrum* $\sigma_{ap}(T)$ *of* T if there exists a sequence $(x_n)_{n \in \mathbb{N}}$ of
vectors from X such that $\|x_n\| = 1$ for all $n \in \mathbb{N}$, and $Tx_n - \lambda x_n \xrightarrow{n \to \infty} \mathbf{0}$ in X.
Prove that $\sigma_{ap}(T) \subset \sigma(T)$.
(2) Let $\Lambda \in CL(\ell^2)$ be the diagonal operator corresponding to a convergent (and
hence bounded) sequence $(\lambda_n)_{n \in \mathbb{N}}$. Prove that $\lim_{n \to \infty} \lambda_n \in \sigma_{ap}(\Lambda)$.

Exercise 2.35. (Point spectrum of the position operator).
Let X be a normed space, and let $A : D_A \to X$ be an "unbounded operator[10]", where the domain D_A is a subspace of X. Then the *point spectrum*
of the unbounded operator A is defined in an analogous manner as before:
$\sigma_p(A) := \{\lambda \in \mathbb{C} : \text{there exists an } x \in D_A \setminus \{\mathbf{0}\} \text{ such that } Ax = \lambda x\}$.
Now consider the *position operator* $Q : D_Q \to L^2(\mathbb{R})$, arising in Quantum Mechanics, where $D_Q := \{\Psi \in L^2(\mathbb{R}) : (x \mapsto x\Psi(x)) =: Q\Psi \in L^2(\mathbb{R})\}$, and
$(Q\Psi)(x) := x\Psi(x)$, for almost all $x \in \mathbb{R}$, and all $\Psi \in D_Q$.
Show that $\sigma_p(Q) = \emptyset$.
Remark. So Q has no eigenvectors in $D_Q \subset L^2(\mathbb{R})$. However, when we learn
elementary distribution theory later on in Chapter 6, we'll see that $x\delta_\lambda = \lambda \delta_\lambda$
for all $\lambda \in \mathbb{R}$, where δ_λ is the "Dirac distribution" with support at $\lambda \in \mathbb{R}$. See
Example 6.11 on page 251.

---
[9] See for example [Taylor and Lay (1980), page 279, Theorem 3.4].
[10] By an *unbounded operator*, we mean a linear transformation that is not continuous.

2.7 (∗) Dual space and the Hahn-Banach Theorem

Definition 2.12. (Dual space of a normed space).
Let X be a normed space over $\mathbb{K}$. Then the normed space $CL(X, \mathbb{K})$, equipped with the operator norm, is called the *dual space of X*. One denotes the dual space of X simply by X'. Elements of the dual space are sometimes called *bounded linear functionals*.

Recall that a consequence of Theorem 2.8 (page 78) was Corollary 2.2, which says that X' is *always* a Banach space, even if X isn't. This is because $\mathbb{K} = \mathbb{R}$ or $\mathbb{C}$ are both Banach spaces.

Given a concrete X, like $\mathbb{R}^d$ or ℓ^p, it is sometimes possible to "recognize" X', that is to establish a (normed space) isomorphism from X' to some other Banach space, for example:
$$(\mathbb{R}^d)' \simeq \mathbb{R}^d, \text{ or } (\ell^p)' \simeq \ell^q \text{ (if } 1 < p < \infty, \text{ and where } \frac{1}{p} + \frac{1}{q} = 1).$$
Such results are called *representation theorems*, and we will see a few such results now, and also later on in the chapter on Hilbert spaces (Chapter 4), when we will learn about the *Riesz Representation Theorem*, page 189.

Theorem 2.14. *For $1 \leq p < \infty$, $(\ell^p)' \simeq \ell^q$, where $\frac{1}{p} + \frac{1}{q} = 1$.*

(Here the understanding is that if $p = 1$, then $q = \infty$.)

Proof. (Sketch). We consider $\mathbb{K} = \mathbb{R}$ for simplicity. Let $1 < p < \infty$.
By Hölder's Inequality, $|a_1 b_1 + \cdots + a_n b_n| \leq \|(a_1, \cdots, a_n)\|_p \|(b_1, \cdots, b_n)\|_q$, with equality if $(a_1^p, \cdots, a_n^p)$ is a multiple of $(b_1^q, \cdots, b_n^q)$. Let $T \in CL(\ell^p, \mathbb{R})$. Let $\mathbf{e}_k \in \ell^p$ be the sequence $(0, \cdots, 0, 1, 0, \cdots)$ with kth term 1, and all others 0. Fix $n \in \mathbb{N}$. Let $\mathbf{a} = (a_1, \cdots, a_n, 0, \cdots) \in \ell^p$ be such that $(a_1^p, \cdots, a_n^p)$ is a multiple of $((T\mathbf{e}_1)^q, \cdots, (T\mathbf{e}_n)^q)$ (i.e., $a_k := (T\mathbf{e}_k)^{p/q}$, $k = 1, \cdots, n$).
Then $\|T\| \geq \dfrac{|T(\mathbf{a})|}{\|\mathbf{a}\|_p} = \dfrac{|a_1 T(\mathbf{e}_1) + \cdots + a_n T(\mathbf{e}_n)|}{\|\mathbf{a}\|_p} = \|(T\mathbf{e}_1, \cdots, T\mathbf{e}_n)\|_q.$
Passing the limit $n \to \infty$, we get $(T\mathbf{e}_1, T\mathbf{e}_2, T\mathbf{e}_3, \cdots) \in \ell^q$. So we get a continuous linear transformation $T \overset{\iota}{\mapsto} (T\mathbf{e}_1, T\mathbf{e}_2, T\mathbf{e}_3, \cdots) : CL(\ell^p, \mathbb{R}) \to \ell^q$. It can be checked that this map ι is injective and surjective. As ι is bijective, it is an isomorphism.

If $p = 1$, then let us now show that $(\ell^1)' \simeq \ell^\infty$. This is easier to see, since if $T \in CL(\ell^1, \mathbb{R})$, then we get immediately that for all k, $|T\mathbf{e}_k| = \dfrac{|T\mathbf{e}_k|}{\|\mathbf{e}_k\|_1} \leq \|T\|$, giving $(T\mathbf{e}_k)_{k \in \mathbb{N}} \in \ell^\infty$. □

Remark 2.6. (The dual space $(\ell^\infty)' \not\simeq \ell^1$.)
If $\mathbf{a} = (a_n)_{n \in \mathbb{N}} \in \ell^1$, then define the functional $\varphi_\mathbf{a} \in CL(\ell^\infty, \mathbb{R}) = (\ell^\infty)'$ by

$$\varphi_\mathbf{a}(\mathbf{b}) = \sum_{n=1}^\infty a_n b_n, \quad \mathbf{b} = (b_n)_{n \in \mathbb{N}} \in \ell^\infty.$$

Then $\mathbf{a} \mapsto \varphi_\mathbf{a} : \ell^1 \to (\ell^\infty)'$ is an injective linear transformation. It is continuous since $|\varphi_\mathbf{a}(\mathbf{b})| \leq \|\mathbf{b}\|_\infty \|\mathbf{a}\|_1$ for all $\mathbf{b} \in \ell^\infty$, giving $\|\varphi_\mathbf{a}\| \leq \|\mathbf{a}\|_1$. However it is not surjective, and this can be shown by using the Hahn-Banach Theorem (see Theorem 2.15 on page 108), which says that a continuous linear functional on a subspace of a normed space can be extended to the whole normed space while preserving the operator norm of the functional. To see how this gives the non-surjectivity of the map $\mathbf{a} \mapsto \varphi_\mathbf{a} : \ell^1 \to (\ell^\infty)'$ above, let us consider the subspace $c \subset \ell^\infty$ of all convergent subsequences, and the "limit functional" $\lambda : c \to \mathbb{R}$, given by

$$\lambda(\mathbf{a}) = \lim_{n \to \infty} a_n, \quad \mathbf{a} = (a_n)_{n \in \mathbb{N}} \in c.$$

Then $\lambda \in CL(c, \mathbb{R}) = c'$, and $\|\lambda\| = 1$. By the Hahn-Banach Theorem, this functional λ on the subspace c of ℓ^∞ has an extension $\Lambda \in CL(\ell^\infty, \mathbb{R})$. But now we see that Λ can't be $\varphi_\mathbf{a}$ for some $\mathbf{a} \in \ell^1$. Otherwise, with $\mathbf{e}_n \in c \subset \ell^\infty$ being the sequence with nth term 1 and all others 0, we have

$$0 = \lambda(\mathbf{e}_n) = \Lambda(\mathbf{e}_n) = \varphi_\mathbf{a}(\mathbf{e}_n) = a_n,$$

for all n, showing that $\mathbf{a} = \mathbf{0}$, and so $\Lambda = \varphi_\mathbf{a} = \mathbf{0}$, which is clearly false, since $\Lambda(1,1,1,\cdots) = \lambda(1,1,1,\cdots) = 1 \neq 0$! So $(\ell^\infty)'$ is "bigger" than ℓ^1.

Remark 2.7. If $1 \leq p < \infty$, then it can be shown that

$$\left(L^p[a,b]\right)' \simeq L^q[a,b],$$

where $\dfrac{1}{p} + \dfrac{1}{q} = 1$.

Exercise 2.36. Consider the subspace $c_0 \subset \ell^\infty$ consisting of all sequences that converge to 0. Prove that $\ell^1 \simeq (c_0)'$.

Exercise 2.37. (∗) (Dual of $C[a,b]$). In this exercise we will learn a representation of the dual space of $C[a,b]$. A function $\boldsymbol{\mu} : [a,b] \to \mathbb{R}$ is said to be of *bounded variation on $[a,b]$* if its *total variation* $\mathrm{var}(\boldsymbol{\mu})$ on $[a,b]$ is finite, where

$$\mathrm{var}(\boldsymbol{\mu}) = \sup_{P \in \mathcal{P}} \sum_{k=1}^n |\boldsymbol{\mu}(t_k) - \boldsymbol{\mu}(t_{k-1})|.$$

Here $\mathcal{P}$ is the set of all *partitions* of $[a,b]$. A *partition* of $[a,b]$ is a finite set $P = \{t_0, t_1, \cdots, t_{n-1}, t_n\}$ with $t_0 := a < t_1 < \cdots < t_{n-1} < b =: t_n$.

(1) Show that the set of all functions of bounded variations on $[a,b]$, with the usual pointwise operations forms a vector space, denoted $BV[a,b]$.

Define $\|\cdot\| : BV[a,b] \to [0,+\infty)$ by $\|\boldsymbol{\mu}\| := |\boldsymbol{\mu}(a)| + \mathrm{var}(\boldsymbol{\mu})$, for $\boldsymbol{\mu} \in BV[a,b]$.

(2) Prove that $\|\cdot\|$ is a norm on $BV[a,b]$.

The *Riemann-Stieltjes integral*: Let $\mathbf{x} \in C[a,b]$ and $\boldsymbol{\mu} \in BV[a,b]$. For a partition of $[a,b]$, say $P = \{t_0, t_1, \cdots, t_{n-1}, t_n\}$, let δ_P be the length of a largest interval $[t_{j-1}, t_j]$, that is, $\delta_P := \max\{t_1 - t_0, \cdots, t_n - t_{n-1}\}$, and set

$$S(P) := \sum_{k=1}^{n} \mathbf{x}(t_k)\big(\boldsymbol{\mu}(t_k) - \boldsymbol{\mu}(t_{k-1})\big).$$

Then it can be shown that there exists a unique real number, denoted by

$$\int_a^b \mathbf{x}(t)d\boldsymbol{\mu},$$

called the *Riemann-Stieltjes integral of $\mathbf{x}$ over $[a,b]$ with respect to $\boldsymbol{\mu}$*, such that for every $\epsilon > 0$ there is a $\delta > 0$ such that if P is a partition of $[a,b]$ satisfying $\delta_P < \delta$, then

$$\left| \int_a^b \mathbf{x}(t)d\boldsymbol{\mu} - S(P) \right| < \epsilon.$$

The usual linearity of the integral (as with the ordinary Riemann integral) holds:

$$\int_a^b \mathbf{x}_1(t) + \mathbf{x}_2(t) d\boldsymbol{\mu} = \int_a^b \mathbf{x}_1(t) d\boldsymbol{\mu} + \int_a^b \mathbf{x}_2(t) d\boldsymbol{\mu} \quad (\mathbf{x}_1, \mathbf{x}_2 \in C[a,b]),$$

$$\int_a^b \alpha \mathbf{x}(t) d\boldsymbol{\mu} = \alpha \int_a^b \mathbf{x}(t) d\boldsymbol{\mu} \quad (\alpha \in \mathbb{R}, \ \mathbf{x} \in C[a,b]).$$

(3) Prove that $\left| \int_a^b \mathbf{x}(t) d\boldsymbol{\mu} \right| \leq \|\mathbf{x}\|_\infty \mathrm{var}(\boldsymbol{\mu})$, where $\mathbf{x} \in C[a,b]$ and $\boldsymbol{\mu} \in BV[a,b]$.

(4) Conclude that every $\boldsymbol{\mu} \in BV[a,b]$ gives rise to a $\varphi_{\boldsymbol{\mu}} \in CL\big(C[a,b], \mathbb{R}\big)$,

$$\mathbf{x} \stackrel{\varphi_{\boldsymbol{\mu}}}{\mapsto} \int_a^b \mathbf{x}(t) d\boldsymbol{\mu},$$

and that $\|\varphi_{\boldsymbol{\mu}}\| \leq \mathrm{var}(\boldsymbol{\mu})$.

The following converse result was proved by F. Riesz: For all $\varphi \in CL\big(C[a,b], \mathbb{R}\big)$, there exists a $\boldsymbol{\mu} \in BV[a,b]$ such that

$$\text{for all } \mathbf{x} \in C[a,b], \ \varphi(\mathbf{x}) = \int_a^b \mathbf{x}(t) d\boldsymbol{\mu},$$

and $\|\varphi\| = \mathrm{var}(\boldsymbol{\mu})$. In other words, every element $\big(C[a,b]\big)'$ can be represented by a Riemann-Stieltjes integral.

(5) For the functional $\mathbf{x} \mapsto \mathbf{x}(a)$ on $C[a,b]$, find a corresponding $\boldsymbol{\mu} \in BV[a,b]$.

Dual spaces are important because, among other things, they allow us to define dual operators. Here is the definition.

Definition 2.13. (Dual operator).

Let X, Y be normed spaces, and $T \in CL(X,Y)$. We define the *dual operator* (of T), $T' \in CL(Y', X')$, by $(T'\psi)(x) = \psi(Tx)$, for all $x \in X$ and $\psi \in Y'$.

Several things need to be checked here:

(1) For $\psi \in Y'$, does $T'\psi$ belong to X'?
(2) Does $T' \in CL(Y', X')$?

Let us begin with (1). If $\psi' \in Y'$, then we have that
(L1) for all $x_1, x_2 \in X$,
$$(T'\psi)(x_1 + x_2) = \psi\big(T(x_1 + x_2)\big) = \psi(Tx_1 + Tx_2)$$
$$= \psi(Tx_1) + \psi(Tx_2) = (T'\psi)(x_1) + (T'\psi)(x_2),$$
(L2) for all $\alpha \in \mathbb{K}$ and $x \in X$,
$$(T'\psi)(\alpha x) = \psi\big(T(\alpha x)\big) = \psi\big(\alpha(Tx)\big) = \alpha\psi(Tx) = \alpha\big((T'\psi)(x)\big).$$
Hence $T'\psi \in L(Y, \mathbb{K})$. Moreover $T'\psi$ is continuous because for all $x \in X$,
$$|(T'\psi)(x)| = |\psi(Tx)| \leq \|\psi\|\|Tx\| \leq \|\psi\|\|T\|\|x\|. \tag{2.7}$$
Now let's check (2), that is, that $T' \in CL(Y', X')$. We have
(L1) for all $\psi_1, \psi_2 \in Y'$, for all $x \in X$,
$$\big(T'(\psi_1 + \psi_2)\big)(x) = (\psi_1 + \psi_2)(Tx) = \psi_1(Tx) + \psi_2(Tx)$$
$$= (T'\psi_1)(x) + (T'\psi_2)(x) = \big(T'(\psi_1) + T'(\psi_2)\big)(x),$$
and so $T'(\psi_1 + \psi_2) = T'(\psi_1) + T'(\psi_2)$,
(L2) for all $\alpha \in \mathbb{K}$, for all $\psi \in Y'$, for all $x \in X$,
$$\big(T'(\alpha\psi)\big)(x) = (\alpha\psi)(Tx) = \alpha\big(\psi(Tx)\big)$$
$$= \alpha\big((T'\psi)(x)\big) = \big(\alpha(T'\psi)\big)(x),$$
and so $T'(\alpha\psi) = \alpha(T'\psi)$.

Thus T' is linear. It is also continuous, because (2.7) gives $\|T'\psi\| \leq \|\psi\|\|T\|$ for all ψ, that is $T' \in CL(Y', X')$ and $\|T'\| \leq \|T\|$.

Example 2.21. Consider $\mathbf{x} \overset{D}{\mapsto} \mathbf{x}' : C^1[0,1] \to C[0,1]$, $\mathbf{x} \in C^1[0,1]$. Then $D' : \big(C[0,1]\big)' \to \big(C^1[0,1]\big)'$ is given by $(D'\psi)(\mathbf{x}) = \psi(D\mathbf{x}) = \psi(\mathbf{x}')$, $\psi \in \big(C[0,1]\big)'$, $\mathbf{x} \in C^1[0,1]$. But $\big(C[0,1]\big)' \subset BV[0,1]$, and so every $\psi \in \big(C[0,1]\big)'$ can be represented by some element $\boldsymbol{\mu}_\psi \in BV[0,1]$, so that
$$\psi(\mathbf{y}) = \int_0^1 \mathbf{y}(t) d\boldsymbol{\mu}_\psi, \quad \mathbf{y} \in C[0,1].$$
Thus if $\psi \in \big(C[0,1]\big)'$, then
$$(D'\psi)(\mathbf{x}) = \int_0^1 \mathbf{x}'(t) d\boldsymbol{\mu}_\psi, \quad \mathbf{x} \in C^1[0,1],$$
where $\boldsymbol{\mu}_\psi \in BV[0,1]$ is such that $\psi(\mathbf{y}) = \int_0^1 \mathbf{y}(t) d\boldsymbol{\mu}_\psi$, $\mathbf{y} \in C[0,1]$. ◇

Sometimes problems for an operator can be simplified by looking at the dual operator, making the consideration of dual spaces and dual operators a useful endeavour.

Remark 2.8. (Dual versus adjoint operators). When we learn about Hilbert spaces, we will learn about the notion of the *adjoint* $T^* \in CL(Y, X)$ of an operator $T \in CL(X, Y)$, where X, Y are Hilbert spaces, and we can use the Riesz Representation Theorem (which we will also learn there) to represent elements of Y', X' by elements of Y, X. (The next sentence should be read after the discussion of the adjoint operator and the Riesz Representation Theorem.) If X, Y are Hilbert spaces, and $T \in CL(X, Y)$, then for $Y \ni y_\psi \equiv \psi \in Y'$, we have for all $x \in X$ that

$$(T'\psi)(x) = \psi(Tx) = \langle Tx, y_\psi \rangle = \langle y_\psi, Tx \rangle^*$$
$$= \langle T^* y_\psi, x \rangle^* = \langle x, T^* y_\psi \rangle = (T^* y_\psi)(x),$$

where we identified $T^* y_\psi \in X$ with the functional $x \mapsto \langle x, T^* y_\psi \rangle : X \to \mathbb{K}$ in X'. In this sense the notions of adjoint and dual "coincide" in this context of operators on Hilbert spaces.

Hahn-Banach Theorem[11]. Finally, we will learn a fundamental result, known as the Hahn-Banach Theorem, which says that X' always contains sufficiently many elements to separate points of X: for $x \neq y$ in X, there exists a $\varphi \in X'$ such that $\varphi(x) \neq \varphi(y)$. In this sense, the elements of X' play the role of "coordinates" for the points of X (which is the kind of thinking one is used to in elementary linear algebra when $X = \mathbb{K}^d$).

Theorem 2.15. (Hahn-Banach).
Let (1) X *be a normed space,*
(2) $Y \subset X$ *be a linear subspace,*
(3) $\varphi \in CL(Y, \mathbb{K})$.
Then there exists a $\Phi \in CL(X, \mathbb{K})$ *such that* $\Phi|_Y = \varphi$ *and* $\|\Phi\| = \|\varphi\|$.

In other words: "Every continuous linear functional on a subspace Y of a normed space X possesses a norm-preserving extension to the entire normed space X".

Before proving the Hahn-Banach Theorem, we will now list a few important consequences one obtains from it.

Corollary 2.6. *Let X be a normed space and $x_0 \in X$. Then there exists an element $\Phi \in X'$ such that $\Phi(x_0) = \|x_0\|$ and $\|\Phi\| = 1$.*

---
[11] Named after the mathematicians Hans Hahn and Stefan Banach.

Proof. Let $Y := \operatorname{span}\{x_0\}$ and $\varphi : Y \to \mathbb{K}$ be given by $\varphi(y) = \alpha \|x_0\|$ for $y = \alpha x_0 \in Y$, $\alpha \in \mathbb{K}$. Then φ is linear. Moreover, φ is a continuous map because $|\varphi(y)| = |\varphi(\alpha x_0)| = |\alpha| \|x_0\| = \|\alpha x_0\| = \|y\|$, that is, $|\varphi(y)| = \|y\|$ for all $y \in Y$. Hence $\|\varphi\| = 1$. By Hahn-Banach there now exists an extension Φ with the desired property. $\square$

As mentioned earlier, once we have the Hahn-Banach Theorem, one has the ability of distinguishing elements of X using elements from X'. This is shown in the two corollaries below.

Corollary 2.7. *Let x and y be elements in a normed space X with $x \neq y$. Then there exists a $\Phi \in X'$ such that $\Phi(x) \neq \Phi(y)$. (In other words, X' separates the points of X.)*

Proof. Take $\Phi \in X'$ with $\Phi(x) - \Phi(y) = \Phi(x - y) = \|x - y\| \neq 0$. $\square$

Exercise 2.38. Let X be a complex normed space and $x_* \in X \setminus \{0\}$. Show that there exists a $\varphi_* \in CL(X, \mathbb{C})$ such that $\varphi_*(x_*) \neq 0$.

Remark 2.9. ($CL(X,Y)$ Banach $\Rightarrow$ Y Banach).
Fix any nonzero $x_* \in X$. By Exercise 2.38, there exists a $\varphi \in CL(X, \mathbb{C})$, such that $\varphi(x_*) \neq 0$. Let $(y_n)_{n \in \mathbb{N}}$ be a Cauchy sequence in Y. For $n \in \mathbb{N}$, define $T_n \in CL(X, Y)$ by

$$T_n x = \frac{\varphi(x)}{\varphi(x_*)} y_n, \quad x \in X.$$

Then using the linearity of φ, it follows that T_n is linear. Also, T_n is continuous because

$$\forall x \in X, \ \|T_n x\| = \frac{|\varphi(x)|}{|\varphi(x_n)|} \|y_n\| \leq \frac{\|y_n\| \|\varphi\|}{|\varphi(x_*)|} \|x\|.$$

A similar computation also gives that for $n, m \in \mathbb{N}$,

$$\|T_n - T_m\| \leq \frac{\|y_n - y_m\| \|\varphi\|}{|\varphi(x_*)|},$$

showing that $(T_n)_{n \in \mathbb{N}}$ is a Cauchy sequence (since $(y_n)_{n \in \mathbb{N}}$ is Cauchy). As $CL(X,Y)$ is Banach, the Cauchy sequence $(T_n)_{n \in \mathbb{N}}$ is convergent, with limit, say, $T \in CL(X,Y)$. But for $x \in X$,

$$\|T_n x - T x\| \leq \|T_n - T\| \|x\| \stackrel{n \to \infty}{\longrightarrow} 0,$$

and so we have that for all $x \in X$, $(T_n x)_{n \in \mathbb{N}}$ converges to Tx. In particular, with $x = x_*$, $(T_n x_*)_{n \in \mathbb{N}} = (y_n)_{n \in \mathbb{N}}$ converges to $T x_* = y$. Hence Y is a Banach space!

Since X' is itself a normed space, we know that X' too has a dual space $(X')' =: X''$, and X'' called the *bidual of* X. For $x \in X$, now consider the map $\varphi_x : X' \to \mathbb{K}$, given by

$$\psi \stackrel{\varphi_x}{\mapsto} \psi(x) : X' \to \mathbb{K}.$$

It is clear that φ_x is linear. Moreover, it is continuous too, since

$$|\varphi_x(\psi)| = |\psi(x)| \leqslant \|\psi\|\|x\|, \quad \psi \in X'.$$

We thus see that $\|\varphi_x\| \leqslant \|x\|$.

If $x = \mathbf{0}$, the zero vector in X, we have $\varphi_x = \mathbf{0}$, the zero linear transformation in $CL(X', \mathbb{K})$. So $\|\varphi_x\| = 0 = \|x\|$ in this case.

If $x \neq \mathbf{0}$, then from Corollary 2.6 it follows that there exists a $\psi \in X' \backslash \{\mathbf{0}\}$ for which $|\psi(x)| = \|x\|$ and $\|\psi\| = 1$. So we get the reverse inequality $\|\varphi_x\| \geqslant \|x\|$ too. Hence $\|\varphi_x\| = \|x\|$ in this case as well.

So we have the following third consequence of the Hahn-Banach Theorem:

Corollary 2.8. *Let X be a normed space and $x \in X$. Then the map φ_x on X', given by $\psi \stackrel{\varphi_x}{\mapsto} \psi(x) : X' \to \mathbb{K}$ has the operator norm $\|\varphi_x\| = \|x\|$.*

Thus map $x \mapsto \varphi_x : X \mapsto X''$ is a *linear isometric embedding* of X in X''. If we consider the elements of X as bounded linear functionals on X' (by identifying x with φ_x), then:

$$X \text{``} \subset \text{''} X'' \tag{2.8}$$

where the norm of x on X agrees with the norm of φ_x in X''. Sometimes, the map $x \mapsto \varphi_x$ from X into X'' is also surjective. In that case, the space X is called *reflexive* and the inclusion (2.8) is replaced by the equality:

$$X \text{``} = \text{''} X''.$$

For proving the Hahn-Banach theorem, we will need a few preliminaries. We will first prove the theorem in the case $\mathbb{K} = \mathbb{R}$, and then show how the result for the case $\mathbb{K} = \mathbb{C}$ can be derived from the real case.

In the following lemma, we consider a normed space X over $\mathbb{R}$, and instead of a norm, we consider a more general function $p : X \to \mathbb{R}$ such that

$$\text{for all } x, y \in X, \; p(x+y) \leqslant p(x) + p(y) \tag{2.9}$$

$$\text{for all } \alpha \in [0, \infty) \text{ and } x \in X, \; p(\alpha x) = \alpha p(x), \tag{2.10}$$

that is, a subadditive and positive-homogeneous functional.

Lemma 2.3. (Hahn-Banach Lemma).
Let X be a normed space over $\mathbb{R}$ and $p : X \to \mathbb{R}$ satisfy (2.9) and (2.10). Furthermore, let $Y \subset X$ be a subspace and $\varphi : Y \to \mathbb{R}$ be a linear map such that
$$\text{for all } y \in Y, \ \varphi(y) \leq p(y). \tag{2.11}$$
Then there exists a linear map $\Phi : X \to \mathbb{R}$ such that $\Phi|_Y = \varphi$, and
$$\text{for all } x \in X, \ \Phi(x) \leq p(x). \tag{2.12}$$
(That is, there exists a linear extension of φ to X preserving the estimate.)

Proof. (∗) This is a rather technical proof, but the idea of the proof is to extend φ "one dimension at a time". Let $x_0 \in X \backslash Y$. Every vector $x \in Y + (\text{span } \{x_0\})$ has a unique decomposition $x = \alpha x_0 + y$, with $y \in Y$ and $\alpha \in \mathbb{R}$. An extension Φ of φ to $Y + (\text{span } \{x_0\})$ is given by $\Phi(x) = \alpha r + \varphi(y)$, where r, which ought to be $\Phi(x_0)$, will be chosen so that (2.12) holds, that is:
$$\text{for all } \alpha \in \mathbb{R}, \text{ and all } y \in Y, \ \alpha r + \varphi(y) \leq p(\alpha x_0 + y). \tag{2.13}$$
Owing to the positive homogeneity of p, it is sufficient to choose r such that (2.13) is satisfied with $\alpha = 1$ and $\alpha = -1$:
$$\text{for all } y \in Y, \ r + \varphi(y) \leq p(x_0 + y), \text{ and} \tag{2.14}$$
$$\text{for all } y \in Y, \ -r + \varphi(y) \leq p(-x_0 + y). \tag{2.15}$$
Indeed, once these hold, then multiplication with $t > 0$ yields
$$\pm tr + \varphi(ty) \leq tp(\pm x_0 + y) = p(\pm tx_0 + ty),$$
which, in light of the fact that every element from Y can be written in the form ty with $y \in Y$, gives (2.13) with $\alpha = \pm t \neq 0$. For $\alpha = 0$, (2.13) is already satisfied according to the hypothesis. Now the inequalities (2.14) and (2.15) are equivalent to the statement:
$$\text{for all } y, z \in Y, \ \varphi(y) - p(-x_0 + y) \leq r \leq -\varphi(z) + p(x_0 + z).$$
But there exists such a r precisely if all numbers $\varphi(y) - p(-x_0 + y)$, with $y \in Y$, lie to the left of all numbers $-\varphi(z) + p(x_0 + z)$, with $z \in Y$, that is,
$$\text{for all } y, z \in Y, \ \varphi(y) - p(-x_0 + y) \leq -\varphi(z) + p(x_0 + z), \tag{2.16}$$
that is, if for all $y, z \in Y$, $\varphi(y) + \varphi(z) \leq p(-x_0 + y) + p(x_0 + z)$. But this is indeed the case since we have for all $y, z \in Y$ that
$$\varphi(y) + \varphi(z) = \varphi(y + z) \leq p(y + z) = p(-x_0 + y + x_0 + z)$$
$$\leq p(-x_0 + y) + p(x_0 + z).$$

Now from (2.16) it now follows that:
$$-\infty < \sup_{y\in Y}\big(\varphi(y) - p(-x_0 + y)\big) \leqslant \inf_{z\in Y}\big(-\varphi(z) + p(x_0 + z)\big) < +\infty$$
and it is sufficient to choose, for instance, $r = \sup_{y\in Y}\big(\varphi(y) - p(-x_0 + y)\big)$.
(In general, the sup and the inf here are unequal, and we can choose r arbitrarily from an interval.) Now the number r also satisfies (2.14) and (2.15) and thus from (2.13), we have obtained an extension to $Y + (\text{span}\ \{x_0\})$ such that (2.12) holds.

Now the idea is that we extend φ one dimension at a time in order to get an extension to the space X. If X were finite dimensional, then it is clear that this can be done. After $\dim X - \dim Y$ steps we will have obtained a linear transformation $\Phi : X \to \mathbb{R}$ that satisfies (2.12).

In the general case, the proof goes through, in essentially the same manner, by successive one-dimensional extensions, but we won't be able to get an extension to X in finitely many steps. In order to complete the process, we will use Zorn's Lemma.

Zorn's Lemma

Zorn's Lemma says that a partially ordered set P with the property that every chain has an upper bound in P possesses a maximal element. The terms are explained below.

A *partial order* $\leqslant$ on a set P is a relation on P satisfying
- (transitivity) for all $x, y, z \in P$, $x \leqslant y$, $y \leqslant z \Rightarrow x \leqslant z$,
- (antisymmetry) for all $x, y \in P$, $x \leqslant y$, $y \leqslant x \Rightarrow x = y$,
- (reflexivity) for all $x \in P$, $x \leqslant x$.

A set with a partial order is called a *partially ordered set*.
A familiar example is $\mathbb{R}$ with the usual $\leqslant$ relation, but the situation can be much more general: for example, consider $\mathbb{R}^2$ with the order: $(a, b) \leqslant_{\mathbb{R}^2} (c, d)$ if $a \leqslant c$ and $b \leqslant d$. This latter example justifies the terminology *partial*. Indeed, $\leqslant_{\mathbb{R}^2}$ is not a *total* order because not every pair of elements can be compared with $\leqslant_{\mathbb{R}^2}$: we have neither $(0, 1) \leqslant_{\mathbb{R}^2} (1, 0)$ nor $(1, 0) \leqslant_{\mathbb{R}^2} (0, 1)$.

A subset C of P is said to be *bounded above* if there exists an element $u \in P$ such that $x \leqslant u$ for all $x \in C$. The element $u \in P$ is then called an *upper bound of C*.

A subset C of P is said to be *chain* if for all $x, y \in C$, there holds

> $x \leqslant y$ or $y \leqslant x$. Thus on a chain C, $\leqslant$ forms a *total* order since any two elements in C can be compared with $\leqslant$. The set $\mathbb{R}^2$ with the above order $\leqslant_{\mathbb{R}^2}$ is not a chain, since neither $(0,1) \leqslant_{\mathbb{R}^2} (1,0)$ nor $(1,0) \leqslant_{\mathbb{R}^2} (0,1)$. However, the diagonal $\{(x,x) : x \in \mathbb{R}\}$ is a chain.
>
> An element $m \in P$ is called *maximal* if whenever $x \in P$ and $m \leqslant x$ we have that $x = m$.
>
> Zorn's Lemma (named after the mathematician Max Zorn) is an axiom in Set Theory. It can be shown that it is equivalent with the *Axiom of Choice*: for every family A_i, $i \in I$, of nonempty sets A_i, there exists a map $I \ni i \mapsto x_i \in A_i$.

In order to apply Zorn's Lemma to complete the proof of the Hahn-Banach Lemma, we proceed as follows.

Consider the set P of all pairs (Z, ψ), where Z is a subspace of X with $Y \subset Z \subset X$, and $\psi : Z \to \mathbb{R}$ is a linear transformation extending φ such that $\psi(z) \leqslant p(z)$ for all $z \in Z$.

We define the partial order $\leqslant$ on P by defining $(Z, \psi) \leqslant (Z', \psi')$ if $Z \subset Z'$ and $\psi = \psi'|_Z$. Then every chain in P has an upper bound, as explained below.

If C is a chain in P, then we can construct an upper bound (Z_C, ψ_C) of C as follows: Let Z_C be the union of all subspaces Z, with $(Z, \psi) \in C$ and let ψ_C be the common extension of the linear transformations ψ. More precisely, for $z \in Z_C$, there exists a $(Z, \psi) \in C$ such that $z \in Z$, and we define $\psi_C(z) = \psi(z)$. This definition of $\psi_C(z)$ is independent of the choice of (Z, ψ). Indeed, if (Z', ψ') also belongs to C, and $z \in Z'$, then we have $(Z, \psi) \leqslant (Z', \psi')$ or $(Z', \psi') \leqslant (Z, \psi)$, and so ψ is the restriction of ψ' or vice versa. In either case, we have $\psi(z) = \psi'(z)$. The map $\psi_C : Z_C \to \mathbb{R}$ so defined is linear: Indeed, if z, z' belong to Z_C, then there exists a (Z, ψ) such that $z \in Z$ and there exists a (Z', ψ') such that $z' \in Z'$. We have $Z \subset Z'$ or $Z' \subset Z$. Suppose that $Z' \subset Z$. Then also $z' \in Z$ so that for $\alpha, \alpha' \in \mathbb{R}$, we have $\alpha z + \alpha' z' \in Z \subset Z_C$, and so it follows that Z_C is subspace of X, and $\psi_C(\alpha z + \alpha' z') = \psi(\alpha z + \alpha' z') = \alpha \psi(z) + \alpha' \psi(z') = \alpha \psi_C(z) + \alpha' \psi_C(z')$. Finally, ψ_C satisfies the inequality $\psi_C(z) \leqslant p(z)$, $z \in Z_C$, since indeed $\psi(z) \leqslant p(z)$ for all $z \in Z$, for all $(Z, \psi) \in C$. Thus we see that (Z_C, ψ_C) belongs to P and that $(Z, \psi) \leqslant (Z_C, \psi_C)$ for all $(Z, \psi) \in C$. This completes the proof that every chain in P has an upper bound.

By Zorn's Lemma, P has a maximal element (Z_*, Φ). Then $Z_* = X$. Indeed, if $Z_* \subsetneq X$, then there exists an $x_* \in X \setminus Z_*$, and then from the first part of the proof of the Hahn-Banach Lemma, it follows that we can extend Φ to $Z_* + (\text{span}\{x_*\})$ with the same estimate given by p, contradicting the maximality of (Z_*, Φ). Thus we have a linear $\Phi : X \to \mathbb{R}$ that extends $\varphi : Y \to \mathbb{R}$, while satisfying the estimate (2.12). This completes the proof of the Hahn-Banach Lemma! $\square$

We will now apply this Hahn-Banach *Lemma* to prove the Hahn-Banach *Theorem*, first of all in the case when $\mathbb{K} = \mathbb{R}$.

Proof. (Of the Hahn-Banach Theorem; real case.)
Let $\varphi : Y \to \mathbb{R}$ be a continuous linear transformation. Then we have:
$$\text{for all } y \in Y, \ |\varphi(y)| \leq \|\varphi\| \|y\|. \tag{2.17}$$
Now we apply the Hahn-Banach Lemma with $p(x) := \|\varphi\| \|x\|$, $x \in X$. From (2.17), we certainly have for all $y \in Y$, $\varphi(y) \leq |\varphi(y)| \leq \|\varphi\| \|x\| = p(y)$. Thus, by the Hahn-Banach Lemma, there exists a linear map $\Phi : X \to \mathbb{R}$, extending φ to X, that moreover satisfies the estimate that for all $x \in X$, $\Phi(x) \leq p(x) = \|\varphi\| \|x\|$. Replacing x by $-x$, we obtain $-\Phi(x) \leq \|\varphi\| \|x\|$, and so for all $x \in X$, $|\Phi(x)| \leq \|\varphi\| \|x\|$. Hence it follows that Φ is continuous and that $\|\Phi\| \leq \|\varphi\|$. Since φ is the restriction of Φ, we have, on the other hand, also that
$$\|\varphi\| = \sup_{\substack{y \in Y \\ \|y\| \leq 1}} \|\varphi y\| = \sup_{\substack{y \in Y \\ \|y\| \leq 1}} \|\Phi y\| \leq \sup_{\substack{x \in X \\ \|x\| \leq 1}} \|\Phi x\| = \|\Phi\|.$$
This proves the Hahn-Banach Theorem in the real case. $\square$

The proof for complex scalars can be derived from the real case. We remark that real versions of the Hahn-Banach Theorem were first proved independently by Hahn and by Banach. The complex version was given by Bohnenblust and Sobcyzk, following the ideas of Murray.

Proof. (Of the Hahn-Banach Theorem; complex case.)
Let X be a normed space over $\mathbb{C}$. By restricting the multiplication with scalars to real numbers, we obtain a normed space over $\mathbb{R}$, which we denote simply by $X_\mathbb{R}$. If $\Phi : X \to \mathbb{C}$ is a linear transformation, then $\Phi_\mathbb{R} : X_\mathbb{R} \to \mathbb{R}$, given by $\Phi_\mathbb{R}(x) = \text{Re}(\Phi(x))$, $x \in X_\mathbb{R}$, is also a linear transformation. We now observe below that Φ is completely determined by its "real part" $\Phi_\mathbb{R}$. For complex $z = a + ib$, with $a, b \in \mathbb{R}$, we have $iz = -b + ia$, and hence $\text{Im}(z) = -\text{Re}(iz)$. So $\text{Im}(\Phi(x)) = -\text{Re}(i\Phi(x)) = -\text{Re}(\Phi(ix)) = -\Phi_\mathbb{R}(ix)$. Thus
$$\Phi(x) = \text{Re}(\Phi(x)) + i\text{Im}(\Phi(x)) = \Phi_\mathbb{R}(x) - i\Phi_\mathbb{R}(ix). \tag{2.18}$$

Now if $\Phi_\mathbb{R} : X_\mathbb{R} \to \mathbb{R}$ is $\mathbb{R}$-linear, then the right-hand side expression of (2.18) determines a $\mathbb{C}$-linear map $\Phi : X \to \mathbb{C}$:

(1) It is clear that Φ is $\mathbb{R}$-linear.
(2) We have $\Phi(ix) = \Phi_\mathbb{R}(ix) - i\Phi_\mathbb{R}(-x) = i\big(\Phi_\mathbb{R}(x) - i\Phi_\mathbb{R}(ix)\big) = i\Phi(x)$. Since every complex number is of the form $a + ib$, with $a, b \in \mathbb{R}$, it follows from here and the above part (1) that Φ is also $\mathbb{C}$-linear.

Finally we show that Φ continuous if and only if its real part $\Phi_\mathbb{R}$ is continuous, and moreover, $\|\Phi\| = \|\Phi_\mathbb{R}\|$.

(1) If Φ is continuous, then since $|\Phi_\mathbb{R}(x)| = |\text{Re}(\Phi(x))| \leqslant |\Phi(x)| \leqslant \|\Phi\|\|x\|$, we have that $\Phi_\mathbb{R}$ is continuous and moreover $\|\Phi_\mathbb{R}\| \leqslant \|\Phi\|$.
(2) Now suppose that $\Phi_\mathbb{R}$ is continuous and that Φ is given by (2.18). For $x \in X$, let $\theta \in \mathbb{R}$ be such that $\Phi(x) = e^{i\theta}|\Phi(x)|$. As $|\Phi(x)|$ is real,

$$|\Phi(x)| = e^{-i\theta}\Phi(x) = \Phi(e^{-i\theta}x) = \text{Re}\big(\Phi(e^{-i\theta}x)\big) = \Phi_\mathbb{R}(e^{-i\theta}x)$$
$$\leqslant \|\Phi_\mathbb{R}\|\|e^{-i\theta}x\| = \|\Phi_\mathbb{R}\|\|x\|,$$

so that Φ is continuous, and moreover $\|\Phi\| \leqslant \|\Phi_\mathbb{R}\|$.

The proof of the Hahn-Banach theorem in the complex case can now be completed as follows. Let $\varphi \in CL(Y, \mathbb{C})$ and let $\varphi_\mathbb{R} \in CL(Y_\mathbb{R}, \mathbb{R})$ be the real part of φ. Then there exists an extension $\Phi_\mathbb{R} \in CL(X_\mathbb{R}, \mathbb{R})$ of $\varphi_\mathbb{R}$ to $X_\mathbb{R}$ with $\|\Phi_\mathbb{R}\| = \|\varphi_\mathbb{R}\|$. Let $\Phi \in CL(X, \mathbb{C})$ defined by (2.18). Then Φ is an extension of φ, and $\|\Phi\| = \|\Phi_\mathbb{R}\| = \|\varphi_\mathbb{R}\| = \|\varphi\|$. $\square$

Exercise 2.39. (Hamel basis). Let X be a vector space over any field $\mathbb{F}$. Show that there exists a subset $B \subset X$ such that B is linearly independent, and span $B = X$. Such a set is called a *Hamel basis of X*.

Exercise 2.40. Let X, Y be vector spaces over a field $\mathbb{F}$. Show that any function $f : B \to Y$ defined on a Hamel basis B of X can be extended to a linear transformation $F : X \to Y$, that is, $F|_B = f$. *Hint:* Every vector in X can be uniquely expressed as a linear combination of vectors from B.

Exercise 2.41. Let X be an infinite dimensional normed space, and let Y be a nontrivial normed space. Prove that there exists a linear transformation from X to Y which is not continuous.

Exercise 2.42. $\mathbb{R}$ is a vector space over $\mathbb{Q}$, and hence has a Hamel basis B. Prove that B is necessarily uncountable.

Exercise 2.43. (Additive discontinuous $F : \mathbb{R} \to \mathbb{R}$). Show that there exists a function $F : \mathbb{R} \to \mathbb{R}$ such that for all $x, y \in \mathbb{R}$, $F(x + y) = F(x) + F(y)$, but F is not continuous on $\mathbb{R}$.

Exercise 2.44. (∗)(Banach limits).
Consider the subspace c of ℓ^∞ comprising convergent sequences.
Let $l : c \to \mathbb{K}$ be the limit functional given by $l(x_n)_{n\in\mathbb{N}} = \lim\limits_{n\to\infty} x_n$, $(x_n)_{n\in\mathbb{N}} \in c$.

(1) Show that l is an element in the dual space $CL(c, \mathbb{K})$ of c, when c is equipped with the induced norm from ℓ^∞.

Let $Y \subset \ell^\infty$ be given by $Y = \left\{ (x_n)_{n\in\mathbb{N}} \in \ell^\infty : \lim\limits_{n\to\infty} \dfrac{x_1 + \cdots + x_n}{n} \text{ exists} \right\}$.

(2) Show that Y is a subspace of ℓ^∞.

(3) Prove that for all $\mathbf{x} \in \ell^\infty$, $\mathbf{x} - S\mathbf{x} \in Y$, where $S : \ell^\infty \to \ell^\infty$ denotes the left shift operator: $S(x_1, x_2, x_3, \cdots) = (x_2, x_3, x_4, \cdots)$, $(x_n)_{n\in\mathbb{N}} \in \ell^\infty$.

(4) Prove that $c \subset Y$.

(5) Show that there exists a $L \in CL(\ell^\infty, \mathbb{K})$ such that $L|_c = l$ and $LS = L$.

This gives a generalisation of the concept of a limit, and the number $L\mathbf{x}$ is called a *Banach limit* of a (possibly divergent!) sequence $\mathbf{x} \in \ell^\infty$.

Hint: First observe that $L_0 : Y \to \mathbb{K}$ defined by
$$L_0(x_n)_{n\in\mathbb{N}} = \lim_{n\to\infty} \frac{x_1 + \cdots + x_n}{n}, \quad (x_n)_{n\in\mathbb{N}} \in Y,$$
is an extension of the functional l from c to Y. Now use the Hahn-Banach Theorem to extend L_0 from Y to ℓ^∞.

(6) Find the Banach limit of the divergent sequence $((-1)^n)_{n\in\mathbb{N}}$.

Notes

The proof of the Open Mapping Theorem given in §2.5, and the proof of the Hahn-Banach Theorem given in §2.7 are based on [Thomas (1997)].

Chapter 3

Differentiation

In the previous chapter we studied continuity of operators from a normed space X to a normed space Y. In this chapter, we will study differentiation: we will define the (Fréchet) derivative of a map $f : X \to Y$ at a point $x_0 \in X$. Roughly speaking, the derivative $f'(x_0)$ of f at a point x_0 will be a continuous linear transformation $f'(x_0) : X \to Y$ that provides a linear approximation of f in the vicinity of x_0.

As an application of the notion of differentiation, we will indicate its use in solving optimisation problems in normed spaces, for example for real valued maps living on $C^1[a,b]$. At the end of the chapter, we'll apply our results to the concrete case of solving the optimisation problem

$$(\text{P}) : \begin{cases} \text{minimise} & \int_a^b F(t, \mathbf{x}(t), \mathbf{x}'(t))\,dt \\ \text{subject to} & \mathbf{x} \in C^1[a,b], \\ & \mathbf{x}(a) = y_a \text{ and } \mathbf{x}(b) = y_b. \end{cases}$$

Setting the derivative of a relevant functional, arising from (P), equal to the zero linear transformation, we get a condition for an extremal curve $\mathbf{x}_*$, called the *Euler-Lagrange equation*. Thus, instead of an *algebraic* equation obtained for example in the minimisation of a polynomial $p : \mathbb{R} \to \mathbb{R}$ using ordinary calculus, now, for the problem (P), the Euler-Lagrange equation is a *differential* equation. The solution of this differential equation is then the sought after function $\mathbf{x}_*$ that solves the optimisation problem (P).

At the end of this chapter, we will also briefly see an application of the language developed in this chapter to Classical Mechanics, where we will describe the Lagrangian equations and the Hamiltonian equations for simple mechanical systems. This will also serve as a stepping stone to a discussion of Quantum Mechanics in the next chapter.

3.1 Definition of the derivative

Let us first revisit the situation in ordinary calculus, where $f : \mathbb{R} \to \mathbb{R}$, and let us rewrite the definition of the derivative of f at $x_0 \in \mathbb{R}$ in a manner that lends itself to generalisation to the case of maps between normed spaces. Recall that for a function $f : \mathbb{R} \to \mathbb{R}$, the derivative at a point x_0 is the approximation of f around x_0 by a straight line.

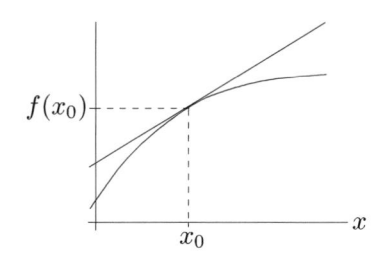

Let $f : \mathbb{R} \to \mathbb{R}$ and let $x_0 \in \mathbb{R}$. Then f is said to be differentiable at x_0 with derivative $f'(x_0) \in \mathbb{R}$ if

$$\lim_{x \to x_0} \frac{f(x) - f(x_0)}{x - x_0} = f'(x_0),$$

that is, for every $\epsilon > 0$, there exists a $\delta > 0$ such that whenever $x \in \mathbb{R}$ satisfies $0 < |x - x_0| < \delta$, there holds that

$$\left| \frac{f(x) - f(x_0)}{x - x_0} - f'(x_0) \right| < \epsilon,$$

that is,

$$\frac{|f(x) - f(x_0) - f'(x_0)(x - x_0)|}{|x - x_0|} < \epsilon.$$

If we now imagine f instead to be a map from a normed space X to another normed space Y, then bearing in mind that the norm is a generalisation of the absolute value in $\mathbb{R}$, we may try mimicking the above definition and replace the denominator $|x - x_0|$ above by $\|x - x_0\|$, and the numerator absolute value can be replaced by the norm in Y, since $f(x) - f(x_0)$ lives in Y. But what object must there be in the box below?

$$\frac{\| f(x) - f(x_0) - \boxed{f'(x_0)}(x - x_0) \|_Y}{\|x - x_0\|_X} < \epsilon.$$

Since $f(x), f(x_0)$ live in Y, we expect the term $f'(x_0)(x - x_0)$ to be also in Y. As $x - x_0$ is in X, $f'(x_0)$ should take this into Y. So we see that

Differentiation

it is natural that we should not expect $f'(x_0)$ to be a number (as was the case when $X = Y = \mathbb{R}$), but rather it we expect it should be a certain mapping from X to Y. We will in fact want it to be a continuous linear transformation from X to Y. Why? We will see this later, but a short answer is that with this definition, we can prove analogous theorems from ordinary calculus, and we can use these theorems in applications to solve (e.g. optimisation) problems. After this rough motivation, let us now see the precise definition.

Definition 3.1. (Derivative).
Let X, Y be normed spaces, $f : X \to Y$ be a map, and $x_0 \in X$.
Then f is said to be *differentiable at* x_0 if there exists a continuous linear transformation $L : X \to Y$ such that for every $\epsilon > 0$, there exists a $\delta > 0$ such that whenever $x \in X$ satisfies $0 < \|x - x_0\| < \delta$, we have

$$\frac{\|f(x) - f(x_0) - L(x - x_0)\|}{\|x - x_0\|} < \epsilon.$$

(If f is differentiable at x_0, then it can be shown that there can be at most one continuous linear transformation L such that the above statement holds. We will prove this below in Theorem 3.1, page 124.)
The unique continuous linear transformation L is denoted by $f'(x_0)$, and is called the *derivative of f at x_0*.
If f is differentiable at every point $x \in X$, then f is said to be *differentiable*.

Before we see some simple illustrative examples on the calculation of the derivative, let us check that this is a genuine extension of the notion of differentiability from ordinary calculus. Over there the concept of derivative was very simple, and $f'(x_0)$ was just a number. Now we will see that over there too, it was actually a continuous linear transformation, but it just so happens that any continuous linear transformation from $\mathbb{R}$ to $\mathbb{R}$ is simply given by multiplication by a fixed number. We explain this below.

Coincidence of our new definition with the old definition when we have $X = Y = \mathbb{R}$, $f : \mathbb{R} \to \mathbb{R}$, $x_0 \in \mathbb{R}$.

(1) Differentiable in the old sense $\Rightarrow$ differentiable in the new sense.
Let $\lim_{x \to x_0} \dfrac{f(x) - f(x_0)}{x - x_0}$ exist and be the number $f'_{\text{old}}(x_0) \in \mathbb{R}$.
Define the map $L : \mathbb{R} \to \mathbb{R}$ by $L(v) = f'_{\text{old}}(x_0) \cdot v$, $v \in \mathbb{R}$.
Then L is a linear transformation as verified below.

(L1) For every $v_1, v_2 \in \mathbb{R}$,
$$L(v_1 + v_2) = f'_{\text{old}}(x_0) \cdot (v_1 + v_2) = f'_{\text{old}}(x_0) \cdot v_1 + f'_{\text{old}}(x_0) \cdot v_2$$
$$= L(v_1) + L(v_2).$$

(L2) For every $\alpha \in \mathbb{R}$ and every $v \in V$,
$$L(\alpha \cdot v) = f'_{\text{old}}(x_0) \cdot (\alpha \cdot v) = \alpha \cdot (f'_{\text{old}}(x_0) \cdot v) = \alpha L(v).$$

L is continuous since $|L(v)| = |f'_{\text{old}}(x_0) \cdot v| = |f'_{\text{old}}(x_0)||v|$ for all $v \in \mathbb{R}$. We know that
$$\lim_{x \to x_0} \frac{f(x) - f(x_0)}{x - x_0} = f'_{\text{old}}(x_0),$$
that is, for every $\epsilon > 0$, there exists a $\delta > 0$ such that whenever $x \in \mathbb{R}$ satisfies $0 < |x - x_0| < \delta$, we have
$$\left| \frac{f(x) - f(x_0)}{x - x_0} - f'_{\text{old}}(x_0) \right| = \frac{|f(x) - f(x_0) - f'_{\text{old}}(x_0) \cdot (x - x_0)|}{|x - x_0|} < \epsilon,$$
that is, $\dfrac{|f(x) - f(x_0) - L(x - x_0)|}{|x - x_0|} < \epsilon$.

So f is differentiable in the new sense too, and $f'_{\text{new}}(x_0) = L$, that is, we have $\big(f'_{\text{new}}(x_0)\big)(v) = f'_{\text{old}}(x_0) \cdot v$, $v \in \mathbb{R}$.

(2) Differentiable in the new sense $\Rightarrow$ differentiable in the old sense.
Suppose there is a continuous linear transformation $f'_{\text{new}}(x_0) : \mathbb{R} \to \mathbb{R}$ such that for every $\epsilon > 0$, there exists a $\delta > 0$ such that whenever $x \in \mathbb{R}$ satisfies $0 < |x - x_0| < \delta$, we have
$$\frac{|f(x) - f(x_0) - \big(f'_{\text{new}}(x_0)\big)(x - x_0)|}{|x - x_0|} < \epsilon.$$
Define $f'_{\text{old}}(x_0) := \big(f'_{\text{new}}(x_0)\big)(1) \in \mathbb{R}$. Then if $x \in \mathbb{R}$, we have
$$\big(f'_{\text{new}}(x_0)\big)(x - x_0) = \big(f'_{\text{new}}(x_0)\big)\big((x - x_0) \cdot 1\big)$$
$$= (x - x_0) \cdot \big(f'_{\text{new}}(x_0)\big)(1) = f'_{\text{old}}(x_0) \cdot (x - x_0).$$
So there exists a number, namely $f'_{\text{old}}(x_0)$, such that for every $\epsilon > 0$, there exists a $\delta > 0$ such that whenever $x \in \mathbb{R}$ satisfies $0 < |x - x_0| < \delta$,
$$\left| \frac{f(x) - f(x_0)}{x - x_0} - f'_{\text{old}}(x_0) \right| = \frac{|f(x) - f(x_0) - f'_{\text{old}}(x_0) \cdot (x - x_0)|}{|x - x_0|}$$
$$= \frac{|f(x) - f(x_0) - \big(f'_{\text{new}}(x_0)\big)(x - x_0)|}{|x - x_0|} < \epsilon.$$
Consequently, f is differentiable at x_0 in the old sense, and furthermore, $f'_{\text{old}}(x_0) = \big(f'_{\text{new}}(x_0)\big)(1)$.

The derivative as a local linear approximation. We know that in ordinary calculus, for a function $f : \mathbb{R} \to \mathbb{R}$ that is differentiable at $x_0 \in \mathbb{R}$, the number $f'(x_0)$ has the interpretation of being the slope of the tangent line to the graph of the function at the point $(x_0, f(x_0))$, and the tangent line itself serves as a local linear approximation to the graph of the function. Imagine zooming into the point $(x_0, f'(x_0))$ using lenses of greater and greater magnification: then there is little difference between the graph of the function and the tangent line. We now show that also in the more general set-up when f is a map from a normed space X to a normed space Y, that is differentiable at a point $x_0 \in X$, $f'(x_0)$ can be interpreted as giving a local linear approximation to the mapping f near the point x_0, and we explain this below. Let $\epsilon > 0$. Then we know that for all x close enough to x_0 and distinct from x_0, we have

$$\frac{\|f(x) - f(x_0) - f'(x_0)(x - x_0)\|}{\|x - x_0\|} < \epsilon,$$

that is, $\|f(x) - f(x_0) - f'(x_0)(x - x_0)\| < \epsilon \|x - x_0\|$. So for all x close enough to x_0,

$$\|f(x) - f(x_0) - f'(x_0)(x - x_0)\| \approx 0,$$

that is, $f(x) - f(x_0) - f'(x_0)(x - x_0) \approx \mathbf{0} \in X$, and upon rearranging,

$$f(x) - f(x_0) \approx f'(x_0)(x - x_0).$$

The above says that near x_0, $f(x) - f(x_0)$ looks like the action of the *linear transformation $f'(x_0)$* acting on $x - x_0$. We will keep this important message in mind because it will help us calculate the derivative in concrete examples. Given an f, for which we need to find $f'(x_0)$, our starting point will always be to start with calculating $f(x) - f(x_0)$ and trying to guess what linear transformation L would give that $f(x) - f(x_0) \approx L(x - x_0)$ for x near x_0. So we would start by writing $f(x) - f(x_0) = L(x - x_0) + \text{error}$, and then showing that the error term is mild enough so that the derivative definition can be verified. We will soon see this in action below, but first let us make an important remark.

Remark 3.1. In our definition of the derivative, why do we insist the derivative $f'(x_0)$ of $f : X \to Y$ at $x_0 \in X$ should be a *continuous* linear transformation—that is, why not settle just for it being a linear transformation (without demanding continuity)? The answer to this question is tied to wanting

$$\boxed{\text{differentiability at } x_0} \Rightarrow \boxed{\text{continuity at } x_0.}$$

We know this holds with the usual derivative concept in ordinary calculus when $f : \mathbb{R} \to \mathbb{R}$. If we want this property to hold also in our more general setting of normed spaces, then just having $f'(x_0)$ as a linear transformation won't do, but in addition we also need the continuity.

On the other hand, for solving optimisation problems, even if one doesn't have differentiability at a point implying continuity at the point, one can prove useful optimisation theorems using the weaker notion of the derivative. The weaker notion is called the Gateaux derivative[1], while our stronger notion is the Fréchet derivative. As we'll only use the Fréchet derivative, we refer to our "Fréchet derivative" simply as "derivative."

Example 3.1. Let X, Y be normed spaces, and let $T : X \to Y$ be a continuous linear transformation. We ask:

Is T differentiable at $x_0 \in X$? If so, what is $T'(x_0)$?

Let us do some rough work first. We would like to fill the question mark in the box below with a continuous linear transformation so that

$$T(x) - T(x_0) \approx \boxed{?}(x - x_0),$$

for x close to x_0. But owing to the linearity of T, we know that for all $x \in X$,

$$T(x) - T(x_0) = T(x - x_0),$$

and (the right-hand side) T is already a continuous linear transformation. So we make a *guess* that $T'(x_0) = T$! Let us check this now.

Let $\epsilon > 0$. Choose any $\delta > 0$, for example, $\delta = 1$. Then whenever $x \in X$ satisfies $0 < \|x - x_0\| < \delta = 1$, we have

$$\frac{\|T(x) - T(x_0) - T(x - x_0)\|}{\|x - x_0\|} = \frac{\|\mathbf{0}\|}{\|x - x_0\|} = \frac{0}{\|x - x_0\|} = 0 < \epsilon.$$

Hence $T'(x_0) = T$. Note that as the choice of x_0 was arbitrary, we have in fact obtained that for all $x \in X$, $T'(x) = T$! This is analogous to the observation in ordinary calculus that a linear function $x \mapsto m \cdot x$ has the same slope at all points, namely the number m. ◇

Example 3.2. Consider $f : C[a,b] \to \mathbb{R}$,

$$f(\mathbf{x}) = \int_a^b \big(\mathbf{x}(t)\big)^2 dt, \quad \mathbf{x} \in C[a,b].$$

Let $\mathbf{x}_0 \in C[a,b]$. What is $f'(\mathbf{x}_0)$?

---
[1] See for example [Luenberger (1969)].

As before, we begin with some rough work to make a guess for $f'(\mathbf{x}_0)$ and we seek a continuous linear transformation L so that for $\mathbf{x} \in C[a,b]$ near $\mathbf{x}_0$, $f(\mathbf{x}) - f(\mathbf{x}_0) \approx L(\mathbf{x} - \mathbf{x}_0)$. We have

$$f(\mathbf{x}) - f(\mathbf{x}_0) = \int_a^b \left((\mathbf{x}(t))^2 - (\mathbf{x}_0(t))^2\right) dt$$

$$= \int_a^b \left(\mathbf{x}(t) + \mathbf{x}_0(t)\right)\left(\mathbf{x}(t) - \mathbf{x}_0(t)\right) dt$$

$$\approx \int_a^b \left(\mathbf{x}_0(t) + \mathbf{x}_0(t)\right)\left(\mathbf{x}(t) - \mathbf{x}_0(t)\right) dt$$

$$= \int_a^b 2\mathbf{x}_0(t)\left(\mathbf{x}(t) - \mathbf{x}_0(t)\right) dt = L(\mathbf{x} - \mathbf{x}_0),$$

where $L : C[a,b] \to \mathbb{R}$ is given by

$$L(\mathbf{h}) = \int_a^b 2\mathbf{x}_0(t)\mathbf{h}(t) dt, \quad \mathbf{h} \in C[a,b].$$

This L is a continuous linear transformation, since it is a special case of Example 2.11, page 72 (when $\mathbf{A} := 2 \cdot \mathbf{x}_0$ and $\mathbf{B} = 0$). Let us now check that the "ϵ-δ definition of differentiability" holds with this L. For $x \in C[a,b]$,

$$f(\mathbf{x}) - f(\mathbf{x}_0) - L(\mathbf{x} - \mathbf{x}_0)$$

$$= \int_a^b \left((\mathbf{x}(t))^2 - (\mathbf{x}_0(t))^2\right) dt - \int_a^b 2\mathbf{x}_0(t)\left(\mathbf{x}(t) - \mathbf{x}_0(t)\right) dt$$

$$= \int_a^b \left(\mathbf{x}(t) + \mathbf{x}_0(t) - 2\mathbf{x}_0(t)\right)\left(\mathbf{x}(t) - \mathbf{x}_0(t)\right) dt$$

$$= \int_a^b \left(\mathbf{x}(t) - \mathbf{x}_0(t)\right)^2 dt,$$

and so

$$|f(\mathbf{x}) - f(\mathbf{x}_0) - L(\mathbf{x} - \mathbf{x}_0)| = \left|\int_a^b \left(\mathbf{x}(t) - \mathbf{x}_0(t)\right)^2 dt\right| \leq \int_a^b |(\mathbf{x} - \mathbf{x}_0)(t)|^2 dt$$

$$\leq \int_a^b \|\mathbf{x} - \mathbf{x}_0\|_\infty^2 dt = (b-a)\|\mathbf{x} - \mathbf{x}_0\|_\infty^2.$$

So if $0 < \|\mathbf{x} - \mathbf{x}_0\|_\infty$, then

$$\frac{|f(\mathbf{x}) - f(\mathbf{x}_0) - L(\mathbf{x} - \mathbf{x}_0)|}{\|\mathbf{x} - \mathbf{x}_0\|_\infty} \leq (b-a)\|\mathbf{x} - \mathbf{x}_0\|_\infty.$$

Let $\epsilon > 0$. Set $\delta := \epsilon/(b-a)$. Then $\delta > 0$ and if $0 < \|\mathbf{x} - \mathbf{x}_0\|_\infty < \delta$,

$$\frac{|f(\mathbf{x}) - f(\mathbf{x}_0) - L(\mathbf{x} - \mathbf{x}_0)|}{\|\mathbf{x} - \mathbf{x}_0\|_\infty} \leq (b-a)\|\mathbf{x} - \mathbf{x}_0\|_\infty < (b-a)\delta = \epsilon.$$

So $f'(\mathbf{x}_0) = L$. In other words, $f'(\mathbf{x}_0)$ is the continuous linear transformation from $C[a,b]$ to $\mathbb{R}$ given by
$$(f'(\mathbf{x}_0))(\mathbf{h}) = \int_a^b 2\mathbf{x}_0(t)\mathbf{h}(t)dt, \quad \mathbf{h} \in C[a,b].$$
So as opposed to the ordinary calculus case, one must stop thinking of the derivative as being a mere number, but instead, in the context of maps between normed spaces, the derivative at a point is itself a map, in fact a continuous linear transformation. So the answer to the question

"What is $f'(\mathbf{x}_0)$?"

should always begin with the phrase

"$f'(\mathbf{x}_0)$ is the continuous linear transformation from X to Y given by $\cdots$".

To emphasise this, let us see some particular cases of our calculation of $f'(\mathbf{x}_0)$ above, for specific choices of $\mathbf{x}_0$.

In particular, we have that the derivative of f at the zero function $\mathbf{0}$, namely $f'(\mathbf{0})$, is the zero linear transformation $\mathbf{0} : C[a,b] \to \mathbb{R}$ that sends every $\mathbf{h} \in C[a,b]$ to the number 0: $\mathbf{0}(\mathbf{h}) = 0$, $\mathbf{h} \in C[a,b]$.

Similarly, $f'(\mathbf{1}) = \left(\mathbf{h} \mapsto 2\int_a^b \mathbf{h}(t)dt : C[a,b] \to \mathbb{R}\right)$. $\diamond$

Exercise 3.1. Consider $f : C[0,1] \to \mathbb{R}$ given by $f(\mathbf{x}) := \int_0^1 \mathbf{x}(t^2)dt$, $\mathbf{x} \in C[0,1]$. Let $\mathbf{x}_0 \in C[0,1]$. What is $f'(\mathbf{x}_0)$? What is $f'(\mathbf{0})$?

We now prove that something we had mentioned earlier, but which we haven't proved yet: if f is differentiable at x_0, then its derivative is unique.

Theorem 3.1. *Let X, Y be normed spaces. If $f : X \to Y$ is differentiable at $x_0 \in X$, then there is a unique continuous linear transformation L such that for every $\epsilon > 0$, there is a $\delta > 0$ such that whenever $x \in X$ satisfies $0 < \|x - x_0\| < \delta$, there holds*
$$\frac{\|f(x) - f(x_0) - L(x - x_0)\|}{\|x - x_0\|} < \epsilon.$$

Proof. Suppose that $L_1, L_2 : X \to Y$ are two continuous linear transformations such that for every $\epsilon > 0$, there is a $\delta > 0$ such that whenever $x \in X$ satisfies $0 < \|x - x_0\| < \delta$, there holds
$$\frac{\|f(x) - f(x_0) - L_1(x - x_0)\|}{\|x - x_0\|} < \epsilon, \tag{3.1}$$
$$\frac{\|f(x) - f(x_0) - L_2(x - x_0)\|}{\|x - x_0\|} < \epsilon. \tag{3.2}$$

Suppose that $L_1(h_0) \neq L_2(h_0)$ for some $h_0 \in X$. Clearly $h_0 \neq \mathbf{0}$ (for otherwise $L_1(\mathbf{0}) = 0 = L_2(\mathbf{0})$!). Take $\epsilon = 1/n$ for some $n \in \mathbb{N}$. Then there exists a $\delta_n > 0$ such that whenever $x \in X$ satisfies $0 < \|x - x_0\| < \delta_n$, the inequalities (3.1), (3.2) hold.

With $x := x_0 + \dfrac{\delta_n}{2\|h_0\|} h_0 \in X$, we have that $x \neq x_0$, and

$$\|x - x_0\| = \frac{\delta_n}{2\|h_0\|}\|h_0\| = \frac{\delta_n}{2} < \delta_n.$$

So (3.1), (3.2) hold for this x. The triangle inequality gives

$$\frac{\|L_1(x - x_0) - L_2(x - x_0)\|}{\|x - x_0\|} = \frac{\|L_1(h_0) - L_2(h_0)\|}{\|h_0\|} < 2\epsilon = \frac{2}{n}.$$

Upon rearranging, we obtain $\|L_1(h_0) - L_2(h_0)\| \leq \dfrac{2}{n}\|h_0\|$.

As the choice of $n \in \mathbb{N}$ was arbitrary, it follows that $\|L_1(h_0) - L_2(h_0)\| = 0$, and so $L_1(h_0) = L_2(h_0)$, a contradiction. This completes the proof. $\square$

Exercise 3.2. (Differentiability $\Rightarrow$ continuity).
Let X, Y be normed spaces, $x_0 \in X$, and $f : X \to Y$ be differentiable at x_0. Prove that f is continuous at x_0.

Exercise 3.3. Consider $f : C^1[0, 1] \to \mathbb{R}$ defined by $f(\mathbf{x}) = (\mathbf{x}'(1))^2$, $\mathbf{x} \in C^1[0, 1]$. Is f differentiable? If so, compute $f'(\mathbf{x}_0)$ at $\mathbf{x}_0 \in C^1[0, 1]$.

Exercise 3.4. (Chain rule).
Given distinct x_1, x_2 in a normed space X, define the straight line $\gamma : \mathbb{R} \to X$ passing through x_1, x_2 by $\gamma(t) = (1 - t)x_1 + tx_2$, $t \in \mathbb{R}$.
(1) Prove that if $f : X \to \mathbb{R}$ is differentiable at $\gamma(t_0)$, for some $t_0 \in \mathbb{R}$,
    then $f \circ \gamma : \mathbb{R} \to \mathbb{R}$ is differentiable at t_0 and $(f \circ \gamma)'(t_0) = f'(\gamma(t_0))(x_2 - x_1)$.
(2) Deduce that if $g : X \to \mathbb{R}$ is differentiable and $g'(x) = \mathbf{0}$ at every $x \in X$,
    then g is constant.

3.2 Fundamental theorems of optimisation

From ordinary calculus, we know the following two facts that enable one to solve optimisation problems for $f : \mathbb{R} \to \mathbb{R}$.

Fact 1. If $x_* \in \mathbb{R}$ is a minimiser of f, then $f'(x_*) = 0$.

Fact 2. If $f''(x) \geq 0$ for all $x \in \mathbb{R}$ and $f'(x_*) = 0$,
then x_* is a minimiser of f.

The first fact gives a necessary condition for minimisation (and allows one to narrow the possibilities for minimisers — together with the knowledge of the existence of a minimiser, this is a very useful result since it then tells us that the minimiser x_* has to be one which satisfies $f'(x_*) = 0$). On the other hand, the second fact gives a sufficient condition for minimisation.

Analogously, we will prove the following two results in this section, but now for a real-valued function $f : X \to \mathbb{R}$ on a *normed space* X.

Fact 1. If $x_* \in X$ is a minimiser of f, then $f'(x_*) = \mathbf{0}$.

Fact 2. If f is convex and $f'(x_*) = \mathbf{0}$, then x_* is a minimiser of f.

We mention that there is no loss of generality in assuming that we have a *minimisation* problem, as opposed to a *maximisation* one. This is because we can just look at $-f$ instead of f. (If $f : S \to \mathbb{R}$ is a given function on a set S, then defining $-f : S \to \mathbb{R}$ by $(-f)(x) = -f(x)$, $x \in S$, we see that $x_* \in S$ is a maximiser for f if and only if x_* is a minimiser for $-f$.)

Optimisation: necessity of vanishing derivative

Theorem 3.2. *Let X be a normed space, and let $f : X \to \mathbb{R}$ be a function that is differentiable at $x_* \in X$. If f has a minimum at x_*, then $f'(x_*) = \mathbf{0}$.*

Let us first clarify what $\mathbf{0}$ above means: $\mathbf{0} : X \to \mathbb{R}$ is the continuous linear transformation that sends everything in X to $0 \in \mathbb{R}$: $\mathbf{0}(h) = 0$, $h \in X$.

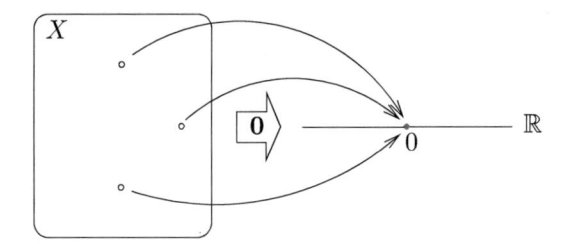

So to say that "$f'(x_*) = \mathbf{0}$" is the same as saying that for all $h \in X$, $\big(f'(x_*)\big)(h) = 0$.

Proof. Suppose that $f'(x_*) \neq \mathbf{0}$. Then there exists a vector $h_0 \in X$ such that $\big(f'(x_*)\big)(h_0) \neq 0$. Clearly this h_0 must be a nonzero vector (because the linear transformation $f'(x_0)$ takes the zero vector in X to the zero vector in $\mathbb{R}$, which is 0). Let $\epsilon > 0$. Then there exists a $\delta > 0$ such that whenever $x \in X$ satisfies $0 < \|x - x_*\| < \delta$, we have

$$\frac{|f(x) - f(x_*) - \big(f'(x_*)\big)(x - x_*)|}{\|x - x_*\|} < \epsilon.$$

Thus whenever $0 < \|x - x_*\| < \delta$, we have

$$\frac{-(f'(x_*))(x - x_*)}{\|x - x_*\|} \leq \frac{\overbrace{f(x) - f(x_*)}^{\geq 0} - (f'(x_*))(x - x_*)}{\|x - x_*\|}$$
$$\leq \frac{|f(x) - f(x_*) - (f'(x_*))(x - x_*)|}{\|x - x_*\|} < \epsilon.$$

Hence whenever $0 < \|x - x_*\| < \delta$, $-\dfrac{(f'(x_*))(x - x_*)}{\|x - x_*\|} < \epsilon$.

Now we will construct a special x using the h_0 from before. Take

$$x := x_* - \underbrace{\left(\frac{\delta}{2} \cdot \frac{(f'(x_*))(h_0)}{|(f'(x_*))(h_0)|} \cdot \frac{1}{\|h_0\|}\right)}_{\text{a scalar}} \cdot h_0 \in X.$$

Then $x \neq x_*$ and $\|x - x_*\| = \dfrac{\delta}{2} \cdot \dfrac{|(f'(x_*))(h_0)|}{|(f'(x_*))(h_0)|} \cdot \dfrac{1}{\|h_0\|} \|h_0\| = \dfrac{\delta}{2} < \delta$.

Using the linearity of $f'(x_0)$, we obtain

$$-\frac{(f'(x_*))(x - x_*)}{\|x - x_*\|} = \frac{\dfrac{\delta}{2} \cdot \dfrac{((f'(x_*))(h_0))^2}{|(f'(x_*))(h_0)|} \cdot \dfrac{1}{\|h_0\|}}{\delta/2} < \epsilon.$$

Thus, $|(f'(x_*))(h_0)| < \epsilon \|h_0\|$. As $\epsilon > 0$ was arbitrary, $|(f'(x_*))(h_0)| = 0$, and so $(f'(x_*))(h_0) = 0$, a contradiction. $\square$

We remark that the condition $f'(x_*) = \mathbf{0}$ is a *necessary* condition for x_* to be a minimiser, but it is not sufficient. This is analogous to the situation to optimisation in $\mathbb{R}$: if we look at $f : \mathbb{R} \to \mathbb{R}$ given by $f(x) = x^3$, $x \in \mathbb{R}$, then with $x_* := 0$, we have that $f'(x_*) = 3x_*^2 = 3 \cdot 0^2 = 0$, but clearly $x_* = 0$ is not a minimiser of f.

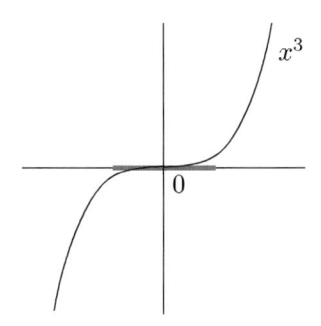

Example 3.3. Let $f : C[0,1] \to \mathbb{R}$ be given by $f(\mathbf{x}) = \big(\mathbf{x}(1)\big)^3$, $\mathbf{x} \in C[0,1]$. Then $f'(\mathbf{0}) = \mathbf{0}$. (Here the $\mathbf{0}$ on the left-hand side is the zero function in $C[0,1]$, while the $\mathbf{0}$ on the right-hand side is the zero linear transformation $\mathbf{0} : C[0,1] \to \mathbb{R}$.) Indeed, given $\epsilon > 0$, we may set $\delta := \min\{\epsilon, 1\}$, and then we have that whenever $\mathbf{x} \in C[0,1]$ satisfies $0 < \|\mathbf{x} - \mathbf{0}\|_\infty < \delta$,

$$\frac{|f(\mathbf{x}) - f(\mathbf{0}) - \mathbf{0}(\mathbf{x} - \mathbf{0})|}{\|\mathbf{x} - \mathbf{0}\|_\infty} = \frac{|\mathbf{x}(1)^3|}{\|\mathbf{x} - \mathbf{0}\|_\infty} \leqslant \frac{\|\mathbf{x}\|_\infty^3}{\|\mathbf{x}\|_\infty} = \|\mathbf{x}\|_\infty^2 < \delta \cdot \delta \leqslant 1 \cdot \epsilon.$$

But $\mathbf{0}$ is not a minimiser for f. For example, with $\mathbf{x} := -\alpha \cdot \mathbf{1} \in C[0,1]$, where $\alpha > 0$, we have $f(\mathbf{x}) = (-\alpha)^3 = -\alpha^3 < 0 = f(\mathbf{0})$, showing that $\mathbf{0}$ is not[2] a minimiser. $\diamondsuit$

Exercise 3.5. Let $f : C[a,b] \to \mathbb{R}$ be given by $f(\mathbf{x}) = \int_0^1 \big(\mathbf{x}(t)\big)^2 dt$, $\mathbf{x} \in C[a,b]$. In Example 3.2, page 122, we showed that if $\mathbf{x}_0 \in C[a,b]$, then $f'(\mathbf{x}_0)$ is given by

$$(f'(\mathbf{x}_0))\mathbf{h} = 2\int_a^b \mathbf{x}_0(t)\mathbf{h}(t)dt, \quad \mathbf{h} \in C[a,b].$$

(1) Find all $\mathbf{x}_0 \in C[a,b]$ for which $f'(\mathbf{x}_0) = \mathbf{0}$.
(2) If we know that $\mathbf{x}_* \in C[a,b]$ is a minimiser for f, what can we say about $\mathbf{x}_*$?

Optimisation: sufficiency in the convex case

We will now show that if $f : X \to \mathbb{R}$ is a convex function, then a vanishing derivative at some point is enough to conclude that the function has a minimum at that point. Thus the condition "$f''(x) \geqslant 0$ for all $x \in \mathbb{R}$" from ordinary calculus when $X = \mathbb{R}$, is now replaced by "f is convex" when X is a general normed space. We will see that in the special case when $X = \mathbb{R}$ (and when f is twice continuously differentiable), convexity is precisely characterised by the second derivative condition above. We begin by giving the definition of a convex function.

Definition 3.2. (Convex set, convex function) Let X be a normed space.
(1) A subset $C \subset X$ is said to be a *convex set* if for every $x_1, x_2 \in C$, and all $\alpha \in (0,1)$, $(1-\alpha) \cdot x_1 + \alpha \cdot x_2 \in C$.
(2) Let C be a convex subset of X. A map $f : C \to \mathbb{R}$ is said to be a *convex function* if for every $x_1, x_2 \in C$, and all $\alpha \in (0,1)$,

$$f\big((1-\alpha) \cdot x_1 + \alpha \cdot x_2\big) \leqslant (1-\alpha)f(x_1) + \alpha f(x_2). \tag{3.3}$$

---
[2] In fact, not even a "local" minimiser because $\|\mathbf{x} - \mathbf{0}\|_\infty = \alpha$ can be chosen as small as we please.

The geometric interpretation of the inequality (3.3), when $X = \mathbb{R}$, is shown below: the graph of a convex function lies above all possible chords.

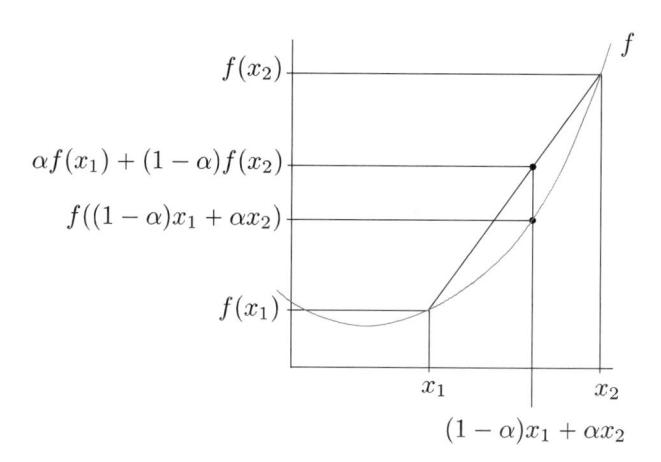

[Figure: graph showing convex function f with points $f(x_1)$, $f(x_2)$, $f((1-\alpha)x_1 + \alpha x_2)$, and $\alpha f(x_1) + (1-\alpha)f(x_2)$ at x_1, $(1-\alpha)x_1 + \alpha x_2$, x_2.]

Exercise 3.6. Let $a < b$, y_a, y_b be fixed real numbers. Show that $S := \{\mathbf{x} \in C^1[a,b] : \mathbf{x}(a) = y_a \text{ and } \mathbf{x}(b) = y_b\}$ is a convex set.

Exercise 3.7. ($\|\cdot\|$ is a convex function).
If X is a normed space, then prove that the norm $x \mapsto \|x\| : X \to \mathbb{R}$ is convex.

Exercise 3.8. (Convex set versus convex function).
Let X be a normed space, C be a convex subset of X, and let $f : C \to \mathbb{R}$. Define the *epigraph* of f by $U(f) = \bigcup_{x \in C} \{(x, y) : y \geq f(x)\} \subset X \times \mathbb{R}$.
Intuitively, we think of $U(f)$ as the "region above the graph of f". Show that f is a convex *function* if and only if $U(f)$ is a convex *set*.

Exercise 3.9. Suppose that $f : X \to \mathbb{R}$ is a convex function on a normed space X. If $n \in \mathbb{N}$, $x_1, \cdots, x_n \in X$, then $f\left(\dfrac{x_1 + \cdots + x_n}{n}\right) \leq \dfrac{f(x_1) + \cdots + f(x_n)}{n}$.

Convexity of functions living in $\mathbb{R}$.

We will now see that for twice differentiable functions $f : (a, b) \to \mathbb{R}$, convexity of f is equivalent to the condition that $f''(x) \geq 0$ for all $x \in (a, b)$. This test will actually help us to show convexity of some functions on spaces like $C^1[a, b]$.

If one were to use the definition of convexity alone, then the verification can be cumbersome. Consider for example the function $f : \mathbb{R} \to \mathbb{R}$ given by $f(x) = x^2$, $x \in \mathbb{R}$. To verify that this function is convex, we note that for $x_1, x_2 \in \mathbb{R}$ and $\alpha \in (0,1)$,

$$\begin{aligned}
f((1-\alpha)x_1 + \alpha x_2) &= ((1-\alpha)x_1 + \alpha x_2)^2 = (1-\alpha)^2 x_1^2 + 2\alpha(1-\alpha)x_1 x_2 + \alpha^2 x_2^2 \\
&= (1-\alpha)x_1^2 + \alpha x_2^2 + ((1-\alpha)^2 - (1-\alpha))x_1^2 + (\alpha^2 - \alpha)x_2^2 \\
&\quad + 2\alpha(1-\alpha)x_1 x_2 \\
&= (1-\alpha)x_1^2 + \alpha x_2^2 - \alpha(1-\alpha)(x_1^2 + x_2^2 - 2x_1 x_2) \\
&= (1-\alpha)x_1^2 + \alpha x_2^2 - \alpha(1-\alpha)(x_1 - x_2)^2 \\
&\leqslant (1-\alpha)x_1^2 + \alpha x_2^2 - 0 = (1-\alpha)f(x_1) + \alpha f(x_2).
\end{aligned}$$

On the other hand, we will now prove the following result.

Theorem 3.3. *Let $f : (a,b) \to \mathbb{R}$ be twice continuously differentiable. Then f is convex if and only if for all $x \in (a,b)$, $f''(x) \geqslant 0$.*

The convexity of $x \mapsto x^2$ is now immediate, as $\dfrac{d^2}{dx^2} x^2 = 2 > 0$, $x \in \mathbb{R}$.

Example 3.4. We have $\dfrac{d^2}{dx^2} e^x = e^x > 0$, $x \in \mathbb{R}$, and so $x \mapsto e^x$ is convex. Consequently, for all $x_1, x_2 \in \mathbb{R}$ and all $\alpha \in (0,1)$, we have the inequality $e^{(1-\alpha)x_1 + \alpha x_2} \leqslant (1-\alpha)e^{x_1} + \alpha e^{x_2}$. $\diamond$

Exercise 3.10. Consider the function $f : \mathbb{R} \to \mathbb{R}$ given by $f(x) = \sqrt{1+x^2}$, $x \in \mathbb{R}$. Show that f is convex.

Proof. (Of Theorem 3.3.) <u>Only if part</u>: Let $x, y \in (a,b)$ and $x < u < y$. Set $\alpha = \dfrac{u-x}{y-x}$. Then $\alpha \in (0,1)$, and $1 - \alpha = \dfrac{y-u}{y-x}$. As f is convex,

$$\frac{y-u}{y-x} f(x) + \frac{u-x}{y-x} f(y) \geqslant f\left(\frac{y-u}{y-x} x + \frac{u-x}{y-x} y\right) = f(u),$$

that is,

$$(y-x)f(u) \leqslant (u-x)f(y) + (y-u)f(x). \tag{3.4}$$

From (3.4), $(y-x)f(u) \leqslant (u-x)f(y) + (y-x+x-u)f(x)$, that is,

$$(y-x)f(u) - (y-x)f(x) \leqslant (u-x)f(y) - (u-x)f(x),$$

and so

$$\frac{f(u) - f(x)}{u - x} \leqslant \frac{f(y) - f(x)}{y - x}. \tag{3.5}$$

From (3.4), we also have $(y-x)f(u) \leq (u-y+y-x)f(y) + (y-u)f(x)$, that is, $(y-x)f(u) - (y-x)f(y) \leq (u-y)f(y) - (u-y)f(x)$, and so
$$\frac{f(y)-f(x)}{y-x} \leq \frac{f(y)-f(u)}{y-u}. \tag{3.6}$$
Combining (3.5) and (3.6), $\dfrac{f(u)-f(x)}{u-x} \leq \dfrac{f(y)-f(x)}{y-x} \leq \dfrac{f(y)-f(u)}{y-u}$.

Passing the limit as $u \searrow x$ and $u \nearrow y$, $f'(x) \leq \dfrac{f(y)-f(x)}{y-x} \leq f'(y)$.

Hence f' is increasing, and so $f''(x) = \lim\limits_{y \searrow x} \dfrac{f'(y)-f'(x)}{y-x} \geq 0$.

Consequently, for all $x \in \mathbb{R}$, $f''(x) \geq 0$.

If part: Since $f''(x) \geq 0$ for all $x \in (a,b)$, it follows that f' is increasing. Indeed by the Fundamental Theorem of Calculus, if $a < x < y < b$, then
$$f'(y) - f'(x) = \int_x^y \underbrace{f''(\xi)}_{\geq 0} d\xi \geq 0.$$

Now let $a < x < y < b$, $\alpha \in (0,1)$, and $u := (1-\alpha)x + \alpha y$. Then $x < u < y$.

By the Mean Value Theorem, $\dfrac{f(u)-f(x)}{u-x} = f'(v)$ for some $v \in (x,u)$.

Similarly, $\dfrac{f(y)-f(u)}{y-u} = f'(w)$ for some $w \in (u,y)$.

As $w > v$, we have $f'(w) \geq f'(v)$, and so
$$\frac{f(y)-f(u)}{(1-\alpha)(y-x)} = \frac{f(y)-f(u)}{y-u} \geq \frac{f(u)-f(x)}{u-x} = \frac{f(u)-f(x)}{\alpha(y-x)}.$$

Rearranging, we obtain
$$(1-\alpha)f(x) + \alpha f(y) \geq (1-\alpha)f(u) + \alpha f(u) = f(u) = f((1-\alpha)x + \alpha y).$$

Thus f is convex. □

Example 3.5. Consider the function $f : C[a,b] \to \mathbb{R}$ given by
$$f(\mathbf{x}) = \int_a^b (\mathbf{x}(t))^2 dt, \quad \mathbf{x} \in C[a,b].$$

Is f convex? We will show below that f is convex, using the convexity of the map $\xi \mapsto \xi^2 : \mathbb{R} \to \mathbb{R}$. Let $\mathbf{x}_1, \mathbf{x}_2 \in C[a,b]$ and $\alpha \in (0,1)$. Then for all $a, b \in \mathbb{R}$, $((1-\alpha)a + \alpha b)^2 \leq (1-\alpha)a^2 + \alpha b^2$. Hence for each $t \in [a,b]$, with $a := \mathbf{x}_1(t)$, $b := \mathbf{x}_2(t)$, we obtain
$$((1-\alpha)\mathbf{x}_1(t) + \alpha \mathbf{x}_2(t))^2 \leq (1-\alpha)(\mathbf{x}_1(t))^2 + \alpha(\mathbf{x}_2(t))^2.$$

Thus
$$f((1-\alpha)\mathbf{x}_1 + \alpha\mathbf{x}_2) = \int_a^b ((1-\alpha)\mathbf{x}_1(t) + \alpha\mathbf{x}_2(t))^2 dt$$
$$\leq \int_a^b \left((1-\alpha)(\mathbf{x}_1(t))^2 + \alpha(\mathbf{x}_2(t))^2\right) dt$$
$$= (1-\alpha)\int_a^b (\mathbf{x}_1(t))^2 dt + \alpha \int_a^b (\mathbf{x}_2(t))^2 dt$$
$$= (1-\alpha)f(\mathbf{x}_1) + \alpha f(\mathbf{x}_2).$$

Consequently, f is convex. ◇

Exercise 3.11. (Convexity of the arc length functional.)
Let $f : C^1[0,1] \to \mathbb{R}$, be given by $f(\mathbf{x}) = \int_0^1 \sqrt{1 + (\mathbf{x}'(t))^2} dt$, $\mathbf{x} \in C^1[0,1]$.
Prove that f is convex.

Example 3.6. Let us revisit Example 0.1, page viii.
There $S := \{\mathbf{x} \in C^1[0,T] : \mathbf{x}(0) = 0 \text{ and } \mathbf{x}(T) = Q\}$, and $f : S \to \mathbb{R}$ was given by
$$f(\mathbf{x}) = \int_0^T (a\mathbf{x}(t) + b\mathbf{x}'(t))\mathbf{x}'(t) dt, \quad \mathbf{x} \in S,$$
where $a, b, Q > 0$ are constants. Let us check that f is convex. The convexity of the map
$$\mathbf{x} \mapsto b\int_0^T (\mathbf{x}'(t))^2 dt : S \to \mathbb{R}$$
follows from the convexity of $\eta \mapsto \eta^2 : \mathbb{R} \to \mathbb{R}$. The map
$$\mathbf{x} \mapsto a\int_0^T \mathbf{x}(t)\mathbf{x}'(t) dt : S \to \mathbb{R}$$
is constant on S because
$$\int_0^T \mathbf{x}(t)\mathbf{x}'(t) dt = \frac{1}{2}\int_0^T \frac{d}{dt}(\mathbf{x}(t))^2 dt = \frac{(\mathbf{x}(T))^2 - (\mathbf{x}(0))^2}{2} = \frac{Q^2 - 0^2}{2} = \frac{Q^2}{2},$$
and so this map is trivially convex. Hence f, being the sum of two convex functions, is convex too. ◇

We now prove the following result on the sufficiency of the vanishing derivative for a minimiser in the case of convex functions.

Theorem 3.4.
Let X be a normed space and $f : X \to \mathbb{R}$ be convex and differentiable. If $x_* \in X$ is such that $f'(x_*) = \mathbf{0}$, then f has a minimum at x_*.

Proof. Suppose that $x_0 \in X$ and $f(x_0) < f(x_*)$. Define $\varphi : \mathbb{R} \to \mathbb{R}$ by $\varphi(t) = f(tx_0 + (1-t)x_*)$, $t \in \mathbb{R}$. The function φ is convex, since if $\alpha \in (0,1)$ and $t_1, t_2 \in \mathbb{R}$, then we have

$$\varphi\big((1-\alpha)t_1 + \alpha t_2\big)$$
$$= f\Big(\big((1-\alpha)t_1 + \alpha t_2\big)x_0 + \big(1 - (1-\alpha)t_1 - \alpha t_2\big)x_*\Big)$$
$$= f\big((1-\alpha)(t_1 x_0 - t_1 x_*) + \alpha(t_2 x_0 - t_2 x_*) + x_*\big)$$
$$= f\big((1-\alpha)(t_1 x_0 - t_1 x_*) + \alpha(t_2 x_0 - t_2 x_*) + (1-\alpha)x_* + \alpha x_*\big)$$
$$= f\big((1-\alpha)(t_1 x_0 - t_1 x_* + x_*) + \alpha(t_2 x_0 - t_2 x_* + x_*)\big)$$
$$= f\big((1-\alpha)(t_1 x_0 + (1-t_1)x_*) + \alpha(t_2 x_0 + (1-t_2)x_*)\big)$$
$$\leq (1-\alpha)f\big(t_1 x_0 + (1-t_1)x_*\big) + \alpha f\big(t_2 x_0 + (1-t_2)x_*\big)$$
$$= (1-\alpha)\varphi(t_1) + \alpha\varphi(t_2).$$

Also, from Exercise 3.4 on page 125, φ is differentiable at 0, and

$$\varphi'(0) = f'(x_*)(x_0 - x_*) = \mathbf{0}(x_0 - x_*) = 0.$$

We have $\varphi(1) = f(x_0) < f(x_*) = \varphi(0)$. By the Mean Value Theorem, there exists a $\theta \in (0,1)$ such that

$$\varphi'(\theta) = \frac{\varphi(1) - \varphi(0)}{1 - 0} < 0 = \varphi'(0).$$

But this is a contradiction because φ is convex (and so φ' must be increasing; see the proof of the "only if" part of Theorem 3.3).
Thus there cannot exist an $x_0 \in X$ such that $f(x_0) < f(x_*)$.
Consequently, f has a minimum at x_*. □

Exercise 3.12. Consider $f : C[0,1] \to \mathbb{R}$, $f(\mathbf{x}) = \int_0^1 (\mathbf{x}(t))^2 dt$, $\mathbf{x} \in C[0,1]$.
Let $\mathbf{x}_0 \in C[0,1]$. From Example 3.2, page 122, $f'(\mathbf{x}_0) : C[0,1] \to \mathbb{R}$ is given by

$$(f'(\mathbf{x}_0))(\mathbf{h}) = 2\int_0^1 \mathbf{x}_0(t)\mathbf{h}(t)dt, \quad \mathbf{h} \in C[0,1].$$

Prove that $f'(\mathbf{x}_0) = \mathbf{0}$ if and only if $\mathbf{x}_0(t) = 0$ for all $t \in [0,1]$.
We have also seen that in Example 3.5, page 131, that f is a convex function.
Find all solutions to the optimisation problem $\begin{cases} \text{minimise } f(\mathbf{x}) \\ \text{subject to } \mathbf{x} \in C[0,1]. \end{cases}$

3.3 Euler-Lagrange equation

Theorem 3.5. *Let*

(1) $\mathbf{x}_* \in S = \{\mathbf{x} \in C^1[a,b] : \mathbf{x}(a) = y_a,\ \mathbf{x}(b) = y_b\}$,

(2) $F : \mathbb{R}^3 \to \mathbb{R},\ (\xi, \eta, \tau) \xmapsto{F} F(\xi, \eta, \tau)$,

(3) $f : S \to \mathbb{R}$ *be given by* $f(\mathbf{x}) = \int_a^b F(\mathbf{x}(t), \mathbf{x}'(t), t)\,dt,\ \mathbf{x} \in S$,

(4) $X := \{\mathbf{h} \in C^1[a,b] : \mathbf{h}(a) = 0,\ \mathbf{x}(bh = 0\}$,

(5) $\widetilde{f} : X \to \mathbb{R}$ *be given by* $\widetilde{f}(\mathbf{h}) = f(\mathbf{x}_* + \widetilde{h}),\ \mathbf{h} \in X$.

Then $\widetilde{f}'(0) = \mathbf{0}$ *if and only if* $\mathbf{x}_* \in S$ *satisfies the* Euler-Lagrange *equation*:

$$\frac{\partial F}{\partial \xi}(\mathbf{x}_*(t), \mathbf{x}'_*(t), t) - \frac{d}{dt}\left(\frac{\partial F}{\partial \eta}(\mathbf{x}_*(t), \mathbf{x}'_*(t), t)\right) = 0 \text{ for all } t \in [a,b].$$

Definition 3.3. Such an $\mathbf{x}_* \in S$, which satisfies the Euler-Lagrange equation, is said to be *stationary for the functional* f.

Note that X defined above in Theorem 3.5 is a vector space, since it is a subspace of $C^1[0,1]$ (Exercise 1.3, page 7), and it inherits the $\|\cdot\|_{1,\infty}$-norm from $C^1[0,1]$. To prove Theorem 3.5, we will need the following result.

Lemma 3.1. ("Fundamental lemma of the calculus of variations").
If $\mathbf{k} \in C[a,b]$ *is such that*

$$\boxed{\text{for all } \mathbf{h} \in C^1[a,b] \text{ with } \mathbf{h}(a) = \mathbf{h}(b) = 0,\ \int_a^b \mathbf{k}(t)\mathbf{h}'(t)\,dt = 0,}$$

then there exists a constant c *such that* $\mathbf{k}(t) = c$ *for all* $t \in [a,b]$.

Of course, if $\mathbf{k} \equiv c$, then by the Fundamental Theorem of Calculus,

$$\int_a^b \mathbf{k}(t)\mathbf{h}'(t)\,dt = \int_a^b c\mathbf{h}'(t)\,dt = c\int_a^b \mathbf{h}'(t)\,dt$$

$$= c\big(\mathbf{h}(b) - \mathbf{h}(a)\big) = c(0 - 0) = 0,$$

for all $\mathbf{h} \in C^1[a,b]$ that satisfy $\mathbf{h}(a) = \mathbf{h}(b) = 0$. The remarkable thing is that the converse is true, namely that the special property in the box forces $\mathbf{k}$ to be a constant.

Proof. Set $c := \dfrac{1}{b-a}\displaystyle\int_a^b \mathbf{k}(t)dt$.

(If $\mathbf{k} \equiv c$, then
$$\frac{1}{b-a}\int_a^b \mathbf{k}(t)dt = \frac{1}{b-a}\cdot c \cdot (b-a) = c;$$
so the c defined above is the constant that $\mathbf{k}$ "is supposed to be".)

Define $\mathbf{h}_0 : [a,b] \to \mathbb{R}$ by $\mathbf{h}_0(t) = \displaystyle\int_a^t \big(\mathbf{k}(\tau) - c\big)d\tau$.

Then (1) $\mathbf{h}_0 \in C^1[a,b]$,

(2) $\mathbf{h}_0(a) = \displaystyle\int_a^a \big(\mathbf{k}(\tau) - c\big)d\tau = 0$,

(3) $\mathbf{h}_0(b) = \displaystyle\int_a^b \big(\mathbf{k}(\tau) - c\big)d\tau = \displaystyle\int_a^b \mathbf{k}(\tau)d\tau - c(b-a) = 0$ (definition of c).

Thus $\displaystyle\int_a^b \mathbf{k}(t)\mathbf{h}_0'(t)dt = 0$. Since $\mathbf{h}_0'(t) = \mathbf{k}(t) - c$, $t \in [a,b]$, we obtain

$$\int_a^b \big(\mathbf{k}(t) - c\big)^2 dt = \int_a^b \big(\mathbf{k}(t) - c\big)\big(\mathbf{k}(t) - c\big) dt = \int_a^b \big(\mathbf{k}(t) - c\big)\mathbf{h}_0'(t) dt$$
$$= \int_a^b \mathbf{k}(t)\mathbf{h}_0'(t) dt - \int_a^b c\mathbf{h}_0'(t) dt = 0 - c\big(\mathbf{h}_0(b) - \mathbf{h}_0(a)\big) = 0.$$

Thus $\mathbf{k}(t) - c = 0$ for all $t \in [a,b]$, and so $\mathbf{k} \equiv c$. $\square$

Proof. (Of Theorem 3.5). We note that $\mathbf{h} \in X$ if and only if $\mathbf{x}_* + \mathbf{h} \in S$.
(If $\mathbf{h} \in X$, then $(\mathbf{x}_* + \mathbf{h})(a) = \mathbf{x}_*(a) + \mathbf{h}(a) = y_a + 0 = y_a$,
$$(\mathbf{x}_* + \mathbf{h})(b) = \mathbf{x}_*(b) + \mathbf{h}(b) = y_b + 0 = y_b,$$
and so $\mathbf{x}_* + \mathbf{h} \in S$.

Vice versa, if $\mathbf{x}_* + \mathbf{h} \in S$, then

$(\mathbf{x}_* + \mathbf{h})(a) = y_a$ implies $y_a + \mathbf{h}(a) = y_a$, and so $\mathbf{h}(a) = 0$,
$(\mathbf{x}_* + \mathbf{h})(b) = y_b$ implies $y_b + \mathbf{h}(b) = y_b$ and so $\mathbf{h}(b) = 0$.

Consequently, $\mathbf{h} \in X$.) Thus $\widetilde{f}$ is well-defined.
What is $\widetilde{f}'(0)$? For $\mathbf{h} \in X$, we have

$$\widetilde{f}(\mathbf{h}) - \widetilde{f}(0) = f(\mathbf{x}_* + \mathbf{h}) - f(\mathbf{x}_*)$$
$$= \int_a^b \Big(F\big(\mathbf{x}_*(t) + \mathbf{h}(t), \mathbf{x}_*'(t) + \mathbf{h}'(t), t\big) - F\big(\mathbf{x}_*(t), \mathbf{x}_*'(t), t\big)\Big)dt.$$

By Taylor's Formula for F, we know that
$$F(\xi_0 + p, \eta_0 + q, \tau_0 + r) - F(\xi_0, \eta_0, \tau_0)$$
$$= p\frac{\partial F}{\partial \xi}(\xi_0, \eta_0, \tau_0) + q\frac{\partial F}{\partial \eta}(\xi_0, \eta_0, \tau_0) + r\frac{\partial F}{\partial \tau}(\xi_0, \eta_0, \tau_0)$$
$$+ \frac{1}{2!}\begin{bmatrix} p & q & r \end{bmatrix} H_F(\xi_0 + \theta p, \eta_0 + \theta q, \tau_0 + \theta r)\begin{bmatrix} p \\ q \\ r \end{bmatrix}$$

for some θ such that $0 < \theta < 1$. We will apply this for each fixed $t \in [a, b]$, with $\xi_0 := \mathbf{x}_*(t)$, $p := \mathbf{h}(t)$, $\eta_0 := \mathbf{x}'_*(t)$, $q := \mathbf{h}'(t)$, $\tau_0 := t$, $r := 0$, and we will obtain a $\theta \in (0, 1)$ for which the above formula works. If I change the t, then I will get a possibly different $\theta \in (0, 1)$. So we have that the θ depends on $t \in [a, b]$. This gives rise to a function $\Theta : [a, b] \to (0, 1)$ so that

$$\widetilde{f}(h) - \widetilde{f}(\mathbf{0})$$
$$= \int_a^b \Big(F\big(\mathbf{x}_*(t) + \mathbf{h}(t), \mathbf{x}'_*(t) + \mathbf{h}'(t), t\big) - F\big(\mathbf{x}_*(t), \mathbf{x}'_*(t), t\big) \Big) dt$$
$$= \int_a^b \Big(\mathbf{h}(t)\frac{\partial F}{\partial \xi}\big(\mathbf{x}_*(t), \mathbf{x}'_*(t), t\big) + \mathbf{h}'(t)\frac{\partial F}{\partial \eta}\big(\mathbf{x}_*(t), \mathbf{x}'_*(t), t\big) + 0 \cdot \frac{\partial F}{\partial \tau}\big(\mathbf{x}_*(t), \mathbf{x}'_*(t), t\big)$$
$$+ \frac{1}{2!}\begin{bmatrix} \mathbf{h}(t) & \mathbf{h}'(t) & 0 \end{bmatrix} H_F\big(\mathbf{x}_*(t) + \Theta(t)\mathbf{h}(t), x'_*(t) + \Theta(t)\mathbf{h}'(t), t\big)\begin{bmatrix} \mathbf{h}(t) \\ \mathbf{h}'(t) \\ 0 \end{bmatrix}\Big) dt$$
$$= \int_a^b \big(\mathbf{A}(t)\mathbf{h}(t) + \mathbf{B}(t)\mathbf{h}'(t)\big) dt + \int_a^b \frac{1}{2!}\begin{bmatrix} \mathbf{h}(t) & \mathbf{h}'(t) & 0 \end{bmatrix} H_F(\mathbf{P}(t))\begin{bmatrix} \mathbf{h}(t) \\ \mathbf{h}'(t) \\ 0 \end{bmatrix} dt,$$

where
$$\mathbf{A}(t) = \frac{\partial F}{\partial \xi}\big(\mathbf{x}_*(t), \mathbf{x}'_*(t), t\big),$$
$$\mathbf{B}(t) = \frac{\partial F}{\partial \eta}\big(\mathbf{x}_*(t), \mathbf{x}'_*(t), t\big), \text{ and}$$
$$\mathbf{P}(t) = \big(\mathbf{x}_*(t) + \Theta(t)\mathbf{h}(t), \mathbf{x}'_*(t) + \Theta(t)\mathbf{h}'(t), t\big)$$

and $H_F(\cdot)$ denotes the Hessian of F:
$$H_F(\mathbf{v}) := \begin{bmatrix} \frac{\partial^2 F}{\partial \xi^2}(\mathbf{v}) & \frac{\partial^2 f}{\partial \xi \partial \eta}(\mathbf{v}) & \frac{\partial^2 f}{\partial \xi \partial \tau}(\mathbf{v}) \\ \frac{\partial^2 f}{\partial \eta \partial \xi}(\mathbf{v}) & \frac{\partial^2 F}{\partial \eta^2}(\mathbf{v}) & \frac{\partial^2 F}{\partial \xi \partial \tau}(\mathbf{v}) \\ \frac{\partial^2 f}{\partial \tau \partial \xi}(\mathbf{v}) & \frac{\partial^2 f}{\partial \tau \partial \eta}(\mathbf{v}) & \frac{\partial^2 f}{\partial \tau^2}(\mathbf{v}) \end{bmatrix}, \quad \mathbf{v} \in \mathbb{R}^3.$$

From the above, we make a guess for $\widetilde{f}'(\mathbf{0})$: define $L : X \to \mathbb{R}$ by

$$L(\mathbf{h}) = \int_a^b \big(\mathbf{A}(t)\mathbf{h}(t) + \mathbf{B}(t)\mathbf{h}'(t)\big)dt, \quad \mathbf{h} \in X.$$

We have seen that L is a continuous linear transformation in Example 2.11, page 72. For $\mathbf{h} \in X$,

$$\begin{aligned}
&|\widetilde{f}(\mathbf{h}) - \widetilde{f}(\mathbf{0}) - L(\mathbf{h} - \mathbf{0})| \\
&= \left|\frac{1}{2}\int_a^b (\mathbf{h}(t))^2 \frac{\partial^2 F}{\partial \xi^2}(\mathbf{P}(t)) + 2\mathbf{h}(t)\mathbf{h}'(t)\frac{\partial^2 F}{\partial \xi \partial \eta}(\mathbf{P}(t)) + (\mathbf{h}'(t))^2 \frac{\partial^2 F}{\partial \eta^2}(\mathbf{P}(t)) dt\right| \\
&\leqslant \frac{1}{2}\int_a^b |\mathbf{h}(t)|^2 \left|\frac{\partial^2 F}{\partial \xi^2}(\mathbf{P}(t))\right| + 2|\mathbf{h}(t)||\mathbf{h}'(t)|\left|\frac{\partial^2 F}{\partial \xi \partial \eta}(\mathbf{P}(t))\right| + |\mathbf{h}'(t)|^2 \left|\frac{\partial^2 F}{\partial \eta^2}(\mathbf{P}(t))\right| dt \\
&\leqslant \frac{1}{2}\int_a^b \|\mathbf{h}\|_\infty^2 \left|\frac{\partial^2 F}{\partial \xi^2}(\mathbf{P}(t))\right| + 2\|\mathbf{h}\|_\infty \|\mathbf{h}'\|_\infty \left|\frac{\partial^2 F}{\partial \xi \partial \eta}(\mathbf{P}(t))\right| + \|\mathbf{h}'\|_\infty^2 \left|\frac{\partial^2 F}{\partial \eta^2}(\mathbf{P}(t))\right| dt \\
&\leqslant \frac{1}{2}\int_a^b \|\mathbf{h}\|_{1,\infty}^2 \left(\left|\frac{\partial^2 F}{\partial \xi^2}(\mathbf{P}(t))\right| + 2\left|\frac{\partial^2 F}{\partial \xi \partial \eta}(\mathbf{P}(t))\right| + \left|\frac{\partial^2 F}{\partial \eta^2}(\mathbf{P}(t))\right|\right) dt = M\|\mathbf{h}\|_{1,\infty}^2,
\end{aligned}$$

where

$$M := \frac{1}{2}\int_a^b \left(\left|\frac{\partial^2 F}{\partial \xi^2}(\mathbf{P}(t))\right| + 2\left|\frac{\partial^2 F}{\partial \xi \partial \eta}(\mathbf{P}(t))\right| + \left|\frac{\partial^2 F}{\partial \eta^2}(\mathbf{P}(t))\right|\right) dt.$$

We note that for each $t \in [a,b]$, the point

$$\mathbf{P}(t) = \big(\mathbf{x}_*(t) + \mathbf{\Theta}(t)\mathbf{h}(t), \mathbf{x}'_*(t) + \mathbf{\Theta}(t)\mathbf{h}'(t), t\big)$$

in $\mathbb{R}^3$ belongs to a ball with centre $\big(\mathbf{x}_*(t), \mathbf{x}'_*(t), t\big)$ and radius $\|\mathbf{h}\|_{1,\infty}$. But $\mathbf{x}_*, \mathbf{x}'_*$ are continuous, and so these centres $\big(\mathbf{x}_*(t), \mathbf{x}'_*(t), t\big)$, for different values of $t \in [a,b]$, lie inside some big compact set in $\mathbb{R}^3$. And if we look at balls with radius, say 1, around these centres, we get a somewhat bigger compact set, say K, in $\mathbb{R}^3$. Since the partial derivatives

$$\frac{\partial^2 F}{\partial \xi^2}, \quad \frac{\partial^2 F}{\partial \xi \partial \eta}, \quad \frac{\partial^2 F}{\partial \eta^2}$$

are all continuous, it follows that their absolute values are bounded on K. Hence M is finite.

Let $\epsilon > 0$, and $\delta := \frac{\epsilon}{M}$. If $\mathbf{h} \in X$ satisfies $0 < \|\mathbf{h} - \mathbf{0}\|_{1,\infty} = \|\mathbf{h}\|_{1,\infty} < \delta$, then

$$\frac{|\widetilde{f}(\mathbf{h}) - \widetilde{f}(\mathbf{0}) - L(\mathbf{h} - \mathbf{0})|}{\|\mathbf{h}\|_{1,\infty}} \leqslant \frac{M\|\mathbf{h}\|_{1,\infty}^2}{\|\mathbf{h}\|_{1,\infty}} = M\|\mathbf{h}\|_{1,\infty} < M\delta = \epsilon.$$

Consequently, $\widetilde{f}'(\mathbf{0}) = L$.

(Only if part). So far we've calculated $\widetilde{f}'(0)$ and found out that it is the continuous linear transformation L. Now suppose that $\widetilde{f}'(0) = L = \mathbf{0}$, that is, for all $\mathbf{h} \in X$, $L\mathbf{h} = 0$, and so

for all $\mathbf{h} \in C^1[a,b]$ with $\mathbf{h}(a) = \mathbf{h}(b) = 0$, $\int_a^b \big(\mathbf{A}(t)\mathbf{h}(t) + \mathbf{B}(t)\mathbf{h}'(t)\big)dt = 0$.

We would now like to use the technical result (Lemma 3.1) we had shown. So we rewrite the above integral and convert the term in the integrand which involves $\mathbf{h}$, into a term involving $\mathbf{h}'$, by using integration by parts:

$$\int_a^b \mathbf{A}(t)\mathbf{h}(t)dt = \mathbf{h}(t)\int_a^t \mathbf{A}(\tau)d\tau\bigg|_a^b - \int_a^b \left(\mathbf{h}'(t)\int_a^t \mathbf{A}(\tau)d\tau\right)dt$$

$$= 0 - \int_a^b \left(\mathbf{h}'(t)\int_a^t \mathbf{A}(\tau)d\tau\right)dt,$$

because $\mathbf{h}(a) = \mathbf{h}(b) = 0$. So for all $\mathbf{h} \in C^1[a,b]$ with $\mathbf{h}(a) = \mathbf{h}(b) = 0$, we have

$$\int_a^b \left(-\int_a^t \mathbf{A}(\tau)d\tau + \mathbf{B}(t)\right)\mathbf{h}'(t)dt = 0.$$

By Lemma 3.1, $-\int_a^t \mathbf{A}(\tau)d\tau + \mathbf{B}(t) = c$, $t \in [a,b]$, for some constant c. By differentiating with respect to t, we obtain

$$\frac{\partial F}{\partial \xi}(\mathbf{x}_*(t), \mathbf{x}'_*(t), t) - \frac{d}{dt}\left(\frac{\partial F}{\partial \eta}(\mathbf{x}_*(t), \mathbf{x}'_*(t), t)\right) = 0 \text{ for all } t \in [a,b].$$

(If part). Now suppose that $\mathbf{x}_*$ satisfies the Euler-Lagrange equation, that is, $\mathbf{A}(t) - \mathbf{B}'(t) = 0$ for all $t \in [a,b]$. For $\mathbf{h} \in X$, we have

$$\int_a^b \mathbf{B}(t)\mathbf{h}'(t)dt = \mathbf{B}(t)\mathbf{h}(t)\bigg|_a^b - \int_a^b \mathbf{B}'(t)\mathbf{h}(t)dt = 0 - \int_a^b \mathbf{B}'(t)\mathbf{h}(t)dt.$$

Thus for all $\mathbf{h} \in X$,

$$L\mathbf{h} = \int_a^b \big(\mathbf{A}(t)\mathbf{h}(t) + \mathbf{B}(t)\mathbf{h}'(t)\big)dt = \int_a^b \big(\mathbf{A}(t)\mathbf{h}(t) - \mathbf{B}'(t)\mathbf{h}(t)\big)dt$$

$$= \int_a^b \underbrace{\big(\mathbf{A}(t) - \mathbf{B}'(t)\big)}_{=0}\mathbf{h}(t)dt = \int_a^b 0 dt = 0.$$

Consequently, $\widetilde{f}'(0) = L = \mathbf{0}$. $\square$

Corollary 3.1.

Let (1) $S = \{\mathbf{x} \in C^1[a,b] : \mathbf{x}(a) = y_a, \mathbf{x}(b) = y_b\}$,

(2) $F : \mathbb{R}^3 \to \mathbb{R}$, $(\xi, \eta, \tau) \stackrel{F}{\mapsto} F(\xi, \eta, \tau)$,

(3) $f : S \to \mathbb{R}$ *be given by* $f(\mathbf{x}) = \int_a^b F(\mathbf{x}(t), \mathbf{x}'(t), t) \, dt$, $\mathbf{x} \in S$.

Then we have:

(a) *If* $\mathbf{x}_*$ *is a minimiser of* f, *then it satisfies the* Euler-Lagrange equation*:*
$$\frac{\partial F}{\partial \xi}(\mathbf{x}_*(t), \mathbf{x}'_*(t), t) - \frac{d}{dt}\left(\frac{\partial F}{\partial \eta}(\mathbf{x}_*(t), \mathbf{x}'_*(t), t)\right) = 0 \text{ for all } t \in [a,b].$$

(b) *If* f *is convex, and* $\mathbf{x}_* \in S$ *satisfies the Euler-Lagrange equation, then* $\mathbf{x}_*$ *is a minimiser of* f.

Proof. Let $X := \{\mathbf{x} \in C^1[a,b] : \mathbf{x}(a) = \mathbf{0}, \mathbf{x}(b) = \mathbf{0}\}$, and $\tilde{f} : X \to \mathbb{R}$ be given by $\tilde{f}(\mathbf{h}) = f(\mathbf{x}_* + \mathbf{h})$, $\mathbf{h} \in X$. Then $\tilde{f}$ is well-defined.

(a) We claim that $\tilde{f}$ has a minimum at $\mathbf{0} \in X$. Indeed, for $\mathbf{h} \in X$, we have
$$\tilde{f}(\mathbf{h}) = f(\mathbf{x}_* + \mathbf{h}) \geqslant f(\mathbf{x}_*) = f(\mathbf{x}_* + \mathbf{0}) = \tilde{f}(\mathbf{0}).$$
So by Theorem 3.2, page 126, $\tilde{f}'(\mathbf{0}) = \mathbf{0}$. From Theorem 3.5, page 134, it follows that $\mathbf{x}_*$ satisfies the Euler-Lagrange equation.

(b) Now let f be convex and $\mathbf{x}_* \in S$ satisfy the Euler-Lagrange equation. By Theorem 3.5, it follows that $\tilde{f}'(\mathbf{0}) = \mathbf{0}$. The convexity of f makes $\tilde{f}$ convex as well. Indeed, if $\mathbf{h}_1, \mathbf{h}_2 \in X$, and $\alpha \in (0,1)$, then
$$\begin{aligned}\tilde{f}((1-\alpha)\mathbf{h}_1 + \alpha\mathbf{h}_2) &= f(\mathbf{x}_* + (1-\alpha)\mathbf{h}_1 + \alpha\mathbf{h}_2) \\ &= f((1-\alpha)\mathbf{x}_* + \alpha\mathbf{x}_* + (1-\alpha)\mathbf{h}_1 + \alpha\mathbf{h}_2) \\ &= f((1-\alpha)(\mathbf{x}_* + \mathbf{h}_1) + \alpha(\mathbf{x}_* + \mathbf{h}_2)) \\ &\leqslant (1-\alpha)f(\mathbf{x}_* + \mathbf{h}_1) + \alpha f(\mathbf{x}_* + \mathbf{h}_2) \\ &= (1-\alpha)\tilde{f}(\mathbf{h}_1) + \alpha\tilde{f}(\mathbf{h}_2).\end{aligned}$$

Recall that in Theorem 3.4, page 132, we had shown that for a convex function, the derivative vanishing at a point implies that that point is a minimiser for the function. Since $\tilde{f}$ is convex, and because $\tilde{f}'(\mathbf{0}) = \mathbf{0}$, $\mathbf{0}$ is a minimiser of $\tilde{f}$. We claim that $\mathbf{x}_*$ is a minimiser of f. Indeed, if $\mathbf{x} \in S$, then $\mathbf{x} = \mathbf{x}_* + (\mathbf{x} - \mathbf{x}_*) = \mathbf{x}_* + \mathbf{h}$, where $\mathbf{h} := \mathbf{x} - \mathbf{x}_* \in X$. Hence
$$f(\mathbf{x}) = f(\mathbf{x}_* + (\mathbf{x} - \mathbf{x}_*)) = \tilde{f}(\mathbf{x} - \mathbf{x}_*) \geqslant \tilde{f}(\mathbf{0}) = f(\mathbf{x}_* + \mathbf{0}) = f(\mathbf{x}_*).$$
This completes the proof. $\square$

Let us revisit Example 0.1, page viii, and solve it by observing that it falls in the class of problems considered in the above result.

Example 3.7. Recall that $S := \{x \in C^1[0,T] : x(0) = 0 \text{ and } x(T) = Q\}$, so that we have $a = 0$, $b = T$, $y_a = 0$ and $y_b = Q$. The cost function $f : S \to \mathbb{R}$ was given by

$$f(\mathbf{x}) = \int_0^T (a\mathbf{x}(t) + b\mathbf{x}'(t))\mathbf{x}'(t)\,dt = \int_0^T F(\mathbf{x}(t), \mathbf{x}'(t), t)\,dt, \quad \mathbf{x} \in S,$$

where $a, b, Q > 0$ are constants, and $F : \mathbb{R}^3 \to \mathbb{R}$ is given by

$$F(\xi, \eta, \tau) = (a\xi + b\eta)\eta, \quad (\xi, \eta, \tau) \in \mathbb{R}^3.$$

So this problem does fall into the class of problems covered by Corollary 3.1. In order to apply the result to solve this problem, we compute

$$\frac{\partial F}{\partial \xi}(\xi, \eta, \tau) = a\eta, \qquad \frac{\partial F}{\partial \eta}(\xi, \eta, \tau) = a\xi + 2b\eta.$$

The Euler-Lagrange equation for $\mathbf{x}_* \in S$ is:

$$\frac{\partial F}{\partial \xi}(\mathbf{x}_*(t), \mathbf{x}'_*(t), t) - \frac{d}{dt}\left(\frac{\partial F}{\partial \eta}(\mathbf{x}_*(t), \mathbf{x}'_*(t), t)\right) = 0$$

that is, $a\mathbf{x}'_*(t) - \dfrac{d}{dt}\bigl(a\mathbf{x}_*(t) + 2b\mathbf{x}'_*(t)\bigr) = 0$ for all $t \in [0,T]$. Thus

$$\frac{d}{dt}(2b\mathbf{x}'_*(t)) = 0, \quad t \in [0,T].$$

By the Fundamental Theorem of Calculus, it follows that there is a constant A such that $\mathbf{x}'_*(t) = A$, $t \in [0,T]$, and integrating again, we obtain a constant B such that $\mathbf{x}_*(t) = At + B$, $t \in [0,T]$. But since $\mathbf{x}_* \in S$, we also have that $\mathbf{x}_*(0) = 0$ and $\mathbf{x}_*(T) = Q$, which we can use to find the constants A, B: $A \cdot 0 + B = 0$, and $A \cdot T + B = Q$, so that $B = 0$ and $A = Q/T$. Consequently, by part (a) of the conclusion in Corollary 3.1, we know that *if $\mathbf{x}_*$ is a minimiser of f, then*

$$\mathbf{x}_*(t) = Q\frac{t}{T}, \quad t \in [0,T].$$

On the other hand, we had checked in Example 3.6 that f is convex. And we know that the $\mathbf{x}_*$ given above satisfies the Euler-Lagrange equation. Consequently, by part (b) of the conclusion in Corollary 3.1, we know that this $\mathbf{x}_*$ is a minimiser. So we have shown, using Corollary 3.1, that

$$\boxed{\mathbf{x}_* \text{ is a minimiser of } f} \quad \Leftrightarrow \quad \boxed{\mathbf{x}_*(t) = Q\frac{t}{T},\ t \in [0,T].}$$

So we have solved our optimal mining question.

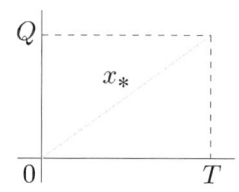

$$\begin{array}{c} Q \\ \\ 0 \end{array} \begin{array}{c} \\ x_* \\ \\ T \end{array}$$

And we now know that the optimal mining operation is given by the humble straight line! ◇

Exercise 3.13. (Euclidean plane).
Let $\mathbb{P}_1 = \mathbb{R}^2$ with $\|(x,t)\|_1 := \sqrt{x^2 + t^2}$ for $(x,t) \in \mathbb{P}_1$.
Set $S := \{\mathbf{x} \in C^1[a,b] : \mathbf{x}(a) = x_a, \mathbf{x}(b) = x_b\}$.
Given $\mathbf{x} \in S$, the map $t \stackrel{\gamma_\mathbf{x}}{\mapsto} \gamma_\mathbf{x}(t) := (\mathbf{x}(t), t) : [a,b] \to \mathbb{P}_1$ is a curve in the Euclidean plane $\mathbb{P}_1$, and we define its arc length by
$$L_1(\gamma_\mathbf{x}) := \int_a^b \|\gamma'_\mathbf{x}(t)\|_1 dt = \int_a^b \sqrt{1 + (\mathbf{x}'(t))^2} dt.$$
Show that the straight line joining (x_a, a) and (x_b, b) has the smallest arc length.

Exercise 3.14. (Galilean spacetime).
Let $\mathbb{P}_0 = \mathbb{R}^2$ with $\|(x,t)\|_0 := \sqrt{0 \cdot x^2 + t^2} = |t|$ for $(x,t) \in \mathbb{P}_0$.
Set $S := \{\mathbf{x} \in C^1[a,b] : \mathbf{x}(a) = x_a, \mathbf{x}(b) = x_b\}$.
Given $\mathbf{x} \in S$, the map $t \stackrel{\gamma_\mathbf{x}}{\mapsto} \gamma_\mathbf{x}(t) := (\mathbf{x}(t), t) : [a,b] \to \mathbb{P}_0$ is a curve in the Euclidean plane $\mathbb{P}_0$, and we define its arc length by
$$L_0(\gamma_\mathbf{x}) := \int_a^b \|\gamma'_\mathbf{x}(t)\|_0 dt.$$
Show that all the curves $\gamma_\mathbf{x}$ joining (x_a, a) and (x_b, b) have the *same* arc length. (If we think of $\mathbb{P}_0$ as the collection of all events (="here and now"), with the coordinates provided by an "inertial frame[3]" choice, then this arc length is the pre-relativistic *absolute time* between the two events (x_a, a) and (x_b, b).)

Exercise 3.15. (Minkowski spacetime).
Let $\mathbb{P}_{-1} = \mathbb{R}^2$ with $\|(x,t)\|_{-1} := \sqrt{|-x^2 + t^2|}$ for $(x,t) \in \mathbb{P}_0$.
Set $S := \{\mathbf{x} \in C^1[a,b] : \mathbf{x}(a) = x_a, \mathbf{x}(b) = x_b, \text{ for all } a \leqslant t \leqslant b, |\mathbf{x}'(t)| < 1\}$.
Given $\mathbf{x} \in S$, the map $t \stackrel{\gamma_\mathbf{x}}{\mapsto} \gamma_\mathbf{x}(t) := (\mathbf{x}(t), t) : [a,b] \to \mathbb{P}_{-1}$ is a curve in the Euclidean plane $\mathbb{P}_{-1}$, and we define its arc length by
$$L_{-1}(\gamma_\mathbf{x}) := \int_a^b \|\gamma'_\mathbf{x}(t)\|_{-1} dt.$$

---
[3] A coordinate system is *inertial* if particles which are "free" that is, not acted upon by any force, move in straight lines with a uniform speed.

Show[4] that among all the curves $\gamma_\mathbf{x}$ joining (x_a, a) and (x_b, b), the straight line has the *largest*(!) arc length.

($\mathbb{P}_{-1}$ can be thought of as the special relativistic spacetime of all events, with the coordinates provided by an "inertial frame" choice. Then the arc length $L(\gamma_\mathbf{x})$ is the *proper time* between the two events (x_a, a) and (x_b, b), which may be thought of as the time recorded by a clock carried by an observer along its worldline $\gamma_\mathbf{x}$. The fact that the straight line has the largest length accounts for the aging of the travelling sibling in the famous *Twin Paradox*. Imagine two twins, say Seeta and Geeta, who are separated at birth, event $(0, 0)$ in an inertial frame, and meet again in adulthood at the event $(0, T)$. Seeta, the meek twin, doesn't move in the inertial frame, described by the straight line γ_S joining the events $(0, 0)$ and $(0, T)$. Meanwhile, the other feisty twin, Geeta, travels in a spaceship (never exceeding the speed of light, 1), with a worldline given by γ_G, starting at $(0, 0)$, and ending to meet the twin at $(0, T)$ as shown.

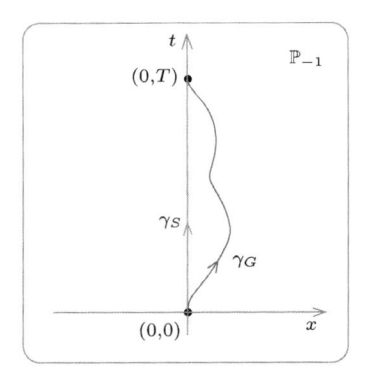

There is no longer any surprise that Seeta has aged far more than Geeta, thanks to our inequality that $L(\gamma_G) < L(\gamma_S)$. Resting is rusting!)

Exercise 3.16. (Euler-Lagrange Equation: vector valued case).
The results in this section can be generalised to the case when f has the form

$$f(\mathbf{x}_1, \cdots, \mathbf{x}_n) = \int_a^b F(\mathbf{x}_1(t), \cdots, \mathbf{x}_n(t), \mathbf{x}_1'(t), \cdots, \mathbf{x}_n'(t), t) dt,$$

where $(\xi_1, \cdots, \xi_n, \eta_1, \cdots, \eta_n, \tau) \mapsto F(\xi_1, \cdots, \xi_n, \eta_1, \cdots, \eta_n, \tau) : \mathbb{R}^{2n+1} \to \mathbb{R}$ is a function with continuous partial derivatives of order $\leqslant 2$, and $\mathbf{x}_1, \cdots, \mathbf{x}_n$ are n continuously differentiable functions of the variable $t \in [a, b]$.

---

[4] Here we tacitly ignore the fact that the set S doesn't quite have the form that we have been considering, since we have the *extra* constraint $|\mathbf{x}'(t)| < 1$ for all ts. Despite this extra condition, a version of Corollary 3.1 holds, mutatis mutandis, with an appropriately adapted proof: instead of $X := \{\mathbf{h} \in C^1[a, b] : \mathbf{h}(a) = 0 = \mathbf{h}(b)\}$, we work in the *open* subset $X_0 := \{\mathbf{h} \in X : |\mathbf{h}'(t)| < 1 \text{ for all } t \in [a, b]\}$ of X. We won't spell out the details here, but we simply use the Euler-Lagrange equation in this exercise.

Then following a similar analysis as before, we obtain n Euler-Lagrange equations to be satisfied by the minimiser $(\mathbf{x}_{1*}, \cdots, \mathbf{x}_{n*})$: for $t \in [a,b]$, and $k \in \{1, \cdots, n\}$,

$$0 = \frac{\partial F}{\partial \xi_k}(\mathbf{x}_{1*}(t), \cdots, \mathbf{x}_{n*}(t), \mathbf{x}'_{1*}(t), \cdots, \mathbf{x}'_{n*}(t), t)$$
$$- \frac{d}{dt}\Big(\frac{\partial F}{\partial \eta_k}(\mathbf{x}_{1*}(t), \cdots, \mathbf{x}_{n*}(t), \mathbf{x}'_{1*}(t), \cdots, \mathbf{x}'_{n*}(t), t)\Big).$$

Let us see an application of this to the planar motion of a body under the action of the gravitational force field (planet around the sun).

If $\mathbf{x}_1(t) = \mathbf{r}(t)$ (the distance to the sun), and $\mathbf{x}_2(t) = \varphi(t)$ (radial angle), then the function to be minimised is

$$f(\mathbf{r}, \varphi) := \int_a^b \Big(\frac{m}{2}\big(\mathbf{r}'(t)\big)^2 + \frac{m}{2}\big(\mathbf{r}(t)\varphi'(t)\big)^2 + \frac{GMm}{\mathbf{r}(t)}\Big)dt.$$

Show that the Euler-Lagrange equations give

$$\mathbf{r}''(t) = \mathbf{r}(t)\big(\varphi'(t)\big)^2 - \frac{GM}{(\mathbf{r}(t))^2}, \text{ and } \frac{d}{dt}\Big(m\big(\mathbf{r}(t)\big)^2\varphi'(t)\Big) = 0.$$

(The latter equation shows that the angular momentum, $L(t) := m\mathbf{r}(t)^2\varphi'(t)$, is conserved, and this gives Kepler's Second Law, saying that a planet sweeps equal areas in equal times.)

Exercise 3.17. (Euler-Lagrange Equation: several independent variables). Suppose that $\Omega \subset \mathbb{R}^d$ is a "region" (an open, path-connected set), and that

$$\mathcal{L}: \quad \mathbb{R}^d \times \mathbb{R} \times \mathbb{R}^d \quad \longrightarrow \mathbb{R}$$
$$(X_1, \cdots, X_d, U, V_1, \cdots, V_d) \longmapsto \mathcal{L}(X_1, \cdots, X_d, U, V_1, \cdots, V_d)$$

is a given C^2 function (called the *Lagrangian density*). We are interested in finding $u \in C^1(\Omega)$ which minimise $I : C^1(\Omega) \to \mathbb{R}$ given by

$$I(u) = \int_\Omega \mathcal{L}(x_1, \cdots, x_d, u, u_{x_1}, \cdots, u_{x_d})dx_1\cdots dx_d$$
$$= \int_\Omega \mathcal{L}\big(\mathbf{x}, u(\mathbf{x}), \nabla u(\mathbf{x})\big)d\mathbf{x}, \quad u \in C^1(\Omega).$$

(Here subscripts indicate respective partial derivatives: for example, $u_{x_1} = \dfrac{\partial u}{\partial x_1}$.)

It can be shown that a necessary condition for u to be a minimiser of I is that it satisfies the *Euler-Lagrange equation* below:

$$\frac{\partial \mathcal{L}}{\partial U}\big(\mathbf{x}, u(\mathbf{x}), \nabla u(\mathbf{x})\big) - \sum_{i=1}^d \frac{\partial}{\partial x_i}\Big(\frac{\partial \mathcal{L}}{\partial V_i}\big(\mathbf{x}, u(\mathbf{x}), \nabla u(\mathbf{x})\big)\Big) = 0.$$

(Note that the Euler-Lagrange equation above is now a Partial Differential Equation (PDE), rather than the Ordinary Differential Equation (ODE) we had met in Theorem 3.5, page 134.)

Let us consider examples of writing the Euler-Lagrange equation.

(1) (Minimal area surfaces).

Consider a smooth surface in $\mathbb{R}^3$ which is the graph of $(x,y) \mapsto u(x,y)$ defined on an open set $\Omega \subset \mathbb{R}^2$.

The area of the surface is given by:
$$I(u) = \iint_\Omega \sqrt{1 + \|\nabla u\|_2^2} \, dxdy, \quad u \in C^2(\Omega).$$
Show that if u is a minimiser, then u must satisfy the PDE
$$(1 + u_x^2)u_{yy} - 2u_x u_y u_{xy} + (1 + u_y^2)u_{xx} = 0.$$
Verify that the following solve this PDE: $u = Ax + By + C$ (plane) and
$$u = \tan^{-1}(y/x) \text{ (helicoid)}.$$
Also, in the case of the helicoid, show that a parametric representation of the surface is given by $x(s,t) = s \cdot \cos t$, $y(s,t) = s \cdot \sin t$, $z(s,t) = t$, by setting $s = \sqrt{x^2 + y^2}$ and $t = \tan^{-1}(y/x)$. Plot the surface[5] with Maple using:
`with(plots): plot3d([s*cos(t), s*sin(t), t], s=-3..3, t=-3..3)`

(2) (Wave equation).

Consider a vibrating string of length 1, whose ends are fixed. If $u(x,t)$ denotes the displacement at position x and time t, where $0 \leq x \leq 1$, then the potential energy at time t is given by
$$\frac{1}{2} \int_0^1 \left(\frac{\partial u}{\partial x}(x,t)\right)^2 dx,$$
and the kinetic energy is $\dfrac{1}{2} \displaystyle\int_0^1 \left(\dfrac{\partial u}{\partial t}(x,t)\right)^2 dx.$

For $u : [0,1] \times [0,T] \to \mathbb{R}$, set
$$I(u) = \frac{1}{2} \int_0^T \int_0^1 \left(\left(\frac{\partial u}{\partial t}(x,t)\right)^2 - \left(\frac{\partial u}{\partial x}(x,t)\right)^2\right) dxdt.$$
Prove that if u_* minimises I, then it satisfies the wave equation
$$\frac{\partial^2 u_*}{\partial t^2} - \frac{\partial^2 u_*}{\partial x^2} = 0.$$
Show that if $f : \mathbb{R} \to \mathbb{R}$ is
- twice continuously differentiable,
- odd ($f(x) = -f(-x)$ for all $x \in \mathbb{R}$), and
- periodic with period 2 (that is, $f(x+2) = f(x)$ for all $x \in \mathbb{R}$),

then u given by $u(x,t) := \dfrac{f(x-t) + f(x+t)}{2}$, is such that

- it solves the wave equation,
- with the boundary conditions $u(0, \cdot) = 0 = u(1, \cdot)$ and
- the initial conditions $u(\cdot, 0) = f$ (position) and $\dfrac{\partial u}{\partial t}(\cdot, 0) \equiv 0$ (velocity).

Interpret the solution graphically.

---

[5]This surface has the least area with a helix as its boundary.

3.4 An excursion in Classical Mechanics

The aim of this section is to apply the Euler-Lagrange equation to illustrate some basic ideas in classical mechanics. Also, this brief discussion will provide some useful background for discussing Quantum Mechanics later on, as an application of Hilbert spaces and their operators.

Newtonian Mechanics. Consider the motion $t \mapsto \mathbf{q}_*(t)$ of a classical point particle of mass m along a straight line. Here $\mathbf{q}_*(t)$ denotes the position of the particle at time t.

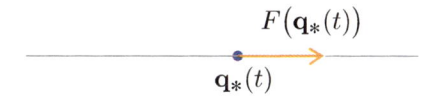

Then the evolution of $\mathbf{q}_*$ is described by Newton's Law, which says that the "mass times the acceleration equals the force acting", that is, if $F(x)$ is the force at position x, then

$$m\mathbf{q}_*''(t) = F(\mathbf{q}_*(t)).$$

Together with the particle's initial position $\mathbf{q}_*(t_i) = q_i$, and initial velocity $\mathbf{q}_*'(t_i) = v_i$, the above equation determines a unique $\mathbf{q}_*$.

Principle of Stationary Action. An alternative formulation of Newtonian Mechanics is given by the "Principle of Stationary[6] Action", which is more useful because it lends itself to generalisations for other types of physical situations, for example in describing the electromagnetic field (when there are no particles). In that sense it is more fundamental as it provides a unifying language.

First, let us define the *potential* $V : \mathbb{R} \to \mathbb{R}$ as follows. Choose any $x_0 \in \mathbb{R}$, and set

$$V(x) = -\int_{x_0}^{x} F(\xi) d\xi.$$

V is thought of as the work done against the force to go from x_0 to x. (Because of the fact that x_0 was chosen arbitrarily, the potential V for a force F is not unique. By the Fundamental Theorem of Calculus, we have

$$\frac{dV}{dx}(x) = F(x), \quad x \in \mathbb{R},$$

---
[6]It is standard to use "Least" rather than "Stationary" because in many cases the action is actually minimised.

and so it can be seen that if V, $\widetilde{V}$ are potentials for F, then as
$$\frac{d}{dx}(V - \widetilde{V}) \equiv 0,$$
there is a constant $c \in \mathbb{R}$ such that $\widetilde{V}(x) = V(x) + c$, $x \in \mathbb{R}$.) We define the kinetic energy of the particle at time t as
$$\frac{1}{2} m \big(\mathbf{q}'(t)\big)^2.$$
Consider for $\mathbf{q} \in C^1[t_i, t_f]$ with $\mathbf{q}(t_i) = x_i$ and $\mathbf{q}(t_f) = x_f$, the *action*
$$A(\mathbf{q}) = \int_{t_i}^{t_f} L\big(\mathbf{q}(t), \mathbf{q}'(t)\big) dt,$$
where L is called the *Lagrangian*, given by $L(x, v) = \dfrac{1}{2} mv^2 - V(x)$. Note that along an imagined trajectory $\mathbf{q}$ of a particle,
$$L\big(\mathbf{q}(t), \mathbf{q}'(t)\big) = \frac{1}{2} m \big(\mathbf{q}'(t)\big)^2 - V\big(\mathbf{q}(t)\big)$$
$$= \text{kinetic energy} - \text{potential energy at time } t.$$

The *Principle of Stationary Action* in Classical Mechanics says that the motion $\mathbf{q}_*$ of the particle moving from position x_i at time t_i to position x_f at time t_f is such that $\widetilde{A}'(\mathbf{0}) = \mathbf{0}$, where $\widetilde{A} : X \to \mathbb{R}$ is given by
$$\widetilde{A}(\mathbf{h}) := A(\mathbf{q}_* + \mathbf{h}), \text{ for } \mathbf{h} \in X := \{\mathbf{h} \in C^1[t_i, t_f] : \mathbf{h}(t_i) = 0 = \mathbf{h}(t_f)\}.$$

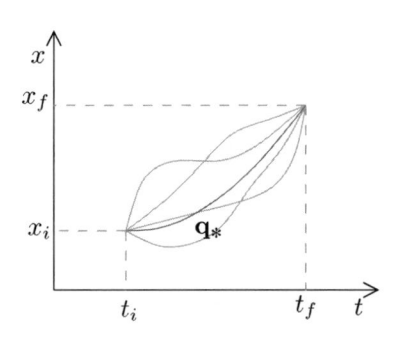

By Theorem 3.5, page 134, the Euler-Lagrange equation is equivalent to $\widetilde{A}'(\mathbf{0}) = \mathbf{0}$, and so the motion $\mathbf{q}_*$ is described by
$$\frac{\partial L}{\partial x}\big(\mathbf{q}_*(t), \mathbf{q}_*'(t)\big) - \frac{d}{dt}\left(\frac{\partial L}{\partial v}\big(\mathbf{q}_*(t), \mathbf{q}_*'(t)\big)\right) = 0,$$

that is,
$$-\frac{dV}{dx}(\mathbf{q}_*(t)) - \frac{d}{dt}\left(\frac{1}{2}m \cdot 2\mathbf{q}'_*(t)\right) = 0.$$

Using $\frac{dV}{dx}(x) = -F(x)$, we obtain Newton's equation of motion,
$$m\mathbf{q}''_*(t) = F(\mathbf{q}_*(t)).$$

Here are a couple of examples.

Example 3.8. (The falling stone).
Let $x \geqslant 0$ denote the height above the surface of the Earth of a stone of mass m. Then its potential energy is given by $V(x) = mgx$. Thus
$$L(x,v) = \frac{1}{2}mv^2 - mgx.$$
Suppose the stone starts from initial height $x_0 > 0$ at time 0, with initial speed 0. Then the height $\mathbf{q}_*(t)$ at time t is described by $m\mathbf{q}''_*(t) = -mg$, that is,
$$\mathbf{q}''_*(t) = -g.$$
Using the initial conditions, we obtain $\mathbf{q}'_*(t) - 0 = -gt$, and so
$$\mathbf{q}_*(t) = x_0 - \frac{1}{2}gt^2, \text{ for } t \geqslant 0.$$
◇

Example 3.9. (The Harmonic Oscillator).
The harmonic oscillator is the simplest oscillating system, where we can imagine a body mass m attached to a spring with spring constant k oscillating about its equilibrium position. For a displacement of x from the equilibrium position of the mass, a force of kx is imparted on the mass by the spring. So
$$V(x) = \frac{1}{2}kx^2, \text{ and } L(x,v) = \frac{1}{2}mv^2 - \frac{1}{2}kx^2.$$
The equation of motion is
$$m\mathbf{q}''_*(t) + k\mathbf{q}_*(t) = 0,$$
describing the displacement $\mathbf{q}_*(t)$ from the equilibrium position at time t. If v_0 is the velocity at time $t = 0$, and the initial position is $\mathbf{q}_*(0) = 0$, then the unique solution is
$$\mathbf{q}_*(t) = v_0\sqrt{\frac{m}{k}}\sin\left(\sqrt{\frac{k}{m}}t\right).$$

(It can be easily verified that this $\mathbf{q}_*$ satisfies the equation of motion $m\mathbf{q}_*''(t) + k\mathbf{q}_*(t) = 0$, as well as the initial conditions $\mathbf{q}_*(0) = 0$ and $\mathbf{q}_*'(0) = v_0$.)

The maximum displacement is
$$x_{\max} = v_0 \sqrt{\frac{m}{k}},$$
and the period of oscillation is $T = 2\pi \sqrt{\dfrac{m}{k}}$. ◇

"Symmetries" of the Lagrangian give rise to "conservation laws":

Law of Conservation of Energy. Since the Lagrangian $L(x,v)$ does not depend on t (that is, it possesses "the symmetry of being invariant under time translations"), we will now see that this results in the Law of Conservation of Energy. Define the *energy* $\mathbf{E}(t)$ along $\mathbf{q}_*$ at time t by

$$\mathbf{E}(t) = \text{kinetic } + \text{ potential energy at time } t$$
$$= \frac{1}{2}m\bigl(\mathbf{q}_*'(t)\bigr)^2 + V\bigl(\mathbf{q}_*(t)\bigr).$$

Then we have
$$\mathbf{E}'(t) = \frac{1}{2}m 2 \mathbf{q}_*'(t)\mathbf{q}_*''(t) + \frac{dV}{dx}\bigl(\mathbf{q}_*(t)\bigr) \cdot \mathbf{q}_*'(t)$$
$$= \bigl(m\mathbf{q}_*''(t) - F(\mathbf{q}_*(t))\bigr) \cdot \mathbf{q}_*'(t) = 0 \cdot \mathbf{q}_*'(t) = 0.$$

Hence the energy $\mathbf{E}$ is constant, that is, it is conserved.

Law of Conservation of Momentum. Now suppose that the Lagrangian does not depend on the position, that is, $L(x,v) = \ell(v)$ for some function ℓ. Define the *momentum* $\mathbf{p}_*(t)$ along $\mathbf{q}_*$ at time t by

$$\mathbf{p}_*(t) = \text{mass} \times \text{velocity} = m\mathbf{q}_*'(t).$$

Then
$$\mathbf{p}_*'(t) = m\mathbf{q}_*''(t) = -\frac{\partial L}{\partial x}\bigl(\mathbf{q}_*(t), \mathbf{q}_*'(t)\bigr) = 0,$$
and so $\mathbf{p}_*$ is constant, that is, the momentum is conserved.

Remark 3.2. (Noether's Theorem).
The above two results are special cases of a much more general result, called *Noether's Theorem*, roughly stating that every differentiable symmetry of the action has a corresponding conservation law. This result is fundamental in theoretical physics. We refer the interested reader to the book [Neuenschwander (2011)].

Example 3.10. (Particle in a Potential Well).
Consider a particle of mass m moving along a line, and which is acted upon by a force

$$F(x) = -\frac{dV}{dx}(x), \quad x \in \mathbb{R},$$

generated by a potential V. The associated Lagrangian is

$$L(x, v) = \frac{1}{2}mv^2 - V(x).$$

Suppose that the motion of the particle is described by $\mathbf{q}_*$ for $t \geq 0$. If

$$E := \frac{1}{2}m(\mathbf{q}'_*(0))^2 + V(\mathbf{q}_*(0)),$$

then for all $t \geq 0$, we have by the Law of Conservation of Energy, that

$$\frac{1}{2}m(\mathbf{q}'_*(t))^2 + V(\mathbf{q}_*(t)) = E,$$

and so $0 \leq (\mathbf{q}'_*(t))^2 = \frac{2}{m}\big(E - V(\mathbf{q}_*(t))\big)$. This implies that $V(\mathbf{q}_*(t)) \leq E$. Hence the particle cannot leave the potential well if $V(x) \to \infty$ as $x \to \pm\infty$.

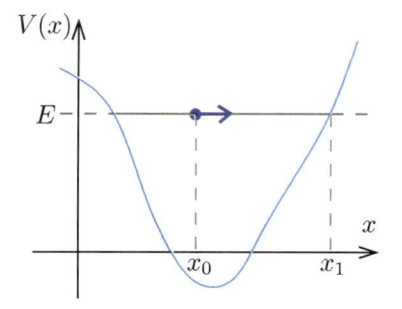

If the velocity of the particle is always positive while moving from initial position x_0 at time $t = 0$ to a position $x > x_0$ at time t, then by integrating,

$$t - 0 = \sqrt{\frac{m}{2}} \int_{x_0}^{x} \frac{1}{\sqrt{E - V(\xi)}} d\xi.$$

If in this manner, the particle reaches x_1, where $E = V(x_1)$ (see the previous picture), then we may ask if the travel time t_1 from x_0 to x_1 is finite. The above expression reveals that $t_1 < \infty$ if and only if

$$\int_{x_0}^{x_1} \frac{1}{\sqrt{E - V(\xi)}} d\xi < \infty.$$

In particular, in the case of the harmonic oscillator, where

$$V(x) = \frac{1}{2}kx^2, \quad x_0 = 0, \quad x_1 = x_{\max} = v_0\sqrt{\frac{m}{k}}, \quad E = \frac{1}{2}mv_0^2,$$

we have that the time of travel from the initial condition x_0 to the maximum displacement $x_{\max}$ is finite, and is given by

$$t_1 = \sqrt{\frac{m}{2}} \int_{x_0}^{x_1} \frac{1}{\sqrt{E - V(\xi)}} d\xi = \sqrt{\frac{m}{2}} \int_0^{v_0\sqrt{\frac{m}{k}}} \frac{1}{\sqrt{\frac{1}{2}mv_0^2 - \frac{1}{2}k\xi^2}} d\xi$$

$$= \sqrt{\frac{m}{2}} \cdot \sqrt{\frac{2}{k}} \cdot \sin^{-1}\left(\frac{v_0\sqrt{\frac{m}{k}}}{\sqrt{\frac{mv_0^2}{k}}}\right) = \frac{1}{4} \cdot 2\pi\sqrt{\frac{m}{k}},$$

which is, as expected, one-fourth of the period of oscillation. ◇

Hamiltonian Mechanics.

The momentum $\mathbf{p}_*$ is defined by $\mathbf{p}_*(t) = m\mathbf{q}'_*(t)$. Since $\dfrac{\partial L}{\partial v} = mv$, we have

$$\mathbf{p}_*(t) = \frac{\partial L}{\partial v}(\mathbf{q}_*(t), \mathbf{q}'_*(t)).$$

The Euler-Lagrange equation, $\dfrac{\partial L}{\partial x}(\mathbf{q}_*(t), \mathbf{q}'_*(t)) - \dfrac{d}{dt}\left(\dfrac{\partial L}{\partial v}(\mathbf{q}_*(t), \mathbf{q}'_*(t))\right) = 0$, can be re-written as

$$\mathbf{p}'_*(t) = \frac{\partial L}{\partial x}(\mathbf{q}_*(t), \mathbf{q}'_*(t)).$$

It turns out that the above two equations can be expressed in a much more symmetrical manner, with the introduction of the *Hamiltonian*,

$$(p, q) \stackrel{H}{\mapsto} H(q, p) := \frac{p^2}{2m} + V(q) : \mathbb{R}^2 \to \mathbb{R},$$

as follows. Note that

$$\frac{\partial H}{\partial p}(q, p) = \frac{p}{m},$$

$$\frac{\partial H}{\partial q}(q, p) = \frac{dV}{dx}(q) = -\frac{\partial L}{\partial x}(q, p).$$

Thus

$$\frac{\partial H}{\partial p}(\mathbf{q}_*(t), \mathbf{p}_*(t)) = \frac{\mathbf{p}_*(t)}{m} = \mathbf{q}'_*(t),$$

$$\frac{\partial H}{\partial q}(\mathbf{q}_*(t), \mathbf{p}_*(t)) = -\frac{\partial L}{\partial x}(\mathbf{q}_*(t), \mathbf{q}'_*(t)) = -\mathbf{p}'_*(t).$$

Differentiation

These two equations are equivalent to $\mathbf{p}_*(t) = m\mathbf{q}'_*(t)$ and the Euler-Lagrange equation. The space $\{(q,p) \in \mathbb{R}^2\}$ is called the *phase plane*, where the position-momentum pairs live. Each point (q,p) in the phase plane is thought of as a possible *state* of the particle. Given an initial state (q_0, p_0), the coupled first order differential equations, describing the evolution of the state, namely the *Hamiltonian equations*

$$\mathbf{q}'_*(t) = \frac{\partial H}{\partial p}\big(\mathbf{q}_*(t), \mathbf{p}_*(t)\big),$$

$$\mathbf{p}'_*(t) = -\frac{\partial H}{\partial q}\big(\mathbf{q}_*(t), \mathbf{p}_*(t)\big),$$

for $t \geq 0$, describe a curve $t \mapsto \big(\mathbf{q}_*(t), \mathbf{p}_*(t)\big)$ in the phase plane, called a *phase plane trajectory*. The collection of all phase plane trajectories, corresponding to various initial conditions, is called the *phase portrait*. The following picture shows the phase portrait for the harmonic oscillator.

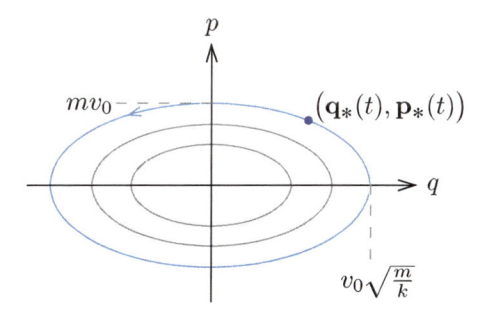

We also observe that the Hamiltonian H, evaluated along a phase plane trajectory $t \mapsto \big(\mathbf{q}_*(t), \mathbf{p}_*(t)\big)$, is

$$H\big(\mathbf{q}_*(t), \mathbf{p}_*(t)\big) = \frac{\big(\mathbf{p}_*(t)\big)^2}{2m} + V\big(\mathbf{q}_*(t)\big) = E(t),$$

the energy, which by the Law of Conservation of Energy, is a constant. So the phase plane trajectories are contained in level sets of the Hamiltonian H. Another proof of this constancy of the function H along phase plane trajectories, based on the Hamiltonian equations, is given below (where we have suppressed writing the argument t):

$$\frac{dH}{dt}(\mathbf{q}_*, \mathbf{p}_*) = \frac{\partial H}{\partial q}(\mathbf{q}_*, \mathbf{p}_*) \cdot \mathbf{q}'_* + \frac{\partial H}{\partial p}(\mathbf{q}_*, \mathbf{p}_*) \cdot \mathbf{p}'_*$$

$$= \frac{\partial H}{\partial q}(\mathbf{q}_*, \mathbf{p}_*) \cdot \frac{\partial H}{\partial p}(\mathbf{q}_*, \mathbf{p}_*) - \frac{\partial H}{\partial p}(\mathbf{q}_*, \mathbf{p}_*) \cdot \frac{\partial H}{\partial q}(\mathbf{q}_*, \mathbf{p}_*) = 0.$$

This sort of a calculation can be used to calculate the time evolution of any "observable" $(q,p) \mapsto F(q,p)$ along phase plane trajectories in the phase plane, as explained in the next paragraph.

Poissonian Mechanics.

All the (mechanical) physical characteristics are functions of the state. For example in our one-dimensional motion of the particle, the coordinate functions $(q,p) \mapsto q$ and $(q,p) \mapsto p$ give, for a state (q,p) of the particle, the position, respectively the momentum, of the particle. Similarly,
$$H(q,p) := \frac{p^2}{2m} + V(q)$$
gives the energy. Motivated by these considerations, we take
$$C^\infty(\mathbb{R}^2) = \{(q,p) \xrightarrow{F} F(q,p) : F \text{ is infinitely many times differentiable}\}$$
as the collection of all *observables*.

We now introduce a binary operation $\{\cdot,\cdot\} : C^\infty(\mathbb{R}^2) \times C^\infty(\mathbb{R}^2) \to C^\infty(\mathbb{R}^2)$, which is connected with the evolution of the mechanical system.

Given two observables F and G in $C^\infty(\mathbb{R}^2)$, define the new observable $\{F,G\} \in C^\infty(\mathbb{R}^2)$, called the *Poisson bracket of F,G*, by
$$\{F,G\}(q,p) = \frac{\partial F}{\partial q}(q,p) \cdot \frac{\partial G}{\partial p}(q,p) - \frac{\partial G}{\partial q}(q,p) \cdot \frac{\partial F}{\partial p}(q,p).$$
The Poisson bracket can be used to express the evolution of an observable F. Suppose that our particle moving along a line, evolves along the phase plane trajectory $(\mathbf{q}_*, \mathbf{p}_*)$ in the phase plane according to Hamilton's equations for a Hamiltonian H. Then the evolution of the observable $F \in C^\infty(\mathbb{R}^2)$ along the trajectory $(\mathbf{q}_*, \mathbf{p}_*)$ is given by (again suppressing t):
$$\frac{dF}{dt}(\mathbf{q}_*, \mathbf{p}_*) = \frac{\partial F}{\partial q}(\mathbf{q}_*, \mathbf{p}_*) \cdot \mathbf{q}'_* + \frac{\partial F}{\partial p}(\mathbf{q}_*, \mathbf{p}_*) \cdot \mathbf{p}'_*$$
$$= \frac{\partial F}{\partial q}(\mathbf{q}_*, \mathbf{p}_*) \cdot \frac{\partial H}{\partial p}(\mathbf{q}_*, \mathbf{p}_*) - \frac{\partial F}{\partial p}(\mathbf{q}_*, \mathbf{p}_*) \cdot \frac{\partial H}{\partial q}(\mathbf{q}_*, \mathbf{p}_*)$$
$$= \{F, H\}(\mathbf{q}_*, \mathbf{p}_*).$$

In particular, if $\{F, H\} = 0$ (as for example is the case when $F = H$!), then F is a conserved quantity.

It can be shown that $C^\infty(\mathbb{R}^2)$ forms a *Lie algebra* with the Poisson bracket, that is, the following properties hold:
$$\{F, G\} = -\{G, F\}, \tag{3.7}$$
$$\{\alpha F + \beta G, H\} = \alpha\{F, H\} + \beta\{G, H\}, \tag{3.8}$$
$$\{F, \{G, H\}\} + \{G, \{H, F\}\} + \{H, \{F, G\}\} = 0 \quad \text{(Jacobi Identity)} \tag{3.9}$$

for $\alpha, \beta \in \mathbb{R}$ and any $F, G, H \in C^\infty(\mathbb{R}^2)$. ($H$ may not be the Hamiltonian!)

We will see in the next chapter, that the role of the *Poisson bracket* in *classical mechanics*,

$$\{F, G\}$$

of observables $F, G \in C^\infty(\mathbb{R}^2)$, is performed by the *commutator*

$$\frac{1}{i\hbar} \cdot [A, B]$$

of observables A, B (which are operators on a Hilbert space H) in *quantum mechanics*.

Exercise 3.18. Prove (3.7)-(3.9).

Exercise 3.19. (Position and Momentum).
Let $Q \in C^\infty(\mathbb{R}^2)$ be the position observable, $(p, q) \overset{Q}{\mapsto} q : \mathbb{R}^2 \to \mathbb{R}$, and $P \in C^\infty(\mathbb{R}^2)$ be the momentum observable $(p, q) \overset{P}{\mapsto} p : \mathbb{R}^2 \to \mathbb{R}$. Show that $\{Q, P\} = 1$.

Chapter 4

Geometry of inner product spaces

In a vector space we can add vectors and multiply vectors by scalars. In a normed space, the vector space is also equipped with a norm, so that we can measure the distance between vectors. The plane $\mathbb{R}^2$ or the space $\mathbb{R}^3$ are examples of normed spaces. However, in the familiar geometry of the plane or of space, we can also measure the *angle* between lines, using the "dot product" of two vectors. Then $\cos\theta = (\vec{x} \cdot \vec{y})/(\|\vec{x}\|\|\vec{y}\|)$. In particular, nonzero vectors $\vec{x}$ and $\vec{y}$ are perpendicular if and only if $\vec{x} \cdot \vec{y} = 0$.

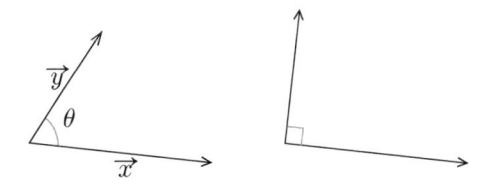

We wish to generalise this notion to abstract spaces, so that we can talk about perpendicularity or orthogonality of vectors even in vector spaces of sequences, functions, etc., and this is what we will do in this chapter, by equipping vector spaces with an "inner product" (analogous to the dot product from $\mathbb{R}^2$, $\mathbb{R}^3$).

Why do we care about orthogonality in more general spaces? There are several reasons for this. It gives a natural generalisation of the familiar $\mathbb{R}^2$, $\mathbb{R}^3$ case, and allows one to solve shortest distance and best approximation problems. Consider the figure on the left in the following picture. We are given a convex set $\mathcal{C}$ and a point $X \notin \mathcal{C}$. Then the closest point P_* in $\mathcal{C}$ to X is characterised by a geometric property involving angles: XP_* makes an *obtuse* angle with P_*Q, where Q is any point in $\mathcal{C}$. Similarly, let us now consider the figure on the right of the following picture, where we are given a plane $\mathcal{P}$ and a point $X \notin \mathcal{P}$. Then the closest point $P_* \in \mathcal{P}$ is the unique point such that XP_* is perpendicular to P_*Q, where Q is any point in $\mathcal{P}$. We will learn about the generalisation of these results in inner product spaces. These results in turn find applications in Fourier Analysis, and in more general "orthonormal basis" expansions. Finally, inner products will also simplify duality theory, and we will learn this later on, when we discuss the Riesz Representation Theorem in Hilbert spaces.

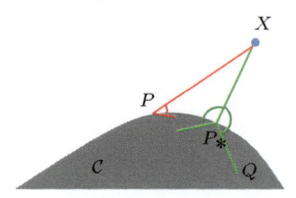

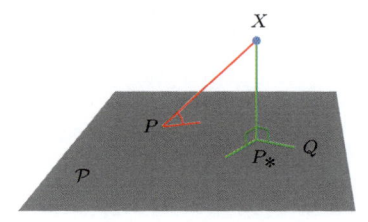

4.1 Inner product spaces

Definition 4.1. (Inner product space; inner product).
Let X be a vector space X over $\mathbb{K}$ ($= \mathbb{R}$ or $\mathbb{C}$). A function $\langle \cdot, \cdot \rangle : X \times X \to \mathbb{K}$ is called an *inner product* on the vector space X over $\mathbb{K}$ if:

(IP1) (Positive definiteness)
For all $x \in X$, $\langle x, x \rangle \geq 0$. If $x \in X$ and $\langle x, x \rangle = 0$, then $x = \mathbf{0}$.

(IP2) (Linearity)
For all $x_1, x_2, y \in X$, $\langle x_1 + x_2, y \rangle = \langle x_1, y \rangle + \langle x_2, y \rangle$.
For all $x, y \in X$ and all $\alpha \in \mathbb{K}$, $\langle \alpha x, y \rangle = \alpha \langle x, y \rangle$.

(IP3) (Conjugate symmetry)
For all $x, y \in X$, $\langle x, y \rangle = \langle y, x \rangle^*$. (Here $\cdot^*$ = complex conjugation).

A vector space X, equipped with an inner product $\langle \cdot, \cdot \rangle : X \times X \to \mathbb{K}$, is called an *inner product space*.

It follows, from (IP2) and (IP3), that the inner product is also *antilinear* with respect to the *second* variable, that is, it is additive, and such that $\langle x, \alpha y \rangle = \alpha^* \langle x, y \rangle$ for all $x, y \in X$ and $\alpha \in \mathbb{K}$. Moreover, for a *real* vector space X, note that (IP3) asserts that for all $x, y \in X$, $\langle x, y \rangle = \langle y, x \rangle$.

We will now see that every inner product space is automatically a normed space, with the norm given by

$$\|x\| := \sqrt{\langle x, x \rangle}, \quad x \in X.$$

We emphasise again that by (IP1), we know that $\langle x, x \rangle \geq 0$ for all $x \in X$, and so, the unique nonnegative real square root $\|x\|$ of $\langle x, x \rangle$ exists. We will call this $\|\cdot\|$, given by $\|x\| = \sqrt{\langle x, x\rangle}$ for $x \in X$, the *norm induced from the inner product on X*. Let us check that $\|\cdot\|$ satisfies (N1), (N2), (N3).

(N1) For all $x \in X$, $\|x\| = \sqrt{\langle x, x \rangle} \geq 0$.
If $x \in X$ and $\|x\| = 0$, then $\langle x, x \rangle = 0$, and by (IP1), $x = \mathbf{0}$.

(N2) For all $x \in X$ and $\alpha \in \mathbb{K}$, we have
$$\|\alpha x\| = \sqrt{\langle \alpha x, \alpha x \rangle} = \sqrt{\alpha \langle x, \alpha x \rangle} = \sqrt{\alpha \alpha^* \langle x, x \rangle} = \sqrt{|\alpha|^2 \|x\|^2} = |\alpha| \|x\|.$$

To show (N3), we will first show the following result, which is an important inequality in its own right.

Theorem 4.1. (Cauchy-Schwarz In/equality).
*If X is an inner product space, then for all $x, y \in X$, $|\langle x, y \rangle| \leq \|x\| \|y\|$.
For $x, y \in X$, $|\langle x, y \rangle| = \|x\| \|y\|$ if and only if x, y are linearly dependent.*

Cauchy-Schwarz

Proof. Let $x \in X$.
If $y = \mathbf{0}$, then $|\langle x, \mathbf{0} \rangle| = |\langle x, 0 \cdot \mathbf{0} \rangle| = |0 \langle x, \mathbf{0} \rangle| = 0 = \|x\| 0 = \|x\| \|\mathbf{0}\|$.
Now consider $y \in X \setminus \{\mathbf{0}\}$ and let $\alpha \in \mathbb{K}$. Then we have
$$\begin{aligned} 0 &\leq \langle x + \alpha y, x + \alpha y \rangle = \langle x, x + \alpha y \rangle + \langle \alpha y, x + \alpha y \rangle \\ &= \langle x, x \rangle + \alpha^* \langle x, y \rangle + \alpha \langle y, x \rangle + |\alpha|^2 \langle y, y \rangle \\ &= \|x\|^2 + 2\mathrm{Re}(\alpha^* \langle x, y \rangle) + |\alpha|^2 \|y\|^2. \end{aligned}$$
Write $\langle x, y \rangle = |\langle x, y \rangle| e^{i\theta}$ for some $\theta \in \mathbb{R}$. Take $\alpha = re^{i\theta}$, where $r \in \mathbb{R}$. Then $|\alpha| = r$, $\alpha^* = re^{-i\theta}$, and
$$\alpha^* \langle x, y \rangle = re^{-i\theta} |\langle x, y \rangle| e^{i\theta} = r |\langle x, y \rangle|.$$
Thus we obtain (from $0 \leq \|x\|^2 + 2\mathrm{Re}(\alpha^* \langle x, y \rangle) + |\alpha|^2 \|y\|^2$) that for all $r \in \mathbb{R}$, $r^2 \|y\|^2 + 2|\langle x, y \rangle| r + \|x\|^2 \geq 0$. Taking
$$r = -\frac{|\langle x, y \rangle|}{\|y\|^2},$$
we obtain $-\dfrac{|\langle x, y \rangle|^2}{\|y\|^2} + \|x\|^2 \geq 0$, that is, $|\langle x, y \rangle| \leq \|x\| \|y\|$.

Next, we'll show that the **equality** holds if and only if x, y are dependent.
(If part).
If $y = \mathbf{0}$, then both sides, $|\langle x, y \rangle|$ and $\|x\|\|y\|$, are equal to 0.
If $y \neq \mathbf{0}$, then since x, y are dependent, there exist $c_1, c_2 \in \mathbb{K}$ with $c_1 \neq 0$, such that $c_1 x + c_2 y = \mathbf{0}$, and so $x = \lambda y$ for some $\lambda \in \mathbb{K}$. Hence
$$|\langle x, y \rangle| = |\langle \lambda y, y \rangle| = |\lambda| \|y\|^2 = |\lambda| \|y\| \|y\| = \|\lambda y\| \|y\| = \|x\| \|y\|.$$
(Only if part).
If $y = \mathbf{0}$, then $0 \cdot x + 1 \cdot y = \mathbf{0}$, and so x, y are linearly dependent.
If $y \neq \mathbf{0}$, then define $\alpha = -\langle x, y \rangle / \|y\|^2$. We have
$$\langle x + \alpha y, x + \alpha y \rangle = \|x\|^2 + 2\mathrm{Re}(\alpha^* \langle x, y \rangle) + |\alpha|^2 \|y\|^2$$
$$= \|x\|^2 + 2\mathrm{Re}\left(-\frac{\langle x, y \rangle^*}{\|y\|^2} \langle x, y \rangle\right) + \frac{|\langle x, y \rangle|^2}{\|y\|^4} \|y\|^2$$
$$= \|x\|^2 - \frac{|\langle x, y \rangle|^2}{\|y\|^2} = 0,$$
and so $1 \cdot x + \alpha \cdot y = \mathbf{0}$, that is, x, y are linearly dependent. $\square$

Now that we've obtained this important inequality, let us complete our verification that every inner product space is a normed space with the induced norm $\|x\| := \sqrt{\langle x, x \rangle}$, $x \in X$. We had already checked (N1) and (N2), and it only remains to verify (N3).
(N3) For $x, y \subset X$, we have
$$\|x + y\|^2 = \langle x + y, x + y \rangle = \|x\|^2 + 2\mathrm{Re}\langle x, y \rangle + \|y\|^2$$
$$\leq \|x\|^2 + 2|\langle x, y \rangle| + \|y\|^2$$
$$\leq \|x\|^2 + 2\|x\|\|y\| + \|y\|^2 \quad \text{(Cauchy-Schwarz)}$$
$$= (\|x\| + \|y\|)^2,$$
and so $\|x + y\| \leq \|x\| + \|y\|$.

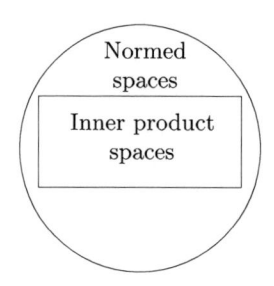

Example 4.1. Let $\mathbb{K} = \mathbb{R}$ or $\mathbb{C}$. Then $\mathbb{K}^d$ is an inner product space with the inner product $\langle \mathbf{x}, \mathbf{y} \rangle := x_1 y_1^* + \cdots + x_d y_d^*$, for $\mathbf{x} = (x_1, \cdots, x_d) \in \mathbb{K}^d$ and $\mathbf{y} = (y_1, \cdots, y_d) \in \mathbb{K}^d$. Note that the induced norm is

$$\|\mathbf{x}\| = \sqrt{\langle \mathbf{x}, \mathbf{x} \rangle} = \sqrt{x_1 x_1^* + \cdots + x_d x_d^*} = \sqrt{|x_1|^2 + \cdots + |x_d|^2} = \|x\|_2,$$

the Euclidean 2-norm $\|\cdot\|_2$. The Cauchy-Schwarz Inequality in the inner product space $\mathbb{C}^d$ gives

$$|x_1 y_1^* + \cdots + x_d y_d^*|^2 \leq (|x_1|^2 + \cdots + |x_d|^2)(|y_1|^2 + \cdots + |y_d|^2)$$

for complex $x_1, \cdots, x_d, y_1, \cdots, y_d$. ◇

Example 4.2. The vector space ℓ^2 of square summable (real or complex) sequences is an inner product space with the inner product

$$\langle \mathbf{x}, \mathbf{y} \rangle = \sum_{n=1}^{\infty} x_n y_n^*,$$

for $\mathbf{x} = (x_n)_{n \in \mathbb{N}}, \mathbf{y} = (y_n)_{n \in \mathbb{N}} \in \ell^2$. The series converges since

$$|x_n y_n^*| = |x_n||y_n| \leq \frac{|x_n|^2 + |y_n|^2}{2}, \quad n \in \mathbb{N}.$$

We have $\|\mathbf{x}\|^2 = \langle \mathbf{x}, \mathbf{x} \rangle = \sum_{n=1}^{\infty} x_n x_n^* = \sum_{n=1}^{\infty} |x_n|^2 = \|x\|_2^2.$

Also, the Cauchy-Schwarz inequality gives

$$\left| \sum_{n=1}^{\infty} x_n y_n^* \right|^2 \leq \left(\sum_{n=1}^{\infty} |x_n|^2 \right)\left(\sum_{n=1}^{\infty} |y_n|^2 \right).$$

◇

Example 4.3. The space of continuous $\mathbb{K}$-valued functions on $[a, b]$ can be equipped with the inner product

$$\langle \mathbf{x}, \mathbf{y} \rangle = \int_a^b \mathbf{x}(t)(\mathbf{y}(t))^* dt,$$

for $\mathbf{x}, \mathbf{y} \in C[a, b]$. The induced norm matches the $\|\cdot\|_2$ norm:

$$\|\mathbf{x}\|^2 = \langle \mathbf{x}, \mathbf{x} \rangle = \int_a^b \mathbf{x}(t)(\mathbf{x}(t))^* dt = \int_a^b |\mathbf{x}(t)|^2 dt = \|\mathbf{x}\|_2^2.$$

The Cauchy-Schwarz inequality gives, for real-valued $\mathbf{x}, \mathbf{y} \in C[a, b]$ that

$$\left(\int_a^b \mathbf{x}(t)\mathbf{y}(t) dt \right)^2 \leq \left(\int_a^b (\mathbf{x}(t))^2 dt \right)\left(\int_a^b (\mathbf{y}(t))^2 dt \right).$$

But we had seen (page 36) that $C[a,b]$ is not a Banach space with the $\|\cdot\|_2$-norm. $L^2[a,b]$ remedies the noncompleteness of $(C[a,b], \|\cdot\|_2)$:

$$L^2[a,b] = \left\{ \mathbf{x} : [a,b] \to \mathbb{K} : \underset{\text{Lebesgue}}{\int_a^b} |\mathbf{x}(t)|^2 dt < \infty \right\},$$

is an inner product space with the inner product

$$\langle [\mathbf{x}], [\mathbf{y}] \rangle = \underset{\text{Lebesgue}}{\int_a^b} \mathbf{x}(t) (\mathbf{y}(t))^* dt,$$

for $[\mathbf{x}], [\mathbf{y}] \in L^2[a,b]$. The induced norm on $L^2[a,b]$ is

$$\|[\mathbf{x}]\| = \sqrt{\underset{\text{Lebesgue}}{\int_a^b} |\mathbf{x}(t)|^2 dt} = \|[\mathbf{x}]\|_2, \quad [\mathbf{x}] \in L^2[a,b],$$

and $L^2[a,b]$ is complete with this norm. $\diamond$

In light of the previous example, it makes sense to introduce the following.

Definition 4.2. (Hilbert space). *An inner product space which is a Banach space with the induced norm is called a* Hilbert space.

Thus $\mathbb{R}^d$, $\mathbb{C}^d$, ℓ^2, $L^2[a,b]$ are all Hilbert spaces (with the aforementioned respective inner products), while $(C[a,b], \|\cdot\|_2)$ isn't. We can ask: What about $(C[a,b], \|\cdot\|_\infty)$? We'd seen it is a Banach space. Is it also a Hilbert space, that is, is the $\|\cdot\|_\infty$-norm induced by *some* inner product? The answer is "no"! And the reason is that an inner product endows the vector space with a certain geometry, and forces the distance function ($\equiv$ norm) to behave in a certain "rigid" manner. To see what we mean, let us show the following result.

Theorem 4.2. (Parallelogram Law).
If X is an inner product space with induced norm $\|\cdot\|$,
then for all $x, y \in X$, $\|x+y\|^2 + \|x-y\|^2 = 2\|x\|^2 + 2\|y\|^2$.

The name is justified, since the identity is a generalisation of the situation in Euclidean $\mathbb{R}^2$, where we know that the sum of the squares of the lengths of the diagonals of a parallelogram equals the sum of the squares of the lengths of the sides of the parallelogram.

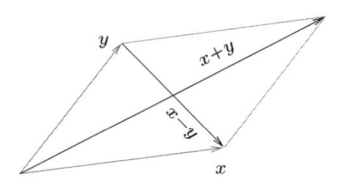

Proof. We have

$$\|x+y\|^2 = \langle x+y, x+y\rangle = \|x\|^2 + \|y\|^2 + 2\operatorname{Re}\langle x,y\rangle, \text{ and} \quad (4.1)$$
$$\|x-y\|^2 = \langle x-y, x-y\rangle = \|x\|^2 + \|y\|^2 - 2\operatorname{Re}\langle x,y\rangle. \quad (4.2)$$

Adding these, we obtain $\|x+y\|^2 + \|x-y\|^2 = 2\|x\|^2 + \|y\|^2$, as wanted. □

Remark 4.1. (The inner product is recovered from the induced norm via the Polarisation Identity). We have, upon subtracting (4.1) and (4.2), that $\operatorname{Re}(\langle x,y\rangle) = \frac{1}{4}\left(\|x+y\|^2 - \|x-y\|^2\right)$. In the case of real scalars:

$$\langle x,y\rangle = \frac{\|x+y\|^2 - \|x-y\|^2}{4}.$$

In the complex case, $\operatorname{Im}(\langle x,y\rangle) = \operatorname{Re}(-i\langle x,y\rangle) = \operatorname{Re}(\langle x,iy\rangle)$, so that

$$\langle x,y\rangle = \frac{\|x+y\|^2 - \|x-y\|^2 + i\|x+iy\|^2 - i\|x-iy\|^2}{4} = \frac{1}{4}\sum_{k=1}^{4} i^k \|x+i^k y\|^2.$$

This is called the *Polarisation Formula*. The inner product is thus uniquely determined by the induced norm.

It can also be shown that if X is a vector space with a norm $\|\cdot\|$ that satisfies the Parallelogram Law, then the norm is induced by the inner product $\langle\cdot,\cdot\rangle$ given by the Polarisation Identity! The interested reader is referred to [Day (1973), page 153].

Remark 4.2. (Pythagoras's Theorem).
If x and y are orthogonal, that is, $\langle x,y\rangle = 0$, then, using (4.1),

$$\|x+y\|^2 = \|x\|^2 + \|y\|^2.$$

Example 4.4. (ℓ^p is an inner product space $\Leftrightarrow p=2$).
With $\mathbf{x} = \mathbf{e}_1 := (1,0,\cdots)$ and $\mathbf{y} = \mathbf{e}_2 := (0,1,0,\cdots)$ in ℓ^p, we have

$$\|\mathbf{x}+\mathbf{y}\|_p^2 + \|\mathbf{x}-\mathbf{y}\|_p^2 = 2^{2/p} + 2^{2/p} = 2^{1+\frac{2}{p}},$$
$$2\|\mathbf{x}\|_p^2 + 2\|\mathbf{y}\|_p^2 = 2\cdot 1 + 2\cdot 1 = 2^2.$$

So we have $\|\mathbf{x}+\mathbf{y}\|_p^2 + \|\mathbf{x}-\mathbf{y}\|_p^2 = 2\|\mathbf{x}\|_p^2 + 2\|\mathbf{y}\|_p^2$ if and only if

$$1 + \frac{2}{p} = 2,$$

that is, if and only if $p=2$. So if $p \neq 2$, the Parallelogram Law fails for the $\|\cdot\|_p$ norm on ℓ^p, showing that this $\|\cdot\|_p$ norm is not induced from any inner product on ℓ^p. We had already seen that the $\|\cdot\|_2$ norm on ℓ^2 is induced from an inner product on ℓ^2. ◇

Exercise 4.1. Show that the supremum norm $\|\cdot\|_\infty$ on $C[0,1]$ is not induced by any inner product on $C[0,1]$.

Exercise 4.2. (Appollonius Identity). Let X be an inner product space. Prove that for all $x, y, z \in X$, $\|z-x\|^2 + \|z-y\|^2 = \dfrac{1}{2}\|x-y\|^2 + 2\left\|z - \dfrac{1}{2}(x+y)\right\|^2$.
This is called the *Appollonius Identity*.
Give a geometric interpretation when $X = \mathbb{R}^2$.

The hierarchy of spaces, with a few key examples, is depicted below.

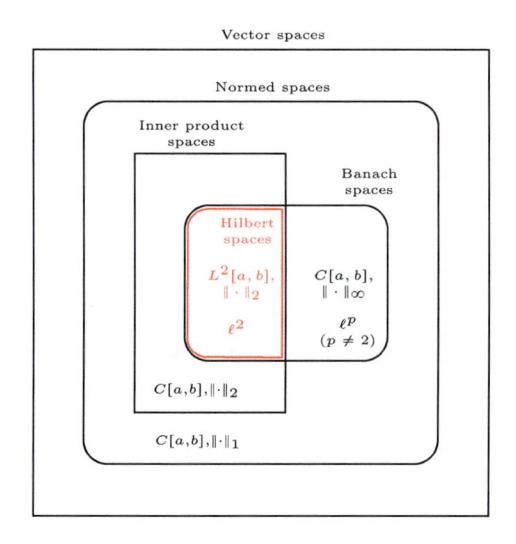

[Diagram: Nested boxes showing Vector spaces ⊃ Normed spaces ⊃ {Inner product spaces, Banach spaces}, with Hilbert spaces = Inner product ∩ Banach. Examples: $L^2[a,b], \|\cdot\|_2$ and ℓ^2 in Hilbert spaces; $C[a,b], \|\cdot\|_\infty$ and ℓ^p ($p \neq 2$) in Banach; $C[a,b], \|\cdot\|_2$ in inner product spaces; $C[a,b], \|\cdot\|_1$ in normed spaces.]

Remark 4.3. Although we've shown a strict containment of normed spaces within vector spaces, we remark that every vector space V can be made into a normed space with a norm devised from a Hamel basis B for V, as follows: for each $x \in V$, there exists an $n \in \mathbb{N}$, scalars $c_1, \cdots, c_n \in \mathbb{K}$, and vectors $b_1, \cdots, b_n \in \mathbb{B}$ such that $x = c_1 b_1 + \cdots + c_n b_n$. Set $\|x\| = |c_1| + \cdots + |c_n|$. Then it can be checked that $\|\cdot\|$ is a well-defined norm on V. But this may not be the "appropriate/natural" norm in the vector space.

Exercise 4.3. (Continuity of $\langle \cdot, \cdot \rangle$).
Let X be an inner product space and suppose that $(x_n)_{n \in \mathbb{N}}$, $(y_n)_{n \in \mathbb{N}}$ are sequences in X which converge to $x, y \in X$, respectively. Show that $(\langle x_n, y_n \rangle)_{n \in \mathbb{N}}$ is convergent in $\mathbb{K}$ with limit $\langle x, y \rangle$.

Exercise 4.4. Prove, using the Cauchy-Schwarz Inequality, that given an ellipse and a circle having equal areas, the perimeter of the ellipse is larger.
Hint: Consider a 90° rotation of the ellipse.

Exercise 4.5. If $A, B \in \mathbb{R}^{m \times n}$, then define $\langle A, B \rangle = \text{tr}(A^\top B)$, where $A^\top$ denotes the transpose of the matrix A. Prove that $\langle \cdot, \cdot \rangle$ defines an inner product on the space of $m \times n$ real matrices. The norm induced by this inner product on $\mathbb{R}^{m \times n}$ is called the *Hilbert-Schmidt norm*. Is $(\mathbb{R}^{m \times n}, \langle \cdot, \cdot \rangle)$ a Hilbert space?

Exercise 4.6.
(1) Let X be an inner product space over $\mathbb{C}$, and let $T \in CL(X)$ be such that for all $x \in X$, $\langle Tx, x \rangle = 0$. Prove that $T = \mathbf{0}$. *Hint:* Consider $\langle T(x+y), x+y \rangle$, and also $\langle T(x+iy), x+iy \rangle$. Finally take $y = Tx$.
(2) On the other hand, consider the continuous linear transformation $T \in CL(\mathbb{R}^2)$ corresponding to an anticlockwise rotation by $90°$: $T = \begin{bmatrix} 0 & -1 \\ 1 & 0 \end{bmatrix}$.
Check that $\langle Tx, x \rangle = 0$ for all $x \in \mathbb{R}^2$, but clearly $T \neq \mathbf{0}$.

Have we got a contradiction to the previous part?

Exercise 4.7. (∗)(Completion of inner product spaces).
If an inner product space $(X, \langle \cdot, \cdot \rangle_X)$ is not complete, then this means that there are some "holes" in it, as there are Cauchy sequences that are not convergent– roughly speaking, the "limits that they are supposed to converge to", do not belong to the space X. One can remedy this situation by filling in these holes, thereby enlarging the space to a larger inner product space $(\overline{X}, \langle \cdot, \cdot \rangle_{\overline{X}})$ in such a manner that:
(C1) X can be identified with a subspace of $\overline{X}$ and $\forall x, y \in X$, $\langle x, y \rangle_X = \langle x, y \rangle_{\overline{X}}$.
(C2) $\overline{X}$ is complete.
Given an inner product space $(X, \langle \cdot, \cdot \rangle_X)$, we now give a construction of an inner product space $(\overline{X}, \langle \cdot, \cdot \rangle_{\overline{X}})$, called the *completion* of X, that has the properties (C1) and (C2).
Let $\mathcal{C}$ be the set of all Cauchy sequences in X.
If $(x_n)_{n \in \mathbb{N}}$, $(y_n)_{n \in \mathbb{N}}$ are in $\mathcal{C}$, then define the relation[1] R on $\mathcal{C}$ as follows:
$$([(x_n)_{n \in \mathbb{N}}], [(y_n)_{n \in \mathbb{N}}]) \in R \Leftrightarrow \lim_{n \to \infty} \|x_n - y_n\|_X = 0.$$
Prove that R is an equivalence relation on $\mathcal{C}$.
Let $\overline{X}$ be the set of equivalence classes of $\mathcal{C}$ under the equivalence relation R. Suppose that the equivalence class of $(x_n)_{n \in \mathbb{N}}$ is denoted by $[(x_n)_{n \in \mathbb{N}}]$.

---

[1] Recall that a *relation* on a set S is simply a subset of the Cartesian product $S \times S$. A relation R on a set S is called an *equivalence relation* if it is reflexive (for all $x \in S$, $(x, x) \in R$), symmetric (if $(x, y) \in R$, then $(y, x) \in R$), and transitive (if $(x, y), (y, z) \in R$, then $(x, z) \in R$). If $x \in S$, then the *equivalence class of* x, denoted by $[x]$, is defined to be the set $\{y \in S : (x, y) \in R\}$. It is easy to see that $[x] = [y]$ if and only if $(x, y) \in R$. Thus equivalence classes are either equal or disjoint. They partition the set S, that is the set can be written as a disjoint union of these equivalence classes.

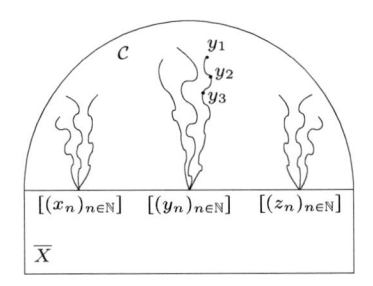

Define vector addition $+ : \overline{X} \times \overline{X} \to \overline{X}$ and scalar multiplication $\cdot : \mathbb{K} \times \overline{X} \to \overline{X}$ by $[(x_n)_{n \in \mathbb{N}}] + [(y_n)_{n \in \mathbb{N}}] = [(x_n + y_n)_{n \in \mathbb{N}}]$ and $\alpha \cdot [(x_n)_{n \in \mathbb{N}}] = [(\alpha x_n)_{n \in \mathbb{N}}]$.

Show that these operations are well-defined. (It can be verified that $\overline{X}$ is a vector space with these operations, but we will accept these straightforward verifications on faith.)

Define $\langle \cdot, \cdot \rangle_{\overline{X}} : \overline{X} \times \overline{X} \to \mathbb{K}$ by $\langle [(x_n)_{n \in \mathbb{N}}], [(y_n)_{n \in \mathbb{N}}] \rangle_{\overline{X}} = \lim\limits_{n \to \infty} \langle x_n, y_n \rangle_X$.

Prove that $\langle \cdot, \cdot \rangle_{\overline{X}}$ is well-defined, and is an inner-product on $\overline{X}$.

Define the map $\iota : X \to \overline{X}$ as follows: if $x \in X$, then $\iota(x) = [(x)_{n \in \mathbb{N}}]$, that is, ι takes x to the equivalence class of the (constant) Cauchy sequence $(x, x, x, \dots)$.

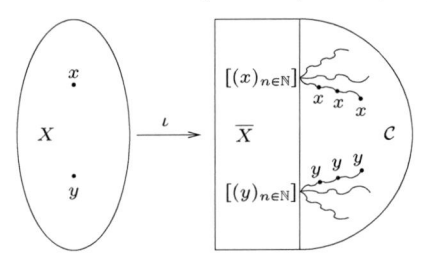

Show that ι is an injective bounded linear transformation (so that X can be identified with a subspace of $\overline{X}$), and that for all x, y in X, $\langle x, y \rangle_X = \langle \iota(x), \iota(y) \rangle_{\overline{X}}$.

We now show that $\overline{X}$ is a Hilbert space. Let $\left([(x_1^{(k)})_{n \in \mathbb{N}}] \right)_{k \in \mathbb{N}}$ be a Cauchy sequence in $\overline{X}$. For each $k \in \mathbb{N}$, let $n_k \in \mathbb{N}$ be such that for all $n, m \geq n_k$, $\|x_n^{(k)} - x_m^{(k)}\|_X < \frac{1}{k}$. Define the sequence $(y_k)_{k \in \mathbb{N}}$ by $y_k = x_{n_k}^{(k)}$, $k \in \mathbb{N}$. We claim that $(y_k)_{k \in \mathbb{N}} \in \mathcal{C}$.

Let $\epsilon > 0$. Choose $N \in \mathbb{N}$ such that $1/N < \epsilon$. Let $K_1 \in \mathbb{N}$ be such that for all $k, l > K_1$, $\|[(x_n^{(k)})_{n\in\mathbb{N}}] - [(x_n^{(l)})_{n\in\mathbb{N}}]\|_{\overline{X}} < \epsilon$, that is, $\lim_{n\to\infty} \|x_n^{(k)} - x_n^{(l)}\|_X < \epsilon$.

Define $K = \max\{N, K_1\}$, and let $k, l > K$. Then for all $n > \max\{n_k, n_l\}$,

$$\|y_k - y_l\|_X = \|y_k - x_n^{(k)} + x_n^{(k)} - x_n^{(l)} + x_n^{(l)} - y_l\|_X$$

$$\leqslant \|y_k - x_n^{k}\|_X + \|x_n^{(k)} - x_n^{(l)}\|_X + \|x_n^{(l)} - y_l\|_X$$

$$\leqslant \frac{1}{k} + \|x_n^{(k)} - x_n^{(l)}\|_X + \frac{1}{l} \leqslant \frac{1}{K} + \|x_n^{(k)} - x_n^{(l)}\|_X + \frac{1}{K}$$

$$< \epsilon + \|x_n^{(k)} - x_n^{(l)}\|_X + \epsilon.$$

So $\|y_k - y_l\|_X \leqslant \epsilon + \lim_{n\to\infty} \|x_n^{(k)} - x_n^{(l)}\|_X + \epsilon < \epsilon + \epsilon + \epsilon = 3\epsilon$.

This shows that $(y_n)_{n\in\mathbb{N}} \in \mathcal{C}$, and so $[(y_n)_{n\in\mathbb{N}}] \in \overline{X}$. We will prove that $([(x_n^{(k)})_{n\in\mathbb{N}}])_{k\in\mathbb{N}}$ converges to $[(y_n)_{n\in\mathbb{N}}] \in \overline{X}$ as $k \to \infty$. Given $\epsilon > 0$, let K_1 be such that $1/K_1 < \epsilon$. As $(y_k)_{k\in\mathbb{N}}$ is a Cauchy sequence, there exists a $K_2 \in \mathbb{N}$ such that for all $k, l > K_2$, $\|y_k - y_l\|_X < \epsilon$. Define $K = \max\{K_1, K_2\}$. Then for all $k > K$ and all $m > \max\{n_k, K\}$, we have

$$\|x_m^{(k)} - y_m\|_X \leqslant \|x_m^{(k)} - x_{n_k}^{(k)}\|_X + \|x_{n_k}^{(k)} - y_m\|_X$$

$$< \frac{1}{k} + \|y_k - y_m\|_X < \frac{1}{K} + \epsilon < \epsilon + \epsilon = 2\epsilon.$$

Hence $\|[(x_m^{(k)})_{m\in\mathbb{N}}] - [(y_m)_{m\in\mathbb{N}}]\|_{\overline{X}} = \lim_{m\to\infty} \|x_m^{(k)} - y_m\|_X \leqslant 2\epsilon$.

Remark 4.4. We had seen that the inner product space $C[a, b]$ with the induced norm $\|\cdot\|_2$ is *not* complete. However, it can be completed by the process discussed above. The completion is isomorphic to $L^2[a, b]$. The new inner product as constructed above is expressible as an integral, namely the *Lebesgue* integral. As we had mentioned earlier, for continuous functions on $[a, b]$, the Lebesgue integral is the same as the Riemann integral, that is, it gives the same value. The L^2 Hilbert spaces arise naturally in Probability Theory (see the remark made in Example 1.12 on page 14). The space of random variables X on a probability space $(\Omega, \mathcal{M}, \mu)$ for which $\mathbb{E}(X^2) < +\infty$, where $\mathbb{E}(\cdot)$ denotes expectation, is a Hilbert space with the inner product $\langle X, Y \rangle = \mathbb{E}(XY)$.

4.2 Orthogonality

Two vectors $\vec{x}$ and $\vec{y}$ in $\mathbb{R}^2$ are *perpendicular/orthogonal*[2] if their dot product $\vec{x} \cdot \vec{y}$ is 0. Since an inner product on a vector space is a generalisation of the notion of dot product, we can analogously talk about perpendicularity/orthogonality of vectors in the general setting of inner product spaces.

---
[2] From Greek, where "ortho" = straight/erect and "gonia" = angle.

Definition 4.3. (Orthogonal vectors; orthonormal set).
Let X be an inner product space.
(1) Vectors $x, y \in X$ are said to be *orthogonal* if $\langle x, y \rangle = 0$.
(2) A subset S of X is said to be an *orthonormal set* if
  (a) for all $x, y \in S$ with $x \neq y$, $\langle x, y \rangle = 0$, and
  (b) for all $x \in S$, $\langle x, x \rangle = 1$.

Example 4.5.

(1) In $\mathbb{R}^d$, $\left\{ \mathbf{e}_1 := \begin{bmatrix} 1 \\ 0 \\ \vdots \\ 0 \end{bmatrix}, \cdots, \mathbf{e}_d := \begin{bmatrix} 0 \\ \vdots \\ 0 \\ 1 \end{bmatrix} \right\}$ is an orthonormal set.

(2) In ℓ^2, $\{\mathbf{e}_i := (0, \cdots, 0, \underset{i^{\text{th}} \text{ term}}{1}, 0, \cdots) : i \in \mathbb{N}\}$ is an orthonormal set.

If δ_{ij} denotes the Kronecker delta, that is,
$$\delta_{ij} = \begin{cases} 1 & \text{if } i = j, \\ 0 & \text{if } i \neq j, \end{cases}$$
then we have $\langle \mathbf{e}_i, \mathbf{e}_j \rangle = \delta_{ij}$.

(3) Consider $C[0, 1]$ with the usual inner product.
For $n \in \mathbb{Z}$, and let $\mathbf{T}_n \in C[0, 1]$ be given by
$$\mathbf{T}_n(t) = e^{2\pi i n t}, \quad t \in [0, 1].$$
Then the set $\{\mathbf{T}_n : n \in \mathbb{Z}\}$ is an orthonormal set. Indeed, if $m, n \in \mathbb{Z}$ and $n \neq m$, then we have
$$\int_0^1 \mathbf{T}_n(t) (\mathbf{T}_m(t))^* dt = \int_0^1 e^{2\pi i n t} e^{-2\pi i m t} dt = \int_0^1 e^{2\pi i (n-m) t} dt$$
$$= \frac{e^{2\pi i (n-m) t}}{2\pi i (n-m)} \bigg|_0^1 = \frac{1 - 1}{2\pi i (n-m)} 0.$$
On the other hand, for all $n \in \mathbb{Z}$,
$$\langle \mathbf{T}_n, \mathbf{T}_n \rangle = \int_0^1 e^{2\pi i n t} e^{-2\pi i n t} dt = \int_0^1 1 dx = 1.$$
So $\{\cdots, e^{-4\pi i t}, e^{-2\pi i t}, e^0 = 1, e^{2\pi i t}, e^{4\pi i t}, \cdots\}$ is an orthonormal set in $C[0, 1]$. ◇

For vectors belonging to the span of orthonormal set, there is a special relation between the coefficients obtained in the relevant linear combinations, the norms and the inner products. Indeed if
$$x = \sum_{k=1}^n \alpha_k u_k$$

and the u_ks are orthonormal, then we have
$$\langle x, u_j \rangle = \Big\langle \sum_{k=1}^{n} \alpha_k u_k, u_j \Big\rangle = \sum_{k=1}^{n} \alpha_k \langle u_k, u_j \rangle = \alpha_j,$$
and
$$\|x\|^2 = \langle x, x \rangle = \Big\langle \sum_{k=1}^{n} \langle x, u_k \rangle u_k, x \Big\rangle = \sum_{k=1}^{n} \langle x, u_k \rangle \langle u_k, x \rangle = \sum_{k=1}^{n} |\langle x, u_k \rangle|^2.$$

Theorem 4.3. (Orthonormality$\Rightarrow$linear independence).
Let X be an inner product space, and $S \subset X$ be an orthonormal set. Then S is linearly independent.

Proof. Let $x_1, \cdots, x_n \in S$ and $\alpha_1, \cdots, \alpha_n \in \mathbb{K}$ be such that
$$\alpha_1 x_1 + \cdots + \alpha_n x_n = \mathbf{0}.$$
For $j \in \{1, \cdots, n\}$, we have
$$0 = \langle \mathbf{0}, x_j \rangle = \langle \alpha_1 x_1 + \cdots + \alpha_n x_n, x_j \rangle = \alpha_1 \langle x_1, x_j \rangle + \cdots + \alpha_n \langle x_n, x_j \rangle = \alpha_j.$$
Consequently, S is linearly independent. □

So every orthonormal set in X is linearly independent. Conversely, given a linearly independent set in X, we can construct an orthonormal set such that the span of this new constructed orthonormal set is the same as the span of the given independent set. We explain this below, and this algorithm is called the *Gram-Schmidt orthonormalisation process*.

Theorem 4.4. (Gram-Schmidt orthonormalisation).
Let $\{x_1, x_2, x_3, \cdots\}$ be a linearly independent subset of an inner product space X. Define
$$u_1 = \frac{x_1}{\|x_1\|} \text{ and}$$
$$u_n = \frac{x_n - \langle x_n, u_1 \rangle u_1 - \cdots - \langle x_n, u_{n-1} \rangle u_{n-1}}{\|x_n - \langle x_n, u_1 \rangle u_1 - \cdots - \langle x_n, u_{n-1} \rangle u_{n-1}\|} \text{ for } n \geq 2.$$
Then $\{u_1, u_2, u_3, \ldots\}$ is an orthonormal set in X and for all $n \in \mathbb{N}$,
$$\operatorname{span}\{x_1, \cdots, x_n\} = \operatorname{span}\{u_1, \cdots, u_n\}.$$

$S_{\text{input}} = \{x_1, x_2, x_3, \cdots\}$ independent set $\rightarrow$ [The Gram-Schmidt orthonormalisation procedure] $\rightarrow$ $S_{\text{output}} = \{u_1, u_2, u_3, \cdots\}$ orthonormal set such that span S_{input} = span S_{output}

Proof. We'll use induction to show that for all $n \in \mathbb{N}$,
$$\text{span}\{x_1, \cdots, x_n\} = \text{span}\{u_1, \cdots, u_n\}.$$

$n = 1$: $\{x_1\}$ being independent, we know that $x_1 \neq \mathbf{0}$. So $u_1 = x_1/\|x_1\|$ is well-defined, and $\|u_1\| = \|x_1\|/\|x_1\| = 1$. Also as $u_1 = x_1/\|x_1\|$ and $x_1 = \|x_1\|u_1$, it follows that $\text{span}\{x_1\} = \text{span}\{u_1\}$.

Suppose that the claim is true for some n, and suppose that $\{x_1, \cdots, x_{n+1}\}$ is a linearly independent set. Then in particular, $\{x_1, \cdots, x_n\}$ is linearly independent too. By the induction hypothesis $\{u_1, \cdots, u_n\}$ is an orthonormal set with $\text{span}\{x_1, \cdots, x_n\} = \text{span}\{u_1, \cdots, u_n\}$. Since $\{x_1, \cdots, x_{n+1}\}$ is a linearly independent set, $x_{n+1} \notin \text{span}\{x_1, \cdots, x_n\} = \text{span}\{u_1, \cdots, u_n\}$. Hence $v := x_{n+1} - (\langle x_{n+1}, u_1 \rangle u_1 + \cdots + \langle x_{n+1}, u_n \rangle u_n) \neq \mathbf{0}$. So the vector $u_{n+1} = v/\|v\|$ is well-defined. Trivially, $\|u_{n+1}\| = 1$. For all $k \leq n$ we have
$$\langle v, u_k \rangle = \langle x_{n+1}, u_k \rangle - \langle x_{n+1}, u_k \rangle \underbrace{\langle u_k, u_k \rangle}_{=1} = 0.$$

So $\langle u_{n+1}, u_k \rangle = 0$ for all $k \leq n$. Hence $\{u_1, \cdots, u_n, u_{n+1}\}$ is an orthonormal set.

Finally, as $u_{n+1} = \dfrac{1}{\|v\|}\Big(x_{n+1} - \underbrace{\boxed{}}_{\substack{\in \text{span}\{u_1,\cdots,u_n\} \\ = \text{span}\{x_1,\cdots,x_n\}}}\Big)$, we have

$$\text{span}\{u_1, \cdots, u_n, u_{n+1}\} = \text{span}\{x_1, \cdots, x_n, u_{n+1}\}$$
$$= \text{span}\{x_1, \cdots, x_n, x_{n+1}\}.$$

By mathematical induction, the proof is complete. $\square$

Exercise 4.8. Let X be an inner product space, and let $(x_n)_{n \in \mathbb{N}}$ be a sequence in X such that $\{x_n : n \in \mathbb{N}\}$ is a linearly independent set. If $(v_n)_{n \in \mathbb{N}}$ is a sequence such that $\{v_n : n \in \mathbb{N}\}$ is an orthonormal set and
$$\text{for all } n \in \mathbb{N}, \text{ span}\{x_1, \cdots, x_n\} = \text{span}\{v_1, \cdots, v_n\},$$
then show that for all $n \in \mathbb{N}$, there exists a scalar α_n with $|\alpha_n| = 1$ and $v_n = \alpha_n u_n$, where $u_1, u_2, u_3, \cdots$ is the sequence of vectors obtained by carrying out the Gram-Schmidt Orthogonalisation Procedure on the sequence $x_1, x_2, x_3, \cdots$.

Example 4.6. (Legendre's Polynomials).
The set of monomials $\{1, t, t^2, t^3, \cdots\}$ form a linearly independent system in $C[-1, 1]$. If $\mathcal{P}_n$ is the subspace of all polynomials of degree at most n, then $\mathcal{P}_n = \text{span}\{1, t, \cdots, t^n\}$. Let us apply the Gram-Schmidt Orthonormalisation Procedure to the set of monomials. This results in a sequence $(\mathbf{u}_n)_{n \in \mathbb{N}}$ of polynomials. Since $\mathbf{u}_n \in \mathcal{P}_n$ and $\mathbf{u}_n$ is orthogonal to every element of

$\mathcal{P}_{n-1}$, it follows that the degree of $\mathbf{u}_n$ is precisely n. Thus there exists a sequence of polynomials $(\mathbf{u}_n)_{n\geq 0}$ such that $\deg \mathbf{u}_n = n$, $n \geq 0$ and

$$\int_{-1}^{1} \mathbf{u}_n(t)\mathbf{u}_m(t)dt = 0 \text{ for } n \neq m.$$

By Exercise 4.8, it follows that *any* sequence $(\mathbf{P}_n)_{n\geq 0}$ with the properties that $\deg \mathbf{P}_n = n$, $n \geq 0$, and

$$\int_{-1}^{1} \mathbf{P}_n(t)\mathbf{P}_m(t)dt = 0 \text{ for } n \neq m$$

is the same as the sequence $(\mathbf{u}_n)_{n\geq 0}$ except for an overall multiplying numerical constant. We'll produce such a sequence $(\mathbf{P}_n)_{n\geq 0}$, called the sequence of *Legendre's Polynomials*. We shall see that $\mathbf{P}_n(1) \neq 0$. Classically, these polynomials $\mathbf{P}_n$s are not normalised by demanding that $\|\mathbf{P}_n\|_2 = 1$, but rather by arranging that $\mathbf{P}_n(1) = 1$. We define $\mathbf{P}_0(t) = 1$ and for $n \in \mathbb{N}$:

$$\mathbf{P}_n(t) = \frac{1}{n!2^n}\left(\frac{d}{dt}\right)^n (t^2-1)^n.$$

This is known as *Rodrigues's Formula*. Then the polynomials $\mathbf{P}_n$, with $n \geq 0$, have the following properties:

(1) $\mathbf{P}_n$ is a polynomial of degree n, and $\mathbf{P}_n(1) = 1$.

(2) The polynomials are orthogonal: $\int_{-1}^{1} \mathbf{P}_n(t)\mathbf{P}_m(t)dt = 0$ for $n \neq m$.

(3) The norm of $\mathbf{P}_n$ is given by $\int_{-1}^{1} \mathbf{P}_n(t)^2 dt = \frac{2}{2n+1}$.

Since $(t^2-1)^n$ is a real polynomial with the leading term t^{2n}, it is clear that $\mathbf{P}_n$ is a real polynomial of degree n.

Let us now calculate $\mathbf{P}_n(1)$. Consider in general the derivative of order $m \leq n$ of the function $(t^2-1)^n$. Thanks to Leibniz's Formula, we have

$$\left(\frac{d}{dt}\right)^m (t^2-1)^n = \left(\frac{d}{dt}\right)^m \left((t-1)^n(t+1)^n\right)$$

$$= \sum_{k=0}^{m} \binom{m}{k}\left(\left(\frac{d}{dt}\right)^k (t-1)^n\right)\left(\left(\frac{d}{dt}\right)^{m-k}(t+1)^n\right).$$

It is clear that if $m = n$, then all terms vanish for $t = 1$, except the one with $k = n$, which yields

$$\binom{n}{n}\left(\frac{d}{dt}\right)^n (t-1)^n \left(\frac{d}{dt}\right)^0 (t+1)^n \bigg|_{t=1} = n!2^n.$$

This gives $\mathbf{P}_n(1) = 1$ if $n \geq 1$.

For $1 \leqslant k \leqslant m < n$, $\left(\dfrac{d}{dt}\right)^k (t-1)^n$ has a factor $t-1$, and so it is 0 for $t=1$.
Also, $\left(\dfrac{d}{dt}\right)^{m-k} (t+1)^n$ is zero for $t=-1$, for $1 \leqslant k \leqslant m < n$.
Consequently,
$$\left(\dfrac{d}{dt}\right)^m (t^2-1)^n \bigg|_{t=\pm 1} = \left(\dfrac{d}{dt}\right)^m \left((t-1)^n (t+1)^n\right) = 0 \text{ for all } m = 0, \cdots, n-1.$$
Now we show that $\mathbf{P}_n$ is orthogonal to $\mathcal{P}_{n-1} = \operatorname{span}\{1, \cdots, t^{n-1}\}$.
After m integrations by parts, we get for $1 \leqslant m < n$:
$$\int_{-1}^{1} \left(\dfrac{d}{dt}\right)^n (t^2-1)^n t^m \, dt = -\int_{-1}^{1} \left(\dfrac{d}{dt}\right)^{n-1} (t^2-1)^n m t^{m-1} \, dt = \cdots$$
$$= (-1)^m m! \int_{-1}^{1} \left(\dfrac{d}{dt}\right)^{n-m} (t^2-1)^n \, dt$$
$$= (-1)^m m! \left(\dfrac{d}{dt}\right)^{n-m-1} (t^2-1)^n \bigg|_{t=-1}^{t=1} = 0.$$
For $m=0$, this follows from the Fundamental Theorem of Calculus.
In particular, it follows that if $m < n$, then $\displaystyle\int_{-1}^{1} \mathbf{P}_n(t) \mathbf{P}_m(t) \, dt = 0$.
So this also for $n \neq m$ by interchanging the roles of n and m.
To calculate the norm of $\mathbf{P}_n$, we use integration by parts to first note that
$$\int_{-1}^{1} \left(\dfrac{d}{dt}\right)^n (t^2-1)^n \left(\dfrac{d}{dt}\right)^n (t^2-1)^n \, dt$$
$$= -\int_{-1}^{1} \left(\dfrac{d}{dt}\right)^{n-1} (t^2-1)^n \left(\dfrac{d}{dt}\right)^{n+1} (t^2-1)^n \, dt = \cdots$$
$$= (-1)^n \int_{-1}^{1} (t^2-1)^n \left(\dfrac{d}{dt}\right)^{2n} (t^2-1)^n \, dt$$
$$= (-1)^n (2n)! \int_{-1}^{1} (t^2-1)^n \, dt = (-1)^n (2n)! \int_{-1}^{1} (t-1)^n (t+1)^n \, dt.$$
The last integral can again be calculated using integration by parts, where the boundary terms containing factors $(1-t)$ and $(t+1)$ become zero:
$$(-1)^n \int_{-1}^{1} (t-1)^n (t+1)^n \, dt$$
$$= \int_{-1}^{1} (1-t)^n (t+1)^n \, dt = \dfrac{n}{n+1} \int_{-1}^{1} (1-t)^{n-1} (t+1)^{n+1} \, dt = \cdots$$
$$= \dfrac{n(n-1)\cdots 1}{(n+1)(n+2)\cdots(2n)} \int_{-1}^{1} (t+1)^{2n} \, dt$$
$$= \dfrac{n(n-1)\cdots 1}{(n+1)(n+2)\cdots(2n)} \dfrac{2^{2n+1}}{2n+1} = \dfrac{(n!)^2}{(2n)!} \dfrac{2^{2n+1}}{2n+1}.$$

Thus we have: $\int_{-1}^{1} (\mathbf{P}_n(t))^2 dt = \frac{1}{(2^n n!)^2}(2n)!\frac{(n!)^2}{(2n)!}\frac{2^{2n+1}}{2n+1} = \frac{2}{2n+1}.$

Using Rodrigues's Formula, we obtain:

$$\mathbf{P}_0(t) = 1$$

$$\mathbf{P}_1(t) = t$$

$$\mathbf{P}_2(t) = \frac{3}{2}t^2 - \frac{1}{2}$$

$$\mathbf{P}_3(t) = \frac{5}{2}t^3 - \frac{3}{2}t$$

$$\mathbf{P}_4(t) = \frac{35}{8}t^4 - \frac{15}{4}t^2 + \frac{3}{8}.$$

We won't treat some less evident properties of Legendre's polynomials, such as the inequality $|\mathbf{P}_n(t)| \leq 1$ for all $t \in [-1,1]$. Legendre's polynomials also appear naturally as a solution to an ODE (see Exercise 4.10), for instance in Physics (when solving the Laplace equation and related PDEs in spherical coordinates—in particular, in Quantum Mechanics for orbital angular momentum calculations). ◇

Exercise 4.9. Show that $\mathbf{P}_n$ is an even function if n is even, and odd if n is odd. In particular: $\mathbf{P}_n(-1) = (-1)^n$.

Exercise 4.10. Show that the Legendre polynomial $\mathbf{P}_n$ solves
$$(1-t^2)\mathbf{x}''(t) - 2t\mathbf{x}'(t) + n(n+1)\mathbf{x}(t) = 0,$$
or equivalently,
$$\frac{d}{dt}\left((1-t^2)\mathbf{x}'\right) + n(n+1)\mathbf{x}(t) = 0.$$
Hint: Let $\mathbf{y}(t) = (t^2-1)^n$. Verify that $(t^2-1)\mathbf{y}'(t) = 2nt\mathbf{y}(t)$, and differentiate this equation $n+1$ times using Leibniz's Formula.

Exercise 4.11. Show that $\mathbf{P}_n$ has n real zeros, and that these belong to $(-1,1)$.
Hint: We have seen that $\left(\frac{d}{dt}\right)^k (t^2-1)^n$ is equal to zero at ± 1 for $0 \leq k \leq n-1$. By induction on k, using Rolle's Theorem show that $\left(\frac{d}{dt}\right)^k (t^2-1)^n$ has at least k zeros in $(-1,1)$.

Exercise 4.12. Recall Exercise 4.5, page 163, where we defined an inner product on $\mathbb{R}^{m \times n}$ by setting $\langle A, B \rangle := \mathrm{tr}(A^\top B)$, $A, B \in \mathbb{R}^{m \times n}$.
Find an orthonormal basis[3] for $(\mathbb{R}^{m \times n}, \langle \cdot, \cdot \rangle)$.

---
[3]In this finite dimensional context, by an *orthonormal basis*, we simply mean a (finite, Hamel) basis of orthonormal vectors.

Exercise 4.13. (Hermite functions and the quantum harmonic oscillator).
(1) For nonnegative integers n, set $\mathbf{H}_n(x) = (-1)^n e^{x^2}(d/dx)^n e^{-x^2}$, $x \in \mathbb{R}$.
These are called the *Hermite polynomials*.
Justify the name "polynomial" as follows. First, note that $\mathbf{H}_0 = 1$.
Next, establish the recursion relation $\mathbf{H}_{n+1}(x) = 2x\mathbf{H}_n(x) - \mathbf{H}_n'(x)$.
Conclude that each $\mathbf{H}_n$ is indeed a polynomial.
Show that the leading term of $\mathbf{H}_n$ is $2^n x^n$. In particular $\mathbf{H}_n$ has degree n.
Determine $\mathbf{H}_0, \mathbf{H}_1, \mathbf{H}_2, \mathbf{H}_3$.
(2) Show that $\varphi_n := e^{-x^2/2}\mathbf{H}_n$, $n \geqslant 0$, belong to $L^2(\mathbb{R})$ and are orthogonal.
What is the L^2-norm of $\mathbf{H}_n$? (φ_n are called *Hermite functions*.)
(3) Show that $(x - d/dx)\varphi_n = \varphi_{n+1}$ for all $n \geqslant 0$.
In particular, $\varphi_n = D^n \varphi_0$, where D is the differential operator $x - d/dx$.
(4) Show that $(x + d/dx)\varphi_n = 2n\varphi_{n-1}$ for all $n \geqslant 1$.
(5) Show that the φ_n satisfy $\big(-(d/dx)^2 + x^2\big)\varphi_n = (2n+1)\varphi_n$, $n \geqslant 0$.
Hint: Expand $(x + d/dx)(x - d/dx)$ using $(d/dx)x\varphi - x(d/dx)\varphi = \varphi$.
(6) Let $a > 0$ and let $\psi_n(x) = \varphi_n(\sqrt{a}x)$, $n \geqslant 0$.
Show that $\big(-(d/dx)^2 + a^2x^2\big)\psi_n = (2n+1)a\psi_n$, $n \geqslant 0$.
Quantum harmonic oscillator: Find values of E_n such that the ψ_n, $n \geqslant 0$, satisfy the Schrödinger equation with $\omega > 0$:
$$\left(-\frac{\hbar^2}{2m}\left(\frac{d}{dx}\right)^2 + \frac{1}{2}m\omega^2 x^2\right)\psi_n = E_n\psi_n.$$

Exercise 4.14. Let $\{u_n : n \in \mathbb{N}\} \subset H$ be an orthonormal set in a Hilbert space H.
Show that $\displaystyle\sum_{n=1}^{\infty} \frac{u_n}{n}$ converges, but not absolutely.

Exercise 4.15. (Bessel's Inequality).
Let $\{u_n : n \in \mathbb{N}\}$ be an orthonormal set in an inner product space X.
Show that for all $x \in X$, $\displaystyle\sum_{n=1}^{\infty} |\langle x, u_n\rangle|^2 \leqslant \|x\|^2$.

Orthogonal complement of a subspace

Definition 4.4. (Orthogonal complement).
Let Y be a subspace of an inner product space X.
The *orthogonal complement* $Y^\perp$ of Y is defined by
$$Y^\perp := \{x \in X : \langle x, y\rangle = 0 \text{ for all } y \in Y\}.$$

Proposition 4.1. *If Y is a subspace of an inner product space X, then $Y^\perp$ is a closed subspace of X.*

Proof. $Y^\perp$ is a subspace of X:
(S1) $\mathbf{0} \in Y^\perp$ since $\langle \mathbf{0}, y \rangle = 0$ for all $y \in Y$.
(S2) If $x_1, x_2 \in Y^\perp$, then $\langle x_1 + x_2, y \rangle = \langle x_1, y \rangle + \langle x_2, y \rangle = 0 + 0 = 0$ for all $y \in Y$, and so $x_1 + x_2 \in Y^\perp$.
(S3) If $\alpha \in \mathbb{K}$ and $x \in Y^\perp$, then $\langle \alpha \cdot x, y \rangle = \alpha \langle x, y \rangle = \alpha \cdot 0 = 0$ for all $y \in Y$, and so $\alpha \cdot x \in Y^\perp$.

Let $(x_n)_{n \in \mathbb{N}}$ be a sequence in $Y^\perp$, which converges in X to $x \in X$. Then for all $y \in Y$, $\langle x, y \rangle = \left\langle \lim_{n \to \infty} x_n, y \right\rangle = \lim_{n \to \infty} \langle x_n, y \rangle = \lim_{n \to \infty} 0 = 0$.
Hence $x \in Y^\perp$. Consequently $Y^\perp$ is closed. □

Example 4.7.
(1) Let $\{\mathbf{e}_1, \mathbf{e}_2, \mathbf{e}_3\}$ be the standard orthonormal basis in $\mathbb{R}^3$ with the usual Euclidean inner product. Then we have $(\operatorname{span}\{\mathbf{e}_1, \mathbf{e}_2\})^\perp = \operatorname{span}\{\mathbf{e}_3\}$.

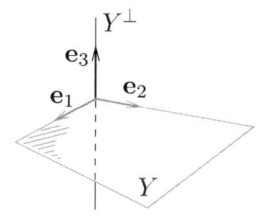

(2) If $\mathbf{e}_1 := (1, 0, \cdots)$ and $\mathbf{e}_2 = (0, 1, 0, \cdots)$ in ℓ^2, then
$$(\operatorname{span}\{\mathbf{e}_1, \mathbf{e}_2\})^\perp = \{(a_n)_{n \in \mathbb{N}} \in \ell^2 : a_1 = a_2 = 0\}. \qquad \diamond$$

Exercise 4.16. Let Y be a subspace of an inner product space X. Show that $Y \cap Y^\perp = \{0\}$.

Exercise 4.17. Let Y be a subspace of an inner product space X.
(1) Prove that $Y \subset (Y^\perp)^\perp$.
(2) Show that if $Y \subset Z$, where Z is another subspace of X, then $Z^\perp \subset Y^\perp$.
(3) Prove that $Y^\perp = \overline{Y}^\perp$, where $\overline{Y}$ denotes the closure of Y.
(4) Show that if Y is dense in X, then $Y^\perp = \{\mathbf{0}\}$.
(5) Let Y_{even} be the subspace of ℓ^2 of all sequences whose oddly indexed terms are zeros. Describe $Y_{\text{even}}^\perp$. Show that $Y_{\text{even}}^{\perp\perp} = Y_{\text{even}}$.
(6) What is $c_{00}^\perp$ in ℓ^2? Show that $c_{00}^{\perp\perp} \neq c_{00}$.
(Thus for a subspace Y, it can happen that $Y \neq Y^{\perp\perp}$.
We'll see later that for a *closed* subspace Y in a Hilbert space H, $Y = Y^{\perp\perp}$.
But for general subspaces Y, we can only say that $Y^{\perp\perp} = \overline{Y}$.)

4.3 Best approximation

Finite dimensional subspace

One reason that orthonormal sets are useful is that it enables us to compute the best approximation in a given finite dimensional subspace to a given vector in an inner product space X. Thus, the following optimisation problem can be solved:

Let $Y = \mathrm{span}\{u_1, \cdots, u_n\}$, where $u_1, \cdots, u_n \in X$ is an orthonormal basis for Y. Given $x \in X$, find $y_* \in Y$ that minimises $\|x - y\|$, subject to $y \in Y$.

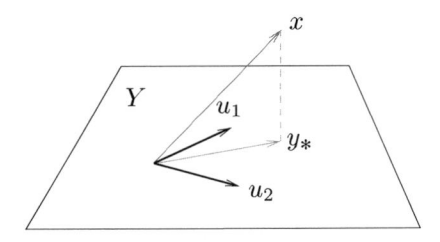

Theorem 4.5.
Let (1) *X be an inner product space,*
    (2) *$x \in X$, and*
    (3) *$Y = \mathrm{span}\{u_1, \cdots, u_n\} \subset X$,*
        where $\{u_1, \ldots, u_n\}$ is an orthonormal basis for Y.
Define $y_* = \sum_{k=1}^{n} \langle x, u_k \rangle u_k.$
Then $y_* \in Y$ is such that for all $y \in Y$, $\|x - y\| \geq \|x - y_*\|$.

Proof.
If $y \in Y$, then there exist scalars $c_1, \cdots, c_n$ such that $y = \sum_{k=1}^{n} c_k u_k$.

So for $j = 1, \cdots, n$, $\langle y, u_j \rangle = \Big\langle \sum_{k=1}^{n} c_k u_k, u_j \Big\rangle = \sum_{k=1}^{n} c_k \langle u_k, u_j \rangle = c_j$.

Then we have

$$\langle x - y_*, y \rangle = \Big\langle x - \sum_{k=1}^{n} \langle x, u_k \rangle u_k, y \Big\rangle = \langle x, y \rangle - \sum_{k=1}^{n} \langle x, u_k \rangle \langle u_k, y \rangle$$

$$= \langle x, y \rangle - \Big\langle x, \sum_{k=1}^{n} \langle y, u_k \rangle u_k \Big\rangle = \langle x, y \rangle - \langle x, y \rangle = 0.$$

In particular, for all $y \in Y$, $\langle x - y_*, y_* - y \rangle = 0$. So for all $y \in Y$,

$$\|x - y\|^2 = \|x - y_* + y_* - y\|^2 = \langle x - y_* + y_* - y, x - y_* + y_* - y \rangle$$
$$= \|x - y_*\|^2 + \|y_* - y\|^2 \geq \|x - y_*\|^2.$$

So for all $y \in Y$, $\|x - y\| \geq \|x - y_*\|$. $\square$

"Least square approximation problems" that arise in applications can be cast as best approximation problems in appropriate inner product spaces. Here is an example, and we'll see more in the exercises.

Example 4.8. Suppose that $\mathbf{f}$ is a continuous function on $[a,b]$ and we want to find a polynomial $\mathbf{p}_*$ of degree at most m such that the 'error'

$$E(\mathbf{p}) = \int_a^b |\mathbf{f}(t) - \mathbf{p}(t)|^2 dt \tag{4.3}$$

is minimised. Let $X = C[a,b]$ with the usual inner product, and let $\mathcal{P}_m$ be the subspace of X comprising all polynomials of degree at most m. Then Theorem 4.5 gives a method of finding such a polynomial $\mathbf{p}_*$. Let us take a concrete case. Let $a = -1$, $b = 1$, $\mathbf{f}(t) = e^t$ for $t \in [-1,1]$, and $m = 2$. As

$$\mathcal{P}_m = \text{span}\{1, t, t^2\},$$

by a Gram-Schmidt orthonormalisation process (Example 4.6, page 168),

$$\mathcal{P}_m = \text{span}\left\{\mathbf{P}_0 = \frac{1}{\sqrt{2}},\ \mathbf{P}_1 = \frac{\sqrt{3}}{\sqrt{2}}t,\ \mathbf{P}_2 = \frac{3\sqrt{10}}{4}\left(t^2 - \frac{1}{3}\right)\right\},$$

where $\mathbf{P}_0, \mathbf{P}_1, \mathbf{P}_2$ are orthogonal, and their span equals $\mathcal{P}_2$. We have

$$\langle \mathbf{f}, \mathbf{P}_0 \rangle = \frac{1}{\sqrt{2}}\left(e - \frac{1}{e}\right),\quad \langle \mathbf{f}, \mathbf{P}_1 \rangle = \frac{\sqrt{6}}{e},\quad \langle \mathbf{f}, \mathbf{P}_2 \rangle = \frac{\sqrt{5}}{\sqrt{2}}\left(e - \frac{7}{e}\right),$$

and so from Theorem 4.5, we obtain

$$\mathbf{p}_*(t) = \frac{1}{2}\left(e - \frac{1}{e}\right) + \frac{3}{e}t + \frac{15}{4}\left(e - \frac{7}{e}\right)\left(t^2 - \frac{1}{3}\right).$$

This polynomial $\mathbf{p}_*$ is a polynomial of degree at most 2 that minimises the L_2-norm error (4.3) on the interval $[-1,1]$ when $\mathbf{f}$ is the exponential function.

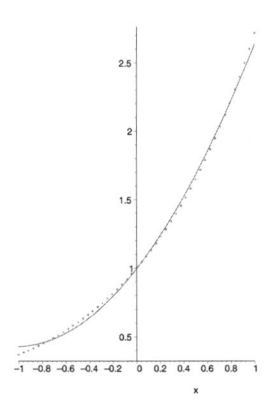

In the picture above, the dots indicate the graph of the exponential function, while the solid line is the graph of the polynomial $\mathbf{p}_*$. ◇

Exercise 4.18. (Least Squares Regression). Suppose we are interested in two variables x and y, which are related via a linear function $y = mx + b$. Suppose that m and b are unknown, but one has measurements $(x_1, y_1), \cdots, (x_n, y_n)$ from experiments. However there are errors, so that the measured values y_i are related to the measured values x_i by $y_i = mx_i + b + e_i$, where e_i is the (unknown) error in measurement i. What are the "best" values of m and b? This is a very common situation occurring in the applied sciences.

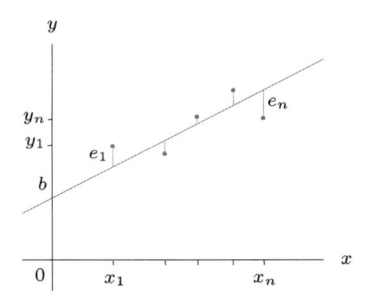

If the m and b we guessed were correct, then most of the errors $e_i := y_i - mx_i + b$ should be reasonably small. So to find the "best" m and b, we should find the m and b that make the e_i's collectively the smallest in some sense. So we introduce the error

$$E(m,b) = \sqrt{\sum_{i=1}^{n} e_i^2} = \sqrt{\sum_{i=1}^{n}(y_i - mx_i - b)^2}$$

and seek m, b such that $E(m,b)$ is minimised. Convert this problem into the setting of a best approximation problem to a vector in a subspace.

Month	Mean temperature (°C)	Inland energy consumption (million tonnes coal equivalent)
January	2.3	9.8
February	3.8	9.3
March	7.0	7.4
April	12.7	6.6
May	15.0	5.7
June	20.9	3.9
July	23.4	3.0
August	20.0	4.1
September	17.9	5.0
October	11.6	6.3
November	5.8	8.3
December	4.7	10.0

The table shows the data on energy consumption and mean temperature in the various months of the year. Draw a scatter chart, and fit a regression line of energy consumption on temperature. What is the intercept, and what does it mean in this case? What is the slope, and how could one interpret it? Use the regression line to forecast the energy consumption for a month with mean temperature 9°C.

Convex set case

What if the subspace Y in the previous section was not necessarily finite dimensional? We will consider this in the next subsection, but first let us consider the far more general case of a *closed convex* set, (not just in an inner product space, but rather) in a Hilbert space.

Recall that a subset C of a vector space X is said to be convex if for all $\alpha \in (0,1)$ and all $x, y \in C$, we have that $(1-\alpha)x + \alpha y \in C$. Now suppose that we are given: a Hilbert space H, a closed convex subset $C \subset H$, and a point $x \in H$. We consider the following optimisation problem:

$$\begin{cases} \text{minimise} & \|x - c\| \\ \text{subject to} & c \in C. \end{cases}$$

In Theorem 4.6 below, we give a characterisation of $c_* \in C$ that solves this optimisation problem.

Theorem 4.6. *Let H be a Hilbert space, C be a closed, convex, nonempty subset of H, and $x \in H$. Then:*

(1) *There exists a unique $c_* \in C$ such that for all $c \in C$, $\|x - c_*\| \leq \|x - c\|$.*

(2) *The point $c_* \in C$ can be characterised by the following property:*

$$\text{for all } c \in C, \ \operatorname{Re}\langle x - c_*, c - c_* \rangle \leq 0. \tag{4.4}$$

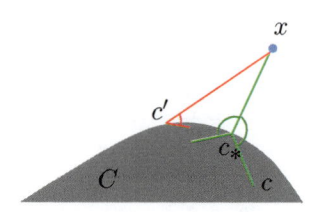

Remark 4.5. (Geometric interpretation of condition (4.4) in Theorem 4.6):
Cauchy-Schwarz gives, for all $x, y \in H\setminus\{0\}$: $-1 \leq \dfrac{\operatorname{Re}\langle x, y\rangle}{\|x\|\|y\|} \leq 1$.
The angle $\theta_{x,y} \in [0, \pi]$ between the nonzero vectors x and y is defined by

$$\theta_{x,y} := \cos^{-1}\frac{\operatorname{Re}\langle x, y\rangle}{\|x\|\|y\|} \in [0, \pi].$$

Thus the condition (4.4) in Theorem 4.6 has the geometric interpretation that the angle between the vectors $x - c_*$ and $c - c_*$ is obtuse, that is, it belongs to $[\pi/2, \pi]$.

Proof.
(1) **Existence:** Let $d = \inf\limits_{y \in C} \|x - c\| \geq 0$ (the distance of x to C).
Then there exists a sequence $(c_n)_{n \in \mathbb{N}}$ in C such that $d = \lim\limits_{n \to \infty} \|x - c_n\|$.
Such a sequence is called a *minimising sequence*.
Now we show that a minimising sequence is a Cauchy sequence.
To do this, we apply the Parallelogram Law to the vectors $x - c_n$ and $x - c_m$, and divide by 4. This gives:

$$\left\|x - \frac{c_n + c_m}{2}\right\|^2 + \frac{1}{4}\|c_n - c_m\|^2 = \frac{1}{2}\Big(\|x - c_n\|^2 + \|x - c_m\|^2\Big). \quad (4.5)$$

Since C is convex, $\dfrac{c_n + c_m}{2} \in C$, so that $\left\|x - \dfrac{c_n + c_m}{2}\right\| \geq d$.
Thus $\dfrac{1}{4}\|c_n - c_m\|^2 \leq \dfrac{1}{2}\Big(\|x - c_n\|^2 + \|x - c_m\|^2\Big) - d^2$. $\quad (*)$
Given $\epsilon > 0$, there is an N such that for all $n > N$, $\big|\|x - c_n\|^2 - d^2\big| < \dfrac{\epsilon^2}{4}$.
Hence, by $(*)$, for $n, m > N$, $\|c_n - c_m\| < \epsilon$. Because H is complete, it follows that the Cauchy sequence $(c_n)_{n \in \mathbb{N}}$ has a limit c_*, and this belongs to C since C is closed. Now we have $d = \lim\limits_{n \to \infty} \|x - c_n\| = \|x - c_*\|$.
Consequently, $\|x - c_*\| = d \leq \inf\limits_{c' \in C} \|x - c'\| \leq \|x - c\|$ for all $c \in C$.

Uniqueness: Assume that $\widetilde{c}_* \in C$ is such that $\|x - \widetilde{c}_*\| \leq \|x - c\|$ for all $c \in C$. Then it follows that $d := \inf_{c \in C} \|x - c\| = \|x - \widetilde{c}_*\|$.

From the previous part, we also know that $\|x - c_*\| = d$.
So it follows that the sequence $c_*, \widetilde{c}_*, c_*, \widetilde{c}_*, \cdots$ is a minimising sequence. But from the above, it follows that this sequence is a Cauchy sequence, and hence convergent. But this sequence has subsequences $c_*, c_*, c_*, \cdots$ and $\widetilde{c}_*, \widetilde{c}_*, \widetilde{c}_*, \cdots$ that converge, respectively, to c_*, and to $\widetilde{c}_*$. Consequently, $c_* = \widetilde{c}_*$.

(2) Suppose that c_* is the unique vector from part (1). Let $c \in C$. Then for $\alpha \in (0,1)$, $\alpha c + (1-\alpha) c_* \in C$, and so
$$\|x - c_*\| \leq \|x - \alpha c - (1-\alpha)c_*\| = \|x - c_* + \alpha(c_* - c)\|.$$

By taking squares, we obtain:
$$\|x - c_*\|^2 \leq \langle x - c_* + \alpha(c_* - c), x - c_* + \alpha(c_* - c) \rangle$$
$$= \|x - c_*\|^2 + \alpha^2 \|c_* - c\|^2 + 2\alpha \operatorname{Re}\langle x - c_*, c_* - c \rangle$$

or $\operatorname{Re}\langle x - c_*, c - c_* \rangle \leq \alpha \|c_* - c\|^2 / 2$. Letting α go to 0, in the limit we obtain $\operatorname{Re}\langle x - c_*, c - c_* \rangle \leq 0$.

Now suppose that $c_* \in C$ is a vector that satisfies (4.4).
We wish to show that for all $c \in C$, $\|x - c_*\| \leq \|x - c\|$. Let $c \in C$.
We write $x - c = x - c_* + c_* - c$. Then
$$\|x - c\|^2 = \langle x - c_* + c_* - c, x - c_* + c_* - c \rangle$$
$$= \|x - c_*\|^2 + \|c_* - c\|^2 - 2\operatorname{Re}\langle x - c_*, c - c_* \rangle$$
$$\geq \|x - c_*\|^2 + \|c_* - c\|^2 - 0 \geq \|x - c_*\|^2.$$

So c_* is the point in C closest to x. $\square$

Exercise 4.19. Let $H = L^2(\mathbb{R})$ be the real Hilbert space of square integrable functions on $\mathbb{R}$, and let $L^2_+(\mathbb{R}) = \{\mathbf{f} \in H : \mathbf{f} \geq 0\}$ (strictly speaking the set of equivalence classes of nonnegative functions). Suppose that $\mathbf{f} \in H$. Show that the element in $L^2_+(\mathbb{R})$ closest to $\mathbf{f}$ is precisely $\mathbf{f}^+ = \max\{\mathbf{f}, \mathbf{0}\}$ (that is, the simplest nonnegative function associated with $\mathbf{f}$!).

Subspace case

A special case of a convex set is a subspace. Thus we can apply Theorem 4.6 to closed subspaces. Note that earlier we had seen the *finite dimensional* subspace case in *inner product spaces*, whereas now we consider the case of (possibly infinite dimensional) subspaces in a *Hilbert space*.

Theorem 4.7.
Let Y be a closed subspace of a Hilbert space H and let $x \in H$. Then there exists a unique $y_* \in Y$ such that for all $y \in Y$, $\|x-y_*\| \leq \|x-y\|$. The point $y_* \in Y$ can be characterised by: for all $y \in Y$, $\langle x - y_*, y\rangle = 0$. (We write $x - y_* \perp Y$.)

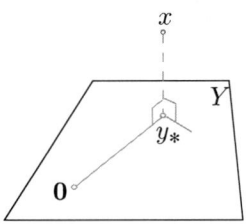

Proof. The existence of $y_* \in Y$ follows from Theorem 4.6. Also, we know that this unique $y_* \in Y$ is such that for all $y' \in Y$, $\operatorname{Re}\langle x - y_*, y' - y_*\rangle \leq 0$. Now let $y \in Y$. Let $\theta \in [0, 2\pi)$ be such that $\langle x - y_*, y\rangle = e^{i\theta}|\langle x - y_*, y\rangle|$. Then with $y' := y_* + e^{i\theta}y$, we have $y' \in Y$ (because Y is a subspace). So
$$0 \geq \operatorname{Re}\langle x - y_*, y' - y_*\rangle = \operatorname{Re}\langle x - y_*, e^{i\theta}y\rangle$$
$$= \operatorname{Re}\left(e^{-i\theta}\langle x - y_*, y\rangle\right) = \operatorname{Re}|\langle x - y_*, y\rangle| = |\langle x - y_*, y\rangle| \geq 0.$$
Hence $\langle x - y_*, y\rangle = 0$ for all $y \in Y$.

Next let $\tilde{y}_* \in Y$ be such that $\langle x - \tilde{y}_*, y\rangle = 0$ for all $y \in Y$. Let $y' \in Y$. Then $y := y' - \tilde{y}_* \in Y$ too (as Y is a subspace). So $\langle x - \tilde{y}_*, y' - \tilde{y}_*\rangle = 0$. In particular, $\operatorname{Re}\langle x - \tilde{y}_*, y' - \tilde{y}_*\rangle = 0$ too. Thus $\tilde{y}_* = y_*$. $\square$

Definition 4.5. (Orthogonal projection). Let Y be a closed subspace of the Hilbert space H, and let $x \in H$. Then the unique $y_* \in Y$ (given by Theorem 4.7) is called the *orthogonal projection of x to Y*. We will denote the orthogonal projection of x to Y by $P_Y x$.

Theorem 4.8. *Let Y be a closed subspace of the Hilbert space H.*
(1) *The map $x \overset{P_Y}{\mapsto} P_Y x : H \to H$ belongs to $CL(H)$.*
(2) $\operatorname{ran} P_Y = Y$.
(3) $\|P_Y\| = 1$ *except if $Y = \{\mathbf{0}\}$, in which case $P_Y = \mathbf{0}$ and $\|P_Y\| = 0$.*
(4) $\ker P_Y = Y^\perp$.
(5) *Every $x \in H$ has a unique decomposition $x = y + z$, with $y \in Y$, $z \in Y^\perp$.*
(6) $P_{Y^\perp} = I - P_Y$.
(7) $P_Y^2 = P_Y$.
(8) P_Y *is symmetric, that is, for all $x, x' \in H$, $\langle P_Y x, x'\rangle = \langle x, P_Y x'\rangle$.*

Proof.
(1) Let $x_1, x_2 \in H$. Then $P_Y x_1, P_Y x_2 \in Y$, and so is their sum. Moreover, for all $y \in Y$,
$$\langle x_1 + x_2 - (P_Y x_1 + P_Y x_2), y \rangle = \langle x_1 - P_Y x_1, y \rangle + \langle x_2 - P_Y x_2, y \rangle$$
$$= 0 + 0 = 0.$$
Thus $P_Y(x_1 + x_2) = P_Y x_1 + P_Y x_2$.
Let $x \in H$ and $\alpha \in \mathbb{K}$. Then $P_Y x \in Y$ and so $\alpha P_Y x \in Y$ too. Moreover, for all $y \in Y$, $\langle \alpha x - \alpha P_Y x, y \rangle = \alpha \langle x - P_Y x, y \rangle = \alpha \cdot 0 = 0$. So $P_Y(\alpha x) = \alpha P_Y x$.
Hence $P_Y \in L(H)$. Next we'll show continuity of P_Y.
For $x \in X$, and all $y \in Y$, $\langle x - P_Y x, y \rangle = 0$, that is, $x - P_Y x \in Y^\perp$. Thus $\|x\|^2 = \|x - P_Y x + P_Y x\|^2 = \|x - P_Y x\|^2 + \|P_Y x\|^2 \geq \|P_Y x\|^2$, showing that $P_Y \in CL(H)$ and that $\|P_Y\| \leq 1$.

(2) For all $x \in H$, $P_Y x \in Y$, and so $\operatorname{ran} P_Y \subset Y$. Now suppose that $y' \in Y$. Then $y' - P_Y y' \in Y$ too. But for all $y \in Y$, we have $\langle y' - P_Y y', y \rangle = 0$. In particular, with $y := y' - P_Y y'$, $\|y' - P_Y y'\|^2 = \langle y' - P_Y y', y \rangle = 0$. Thus $P_Y y' = y'$. Hence $Y \subset \operatorname{ran} P_Y$.
(Alternatively, one could just observe that for $y \in Y$, $P_Y y = y$, since y is the closest member of Y to y, and this shows that $y \in \operatorname{ran} P_Y$.)

(3) If $Y = \{\mathbf{0}\}$, then $P_Y = \mathbf{0}$, and so $\|P_Y\| = 0$.
Suppose that $Y \neq \{\mathbf{0}\}$. Then there exists a $y' \in Y \backslash \{0\}$, and as we had seen above, $P_Y y' = y'$. So $\|P_Y\| \|y'\| \geq \|P_Y y'\| = \|y'\|$, giving $\|P_Y\| \geq 1$. Also, we had established in part (1) that $\|P_Y\| \leq 1$. Thus $\|P_Y\| = 1$.

(4) Let $x \in \ker P_Y$. Then $P_Y x = \mathbf{0}$. For all $y \in Y$,
$$0 = \langle x - P_Y x, y \rangle = \langle x - \mathbf{0}, y \rangle = \langle x, y \rangle,$$
and so $x \in Y^\perp$. Thus $\ker P_Y \subset Y^\perp$.
Let $x \in Y^\perp$. Then for all $y \in Y$, $\langle x - \mathbf{0}, y \rangle = \langle x, y \rangle = 0$. Thus $P_Y x = \mathbf{0}$. Hence $Y^\perp \subset \ker P_Y$ too.

(5) If $x \in X$, then $x - P_Y x \in Y^\perp$. Thus, taking $y := P_Y x$ and $z := x - P_Y x$, we get $x = y + z$ with $y \in Y$ and $z \in Y^\perp$. Also, if $y' \in Y$ and $z' \in Y^\perp$ are such that $x = y' + z'$, then $y + z = y' + z'$, and so we obtain that $Y \ni y - y' = z' - z \in Y^\perp$. But $Y \cap Y^\perp = \{\mathbf{0}\}$. Hence $y - y' = \mathbf{0} = z' - z$, that is, $y = y'$ and $z = z'$.

(6) We have seen that $x = y + z$, where $y = P_Y x \in Y$ and $z = x - P_Y x \in Y^\perp$. As $Y^\perp$ is also a closed subspace and using Exercise 4.17(1) (page 173), it follows from part (5) that $P_{Y^\perp} x = z$. Hence $P_{Y^\perp} = I - P_Y$.

(7) We have $P_Y x = \underbrace{P_Y x}_{\in Y} + \underbrace{0}_{\in Y^\perp}$, and so by part (5), $P_Y(P_Y x) = P_Y x$.

(8) Let $x, x' \in H$. Then we have

$$\begin{aligned}\langle P_Y x, x'\rangle &= \langle P_Y x, P_Y x' + P_{Y^\perp} x'\rangle \\ &= \langle P_Y x, P_Y x'\rangle + \langle P_Y x, P_{Y^\perp} x'\rangle \\ &= \langle P_Y x, P_Y x'\rangle + 0 = \langle P_Y x, P_Y x'\rangle \\ &= \langle x - P_{Y^\perp} x, P_Y x'\rangle \\ &= \langle x, P_Y x'\rangle - \langle P_{Y^\perp} x, P_Y x'\rangle = \langle x, P_Y x'\rangle - 0 \\ &= \langle x, P_Y x'\rangle.\end{aligned}$$

Thus P_Y is symmetric[4]. $\square$

Corollary 4.1. *Let Y be a closed subspace of a Hilbert space H. Then $(Y^\perp)^\perp = Y$.*

Proof. We have $P_{(Y^\perp)^\perp} = I - P_{Y^\perp} = I - (I - P_Y) = P_Y$.
So the ranges of $P_{(Y^\perp)^\perp}$ and $P_{Y^\perp}$ are equal, that is, $(Y^\perp)^\perp = Y$. $\square$

Exercise 4.20. Let Y be a subspace of a Hilbert space H. Show that $Y^{\perp\perp} = \overline{Y}$.

Exercise 4.21. Let $H = L^2(\mathbb{R})$ and let $Y = \{\mathbf{f} \in L^2(\mathbb{R}) : \mathbf{f} = \check{\mathbf{f}} := \mathbf{f}(-\cdot)\}$ be the closed subspace of even functions in $L^2(\mathbb{R})$. Show that $P_Y \mathbf{f} = (\mathbf{f} + \check{\mathbf{f}})/2$. Show that $Y^\perp$ is equal to the space of odd functions in $L^2(\mathbb{R})$. Determine $P_{Y^\perp}$, and using this, verify the fact that every element of $L^2(\mathbb{R})$ has a unique decomposition into the sum of an even and an odd function in $L^2(\mathbb{R})$.

Exercise 4.22. Let H be a Hilbert space. Suppose that $S \in CL(H)$ is such that $S^2 = I$ and S is symmetric (that is, $\langle Sx, y\rangle = \langle x, Sy\rangle$ for all $x, y \in H$). Let $Y = \{x \in H : Sx = x\}$ and let $Z = \{x \in H : Sx = -x\}$. Show that $Z = Y^\perp$, $Y = Z^\perp$, and determine the projections P_Y and P_Z in terms of S.

Exercise 4.23. Let $H = L^2(\mathbb{R})$. Let $A \subset \mathbb{R}$ be a measurable set and let Y_A be the set of elements $\mathbf{f} \in L^2(\mathbb{R})$ such that $\mathbf{f}(x) = 0$ for almost all $x \in \complement A = \mathbb{R}\backslash A$. Show that Y_A is a closed linear subspace, and determine the projection P_A onto Y_A.

Exercise 4.24. Let $D \subset H$ be a subspace of the Hilbert space H. Show that D is dense in H if and only if $D^\perp = \{\mathbf{0}\}$, that is, there does not exist a nonzero $x \in H$ that is orthogonal to D. *Hint:* Use the fact that $\overline{D}$ is a closed linear subspace.

---

[4] After we introduce the notion of the adjoint of an operator, we will see that this fact means that P_Y is self-adjoint, that is, $P_Y^* = P_Y$.

4.4 Generalised Fourier series

Recall that using Zorn's Lemma, we'd seen that every vector space X has a Hamel basis, that is, a set $B \subset X$ such that $\mathrm{span}(B) = X$ (so that every vector in X can be written as a linear combination of elements of B) and (this representation is not "wasteful", and is "efficient" in the sense that) B is linearly independent. However, if X is an inner product space, then the Hamel basis suffers from the following two "flaws":

(1) The notion of a Hamel basis is purely "algebraic", that is, it refers only to the vector space structure of X. But if X is an inner product space, then more structure is available, and it would be great if the basis reflects this structure too.
(2) A Hamel basis of an infinite dimensional Hilbert space is uncountable, and there is no constructive procedure to find it. So it is "inconvenient" in this context.

We'll now see that as opposed to the Hamel basis, there is much nicer concept of *orthonormal basis* in the context of inner product spaces, which does away with the above deficiencies. Indeed, an orthonormal basis has good algebraic properties (being independent) and good analytic properties (since it is an orthonormal set, and hence it is nicely behaved with respect to the inner product). Moreover, in the infinite dimensional case, for most examples one meets in applications, it is a countable and an explicit set. As opposed to Hamel basis, where we insist on using *finite* number of vectors from the Hamel basis at a time to generate vectors (by taking linear combinations), we'll see that an orthonormal basis is more "efficient" in that limits of linear combinations are also allowed for generation of vectors.

Definition 4.6. (Orthonormal basis). Let X be an inner product space. A set $B \subset X$ is called an *orthonormal basis for X* if
  (a) $\mathrm{span}(B)$ is dense in X, and
  (b) B is an orthonormal set.

Example 4.9. Consider $\mathbb{R}^d$ with the usual Euclidean inner product. Then $\left\{ \mathbf{e}_1 := \begin{bmatrix} 1 \\ 0 \\ \vdots \\ 0 \end{bmatrix}, \cdots, \mathbf{e}_d := \begin{bmatrix} 0 \\ \vdots \\ 0 \\ 1 \end{bmatrix} \right\}$ is an orthonormal basis for $\mathbb{R}^d$. ◇

Example 4.10. Consider ℓ^2 with the usual inner product. For $i \in \mathbb{N}$, let $\mathbf{e}_i := (0, \cdots, 0, 1, 0, \cdots)$, the sequence with the ith term equal to 1 and all other terms equal to 0. Then $B := \{\mathbf{e}_i : i \in \mathbb{N}\}$ is an orthonormal set, and $\text{span}(B) = c_{00}$, which is dense in ℓ^2 (Exercise 1.24, page 21). Hence B is an orthonormal basis of ℓ^2. ◇

Exercise 4.25. Show that the polynomials are dense in $C[-1,1]$ with the usual inner product. *Hint*: Weierstrass's Theorem from Exercise 1.26, page 22.

Example 4.11. (Legendre's Polynomials).
Let us revisit Example 4.6, page 168. From Exercise 4.25, it follows that the set of polynomials $\left\{\sqrt{\frac{2n+1}{2}}\mathbf{P}_n : n \geq 0\right\}$ forms an orthonormal basis of $C[-1, 1]$ (and of $L^2[-1, 1]$). ◇

Example 4.12. (Trigonometric Polynomials).
Consider the example in item (3) on page 166. It can be shown that the set $B := \{\mathbf{T}_n : n \in \mathbb{Z}\}$ is an orthonormal basis for $X = C[0, 1]$ (or $L^2[0, 1]$). We had seen earlier that the $\mathbf{T}_n$s are orthonormal. The proof of the density of span B in X is somewhat tedious, and we will not prove this here [5]. ◇

The result below justifies the terminology "basis" in "orthonormal basis".

Theorem 4.9. *Let X be an inner product space with a countable orthonormal basis $\{u_1, u_2, u_3, \cdots\}$. Then:*

(1) *For all $x \in X$, $x = \displaystyle\sum_{n=1}^{\infty} \langle x, u_n\rangle u_n$.*

(2) *For all $x, y \in X$, $\langle x, y\rangle = \displaystyle\sum_{n=1}^{\infty} \langle x, u_n\rangle \langle y, u_n\rangle^*$.*

(3) *For all $x \in X$, $\|x\|^2 = \displaystyle\sum_{n=1}^{\infty} |\langle x, u_n\rangle|^2$.*

If $X = H$ is a Hilbert space with a countable orthonormal basis $\{u_1, u_2, \cdots\}$, then for all $(c_n)_{n \in \mathbb{N}} \in \ell^2$, we have $\displaystyle\sum_{n=1}^{\infty} c_n u_n \in H$.

Remark 4.6.
(1) It is tempting to say "the right-hand side of the decomposition in (1) is a linear combination of the u_ns", but this remark is faulty because a linear combination by definition is a *finite* sum.
And "infinite linear combination" is considered troublesome terminology

---
[5]The interested reader is referred to [Rudin (1976), Theorem 8.16, page 191].

owing to the persistent query about convergence. Instead we'll call the decomposition in part (1) of Theorem 4.9 a *generalised Fourier series*, because of the similarity with the classical Fourier series, which will be explored in Example 4.13 below.

(2) Recall that in elementary (finite dimensional) linear algebra, one learns the following "representation result":

> All d dimensional vector spaces over $\mathbb{K}$ are isomorphic to $\mathbb{K}^d$.

Analogously, the theorem above contains an important representation result, saying that

> All Hilbert spaces with a countable orthonormal basis are isomorphic to ℓ^2.

Indeed, consider the map
$$x \overset{\iota}{\mapsto} \big(\langle x, u_1\rangle, \langle x, u_2\rangle, \langle x, u_1\rangle, \cdots\big) : H \to \ell^2.$$
It can be checked that this map ι is linear, and thanks to the result in Theorem 4.9, also continuous, one-to-one and onto. So it is an element of $CL(H, \ell^2)$, and as the map ι is bijective, it has a continuous inverse $\iota^{-1} \in CL(\ell^2, H)$. Thus ι is an isomorphism. As $\|x\|_H = \|\iota(x)\|_2$, we call ι an *isometric* isomorphism.

Exercise 4.26.
If H is a Hilbert space with a countable orthonormal basis $\{u_1, u_2, u_3, \cdots\}$, then using Theorem 4.9, show that the map $x \overset{\iota}{\mapsto} \big(\langle x, u_1\rangle, \langle x, u_2\rangle, \langle x, u_1\rangle, \cdots\big) : H \to \ell^2$ is indeed an isometric isomorphism.

Proof. (Of Theorem 4.9)
(1) Let $x \in X$, and $\epsilon > 0$. As span$\{u_1, u_2, u_3, \cdots\}$ is dense in X, there exists an $N \in \mathbb{N}$ and scalars $c_1, \cdots, c_N \in \mathbb{K}$ such that
$$\left\| x - \sum_{k=1}^N c_k u_k \right\| < \epsilon.$$
Let $n > N$, and $Y = \text{span}\{u_1, \cdots, u_n\}$. As $\sum_{k=1}^N c_k u_k \in Y$, by Theorem 4.5,
$$\underbrace{\left\| x - \sum_{k=1}^n \langle x, u_k\rangle u_k \right\|}_{y_*} \leqslant \underbrace{\left\| x - \sum_{k=1}^N c_n u_n \right\|}_{\in Y}.$$
Thus $\left(\sum_{k=1}^n \langle x, u_k\rangle u_k\right)_{n \in \mathbb{N}}$ is convergent in X with limit x, that is,
$$x = \sum_{n=1}^\infty \langle x, u_n\rangle u_n.$$

(2) If $x, y \in X$, then
$$x_n := \sum_{k=1}^{n} \langle x, u_k \rangle u_k \xrightarrow{n \to \infty} x, \text{ and } y_n := \sum_{k=1}^{n} \langle y, u_k \rangle u_k \xrightarrow{n \to \infty} y.$$
So
$$\langle x, y \rangle = \Big\langle \lim_{n \to \infty} x_n, \lim_{n \to \infty} y_n \Big\rangle = \lim_{n \to \infty} \langle x_n, y_n \rangle$$
$$= \lim_{n \to \infty} \Big\langle \sum_{k=1}^{n} \langle x, u_k \rangle u_k, \sum_{j=1}^{n} \langle y, u_j \rangle u_j \Big\rangle$$
$$= \lim_{n \to \infty} \sum_{k=1}^{n} \sum_{j=1}^{n} \langle x, u_k \rangle \langle y, u_j \rangle^* \underbrace{\langle u_k, u_j \rangle}_{\delta_{kj}} = \lim_{n \to \infty} \sum_{k=1}^{n} \langle x, u_k \rangle \langle y, u_k \rangle^*.$$
Thus $\langle x, y \rangle = \sum_{n=1}^{\infty} \langle x, u_n \rangle \langle y, u_n \rangle^*$.

(3) This follows from (2) by taking $y = x$.

Now suppose that $X = H$, a Hilbert space.

Let $(c_n)_{n \in \mathbb{N}} \in \ell^2$. Then $\Big(\sum_{k=1}^{n} |c_k|^2 \Big)_{n \in \mathbb{N}}$ is a Cauchy sequence.

We will show that $\Big(\sum_{k=1}^{n} c_k u_k \Big)_{n \in \mathbb{N}}$ is a Cauchy sequence. For $n > m$,

$$\Big\| \sum_{k=1}^{n} c_k u_k - \sum_{k=1}^{m} c_k u_k \Big\|^2 = \Big\| \sum_{k=m+1}^{n} c_k u_k \Big\|^2 = \sum_{k=m+1}^{n} |c_k|^2$$
$$= \Big| \sum_{k=1}^{n} |c_k|^2 - \sum_{k=1}^{m} |c_k|^2 \Big|.$$

Thus $\Big(\sum_{k=1}^{n} c_k u_k \Big)_{n \in \mathbb{N}}$ is a Cauchy sequence in H.

As $X = H$ is a Hilbert space, $\Big(\sum_{k=1}^{n} c_k u_k \Big)_{n \in \mathbb{N}}$ is convergent in H. $\square$

The identity
$$\langle x, y \rangle = \sum_{n=1}^{\infty} \langle x, u_n \rangle \langle y, u_n \rangle^*, \text{ for } x, y \in X,$$
is called *Parseval's Identity*.

Example 4.13. Consider the Hilbert space $C[0,1]$ with the usual inner product, and the orthonormal basis $\{\mathbf{T}_n : n \in \mathbb{N}\}$ from page 166.
By Theorem 4.9, every $\mathbf{x} \in C[0,1]$ has the expansion: $\mathbf{x} = \sum_{n=-\infty}^{\infty} \langle \mathbf{x}, \mathbf{T}_n \rangle \mathbf{T}_n$.
Note that $\langle \mathbf{x}, \mathbf{T}_n \rangle$ is precisely the nth Fourier coefficient of $\mathbf{x}$:

$$\langle \mathbf{x}, \mathbf{T}_n \rangle = \int_0^1 \mathbf{x}(t)\left(e^{2\pi i n t}\right)^* dt = \int_0^1 \mathbf{x}(t) e^{-2\pi i n t} dt =: \hat{\mathbf{x}}_n.$$

Hence the above expansion becomes $\mathbf{x} = \sum_{n=-\infty}^{\infty} \hat{\mathbf{x}}_n e^{2\pi i n \cdot}$.

But we should bear in mind that the above does *not* necessarily mean pointwise convergence! All we know is that the sequence of partial sums s_N, $N \in \mathbb{N}$, given by

$$s_N = \sum_{n=-N}^{N} \langle \mathbf{x}, \mathbf{T}_n \rangle \mathbf{T}_n,$$

converges to $\mathbf{x}$ in the $\|\cdot\|_2$-norm: $\lim_{N \to \infty} \int_0^1 |s_N - \mathbf{x}(t)|^2 dt = 0$. ◇

Exercise 4.27. Using $t \mapsto t \in C[0,1]$, show that $\sum_{n=1}^{\infty} \frac{1}{n^2} = \frac{\pi^2}{6}$.

Exercise 4.28. Show that an inner product space is dense in its completion.

Exercise 4.29. (Isoperimetric Theorem).
Among all simple, closed smooth curves of length L in the plane, the circle encloses the maximum area. This can be proved by proceeding as follows.
Suppose that $s \mapsto (\mathbf{x}(s), \mathbf{y}(s))$ is a representation of the curve, where s is the arc length parameter. We may consider $\mathbf{x}, \mathbf{y}$ as L-periodic functions.

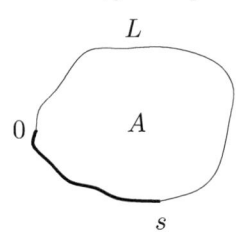

So $\mathbf{x}, \mathbf{y}$ have Fourier series expansions

$$\mathbf{x}(s) = a_0 + \sum_{n=1}^{\infty} a_n \cos\left(\frac{2\pi}{L} n s\right) + b_n \sin\left(\frac{2\pi}{L} n s\right)$$

$$\mathbf{y}(s) = c_0 + \sum_{n=1}^{\infty} c_n \cos\left(\frac{2\pi}{L} n s\right) + d_n \sin\left(\frac{2\pi}{L} n s\right).$$

Then
$$\mathbf{x}'(s) = \sum_{n=1}^{\infty} \frac{2\pi n}{L}\left(-a_n \sin\left(\frac{2\pi}{L}ns\right) + b_n \cos\left(\frac{2\pi}{L}ns\right)\right)$$
$$\mathbf{y}'(s) = \sum_{n=1}^{\infty} \frac{2\pi n}{L}\left(-c_n \sin\left(\frac{2\pi}{L}ns\right) + d_n \cos\left(\frac{2\pi}{L}ns\right)\right).$$

Using Parseval's Identity, it can be shown that
$$\int_0^L \left(\mathbf{x}'(s)\right)^2 ds = \frac{4\pi^2}{L^2} \cdot \frac{L}{2} \sum_{n=1}^{\infty} n^2(a_n^2 + b_n^2)$$
$$\int_0^L \left(\mathbf{y}'(s)\right)^2 ds = \frac{4\pi^2}{L^2} \cdot \frac{L}{2} \sum_{n=1}^{\infty} n^2(c_n^2 + d_n^2).$$

We have $L = \int_0^L ds = \int_0^L \left((\mathbf{x}'(s))^2 + (\mathbf{y}'(s))^2\right) ds = \frac{2\pi^2}{L} \sum_{n=1}^{\infty} n^2(a_n^2 + b_n^2 + c_n^2 + d_n^2).$

So $L^2 = 2\pi^2 \sum_{n=1}^{\infty} n^2(a_n^2 + b_n^2 + c_n^2 + d_n^2).$

The area A is given by $A = \int_0^L \mathbf{x}(s)\mathbf{y}'(s) ds = \frac{2\pi}{L} \cdot \frac{L}{2} \sum_{n=1}^{\infty} n(a_n d_n - b_n c_n).$

Prove that $L^2 - 4\pi A \geq 0$ and that equality holds if and only if
$$a_1 = d_1,\ b_1 = -c_1,\ \text{and } a_n = b_n = c_n = d_n = 0 \text{ for all } n \geq 2,$$
which describes the equation of a circle.

Exercise 4.30. Which of the following two sets is an orthonormal basis for ℓ^2?
$\frac{1}{\sqrt{5}}(1,2,0,\cdots),\ \frac{1}{\sqrt{5}}(0,0,1,2,0,\cdots),\ \frac{1}{\sqrt{5}}(0,0,0,0,1,2,0,\cdots),\cdots.$
$\frac{1}{\sqrt{2}}(1,-1,0,\cdots),\ \frac{1}{\sqrt{2}}(1,1,0,\cdots),\ \frac{1}{\sqrt{2}}(0,0,1,-1,0,\cdots),\ \frac{1}{\sqrt{2}}(0,0,1,1,0,\cdots),\cdots.$

Exercise 4.31. Show that an inner product space with a countable orthonormal basis is separable.

Exercise 4.32. (Nonseparable Hilbert spaces).
Although an analogue of Theorem 4.9 can be shown for spaces with an uncountable orthonormal basis, we won't discuss this. One reason is that most examples met in practice are separable.

A notable exception is the class of *Besicovitch Almost Periodic Functions*, which is the completion of the span X of the exponentials $\{e^{i\lambda x} : \lambda \in \mathbb{R}\}$ under the inner product
$$\langle \mathbf{f}, \mathbf{g} \rangle = \lim_{T \to \infty} \frac{1}{2T} \int_{-T}^{T} \mathbf{f}(x)\big(\mathbf{g}(x)\big)^* dx,\quad \mathbf{f}, \mathbf{g} \in X.$$
We won't check that this does define an inner product. Rather, show that the space of Besicovitch Almost Periodic Functions is not separable by calculating the distance between $e^{i\lambda x}$ and $e^{i\mu x}$ when $\lambda \neq \mu$.

Exercise 4.33.
Let $\{u_i : i \in I\}$ be an orthonormal set in an inner product space X, and let $x \in X$.
Show that $\langle x, u_i \rangle$ is nonzero for at most a countable number of the u_i's.
Hint: Using Bessel's Inequality, for each $n \in \mathbb{N}$, $U_n := \{u_i : |\langle x, u_i \rangle|^2 > \|x\|^2/n\}$
has at most $n - 1$ elements. Conclude that $U := \{u_i : \langle x, u_i \rangle \neq 0\}$ must be at most countable.

4.5 Riesz Representation Theorem

Let H be a Hilbert space, and $y \in H$. Then the map $\varphi_y : H \to \mathbb{C}$ given by

$$x \stackrel{\varphi_y}{\mapsto} \langle x, y \rangle : H \to \mathbb{C}$$

is a continuous linear transformation on H. Indeed, we have:
(L1) For all $x_1, x_2 \in H$,
$$\varphi_y(x_1 + x_2) = \langle x_1 + x_2, y \rangle = \langle x_1, y \rangle + \langle x_2, y \rangle = \varphi_y(x_1) + \varphi_y(x_2).$$
(L2) For all $\alpha \in \mathbb{C}$ and $x \in H$, $\varphi_y(\alpha \cdot x) = \langle \alpha \cdot x, y \rangle = \alpha \langle x, y \rangle = \alpha \varphi_y(x)$.
Continuity: For all $x \in H$, $|\varphi_y(x)| = |\langle x, y \rangle| \leq \|y\| \|x\|$, by the Cauchy-Schwarz Inequality.

Thus φ_y is an element of the dual space $H' = CL(H, \mathbb{C})$.
Now we will see that *every* element of the dual space is of this type!
So H' can be "identified" with H.
This is called the *Riesz Representation Theorem* after (Frigyes)[6] Riesz.

Theorem 4.10. (F. Riesz Representation Theorem).
*If H is a Hilbert space, and $\varphi \in CL(H, \mathbb{C})$,
then there exists a unique $y \in X$ such that $\varphi = \varphi_y$, that is,
for all $x \in H$, $\varphi(x) = \langle x, y \rangle$.*

Proof. Let $Y = \ker \varphi$. Then Y is closed. (Indeed, if $Y \ni x_n \to x$, then $\varphi(x) = \varphi(\lim x_n) = \lim \varphi(x_n) = \lim 0 = 0$.)
If $Y = H$, then we may take $y = \mathbf{0}$.
If $Y \neq H$, then let $x_0 \in H \setminus Y$.
Write $x_0 = P_Y x_0 + P_{Y^\perp} x_0$, where $P_{Y^\perp} x_0 \neq \mathbf{0}$ because $x_0 \notin Y$.
Let us note that $\varphi(P_{Y^\perp} x_0) \neq 0$. Indeed,

$$0 \neq \varphi(x_0) = \varphi(P_Y x_0 + P_{Y^\perp} x_0) = 0 + \varphi(P_{Y^\perp} x_0) = \varphi(P_{Y^\perp} x_0).$$

For $x \in H$, $x_1 := x - \dfrac{\varphi(x)}{\varphi(P_{Y^\perp} x_0)} P_{Y^\perp} x_0 \in \ker \varphi = Y$.

---
[6] His brother, Marcel, is also famous for results in functional analysis.

Set $y := \underbrace{\dfrac{\left(\varphi(P_{Y^\perp}x_0)\right)^*}{\|P_{Y^\perp}x_0\|^2}}_{\text{a scalar}} \underbrace{P_{Y^\perp}x_0}_{\in Y^\perp} \in Y^\perp.$

We have
$$0 = \langle x_1, y\rangle = \left\langle x - \frac{\varphi(x)}{\varphi(P_{Y^\perp}x_0)}P_{Y^\perp}x_0, y\right\rangle$$
$$= \langle x, y\rangle - \frac{\varphi(x)}{\varphi(P_{Y^\perp}x_0)} \cdot \frac{\left(\varphi(P_{Y^\perp}x_0)\right)^{**}}{\|P_{Y^\perp}x_0\|^2}\langle P_{Y^\perp}x_0, P_{Y^\perp}x_0\rangle$$
$$= \langle x, y\rangle - \varphi(x)$$

that is, $\langle x, y\rangle = \varphi(x)$.

Finally, we prove the claimed uniqueness. Suppose that $y, y' \in H$ are such that for all $x \in H$, $\varphi(x) = \langle x, y\rangle = \langle x, y'\rangle$. Then for all $x \in H$, $\langle x, y - y'\rangle = 0$. In particular, taking $x = y - y'$, we obtain
$$\|y - y'\|^2 = \langle y - y', y - y'\rangle = \langle x, y - y'\rangle = 0,$$
and so $y = y'$. □

The above theorem characterises linear functionals on Hilbert spaces: they are precisely inner products with a fixed vector! In this sense we can "identify" the dual space $H' = CL(H, \mathbb{C})$ with H itself, at least set theoretically.

Exercise 4.34. Let H be a complex Hilbert space, $y \in H$, and $\varphi_y \in CL(H, \mathbb{C})$ be given by $\varphi_y(x) = \langle x, y\rangle$.
(1) Show that $\|\varphi_y\|$ in $CL(H, \mathbb{C})$ equals $\|y\|$ as an element of H.
(2) Prove that $\varphi_{iy} = -i\varphi_y$, and so the map $y \mapsto \varphi_y : H \to CL(H, \mathbb{C})$ is not linear when H has nonzero vectors. (Although for real Hilbert spaces, $y \mapsto \varphi_y$ *is* linear.)

4.6 Adjoints of bounded operators

With every operator $T \in CL(H, K)$, where H, K are Hilbert spaces, one can associate another operator, called its adjoint, $T^* \in CL(K, H)$, which is geometrically related to T.

Theorem 4.11. *Let H, K be Hilbert spaces, and $T \in CL(H, K)$. Then there exists a unique $T^* \in CL(K, H)$ such that*
$$\text{for all } h \in H \text{ and } k \in K, \langle Th, k\rangle = \langle h, T^*k\rangle.$$

Definition 4.7. (Adjoint operator). This T^* is called the *adjoint of T*.

Proof.
Step 1: Fix $k \in K$. The map $h \stackrel{\varphi_k}{\mapsto} \langle Th, k \rangle : H \to \mathbb{K}$ is a linear functional on H. We verify this below.

(L1) If $h_1, h_2 \in H$, then
$$\varphi_k(h_1 + h_2) = \langle T(h_1 + h_2), k \rangle = \langle Th_1 + Th_2, k \rangle$$
$$= \langle Th_1, k \rangle + \langle Th_2, k \rangle = \varphi_k(h_1) + \varphi_k(h_2).$$

(L2) If $\alpha \in \mathbb{C}$ and $h \in H$, then
$$\varphi_k(\alpha h) = \langle T(\alpha h), k \rangle = \langle \alpha Th, k \rangle = \alpha \langle Th, k \rangle = \alpha \varphi_k(h).$$

Continuity: For all $h \in H$, $|\varphi_k(h)| = |\langle Th, k \rangle| \leq \|Th\| \|k\| \leq \|T\| \|k\| \|h\|$, and so $\|\varphi_k\| \leq \|T\| \|k\| < +\infty$.

Hence $\varphi_k \in CL(H, \mathbb{C})$, and by the Riesz Representation Theorem, it follows that there exists a unique vector, which we denote by T^*k, such that for all $h \in H$, $\varphi_k(h) = \langle h, T^*k \rangle$, that is, for all $h \in H$, $\langle Th, k \rangle = \langle h, T^*k \rangle$.

In this manner, we get a map $k \stackrel{T^*}{\mapsto} T^*k : K \to H$ from K to H.

We ask: Is $T^* \in CL(K, H)$?

We will show in the next step that the answer is "yes".

Step 2: $T^* \in CL(K, H)$.

(L1) If $k_1, k_2 \in K$, then for all $h \in H$ we have
$$\langle h, T^*(k_1 + k_2) \rangle = \langle Th, k_1 + k_2 \rangle = \langle Th, k_1 \rangle + \langle Th, k_2 \rangle$$
$$= \langle h, T^*k_1 \rangle + \langle h, T^*k_2 \rangle = \langle h, T^*k_1 + T^*k_2 \rangle.$$

Thus $\langle h, T^*(k_1 + k_2) - (T^*k_1 + T^*k_2) \rangle = 0$ for all $h \in H$.

In particular, taking $h = T^*(k_1 + k_2) - (T^*k_1 + T^*k_2)$, we obtain that $T^*(k_1 + k_2) = T^*k_1 + T^*k_2$.

(L2) If $\alpha \in \mathbb{C}$ and $k \in K$, then for all $h \in H$,
$$\langle h, T^*(\alpha k) \rangle = \langle Th, \alpha k \rangle = \alpha^* \langle Th, k \rangle = \alpha^* \langle h, T^*k \rangle = \langle h, \alpha T^*k \rangle.$$

In particular, taking $h = T^*(\alpha k) - \alpha(T^*k)$, we get $T^*(\alpha k) = \alpha(T^*k)$.

This completes the proof of the linearity of T^*.

Continuity: By Exercise 4.34(1), we have $\|\varphi_k\| = \|Tk\|$, $k \in K$.
Thus for all $k \in K$, $\|T^*k\| = \|\varphi_k\| \leq \|T\| \|k\|$.
Consequently, $T^* \in CL(K, H)$ (and $\|T^*\| \leq \|T\|$).

Uniqueness: If $S \in CL(K, H)$ is another operator satisfying
$$\text{for all } h \in H \text{ and } k \in K, \ \langle Th, k \rangle = \langle h, Sk \rangle,$$
then we have for all $h \in H$ and $k \in K$, $\langle h, T^*k \rangle = \langle Th, k \rangle = \langle h, Sk \rangle$, and so $\langle h, T^*k - Sk \rangle = 0$. Let $k \in K$. Taking $h := T^*k - Sk \in H$ in the above, we obtain $T^*k = Sk$. As $k \in K$ was arbitrary, we conclude that $S = T^*$. Consequently T^* is unique. $\square$

Example 4.14. Let $H = \mathbb{C}^n$ and $K = \mathbb{C}^m$ with the usual respective Euclidean inner products. Let $T_A : \mathbb{C}^n \to \mathbb{C}^n$ be the continuous linear transformation corresponding to multiplication by the $n \times n$ matrix A of complex numbers:

$$A = \begin{bmatrix} a_{11} & \cdots & a_{1n} \\ \vdots & & \vdots \\ a_{m1} & \cdots & a_{mn} \end{bmatrix} \in \mathbb{C}^{m \times n}.$$

What is the adjoint $(T_A)^*$? We will show that $(T_A)^* = T_{A^*}$, where A^* is the matrix obtained by transposing the matrix A and by taking the complex conjugates of each of the entries. Thus

$$A^* = \begin{bmatrix} a_{11}^* & \cdots & a_{m1}^* \\ \vdots & & \vdots \\ a_{1n}^* & \cdots & a_{mn}^* \end{bmatrix}.$$

For all $\mathbf{h} = (h_1, \cdots, h_n) \in \mathbb{C}^n$ and all $\mathbf{k} = (k_1, \cdots, k_m) \in \mathbb{C}^m$, we have

$$\langle T_A \mathbf{h}, \mathbf{k} \rangle = \sum_{i=1}^{m} \Big(\sum_{j=1}^{n} a_{ij} h_j \Big) k_i^* = \sum_{i=1}^{m} \sum_{j=1}^{n} a_{ij} h_j k_i^* = \sum_{j=1}^{n} h_j \sum_{i=1}^{m} a_{ij} k_i^*$$

$$= \sum_{j=1}^{n} h_j \Big(\sum_{i=1}^{m} a_{ij}^* k_i \Big)^* = \langle \mathbf{h}, T_{A^*} \mathbf{k} \rangle.$$

Thus $(T_A)^* = T_{A^*}$. $\diamond$

Example 4.15. (Right/left shift operators).
For the right shift operator $R \in CL(\ell^2)$, one has that $R^* = L$.
For $\mathbf{h} = (h_n)_{n \in \mathbb{N}}$ and $\mathbf{k} = (k_n)_{n \in \mathbb{N}}$, we have

$$\langle R\mathbf{h}, \mathbf{k} \rangle = \langle (0, h_1, h_2, \cdots), (k_1, k_2, k_3, \cdots) \rangle = h_1 k_2^* + h_2 k_3^* + \cdots$$
$$= \langle (h_1, h_2, h_3, \cdots), (k_2, k_3, k_4, \cdots) \rangle = \langle \mathbf{h}, L\mathbf{k} \rangle.$$

Thus $R^* = L$.

What about L^*? Clearly $L^* = (R^*)^*$, and from the following result, we can conclude that $(R^*)^* = R$. Thus $L^* = R$. $\diamond$

Theorem 4.12. *Let H, K be Hilbert spaces, $T, S \in CL(H, K)$ and $\alpha \in \mathbb{C}$. Then:*

(1) $(T^*)^* = T$.
(2) $\|T^*\| = \|T\|$.
(3) $(\alpha T)^* = \alpha^* T$.
(4) $(S + T)^* = S^* + T^*$.
(5) *If $T_1 \in CL(H_1, K)$ and $T_2 \in CL(K, H_2)$, then $(T_2 T_1)^* = T_1^* T_2^*$, where H_1, H_2, K are Hilbert spaces.*

Proof.
(1) For all $h \in H$ and $k \in K$, $\langle T^*k, h \rangle = \langle h, T^*k \rangle^* = \langle Th, k \rangle^* = \langle k, Th \rangle$, and so $(T^*)^* = T$.

(2) In the proof of Theorem 4.11, we had seen that $\|T^*\| \leq \|T\|$. Also, $\|T\| = \|(T^*)^*\| \leq \|T^*\|$. Consequently $\|T\| = \|T^*\|$.

(3) For all $h \in H$ and $k \in K$, we have
$$\langle (\alpha T)h, k \rangle = \langle \alpha(Th), k \rangle = \alpha \langle Th, k \rangle = \alpha \langle h, T^*k \rangle = \langle h, \alpha^*(T^*k) \rangle$$
$$= \langle h, (\alpha^* T^*)k \rangle$$
and so $(\alpha T)^* = \alpha^* T^*$.

(4) For all $h \in H$ and $k \in K$, we have
$$\langle (T+S)h, k \rangle = \langle Th, k \rangle + \langle Sh, k \rangle = \langle h, T^*k \rangle + \langle h, S^*k \rangle$$
$$= \langle h, T^*k + S^*k \rangle = \langle h, (T^* + S^*)k \rangle$$
and so $(T+S)^* = T^* + S^*$.

(5) For all $h_1 \in H_1$ and $h_2 \in H_2$, we have
$$\langle T_2 T_1 h_1, h_2 \rangle = \langle T_1 h_1, T_2^* h_2 \rangle = \langle h_1, T_1^*(T_2^* h_2) \rangle = \langle h_1, (T_1^* T_2^*) h_2 \rangle,$$
and so $(T_2 T_1)^* = T_1^* T_2^*$. □

Corollary 4.2.
If H is a Hilbert space and $T \in CL(H)$ is invertible in $CL(H)$, then T^* is invertible in $CL(H)$ and $(T^*)^{-1} = (T^{-1})^*$.

Proof. We have for all $h, k \in H$ that $\langle Ih, k \rangle = \langle h, k \rangle = \langle h, Ik \rangle$, and so $I^* = I$. Since $TT^{-1} = I = T^{-1}T$, by taking adjoints, we obtain that $(TT^{-1})^* = I^* = (T^{-1}T)^*$, that is, $(T^{-1})^* T^* = I = T^*(T^{-1})^*$. Hence T^* is invertible, and its inverse is $(T^{-1})^*$. □

The definition of the adjoint operator T shows that T^* is geometrically related to T. A manifestation of this is the following result.

Theorem 4.13. *Let H, K be Hilbert spaces and $T \in CL(H, K)$. Then:*
(1) $\ker T = (\operatorname{ran} T^*)^\perp$.
(2) $\overline{\operatorname{ran} T} = (\ker T^*)^\perp$.

Proof.
(1) $\ker T \subset (\operatorname{ran} T^*)^\perp$: If $h \in \ker T$, then $Th = \mathbf{0}$.
So for all $k \in K$, $\langle h, T^*k \rangle = \langle Th, k \rangle = \langle \mathbf{0}, k \rangle = 0$. Thus $h \in (\operatorname{ran} T^*)^\perp$.

$(\operatorname{ran} T^*)^\perp \subset \ker T$: Let $h \in (\operatorname{ran}(T^*))^\perp$. Then for all $k \in K$, we have $0 = \langle h, T^*k \rangle = \langle Th, k \rangle$, and in particular, with $k = Th$, we obtain that $\langle Th, Th \rangle = 0$, that is $Th = \mathbf{0}$. So $h \in \ker T$, and $(\operatorname{ran}(T^*))^\perp \subset \ker(T)$. Consequently, $\ker T = (\operatorname{ran} T^*)^\perp$.

(2) From part (1), and using Exercise 4.20, page 182, for $T \in CL(H, K)$, we have that

$$(\ker T)^\perp = ((\operatorname{ran} T^*)^\perp)^\perp = \overline{\operatorname{ran} T^*}.$$

With T replaced by T^*, we obtain $(\ker T^*)^\perp = \overline{\operatorname{ran}(T^*)^*} = \overline{\operatorname{ran} T}$. □

In our discussion on adjoints, we've been working with complex spaces, but analogous results also hold for real spaces.

Example 4.16. Take $H = K = \mathbb{R}^2$.
Let $P = P_Y$ be the projection operator onto $Y = \operatorname{span}\begin{bmatrix} 1 \\ 0 \end{bmatrix}$ (the "x-axis").
Then $P = T_A$, where $A = \begin{bmatrix} 1 & 0 \\ 0 & 0 \end{bmatrix}$. As $A^* = A$, we see that $P^* = P$.

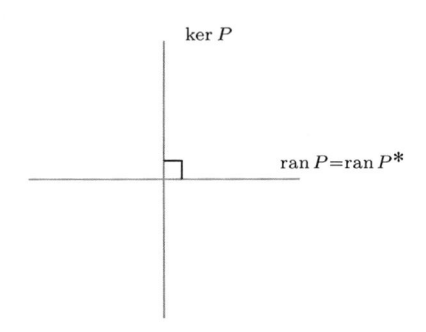

Also, $\ker P = \operatorname{span}\begin{bmatrix} 0 \\ 1 \end{bmatrix}$ and $\operatorname{ran} P = \operatorname{span}\begin{bmatrix} 1 \\ 0 \end{bmatrix}$. ◇

In the above example, we saw that for the projection operator, its adjoint was itself. This is no coincidence!

Theorem 4.14. *Let Y be a closed subspace of a Hilbert space H, and let P_Y be the orthogonal projection operator onto Y (Theorem 4.8, page 180). Then $(P_Y)^* = P_Y$.*

Proof. For all $h, k \in H$, we have

$$\begin{aligned}\langle P_Y h, k\rangle &= \langle P_Y h, P_Y k + P_{Y^\perp} k\rangle = \langle P_Y h, P_Y k\rangle + \langle P_Y h, P_{Y^\perp} k\rangle \\ &= \langle P_Y h, P_Y k\rangle + 0 = \langle P_Y h, P_Y k\rangle = \langle P_Y h, P_Y k\rangle + 0 \\ &= \langle P_Y h, P_Y k\rangle + \langle P_{Y^\perp} h, P_Y k\rangle = \langle P_Y h + P_{Y^\perp} h, P_Y k\rangle \\ &= \langle h, P_Y k\rangle,\end{aligned}$$

and so $(P_Y)^* = P_Y$. $\square$

Definition 4.8. (Self-adjoint or Hermitian operators).
Let H be a Hilbert space. An operator $T \in CL(H)$ is called *self-adjoint* or *Hermitian* if $T^* = T$.

Such operators are often encountered in applications, for example in physics and in PDE theory. We'll see a glimpse of this in our discussion on Quantum Mechanics later in this chapter. Note that if T is a Hermitian operator on a Hilbert space H, then for all $x \in H$, $\langle Tx, x\rangle$ is real because

$$\langle Tx, x\rangle = \langle x, T^* x\rangle = \langle x, Tx\rangle = \langle Tx, x\rangle^*.$$

Exercise 4.35. (Idempotent self-adjoints are orthogonal projections.)
Let H be a Hilbert space, and $P \in CL(H)$ be such that $P^2 = P = P^*$. Prove that $Y := \operatorname{ran} P$ is closed and that $P = P_Y$.

Exercise 4.36.
An operator $T \in CL(H)$ on a Hilbert space H is called *skew-adjoint* if $T = -T^*$. Show that every $T \in CL(H)$ can be written as a sum of self-adjoint operator and a skew-adjoint operator in a unique manner.

Exercise 4.37. Let $(\lambda_n)_{n \in \mathbb{N}}$ be a bounded sequence in $\mathbb{C}$, and consider $\Lambda \in CL(\ell^2)$ given by $\Lambda(a_1, a_2, a_3, \cdots) = (\lambda_1 a_1, \lambda_2 a_2, \lambda_3 a_3, \cdots)$ for all $(a_1, a_2, a_3, \cdots) \in \ell^2$. Determine Λ^*.

Exercise 4.38.
Let $\mathrm{I}: L^2[0,1] \to L^2[0,1]$ be the operator on $L^2[0,1]$ (real version), given by

$$(\mathrm{I}\mathbf{x})(t) = \int_0^t \mathbf{x}(\tau) d\tau, \quad t \in [0,1], \ \mathbf{x} \in L^2[0,1].$$

From Example 2.10 (page 70), with $A(t, \tau) = \begin{cases} 1 & \text{if } t \geq \tau, \\ 0 & \text{otherwise} \end{cases}$, $\mathrm{I} \in CL(L^2[0,1])$. Determine I^*.

Exercise 4.39. Consider the anticlockwise *rotation* through angle θ in $\mathbb{R}^2$, given by the operator $T_A : \mathbb{R}^2 \to \mathbb{R}^2$ corresponding to the matrix $A = \begin{bmatrix} \cos\theta & -\sin\theta \\ \sin\theta & \cos\theta \end{bmatrix}$. Give a geometric interpretation of T_A^*. (Note that $T_A^* = T_A^{-1}$. We call $U \in CL(H)$, where H is a Hilbert space, *unitary* if $UU^* = U^*U = I$.)

Exercise 4.40. Let $\{u_n : n \in \mathbb{N}\}$ be an orthonormal basis for a Hilbert space H. For $n \in \mathbb{N}$, consider the subspace $Y_n = \mathrm{span}\{u_1, \cdots, u_n\}$.
Show that the projection $P_n := P_{Y_n} \in H$ is given by $P_n x = \sum_{k=1}^{n} \langle x, u_k \rangle u_k$, $x \in H$.
Prove that $(P_n)_{n \in \mathbb{N}}$ converges strongly to the identity operator I, that is, for all $x \in H$, $P_n x \xrightarrow{n \to \infty} x$ in H. (Although $\|P_n - I\| = 1$ for all $n \in \mathbb{N}$.)

Exercise 4.41. (Hilbert-Schmidt operators).
Let $B = \{u_n : n \in \mathbb{N}\}$ be an orthonormal basis for a Hilbert space H.
An operator $T \in CL(H)$ is said to be *Hilbert-Schmidt* if $\|T\|_{\mathrm{HS}}^2 := \sum_{n=1}^{\infty} \|Tu_n\|^2 < \infty$.
$\|\cdot\|_{\mathrm{HS}}$ is called the *Hilbert-Schmidt norm* of T.

(1) Show that $\|T\|_{\mathrm{HS}}$ does not depend on the choice of the orthonormal basis. *Hint:* If $B' = \{u'_n : n \in \mathbb{N}\}$ is another orthonormal basis, then show that
$$\sum_{n=1}^{\infty} \|Tu_n\|^2 = \sum_{m=1}^{\infty} \|T^* u'_m\|^2 = \sum_{n=1}^{\infty} \|Tu'_n\|^2$$
by using the fact that for a finite double sum of nonnegative terms, the order of summation may be interchanged.

(2) Show that the collection $S_2(H)$ of all Hilbert-Schmidt operators is a subspace of $CL(H)$, and that $\|\cdot\|_{\mathrm{HS}}$ defines a norm on $S_2(H)$.

(3) Prove that $\|T\| \leq \|T\|_{\mathrm{HS}}$ for all Hilbert-Schmidt $T \in CL(H)$.

The importance of Hilbert-Schmidt operators stems from the fact that they are "compact" (to be studied in the next chapter), and hence they can be approximated by finite rank operators (finite matrices). This is useful in numerical analysis. An important example of Hilbert-Schmidt operators are the integral operators $T_A : L^2[0,1] \to L^2[0,1]$ on $L^2[0,1]$ (real version), given by

$$(T_A \mathbf{x})(t) = \int_0^1 A(t, \tau) \mathbf{x}(\tau) d\tau, \quad t \in [0,1], \ \mathbf{x} \in L^2[0,1],$$

where the kernel $A \in L^2([0,1] \times [0,1])$, that is, $\iint_{[0,1] \times [0,1]} (A(t,\tau))^2 d\tau dt < \infty$.

Indeed, if $\{\mathbf{u}_n : n \in \mathbb{N}\}$ is an orthonormal basis for $L^2[0,1]$, then

$$(T_A \mathbf{u}_n)(t) = \int_0^1 A(t,\tau) \mathbf{u}_n(\tau) d\tau = \langle A(t,\cdot), \mathbf{u}_n \rangle, \quad t \in [0,1],$$

and so

$$\sum_{n=1}^{\infty} \|T_A \mathbf{u}_n\|^2 = \sum_{n=1}^{\infty} \int_0^1 \langle A(t,\cdot), \mathbf{u}_n \rangle^2 dt = \int_0^1 \sum_{n=1}^{\infty} \langle A(t,\cdot), \mathbf{u}_n \rangle^2 dt = \int_0^1 \|A(t,\cdot)\|_2^2 dt$$
$$= \int_0^1 \int_0^1 (A(t,\tau))^2 d\tau dt = \iint_{[0,1] \times [0,1]} (A(t,\tau))^2 d\tau dt.$$

Thus T_A is Hilbert-Schmidt with $\|T_A\|_{\mathrm{HS}} = \left(\iint_{[0,1] \times [0,1]} (A(t,\tau))^2 d\tau dt \right)^{1/2}$.

Exercise 4.42. Let H be a Hilbert space, and let $A \in CL(H)$ be fixed. We define $\Lambda : CL(H) \to CL(H)$ by $\Lambda(T) = A^*T + TA$ for $T \in CL(H)$. Show that $\Lambda \in CL(CL(H))$, and that $\Lambda(T)$ is self-adjoint if T is self-adjoint.

Exercise 4.43. Let H be a Hilbert space. Show that the set of all self-adjoint operators on H is a closed subspace of $CL(H)$.

Some more spectral theory

We will begin with a few elementary exercises, and move on to discuss the spectral theory for self-adjoint operators.

Exercise 4.44. Let H be a Hilbert space and $T \in CL(H)$.
Prove that $\mu \in \rho(T)$ if and only if $\mu^* \in \rho(T^*)$.
What is the spectrum of the right shift operator on ℓ^2?

Exercise 4.45. Let Y be a proper closed subspace of a Hilbert space H, and P_Y be the projection onto Y. If $|\lambda| > \|P_Y\| \geq 0$, then we can write by the Neumann Series Theorem that

$$(\lambda I - P_Y)^{-1} = \frac{1}{\lambda}\left(I - \frac{P_Y}{\lambda}\right)^{-1} = \frac{1}{\lambda}\left(I + \frac{P_Y}{\lambda} + \frac{P_Y^2}{\lambda^2} + \frac{P_Y^3}{\lambda^3} + \cdots\right)$$

$$= \frac{1}{\lambda}\left(I + \frac{P_Y}{\lambda} + \frac{P_Y}{\lambda^2} + \frac{P_Y}{\lambda^3} + \cdots\right) = \frac{1}{\lambda}\left(I + \frac{P_Y}{\lambda}\frac{1}{1 - 1/\lambda}\right)$$

$$= \frac{1}{\lambda}\left(I + \frac{P_Y}{\lambda - 1}\right).$$

Show that for $\lambda \in \mathbb{C}\backslash\{0, 1\}$, $(\lambda I - P_Y)^{-1} = \frac{1}{\lambda}\left(I + \frac{P_Y}{\lambda - 1}\right)$.
What is the spectrum of P_Y? What is its point spectrum?

Exercise 4.46. Let H be a Hilbert space and $U \in CL(H)$ be a unitary operator, that is $UU^* = I = U^*U$. Show that eigenvalues $\lambda \in \sigma_p(A)$ lie on the unit circle with center 0 in the complex plane. Also show that the eigenvectors corresponding to distinct eigenvalues are orthogonal.

Recall that the finite-dimensional spectral theorem from linear algebra states that a Hermitian linear transformation T on $\mathbb{C}^d$ has all eigenvalues real, with corresponding eigenvectors forming an orthogonal basis. A similar result holds for Hermitian continuous linear transformations on a Hilbert space H.

Theorem 4.15. *Let H be a Hilbert space and $T \in CL(H)$ satisfy $T = T^*$. Then $\sigma(T) \subset \mathbb{R}$.*

Proof. We will show that for all $\lambda \in \mathbb{C}\backslash\mathbb{R}$, $\lambda I - T$ is invertible, by showing that it is one-to-one and onto. Set $\lambda = a + ib$, where $a, b \in \mathbb{R}$ and $b \neq 0$.

One-to-one: $\|(\lambda I - T)x\|^2 = \|(aI - T)x\|^2 - 2\mathrm{Re}\big(ib\langle x, (aI-T)x\rangle\big) + \|ibx\|^2$ for $x \in H$. As $(aI-T)^* = aI - T^* = aI - T$, $\langle x, (aI-T)x\rangle$ is real, and the term $\mathrm{Re}(\cdots)$ is 0. So $\|(\lambda I - T)x\|^2 = \|(aI - T)x\|^2 + |b|^2\|x\|^2 \geqslant |b|^2\|x\|^2$. Hence if $(\lambda I - T)x = \mathbf{0}$, then $|b|^2\|x\|^2 = 0$, giving $x = \mathbf{0}$. Thus $\lambda I - T$ is one-to-one whenever $\mathrm{Im}(\lambda) \neq 0$.

Onto: First we note that the range of $\lambda I - T$ is closed. To see this, let $(y_n)_{n\in\mathbb{N}}$ be a sequence of vectors in $\lambda I - T$ that converges in H to y. Then there is a sequence $(x_n)_{n\in\mathbb{N}}$ in H such that $(\lambda I - T)x_n = y_n$, $n \in \mathbb{N}$. But the inequality $\|(\lambda I - T)x\| \geqslant |b|\|x\|$, $x \in H$, established above gives with $x = x_n - x_m$ that $\|y_n - y_m\| \geqslant |b|\|x_n - x_m\|$, $n, m \in \mathbb{N}$. As $(y_n)_{n\in\mathbb{N}}$ is Cauchy, so is $(x_n)_{n\in\mathbb{N}}$, and since H is Hilbert, there is some $x \in H$ to which $(x_n)_{n\in\mathbb{N}}$ converges. But then

$$(\lambda I - T)x = (\lambda I - T)\Big(\lim_{n\to\infty} x_n\Big) = \lim_{n\to\infty}(\lambda I - T)x_n = \lim_{n\to\infty} y_n = y.$$

So y belongs to the range of $\lambda I - T$. Consequently, the range of $\lambda I - T$ is closed. Next,

$$\mathrm{ran}(\lambda I - T) = \overline{\mathrm{ran}(\lambda I - T)} = \big(\ker(\lambda I - T)^*\big)^\perp$$
$$= \big(\ker(\lambda^* I - T)\big)^\perp = \{\mathbf{0}\}^\perp = H.$$

So T is onto. $\square$

Just as in the finite dimensional case, one can also show orthogonality of eigenvectors.

Theorem 4.16.
*Let H be a Hilbert space and $T \in CL(H)$ be such that $T = T^*$.
Then the eigenvectors corresponding to distinct eigenvalues are orthogonal.*

Proof. We already know that $\sigma_p(T) \subset \sigma(T) \subset \mathbb{R}$. Let λ_1, λ_2 be distinct real eigenvalues with corresponding eigenvectors v_1, v_2 respectively. Then

$$\lambda_1\langle v_1, v_2\rangle = \langle \lambda_1 v_1, v_2\rangle = \langle Tv_1, v_2\rangle = \langle v_1, T^*v_2\rangle = \langle v_1, Tv_2\rangle$$
$$= \langle v_1, \lambda_2 v_2\rangle = \lambda_2^*\langle v_1, v_2\rangle = \lambda_2\langle v_1, v_2\rangle.$$

This gives $(\lambda_1 - \lambda_2)\langle v_1, v_2\rangle = 0$, and so $\langle v_1, v_2\rangle = 0$. $\square$

Remark 4.7. In the finite dimensional case, we can construct the Hermitian operator given the knowledge of the eigenvalues and eigenvectors. In the infinite dimensional case, an analogous construction can be given. To see this, let us first revisit the finite-dimensional case, and write a procedure for constructing the operator in a way that lends itself to a generalisation to the infinite-dimensional case.

Arrange the spectrum of the Hermitian $T \in CL(\mathbb{C}^d)$ in increasing order: $\lambda_1 < \cdots < \lambda_k$, and let the corresponding eigenspaces be denoted by $Y_1, \cdots, Y_k$. Thus $Y_j = \ker(\lambda_j I - T)$, $j = 1, \cdots, k$. Then the Y_js are mutually orthogonal, and together, they span $\mathbb{C}^d$. If P_j is the projection onto Y_j, then
$$T = \lambda_1 P_1 + \cdots + \lambda_k P_k,$$
$$I = P_1 + \cdots + P_k,$$
$$P_i P_j = P_j P_i = \delta_{ij} P_i.$$
With $E_0 := \mathbf{0}$, we have
$$E_i = P_1 + \cdots + P_i, \quad i = 1, \cdots, k.$$
Also, $E_i^2 = E_i$ and $E_k = I$. One thinks of the family of projections E_is as "growing from $\mathbf{0}$ to I". We have
$$T = \sum_{i=1}^{k} \lambda_i (E_i - E_{i-1}).$$
This is essentially the content of the finite dimensional spectral theorem for Hermitian linear transformations.

Analogously, for a Hermitian operator T on a Hilbert space H, there exists an increasing family of projection operators $(E_\lambda)_{\lambda \in \mathbb{R}}$ such that $E_\lambda = \mathbf{0}$ if $\lambda < \inf \sigma(T)$, $E_\lambda = I$ if $\lambda > \sup \sigma(T)$, and
$$T = \int_{-\infty}^{\infty} \lambda dE_\lambda.$$
The above integral can be used in the following ordinary Riemann-Stieltjes integral sense to reconstruct T:
$$\langle Tx, y \rangle = \int_{-\infty}^{\infty} \lambda d\langle E_\lambda x, y \rangle, \quad x, y \in H.$$
We refer the reader to [Kreyszig (1978), Theorem 9.9-1, page 505] for details. With the extra assumption of compactness, we'll see this in Chapter 5.

Exercise 4.47. (Cayley transform).
Let H be a Hilbert space, and let $T = T^* \in CL(H)$.
Conclude that $T + iI$ is invertible, and that $(T + iI)^{-1}$ commutes with $T - iI$.
Show that $(T - iI)(T + iI)^{-1} =: U$ is unitary.
Moreover, show that $T = i(I + U)(I - U)^{-1}$.

Exercise 4.48. If A, B are Hermitian operators on a Hilbert space H, we write $A \leqslant B$ if for all $x \in H$, $\langle Ax, x \rangle \leqslant \langle Bx, x \rangle$. Show that if Y, Z are closed subspaces of H, then $P_Y \leqslant P_Z$ if and only if $Y \subset Z$.

4.7 An excursion in Quantum Mechanics

We've seen that the Hilbert space theory grows naturally out of a need to generalise the setting of finite-dimensional Euclidean space. Another motivation for their study comes from Physics, where it provides a framework for Quantum Mechanics. The determinism of Newtonian Physics, even after Einstein's revolution with the special and general theories of relativity, was still the prevalent view at the beginning of the 20th century. However, this started getting challenged by experimental evidence in the quantum world, and the deterministic view of the universe started giving way to the probabilistic description of the behaviour of quantum particles. The two seemingly disparate formulations of the fundamental laws of quantum mechanics, developed in the 1920s, the *Matrix Mechanics* of Heisenberg, versus the *Wave Mechanics* of Schrödinger, were shown to be in fact equivalent, both falling under the theory of linear operators on Hilbert spaces. We will briefly discuss the axiomatics of Quantum Mechanics in order to give a flavour of this application of Hilbert space theory to Physics.

State space and states. Every physical system is assumed to correspond to a complex separable Hilbert space, and is called the *state space*. A *state* of the system is a one-dimensional subspace of H, and is represented by a nonzero vector $\psi \in H$ such that $\|\psi\| = 1$. We note that $e^{i\rho}\psi$, $\rho \in \mathbb{R}$ represents[7] the same state as ψ.

Observables. In classical mechanics, recall that observables are real-valued functions on the state space. On the other hand, in quantum mechanics, observables are self-adjoint operators on the Hilbert space H.

Although we have been considering continuous linear transformations, to include physically realistic examples, one needs to consider "unbounded linear operators", that is, (not necessarily continuous) linear transformations $T : D_T(\subset H) \to H$, where D_T is a subspace of H. A thorough discussion of self-adjointness[8] of unbounded operators will take us too far afield in these elementary notes, and so we will assume continuity in our arguments, but relax this assumption in the discussion of examples! We hope that despite this "schizophrenic" approach, much of the spirit of the application of Hilbert space theory to quantum mechanics will nevertheless be conveyed to an uninitiated reader.

---

[7] The "angular freedom", $e^{i\rho}$, is sometimes called the *phase* of the state.
[8] Among other things, a self adjoint operator $T : D_T(\subset H) \to H$ should first of all be *densely defined* (that is, D_T should be dense in H) and *symmetric* (that is, it should satisfy the symmetric condition that for all $\varphi, \psi \in D_T$, $\langle T\psi, \varphi \rangle = \langle \psi, T\varphi \rangle$).

In the classical case, if the state space is $X = \mathbb{R}^2$, and $F \in C^\infty(\mathbb{R}^2)$ is an observable, then this observable imparts a structure on the state space X by the consideration of the fibres of F, which partition X into disjoint subsets, namely the level sets of F. Indeed, we define the equivalence relation $\sim$ on X by setting $x_1 \sim x_2$ if $F(x_1) = F(x_2)$. Then equivalence classes of this equivalence relation partition X into disjoint subsets, and F labels each subset with a distinct real number λ_i, i in some index set I. The quantum mechanical analogue of a partition into disjoint subsets is a decomposition into orthogonal subspaces, and so the analogue of the classical observable should decompose H into orthogonal subspaces. But this is what self-adjoint operators do (for example, we know this by the spectral theorem at least when $\dim H < \infty$, and the following figure illustrates this).

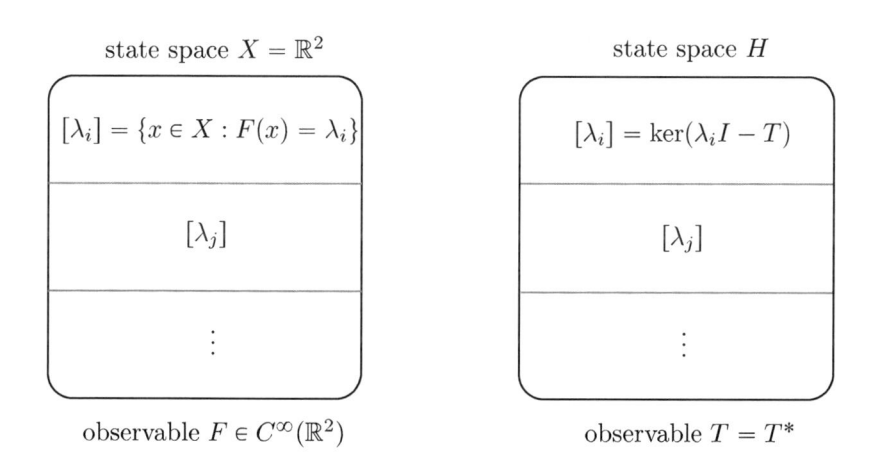

state space $X = \mathbb{R}^2$ state space H

$[\lambda_i] = \{x \in X : F(x) = \lambda_i\}$ $[\lambda_i] = \ker(\lambda_i I - T)$

$[\lambda_j]$ $[\lambda_j]$

$\vdots$ $\vdots$

observable $F \in C^\infty(\mathbb{R}^2)$ observable $T = T^*$

In classical mechanics, the result of an experiment measuring an observable F on a classical system in a state x is the real number $F(x)$ which belongs to the range of the observable at hand. In quantum mechanics, the result of an experiment measuring an observable T on a quantum system in state ψ is an eigenvalue of T. Which eigenvalue of T? The answer is no longer deterministic, but rather probabilistic. The act of measuring the observable T "collapses" the state ψ of the system into a state φ, where φ is a normalised eigenvector of T. The probability with which this happens is $|\langle \psi, \varphi \rangle|^2$. If the state has collapsed to φ, and $T\varphi = \lambda\varphi$ then the measured value of the observable T is λ. If a quantum mechanical system is in a state φ, where φ is an eigenvector of the observable T, then what is the probability that the state stays in the same state φ after T is measured?

The answer is $|\langle \varphi, \varphi \rangle|^2 = \|\varphi\|^2 = 1$. This feature of Quantum Mechanics, where the result of an experiment is not deterministic, but random, and the fact that the state of the system is in general no longer the same as that prior to the measurement is one of the key differences between classical and quantum mechanics.

We now introduce some terminology from probability theory. The *expected value of an observable T of a system which is in a state ψ* is given by

$$\langle T \rangle_\psi := \langle T\psi, \psi \rangle.$$

Note that since T is self-adjoint, this is a real number. This number is to be interpreted as follows: if we do multiple experiments on a quantum system in the *same* state ψ, and we measure the observable T in each instance, this is the average of the measured value we would obtain. The *root mean square deviation of an observable T of a system which is in a state ψ* is given by

$$\Delta_\psi T := \sqrt{\langle (T - \langle T \rangle_\psi I)^2 \rangle_\psi}.$$

Let us simplify this somewhat. First note that $T - \langle T \rangle_\psi I$ is self adjoint, and so is its square. So the quantity under the square root is well-defined and moreover, it is nonnegative because

$$\langle (T - \langle T \rangle_\psi I)^2 \rangle_\psi = \langle (T - \langle T \rangle_\psi I)^2 \psi, \psi \rangle = \|(T - \langle T \rangle_\psi I)\psi\|^2 \geq 0.$$

Moreover, we have

$$\begin{aligned}
(\Delta_\psi T)^2 &= \langle (T - \langle T \rangle_\psi I)^2 \rangle_\psi \\
&= \langle (T - \langle T \rangle_\psi I)\psi, (T - \langle T \rangle_\psi I)\psi \rangle \\
&= \langle T\psi, T\psi \rangle - 2\mathrm{Re}\langle T\psi, \langle T \rangle_\psi I\psi \rangle + \langle \langle T \rangle_\psi I\psi, \langle T \rangle_\psi I\psi \rangle \\
&= \langle T^2 \psi, \psi \rangle - 2\langle T \rangle_\psi \langle T\psi, \psi \rangle + \langle T \rangle_\psi^2 \langle \psi, \psi \rangle \\
&= \langle T^2 \rangle_\psi - 2\langle T \rangle_\psi \langle T \rangle_\psi + \langle T \rangle_\psi^2 \cdot 1 \\
&= \langle T^2 \rangle_\psi - \langle T \rangle_\psi^2.
\end{aligned}$$

Note that if ψ is a normalised eigenvector of T with eigenvalue λ, then $\langle T \rangle_\psi = \lambda$, and $\langle T^2 \rangle_\psi = \lambda^2$, so that $\Delta_\psi T = 0$. Vice versa, if $\Delta_\psi T = 0$, then $(T - \langle T \rangle_\psi I)\psi = \mathbf{0}$, that is, $T\psi = \langle T \rangle_\psi \psi$, so that ψ is an eigenvector of T corresponding to the eigenvalue $\langle T \rangle_\psi$. So we've shown that a state ψ is an eigenvector of an observable T if and only if $\Delta_\psi T = 0$ ("dispersion-free"). Such states are called *pure states of the observable T*.

Example 4.17. (Position and momentum operators). Let $H = L^2(\mathbb{R})$, and define the *position operator* $Q : D_Q(\subset L^2(\mathbb{R})) \to L^2(\mathbb{R})$ by
$$(Q\Psi)(x) = x\Psi(x), \ x \in \mathbb{R}, \ \Psi \in D_Q := \{\Psi \in L^2(\mathbb{R}) : x \mapsto x\Psi(x) \in L^2(\mathbb{R})\}.$$
Then Q is a linear transformation. Moreover, for $\Psi_1, \Psi_2 \in D_Q$, we have
$$\langle Q\Psi_1, \Psi_2 \rangle = \int_{-\infty}^{\infty} x\Psi_1(x)\big(\Psi_2(x)\big)^* dx = \int_{-\infty}^{\infty} \Psi_1(x)\big(x\Psi_2(x)\big)^* dx = \langle \Psi_1, Q\Psi_2 \rangle.$$
The expected value of Q for a system in a state Ψ is
$$\langle Q \rangle_\Psi = \langle Q\Psi, \Psi \rangle = \int_{-\infty}^{\infty} x|\Psi(x)|^2 dx.$$
So if we think of $\mathbb{R}$ as space, then $|\Psi|^2$ can be thought of as the probability density function for the random variable of the result of the experiment of measuring the position of the particle in a state Ψ. The probability of finding the particle in a region $(a, b) \subset \mathbb{R}$ is then
$$\int_a^b |\Psi(x)|^2 dx.$$
The *momentum operator* $P : D_P(\subset L^2(\mathbb{R})) \to L^2(\mathbb{R})$ is given by
$$(P\Psi)(x) = -i\hbar\frac{d}{dx}\Psi(x), \ x \in \mathbb{R}, \ \Psi \in D_P := \{\Psi \in L^2(\mathbb{R}) : \Psi' \in L^2(\mathbb{R})\}.$$
Then P is a linear transformation. Moreover, for $\Psi_1, \Psi_2 \in D_P$, we have, using integration by parts, that
$$\langle P\Psi_1, \Psi_2 \rangle = \int_{-\infty}^{\infty} -i\hbar\Psi_1'(x)\big(\Psi_2(x)\big)^* dx$$
$$= -i\hbar\Psi_1(x)\Psi_2(x)\Big|_{-\infty}^{\infty} - \int_{-\infty}^{\infty} \big(-i\hbar\Psi_1(x)\big)\big(\Psi_2'(x)\big)^* dx$$
$$= 0 + \int_{-\infty}^{\infty} \Psi_1(x)\big(-i\hbar\Psi_2'(x)\big)^* dx \quad \text{(Exercise 4.49)}$$
$$= \langle \Psi_1, P\Psi_2 \rangle.$$

We had seen in Exercise 2.35, page 103, that Q has no eigenvectors. The momentum operator doesn't have any eigenvectors either, since if $\Psi \in L^2(\mathbb{R}) \setminus \{\mathbf{0}\}$ is such that $P\Psi = \lambda\Psi$ for some $\lambda \in \mathbb{R}$, then
$$\frac{d}{dx}\Psi(x) = \frac{i\lambda}{\hbar}\Psi(x), \quad x \in \mathbb{R},$$
giving $\Psi(x) = e^{i\lambda x/\hbar}\Psi(0)$, and as $\Psi \neq \mathbf{0}$, $\Psi(0) \neq 0$. But then
$$\|\Psi\|_2^2 = \int_{-\infty}^{\infty} |\Psi(x)|^2 dx = \int_{-\infty}^{\infty} |e^{i\lambda x/\hbar}\Psi(0)|^2 dx = |\Psi(0)|^2 \int_{-\infty}^{\infty} 1 dx = \infty,$$
a contradiction to $\Psi \in L^2(\mathbb{R})$! $\diamondsuit$

Exercise 4.49.
(1) Let $f : [0, \infty) \to [0, \infty)$ be differentiable and $\int_0^\infty f(x)dx$, $\int_0^\infty f'(x)dx$ exist. Show that $\lim_{x \to \infty} f(x) = 0$.
(2) Prove that if $\Psi, \Psi' \in L^2(\mathbb{R})$, then $\lim_{x \to \pm\infty} \Psi(x) = 0$.

Heisenberg Uncertainty Principle. The famous Heisenberg Uncertainty Principle that one learns in elementary physics courses states that the position and momentum of a quantum particle cannot be measured simultaneously, with their respective dispersions satisfying $(\sigma_x)(\sigma_p) \geq \hbar/2$, where $\hbar$ is the reduced Planck's constant. (This famously resolves the question of the stability of the atom, namely why the negatively charged electrons don't just radiate electromagnetic energy and fall into the positively charged nucleus. The uncertainty principle guarantees that when the electron tries to do so, it would be confined in a smaller space, thus reducing the dispersion in position, and increasing the dispersion in momentum, and hence creating a tendency in the electron to not want to do so! So it settles in a happy cloud around the nucleus, giving the atom its "size".) We will now show an Uncertainty Principle not just for the position-momentum pair of observables, but for any pair, and this is just an application of the Cauchy-Schwarz Inequality!

Theorem 4.17. (Uncertainty Principle).
Let A, B be self-adjoint operators on a Hilbert space H and $\psi \in H$. Then
$$(\Delta_\psi A)(\Delta_\psi B) \geq \frac{|\langle [A, B] \rangle_\psi|}{2},$$
where $[A, B] := AB - BA$ is the commutator of A, B.

Proof. We have
$$|\langle [A,B]\rangle_\psi|^2 = |\langle (AB-BA)\psi, \psi\rangle|^2 = |\langle AB\psi, \psi\rangle - \langle BA\psi, \psi\rangle|^2$$
$$= |\langle AB\psi, \psi\rangle - \langle \psi, AB\psi\rangle|^2 = |\langle AB\psi, \psi\rangle - \langle AB\psi, \psi\rangle^*|^2$$
$$= (2\mathrm{Im}\langle AB\psi, \psi\rangle)^2, \qquad (4.6)$$

where $\mathrm{Im}(\cdot)$ denotes the imaginary part of a complex number. If $a := \langle A\rangle_\psi$ and $b := \langle B\rangle_\psi$, then $a, b \in \mathbb{R}$, and we have
$$[A - aI, B - bI] = (A - aI)(B - bI) - (B - bI)(A - aI)$$
$$= AB - aB - bA + abI - BA + aB + bA - abI$$
$$= AB - BA = [A, B].$$

Thus we have

$$\begin{aligned}|\langle[A,B]\rangle_\psi|^2 &= |\langle[A-aI,B-bI]\rangle_\psi|^2 \\ &= (2\operatorname{Im}\langle(A-aI)(B-bI)\psi,\psi\rangle)^2 \quad \text{(using (4.6))} \\ &= (2\operatorname{Im}\langle(B-bI)\psi,(A-aI)\psi\rangle)^2 \\ &\leqslant 4\|(A-aI)\psi\|^2\|(B-bI)\psi\|^2 \quad \text{(Cauchy-Schwarz)} \\ &= 4(\Delta_\psi A)^2(\Delta_\psi B)^2.\end{aligned}$$

This completes the proof. □

Example 4.18. (Position-momentum uncertainty).
For $\Psi \in D_P \cap D_Q$ (for example Ψ can be any compactly supported smooth function), we have

$$[P,Q]\Psi = (PQ-QP)\Psi = -i\hbar\bigl((x\Psi)' - x(\Psi')\bigr) = -i\hbar\bigl(x\Psi' + \Psi - x\Psi'\bigr) = -i\hbar\Psi,$$

and so $(\Delta_\Psi P)(\Delta_\Psi Q) \geqslant \dfrac{\hbar}{2}$. ◇

Exercise 4.50. Show that the commutator of two observables is skew-adjoint.

Exercise 4.51. (Jacobi Identity).
Show that for observables A,B,C, $[A,[B,C]] + [B,[C,A]] + [C,[A,B]] = 0$.

Exercise 4.52. Show that for $n \in \mathbb{N}$, $[Q^n, P] = i\hbar n Q^{n-1}$.

The uncertainty relation of the position-momentum type given in Example 4.18 is obtained for any two observables satisfying the commutation relation $[A,B] = i\hbar I$, and such observables are said to be *complementary*. The need for considering unbounded operators in Quantum Mechanics is thus further supported by Exercise 2.22, page 86, from which it follows that there do not exist continuous linear transformations A,B on H such that $[A,B] = -i\hbar I$. Thus the beloved continuity of mathematicians appears quite at odds with the "unbounded" requirements of physicists for realistic applications. Nevertheless, $CL(H)$ still has its role to play in quantum mechanics, for these unbounded operators do "generate" groups of unitary operators, which play a role in dynamics, to be discussed below.

Quantum dynamics and Schrödinger's equation. In classical mechanics, the evolution equations for the state $(\mathbf{q}_*(t), \mathbf{p}_*(t))$ in Hamiltonian form are:

$$\frac{d\mathbf{q}_*}{dt}(t) = \frac{\partial \mathrm{H}}{\partial p}(\mathbf{q}_*(t), \mathbf{p}_*(t))$$
$$\frac{d\mathbf{p}_*}{dt}(t) = -\frac{\partial \mathrm{H}}{\partial q}(\mathbf{q}_*(t), \mathbf{p}_*(t)),$$

where $(q, p) \mapsto \mathrm{H}(q, p) : \mathbb{R}^2 \to \mathbb{R}$ is the Hamiltonian observable[9]. With

$$q \leftrightsquigarrow Q,$$
$$p \leftrightsquigarrow P,$$

we get a quantum mechanical Hamiltonian observable $\mathrm{H}(Q, P)$, and the evolution equation for the state ψ is *Schrödinger's equation*,

$$\frac{d}{dt}\psi(t) = -\frac{i}{\hbar}\mathrm{H}\psi(t).$$

Here we assume that H is time-independent. What does the solution look like? We elaborate on this in the following paragraph.

Just as real numbers are related to unit-modulus complex numbers via the exponential map $\tau \mapsto e^{i t\tau}$, $t \in \mathbb{R}$, (even unbounded) self-adjoint operators are related to (bounded) unitary ones. One has the following result due to Stone from 1932:

Theorem 4.18. *Every self-adjoint operator T on a Hilbert space H generates a strongly continuous one-parameter group of unitary operators e^{itT}, $t \in \mathbb{R}$, on H. Conversely, every such one-parameter group $(U(t))_{t\in\mathbb{R}}$ is generated by a unique self-adjoint operator, given by*

$$T\psi = \lim_{t \to 0} \frac{U(t)\psi - U(0)\psi}{it}, \quad \psi \in D_T,$$
$$D_T := \left\{\psi \in H : \lim_{t \to 0} \frac{U(t)\psi - U(0)\psi}{it} \text{ exists}\right\}.$$

If $\psi(0) \in D_\mathrm{H}$, then the solution to Schrödinger's equation is given by

$$\psi(t) = e^{-it\mathrm{H}/\hbar}\psi(0), \quad t \in \mathbb{R}.$$

The evolution of the expected value of an observable T is given by

$$\frac{d}{dt}\langle T \rangle_{\psi(t)} = \frac{d}{dt}\langle T\psi(t), \psi(t)\rangle = \langle T\psi'(t), \psi(t)\rangle + \langle T\psi(t), \psi'(t)\rangle$$
$$= -\frac{i}{\hbar}\langle T\mathrm{H}\psi(t), \psi(t)\rangle + \frac{i}{\hbar}\langle T\psi(t), \mathrm{H}\psi(t)\rangle$$
$$= -\frac{i}{\hbar}\langle T\mathrm{H}\psi(t), \psi(t)\rangle + \frac{i}{\hbar}\langle \mathrm{H}T\psi(t), \psi(t)\rangle$$
$$= -\frac{i}{\hbar}\langle (T\mathrm{H} - \mathrm{H}T)\psi(t), \psi(t)\rangle = \left\langle -\frac{i}{\hbar}[T, \mathrm{H}]\psi(t), \psi(t)\right\rangle$$
$$= \left\langle \frac{1}{i\hbar}[T, \mathrm{H}]\right\rangle_{\psi(t)},$$

---

[9] We use upright H for the Hamiltonian here in order to avoid confusion with the letter H used for the underlying Hilbert space in our quantum mechanical system.

that is,
$$\frac{d}{dt}\langle T\rangle_{\psi(t)} = \left\langle \frac{1}{i\hbar}[T,\mathrm{H}]\right\rangle_{\psi(t)},$$
which is reminiscent of the Poissonian equation in classical mechanics for the evolution of an observable F:
$$\frac{dF}{dt} = \{F,\mathrm{H}\}.$$
Thus the role of the Poisson bracket $\{\cdot,\cdot\}$ in Classical Mechanics is replaced by commutator
$$\frac{1}{i\hbar}[\cdot,\cdot]$$
in Quantum Mechanics.

Exercise 4.53. (Quantisation).
In our discussion of Classical Mechanics in Chapter 3, recall that in one space dimension, the position observable $Q \in C^\infty(\mathbb{R}^2)$ was given by
$$(p,q) \stackrel{Q}{\mapsto} q : \mathbb{R}^2 \to \mathbb{R},$$
and the momentum observable $P \in C^\infty(\mathbb{R}^2)$ was given by
$$(p,q) \stackrel{P}{\mapsto} p : \mathbb{R}^2 \to \mathbb{R}.$$
Show that the classical Poisson bracket $\{Q^2, P^2\} = 4QP$, but that this *does not* correspond to the quantum mechanical commutator $[Q^2, P^2]/(i\hbar)$ of the quantum mechanical observables Q^2 and P^2.

By rewriting the classical Poisson bracket $\{Q^2, P^2\} = 4QP = 2(QP+PQ)$ (which is allowed, since the order or writing does not matter classically), this expression does correspond to the quantum mechanical case.

Check that QP is not self-adjoint, but that $QP + PQ$ satisfies the symmetry condition $\langle (PQ+QP)\Phi, \Psi\rangle = \langle \Phi, (PQ+QP)\Psi\rangle$ for all Ψ, Φ.

(The process of replacing classical observables by quantum mechanical ones is called "quantisation", and this example shows that there is no unique way of doing this owing to the lack of commutativity of the quantum mechanical observables: a helpful device available is that one should rearrange the Poisson bracket in a manner such that when quantised, it results in a self-adjoint operator.)

Exercise 4.54. (Probability is conserved).
Let $t \mapsto \psi(t)$ be the solution to Schrödinger's equation for some selfadjoint Hamiltonian H. Show that if $\|\psi(0)\|^2 = 1$, then $\|\psi(t)\|^2 = 1$ for all $t \in \mathbb{R}$.

Exercise 4.55. The one-dimensional Schrödinger equation in $H = L^2(\mathbb{R})$ is
$$i\hbar \frac{\partial \Psi}{\partial t} = -\frac{\hbar^2}{2m}\frac{\partial^2 \Psi}{\partial x^2} + V(x)\cdot \Psi,$$
where V is the potential energy.
One method to find solutions is to assume that variables separate, that is, the solution has the form $\Psi(x,t) = X(x)T(t)$.

Substituting $\Psi(x,t) = X(x)T(t)$ in the Schrödinger equation gives
$$\frac{i\hbar T'(t)}{T(t)} = \frac{-\frac{\hbar^2}{2m}X''(x) + V(x)X(x)}{X(x)}.$$
As the left-hand side is constant along vertical lines in the (x,t)-plane, and the right-hand side is constant along horizontal lines, it follows that their common value is a constant function in the (x,t)-plane, with constant value, say E (for "energy"). So we obtain the following equation for T,
$$T'(t) = \frac{-iE}{\hbar}T(t),$$
which has the solution $T(t) = C\exp\left(\frac{-iEt}{\hbar}\right) = C\left(\cos\left(\frac{Et}{\hbar}\right) - i\sin\left(\frac{Et}{\hbar}\right)\right)$.

The equation for X is $-\frac{\hbar^2}{2m}X''(x) + (V(x) - E)X(x) = 0$.

Consider a free particle of mass m confined to the interval $0 < x < \pi$ (so that $V \equiv 0$ in $(0,\pi)$), and suppose that $\Psi(0,t) = \Psi(\pi,t) = 0$ for all t. (Imagine the particle to be in an "infinite potential well".)

Show that this problem has a nontrivial solution if and only if
$$E = \frac{n^2\hbar^2}{2m}, \quad n = 1, 2, 3, \cdots.$$
Sketch the probability density function $|\Psi|^2$ when $n = 1, 2$, and compute the probability that the particle is in the interval $[0, 1/4]$ in each case.

Notes

Example 4.6 and Theorem 4.6 are based on [Thomas (1997)].

Chapter 5

Compact operators

In this chapter, we study a special class of linear operators, called compact operators.

Why should we study compact operators? One important reason is that they can be approximated by finite rank operators. So they play an important role in the numerical approximation of solutions to operator equations. We had seen that if H is an infinite-dimensional Hilbert space and $A \in CL(H)$ with $\|A\| < 1$, then given $y \in H$, there is a unique $x \in H$ such that

$$(I - A)x = y,$$

which is given by the Neumann series

$$x = (I - A)^{-1}y = (I + A + A^2 + A^3 + \cdots)y.$$

But all of the operators $A, A^2, A^3, \cdots$ act on the infinite-dimensional H, so that computing these powers may not at all be feasible, and the convergence of the series may be "slow" (see Exercise 2.23, page 87). But now imagine that we can approximate A by finite matrices A_n and consider instead

$$(I - A_n)x_n = y_n$$

where the $y_n \to y$ as $n \to \infty$, and the unknown x_n are obtained by solving the finite linear algebraic equation $(I - A_n)x_n = y_n$. Then we can easily compute $x_n = (I - A_n)^{-1}y_n$, and if $x_n \to x$, then we are able to determine x approximately. This wishful thinking can be made a reality if A is "compact", as we shall see later on in this chapter when we learn Theorem 5.6 (page 218) on Galerkin approximations.

We begin by giving the definition of a compact operator.

5.1 Compact operators

Definition 5.1. (Compact operators). Let X, Y be normed spaces. A linear transformation $T : X \to Y$ is said to be *compact* if

> for every bounded sequence $(x_n)_{n \in \mathbb{N}}$ contained in X, $(Tx_n)_{n \in \mathbb{N}}$ has a convergent subsequence.

We will denote the set of all compact operators from X to Y by $K(X, Y)$.

Why do we call such operators compact? The following result answers this question. Recall firstly that a closed and bounded set in an *infinite* dimensional normed space may fail to be compact (Example 1.26, page 48, and Example 1.28, page 50). So if $T \in CL(X, Y)$, and B is the closed unit ball in X with centre $\mathbf{0}$, then although we know that $\overline{T(B)}$ is closed and bounded, it needn't be compact. However, *compact* operators T are special in the sense that $\overline{T(B)}$ is guaranteed to be compact!

Theorem 5.1.
Let X, Y be normed spaces, and $T : X \to Y$ be a linear transformation. Then the following are equivalent:
(1) T is compact.
(2) The closure $\overline{T(B)}$ of the image under T of the closed unit ball, $B := \{x \in X : \|x\| \leq 1\}$, is compact.

Proof.
(1) $\Rightarrow$ (2): Let $(z_n)_{n \in \mathbb{N}}$ be a sequence in $\overline{T(B)}$. Then there is a sequence $(y_n)_{n \in \mathbb{N}}$ in $T(B)$ such that

$$\|y_n - z_n\| < \frac{1}{n} \text{ for all } n \in \mathbb{N}. \tag{5.1}$$

Let $y_n = Tx_n$, $x_n \in B$, $n \in \mathbb{N}$. Since for all n we have $\|x_n\| \leq 1$, and because T is compact, $(y_n)_{n \in \mathbb{N}}$ has a convergent subsequence, say $(y_{n_k})_{k \in \mathbb{N}}$, converging to, say y. As each $y_n \in T(B)$, we have $y \in \overline{T(B)}$. From (5.1), $(z_{n_k})_{k \in \mathbb{N}}$ converges to y too. Hence $\overline{T(B)}$ is compact.

(2) $\Rightarrow$ (1): Let $(x_n)_{n \in \mathbb{N}}$ be a bounded sequence in X, and let $M > 0$ be such that for all $n \in \mathbb{N}$, $\|x_n\| \leq M$. Then $(x_n/M)_{n \in \mathbb{N}}$ is in B, and $(T(x_n)/M)_{n \in \mathbb{N}}$ is in $T(B) \subset \overline{T(B)}$. As $\overline{T(B)}$ is compact, $(T(x_n)/M)_{n \in \mathbb{N}}$ has a convergent subsequence. Thus $(Tx_n)_{n \in \mathbb{N}}$ has a convergent subsequence. Consequently, T is compact. $\square$

5.2 The set $K(X, Y)$ of all compact operators

Corollary 5.1.
Every compact operator is continuous, that is $K(X, Y) \subset CL(X, Y)$.

Proof. Let $B := \{x \in X : \|x\| \leq 1\}$. If T is compact, then $\overline{T(B)}$ is compact, and in particular, bounded. So $T(B) \subset \overline{T(B)}$ is bounded too. So there is some $M > 0$ such that $\|Tz\| \leq M$ for all $z \in B$. But this gives $\|Tx\| \leq M\|x\|$ for all $x \in X$. (This is trivially true when $x = \mathbf{0}$, and if $x \neq \mathbf{0}$, then by taking $z = x/\|x\| \in B$, we have $\|Tz\| \leq M$, yielding the desired inequality.) So $T \in CL(X, Y)$. □

Is it true that $K(X, Y) = CL(X, Y)$? No:

Example 5.1. (Not all continuous linear transformations are compact). Let X be any infinite dimensional inner product space, for example ℓ^2. We will show that the identity operator $I \in CL(X)$ is not compact.

Let $\{u_1, u_2, u_3, \cdots\}$ be any orthonormal set in X. (Start with any countably infinite independent set, and use Gram-Schmidt.) Then $\|u_n\| = 1$ for all $n \in \mathbb{N}$, and so the sequence $(u_n)_{n \in \mathbb{N}}$ is bounded. However, the sequence $(Iu_n)_{n \in \mathbb{N}}$ has no convergent subsequence, since for all $n, m \in \mathbb{N}$ with $n \neq m$, we have $\|Iu_n - Iu_m\|^2 = \|u_n - u_m\|^2 = 1 + 0 + 0 + 1 = 2$, and this can't be made as small as we please. Hence I is not compact, but is continuous. ◇

In contrast to the above, it turns out that all finite rank operators are compact. Recall that an operator T is called a *finite rank operator* if its range, $\text{ran}(T)$, is a finite-dimensional vector space. For ease of exposition, we will just prove this when Y is an inner product space.

Theorem 5.2. *Let X be a normed space and Y be an inner product space. Suppose that $T \in CL(X, Y)$ is such that $\text{ran}(T)$ is finite dimensional. Then T is compact.*

Proof. Let $\{u_1, \cdots, u_m\}$ be an orthonormal basis for $\text{ran}(T)$. Let $(x_n)_{n \in \mathbb{N}}$ be a bounded sequence in X. Suppose that $M > 0$ is such that $\|x_n\| \leq M$ for all $n \in \mathbb{N}$. We want to show that $(Tx_n)_{n \in \mathbb{N}}$ has a convergent subsequence. (We will show that

$$\langle Tx_{n_k}, u_1 \rangle \xrightarrow{k \to \infty} \alpha_1, \cdots, \langle Tx_{n_k}, u_m \rangle \xrightarrow{k \to \infty} \alpha_m \text{ and that } Tx_{n_k} \to \sum_{\ell=1}^{m} \alpha_\ell u_\ell$$

for some subsequence $(x_{n_k})_{k \in \mathbb{N}}$.) For all $n \in \mathbb{N}$ and each $\ell \in \{1, \ldots, m\}$,
$$|\langle Tx_n, u_\ell \rangle| \leq \|Tx_n\|\|u_\ell\| \leq \|T\|\|x_n\| \cdot 1 \leq \|T\|M.$$

$(x_n)_{n\in\mathbb{N}}$ has some subsequence $(x_n^{(1)})_{n\in\mathbb{N}}$ such that $\langle Tx_n^{(1)}, u_1\rangle \stackrel{n\to\infty}{\longrightarrow} \alpha_1$.
$(x_n^{(1)})_{n\in\mathbb{N}}$ has some subsequence $(x_n^{(2)})_{n\in\mathbb{N}}$ such that $\langle Tx_n^{(2)}, u_2\rangle \stackrel{n\to\infty}{\longrightarrow} \alpha_2$.
...
$(x_n^{(m-1)})_{n\in\mathbb{N}}$ has some subsequence $(x_n^{(m)})_{n\in\mathbb{N}}$ such that $\langle Tx_n^{(m)}, u_m\rangle \stackrel{n\to\infty}{\longrightarrow} \alpha_m$.

Claim: $(Tx_n^{(m)})_{n\in\mathbb{N}}$ converges to $\alpha_1 u_1 + \cdots + \alpha_m u_m$.

As
$$\left\|Tx_n^{(m)} - (\alpha_1 u_1 + \cdots + \alpha_m u_m)\right\|^2 = \left\|\sum_{\ell=1}^m \langle Tx_n^{(m)}, u_\ell\rangle u_\ell - \sum_{\ell=1}^m \alpha_\ell u_\ell\right\|^2$$
$$= \left\|\sum_{\ell=1}^m (\langle Tx_n^{(m)}, u_\ell\rangle - \alpha_\ell) u_\ell\right\|^2$$
$$= \sum_{\ell=1}^m |\langle Tx_n^{(m)}, u_\ell\rangle - \alpha_\ell|^2 \stackrel{n\to\infty}{\longrightarrow} 0,$$

it follows that $(Tx_n^{(m)})_{n\in\mathbb{N}}$ is a convergent subsequence of the sequence $(Tx_n)_{n\in\mathbb{N}}$. Consequently T is compact. □

In elementary linear algebra, not only were all linear transformations from $\mathbb{C}^n$ to $\mathbb{C}^m$ continuous, they were even compact!

Example 5.2. ($L(\mathbb{C}^n, \mathbb{C}^m) = CL(\mathbb{C}^n, \mathbb{C}^m) = K(\mathbb{C}^n, \mathbb{C}^m)$).
If $A \in \mathbb{C}^{n\times m}$, then $T_A \in CL(\mathbb{C}^n, \mathbb{C}^m)$ given by $T_A\mathbf{x} = A\mathbf{x}$, $\mathbf{x} \in \mathbb{C}^n$, is finite-rank because $\mathrm{ran}\, T_A \subset \mathbb{C}^m$, and so T_A is compact. In particular, the identity map $I : \mathbb{C}^d \to \mathbb{C}^d$ is compact. ◇

We had seen that $K(X,Y) \subset CL(X,Y)$. But $CL(X,Y)$ is a vector space, with the usual pointwise operations. So it is natural to ask if $K(X,Y)$ is a *subspace* of $CL(X,Y)$. The answer is "yes", and this is what we show next.

Theorem 5.3. ($K(X,Y) \underset{\text{subspace}}{\subset} CL(X,Y)$).
If X, Y are normed spaces, then $K(X,Y)$ is a subspace of $CL(X,Y)$.

Proof.

(S1) $\mathbf{0}$ is compact since $(\mathbf{0}x_n)_{n\in\mathbb{N}} = (\mathbf{0})_{n\in\mathbb{N}}$ is convergent for all bounded sequences $(x_n)_{n\in\mathbb{N}}$ in X.

(S2) If T, S are compact, and $(x_n)_{n\in\mathbb{N}}$ is bounded, then $(T_n)_{n\in\mathbb{N}}$ has some subsequence $(Tx_{n_k})_{k\in\mathbb{N}}$ that is convergent, and $(Sx_{n_k})_{k\in\mathbb{N}}$ has some subsequence $(Sx_{n_{k_\ell}})_{\ell\in\mathbb{N}}$ that is convergent. So $\big((T+S)x_{n_{k_\ell}}\big)_{\ell\in\mathbb{N}}$ is convergent. Thus $T + S$ is compact.

(S3) If T, S are compact, $\alpha \in \mathbb{K}$ and $(x_n)_{n\in\mathbb{N}}$ is bounded, then $(T_n)_{n\in\mathbb{N}}$ has some subsequence $(Tx_{n_k})_{k\in\mathbb{N}}$ that is convergent, and so it follows that $\big((\alpha T)x_{n_k}\big)_{k\in\mathbb{N}} = \big(\alpha(Tx_{n_k})\big)_{k\in\mathbb{N}}$ is convergent. Thus αT is compact. □

Compact operators

Since $CL(X,Y)$ is a normed space, we can even ask if $K(X,Y)$ is a *closed* subspace of $CL(X,Y)$. We now show that *if Y is a Banach space*, then $K(X,Y)$ is a closed subspace of $CL(X,Y)$, or briefly:

"Limits of compact operators are compact."

Theorem 5.4. *Let X be a normed space, Y a Banach space, and $(T_n)_{n\in\mathbb{N}}$ be a sequence in $K(X,Y)$ that converges in $CL(X,Y)$ to $T \in CL(X,Y)$. Then T is compact.*

Proof. Suppose that $(x_n)_{n\in\mathbb{N}}$ is a bounded sequence in X, and let $M > 0$ be such that for all $n \in \mathbb{N}$, $\|x_n\| \leq M$. Since T_1 is compact, $(T_1 x_n)_{n\in\mathbb{N}}$ has a convergent subsequence $(T_1 x_n^{(1)})_{n\in\mathbb{N}}$, say. Again, since $(x_n^{(1)})_{n\in\mathbb{N}}$ is a bounded sequence, and T_2 is compact, $(T_2 x_n^{(1)})_{n\in\mathbb{N}}$ has a convergent subsequence, say $(T_2 x_n^{(2)})_{n\in\mathbb{N}}$. We continue in this manner to obtain the following:

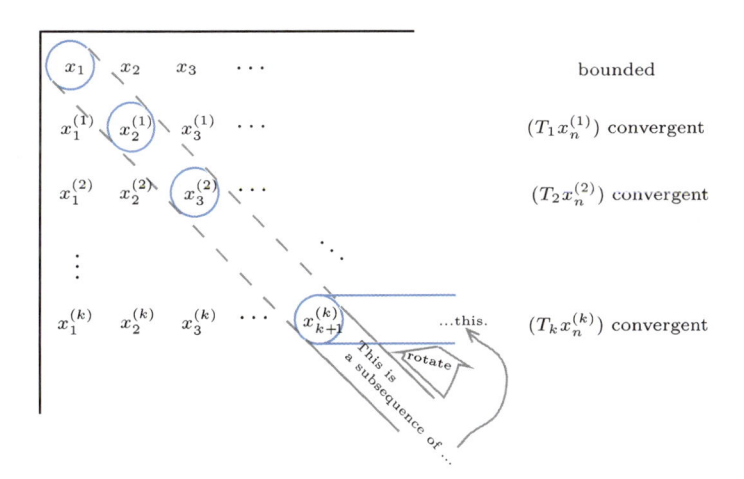

Consider the diagonal sequence $x_1, x_2^{(1)}, x_3^{(2)}, \cdots$.
By meditating on the above picture, one can convince oneself that

$$x_{k+1}^{(k)}, x_{k+2}^{(k+1)}, x_{k+3}^{(k+2)}, \cdots \text{ is a subsequence of } x_{k+1}^{(k)}, x_{k+2}^{(k)}, x_{k+3}^{(k)}, \cdots.$$

As $T_k x_1^{(k)}, T_k x_2^{(k)}, T_k x_3^{(k)}, \cdots$ converges, its subsequence,

$$T_k x_{k+1}^{(k)}, T_k x_{k+2}^{(k+1)}, T_k x_{k+3}^{(k+2)}, \cdots$$

converges too, and so $(T_k x_{n+1}^{(n)})_{n\in\mathbb{N}}$ converges.

For $n, m \in \mathbb{N}$, we have

$$\|Tx_{n+1}^{(n)} - Tx_{m+1}^{(m)}\|$$
$$\leq \|Tx_{n+1}^{(n)} - T_k x_{n+1}^{(n)}\| + \|T_k x_{n+1}^{(n)} - T_k x_{m+1}^{(m)}\| + \|T_k x_{m+1}^{(m)} - Tx_{m+1}^{(m)}\|$$
$$\leq \|T - T_k\| \|x_{n+1}^{(n)}\| + \|T_k x_{n+1}^{(n)} - T_k x_{m+1}^{(m)}\| + \|T_k - T\| \|x_{m+1}^{(m)}\|$$
$$\leq 2M \cdot \underbrace{\|T - T_k\|}_{\substack{\text{can be made small} \\ \text{by choosing } k \text{ big}}} + \underbrace{\|T_k x_{n+1}^{(n)} - T_k x_{m+1}^{(m)}\|}_{\substack{\text{can be made small} \\ \text{by choosing } n, m \text{ big}}}.$$

Hence $(Tx_{n+1}^{(n)})_{n \in \mathbb{N}}$ is a Cauchy sequence in Y and since Y is complete, it converges in Y. So, starting from the bounded sequence $(x_n)_{n \in \mathbb{N}}$ in X, we have found a subsequence $(Tx_{n+1}^{(n)})_{n \in \mathbb{N}}$ of the sequence $(Tx_n)_{n \in \mathbb{N}}$, that converges in Y. Consequently, T is compact. □

Corollary 5.2. *Let X be a normed space, Y a Hilbert space, and $(T_n)_{n \in \mathbb{N}}$ be a sequence of finite rank operators in $CL(X,Y)$ that converges in $CL(X,Y)$ to $T \in CL(X,Y)$. Then T is compact.*

Example 5.3. (When is a diagonal operator on ℓ^2 compact?)
Let $X = Y = \ell^2$, $(\lambda_n)_{n \in \mathbb{N}}$ be a bounded in $\mathbb{K}$, and $\Lambda \in CL(\ell^2)$ be "given by"

$$\Lambda = \begin{bmatrix} \lambda_1 & & & \\ & \lambda_2 & & \\ & & \lambda_3 & \\ & & & \ddots \end{bmatrix}.$$

Then we had seen in Exercise 2.17 (page 76) that $\|\Lambda\| = \sup_{n \in \mathbb{N}} |\lambda_n|$.

Claim: Λ is compact if and only if $\lim_{n \to \infty} \lambda_n = 0$.

(If part): Consider for $n \in \mathbb{N}$, the operators $\Lambda_n \in CL(\ell^2)$ given by

$$\Lambda_n = \begin{bmatrix} \lambda_1 & & & & \\ & \ddots & & & \\ & & \lambda_n & & \\ & & & 0 & \\ & & & & \ddots \end{bmatrix}.$$

Each Λ_n is a finite rank operator because $\operatorname{ran} \Lambda_n \subset \operatorname{span}\{\mathbf{e}_1, \cdots, \mathbf{e}_n\}$, where $\mathbf{e}_k$ is the sequence with kth term equal to 1, and all others equal to 0. Hence

Λ_n is compact. Then

$$\|\Lambda - \Lambda_n\| = \left\| \begin{bmatrix} 0 & & & & \\ & \ddots & & & \\ & & 0 & & \\ & & & \lambda_{n+1} & \\ & & & & \ddots \end{bmatrix} \right\| = \sup_{k:k>n} |\lambda_k| \overset{n\to\infty}{\longrightarrow} 0.$$

Consequently, Λ, being the uniform limit of a sequence of compact operators, is compact.

(Only if part): Suppose that Λ is compact, but it is not the case that $(\lambda_n)_{n\in\mathbb{N}}$ is convergent with limit 0, that is,

$$\neg\Big(\forall \epsilon > 0,\ \exists N \text{ such that } \forall n > N,\ |\lambda_n - 0| < \epsilon\Big)$$

that is,

$$\exists \epsilon > 0 \text{ such that } \forall N \in \mathbb{N},\ \exists n > N \text{ such that } |\lambda_n| \geq \epsilon.$$

Taking $N = 1$, there exists $n_1 > 1$ such that $|\lambda_{n_1}| \geq \epsilon$.
Taking $N = n_1$, there exists $n_2 > n_1$ such that $|\lambda_{n_2}| \geq \epsilon$.
...

Proceeding in this manner, we can construct inductively a subsequence $(\lambda_{n_k})_{k\in\mathbb{N}}$ of $(\lambda_n)_{n\in\mathbb{N}}$ such that for all $k \in \mathbb{N}$, $|\lambda_{n_k}| \geq \epsilon$. Now consider the bounded sequence $(\mathbf{e}_{n_k})_{k\in\mathbb{N}}$ in ℓ^2. We have $(\Lambda \mathbf{e}_{n_k})_{k\in\mathbb{N}} = (\lambda_{n_k} \mathbf{e}_{n_k})_{k\in\mathbb{N}}$. But for all $k \neq j$, $\|\lambda_{n_k} \mathbf{e}_{n_k} - \lambda_{n_j} \mathbf{e}_{n_j}\|^2 = |\lambda_{n_j}|^2 + |\lambda_{n_j}|^2 \geq \epsilon^2 + \epsilon^2 = 2\epsilon^2$. This shows that $(\lambda_{n_k} \mathbf{e}_{n_k})_{k\in\mathbb{N}}$ has no convergent subsequence, contradicting the compactness of Λ. $\diamond$

Exercise 5.1. (Hilbert Schmidt operators are compact.)
Let H be a Hilbert space with an orthonormal basis $\{u_1, u_2, u_3, \cdots\}$.
Let $T \in CL(H)$ be Hilbert-Schmidt, that is, $\sum_{n=1}^{\infty} \|Tu_n\|^2 < +\infty$.

(1) If $m \in \mathbb{N}$, then define $T_m : H \to H$ by $T_m x = \sum_{n=1}^{m} \langle x, u_n \rangle T u_n$, $x \in H$.

Prove that $T_m \in CL(H)$ and that $\|(T - T_m)x\|^2 \leq \|x\|^2 \sum_{n=m+1}^{\infty} \|Tu_n\|^2$.

Hint: $\|(T - T_m)x\|^2 = \Big\| \sum_{n=m+1}^{\infty} \langle x, u_n \rangle T u_n \Big\|^2 \leq \Big(\sum_{n=m+1}^{\infty} |\langle x, u_n \rangle| \|Tu_n\| \Big)^2$

and use the Cauchy-Schwarz inequality in ℓ^2.

(2) Show that every Hilbert-Schmidt operator T is compact.
Hint: Using (1), conclude that T is the limit in $CL(H)$ of the sequence of finite rank operators T_m, $m \in \mathbb{N}$.

Exercise 5.2. Let H be a Hilbert space, and $x_0, y_0 \in H$ be fixed. Define $x_0 \otimes y_0 : H \to H$ by $(x_0 \otimes y_0)(x) = \langle x, y_0 \rangle x_0$, $x \in H$.
(1) Show that $x_0 \otimes y_0 \in CL(H)$ and that $\|x_0 \otimes y_0\| \leqslant \|x_0\| \|y_0\|$.
(2) Is $x_0 \otimes y_0$ compact?
(3) Let $A, B \in CL(H)$. Show that $A(x_0 \otimes y_0)B = (Ax_0) \otimes (B^* y_0)$.

Recall that $CL(H)$ has the structure of a complex algebra with multiplication of $T, S \in CL(H)$ taken as composition $T \circ S \in CL(H)$. What is the relation of $K(H)$ as a subset of $CL(H)$ with respect to this operation of multiplication? The answer is that $K(H)$ forms an "ideal" in $CL(H)$.

Definition 5.2. (Ideal in an algebra).
An *ideal* I of an algebra R is a subset I of R having the properties:
  (I1) $0 \in I$.
  (I2) If $a, b \in I$, then $a + b \in I$.
  (I3) If $a \in I$ and $r \in R$, then $ar \in I$ and $ra \in I$.

For example, if $R = \mathbb{Z}$, the set of all integers, then $I = 2\mathbb{Z}$, the set of all even integers, is an ideal in R. In algebra, ideals are important, since they serve as kernels of algebra homomorphisms.

Theorem 5.5. *Let H be a Hilbert space. Then we have:*
(1) *If $T \in K(H)$ is compact and $S \in CL(H)$, then TS is compact.*
(2) *If $T \in CL(H)$ is compact, then T^* is compact.*
(3) *If $T \in CL(H)$ is compact and $S \in K(H)$, then ST is compact.*

Proof.
(1) Let $(x_n)_{n \in \mathbb{N}}$ be a bounded sequence in H. Suppose $M > 0$ is such that $\|x_n\| \leqslant M$ for all $n \in \mathbb{N}$. Since $S \in CL(H)$, it follows that $(Sx_n)_{n \in \mathbb{N}}$ is also a bounded sequence ($\|Sx_n\| \leqslant \|S\| \|x_n\| \leqslant \|S\| M$). As T is compact, $(T(Sx_n))_{n \in \mathbb{N}} = (TSx_n)_{n \in \mathbb{N}}$ has a convergent subsequence. Thus TS is compact.

(2) As $T \in K(H)$ and $T^* \in CL(H)$, by part (1) above, TT^* is compact. Let $(x_n)_{n \in \mathbb{N}}$ be a bounded sequence in X and $\|x_n\| \leqslant M$ for all n. Then $(TT^* x_n)_{n \in \mathbb{N}}$ has some convergent subsequence, say $(TT^* x_{n_k})_{k \in \mathbb{N}}$. Hence, given an $\epsilon > 0$,
$$0 \leqslant \|T^* x_{n_k} - T^* x_{n_\ell}\|^2 = \langle TT^*(x_{n_k} - x_{n_\ell}), (x_{n_k} - x_{n_\ell}) \rangle$$
$$\leqslant \underbrace{\|TT^*(x_{n_k} - x_{n_\ell})\|}_{< \frac{\epsilon}{2M} \text{ for big } k, \ell} \underbrace{\|x_{n_k} - x_{n_\ell}\|}_{\leqslant 2M}.$$
So $(T^* x_{n_k})_{k \in \mathbb{N}}$ is a Cauchy sequence, and as H is a Hilbert space, it is convergent. Consequently, T^* is compact.

(3) Since T is compact, by part (2), it follows that T^* is also compact. Moreover, as $S^* \in CL(H)$, we have T^*S^* is compact, using part (1). From part (2) again, we get $(T^*S^*)^* = S^{**}T^{**} = ST$ is compact. □

Summary: The set $K(H)$ is a closed ideal of $CL(H)$.

Example 5.4. (Compact operators on infinite dimensional Hilbert spaces are never invertible.) Let H be an infinite dimensional Hilbert space, and $T \in K(H)$. If T were invertible in $CL(H)$, then $T^{-1} \in CL(H)$, so that $I = TT^{-1} \in K(H)$, which is false, since we had seen that the identity operator on an infinite dimensional Hilbert space is not compact. ◇

Exercise 5.3. Let $T \in CL(H)$, where H is an infinite-dimensional Hilbert space.
 (1) Give an example of H and T such that T^2 is compact, but T isn't.
 (2) Show that if T is self-adjoint and T^2 is compact, then T is compact.

Exercise 5.4. Determine if the following statements are true for all $S, T \in CL(H)$, where H is an infinite dimensional Hilbert space.
 (1) If S and T are compact, then $S + T$ is compact.
 (2) If $S + T$ is compact, then S or T is compact.
 (3) If S or T is compact, then ST is compact.
 (4) If ST is compact, then S is compact or T is compact.

Exercise 5.5. Let H be a Hilbert space. Let $A \in CL(H)$ be fixed. We define $\Lambda \in CL(CL(H))$ by $\Lambda(T) = A^*T + TA$, $T \in CL(H)$. Show that the subspace $K(H)$ of $CL(H)$ is Λ-invariant, that is, $\Lambda K(H) \subset K(H)$.

5.3 Approximation of compact operators

Compact operators play an important role in numerical analysis since they can be approximated by finite rank operators. This means that when we want to solve an operator equation involving a compact operator, then we can replace the compact operator by a sufficiently good finite-rank approximation, reducing the operator equation to an equation involving finite matrices. The solution can then be found using linear algebra. In this section we will prove Theorems 5.6 and 5.7, which form the basis of the Galerkin Method in numerical analysis.

Consider the equation $(I - K)x = y$, where K is a given operator on a Hilbert space H, $y \in H$ is a given vector, and $x \in H$ is the unknown. Suppose we consider instead the equation $(I - K_0)x_0 = y_0$, where K_0 is close to K, and y_0 is close to y. The following result describes how big

$\|x - x_0\|$ can get.
$$(I-K)\ x\ =\ y$$
$$\wr\quad ?\quad \wr$$
$$(I-K_0)\ x_0\ =\ y_0$$
$$?$$

Theorem 5.6.
Let
  (1) H be a Hilbert space,
  (2) $K \in CL(H)$ be such that $I - K$ is invertible in $CL(H)$,
  (3) $K_0 \in CL(H)$ be such that $\epsilon := \|(K - K_0)(I - K)^{-1}\| < 1$.
Then for every $y, y_0 \in H$, there exist unique $x, x_0 \in X$ such that

(a) $(I - K)x = y$,

(b) $(I - K_0)x_0 = y_0$, and

(c) $\|x - x_0\| \leq \dfrac{(I - K)^{-1}}{1 - \epsilon}(\epsilon\|y\| + \|y - y_0\|)$.

Note that from part (c) we see that the upper estimate on $\|x - x_0\|$ is small when $y \approx y_0$ and $K \approx K_0$. So the result is telling us that *if* we have a scheme of approximating the operator K and the vector y, then we can solve the equation
$$(I - K)x = y$$
approximately by solving instead the equation
$$(I - K_0)x_0 = y_0,$$
and moreover, we have a handle on how large the error $\|x - x_0\|$ can get. Later on, in Theorem 5.7 we will see that for *compact* operators K, such an approximating scheme for producing K_0 does exist.

Proof. As $\|(K - K_0)(I - K)^{-1}\| < 1$, by the Neumann Series Theorem, we have $I + (K - K_0)(I - K)^{-1}$ is invertible, and so
$$I - K_0 = I - K + K - K_0 = \underbrace{\left(I + (K - K_0)(I - K)^{-1}\right)}_{\text{invertible}} \underbrace{(I - K)}_{\text{invertible}}$$
is invertible as well. Moreover,
$$\|(I - K_0)^{-1}\| = \left\|(I - K)^{-1}\left(I + (K - K_0)(I - K)^{-1}\right)^{-1}\right\|$$
$$\leq \|(I - K)^{-1}\|\left\|\left(I + (K - K_0)(I - K)^{-1}\right)^{-1}\right\|$$
$$\leq \frac{\|(I - K)^{-1}\|}{1 - \|(K - K_0)(I - K)^{-1}\|} = \frac{\|(I - K)^{-1}\|}{1 - \epsilon}.$$

Furthermore, we have

$$\begin{aligned}(I - K)^{-1} - (I - K_0)^{-1} &= (I - K_0)^{-1}\big((I - K_0)(I - K)^{-1} - I\big) \\ &= (I - K_0)^{-1}\big((I - K_0) - (I - K)\big)(I - K)^{-1} \\ &= (I - K_0)^{-1}(K - K_0)(I - K)^{-1},\end{aligned}$$

and so

$$\begin{aligned}\|(I - K)^{-1} - (I - K_0)^{-1}\| &\leq \|(I - K_0)^{-1}\|\|(K - K_0)(I - K)^{-1}\| \\ &\leq \frac{\|(I - K)^{-1}\|}{1 - \epsilon}\epsilon.\end{aligned}$$

Let $y, y_0 \in X$. Since $I - K$ and $I - K_0$ are invertible, there are unique $x, x_0 \in X$ such that $(I - K)x = y$ and $(I - K_0)x_0 = y_0$. Also,

$$\begin{aligned}x - x_0 &= (I - K)^{-1}y - (I - K_0)^{-1}y_0 \\ &= \big((I - K)^{-1} - (I - K_0)^{-1}\big)y + (I - K_0)^{-1}(y - y_0),\end{aligned}$$

and so $\|x - x_0\| \leq \dfrac{\epsilon\|(I - K)^{-1}\|}{1 - \epsilon}\|y\| + \dfrac{\|(I - K)^{-1}\|}{1 - \epsilon}\|y - y_0\|$, as desired. □

Question: If K is compact, $y \in H$, then how do we find approximations K_0 to K and y_0 to y?

Answer: Via projections.

Theorem 5.7. (Galerkin approximation).
Let
(1) *H be a Hilbert space,*
(2) *K be a compact operator on H,*
(3) *$(P_n)_{n \in \mathbb{N}}$ be a sequence of projections ($P_n^2 = P_n = P_n^* \in CL(H)$) of finite rank such that P_n converges strongly to I (for all $x \in H$, $\lim_{n \to \infty} P_n x = x$).*

Then $P_n K P_n \xrightarrow{n \to \infty} K$ in $CL(H)$.

We remark that a mere *strong* convergence assumption results in *uniform* convergence, and this miracle happens since we have a *compact* operator at hand. We also remark that a standard way of producing such a sequence of projections is via choosing an orthonormal basis $\{u_1, u_2, u_3, \cdots\}$ for H, and then we can take P_n to be the projection onto the closed finite dimensional subspace $Y = \mathrm{span}\{u_1, \cdots, u_n\}$:

$$P_n x := P_Y x = \sum_{k=1}^n \langle x, u_n\rangle u_n, \quad x \in H.$$

Proof. We'll prove the following claims:
(1) $P_n K \overset{n\to\infty}{\longrightarrow} K$ in $CL(H)$ (projection approximation),
(2) $KP_n \overset{n\to\infty}{\longrightarrow} K$ in $CL(H)$ (Sloan approximation),
(3) $P_n K P_n \overset{n\to\infty}{\longrightarrow} K$ in $CL(H)$ (Galerkin approximation).

(1): For all $x \in H$, we have
$$\|P_n x\|^2 = \langle P_n x, P_n x\rangle = \langle x, P_n^* P_n x\rangle = \langle x, P_n^2 x\rangle = \langle x, P_n x\rangle \leqslant \|P_n x\|\|x\|,$$
and so $\|P_n x\| \leqslant \|x\|$ for all x, that is, $\|P_n\| \leqslant 1$. Suppose that it is not the case that $P_n K - K$ converges to $\mathbf{0}$ in $CL(H)$ as $n \to \infty$. This means that
$$\neg\Big(\forall \epsilon > 0 \ \exists N \in \mathbb{N} \text{ such that } \forall n > N, \ \|P_n K - K\| \leqslant \epsilon\Big).$$
Thus there exists an $\epsilon > 0$ such that for all $N \in \mathbb{N}$, there exists an $n > N$, such that $\|P_n K - K\| > \epsilon$.

Hence there exists an $\epsilon > 0$ such that for all $N \in \mathbb{N}$, there exists an $n > N$, such that $\sup\{\|(P_n K - K)x\| : x \in H, \ \|x\| \leqslant 1\} > \epsilon$.

So there exists an $\epsilon > 0$ such that for all $N \in \mathbb{N}$, there exists an $n > N$, such that there exists an $x \in H$ with $\|x\| \leqslant 1$, but $\|(P_n K - K)x\| > \epsilon$.

The last statement allows us to construct a sequence $(x_{n_k})_{k\in\mathbb{N}}$ in X such that $\|x_{n_k}\| \leqslant 1$ and $\|(P_{n_k} K - K)x_{n_k}\| > \epsilon$ as follows.

Taking $N = 1$, there exists an $n_1 > 1$ and an $x_{n_1} \in H$ with $\|x_{n_1}\| \leqslant 1$ but $\|(P_{n_1} K - K)x_{n_1}\| > \epsilon$.

Taking $N = n_1$, there exists an $n_2 > n_1$ and an $x_{n_2} \in H$ with $\|x_{n_2}\| \leqslant 1$ but $\|(P_{n_2} K - K)x_{n_2}\| > \epsilon$.

Taking $N = n_3$, there exists an $n_3 > n_2$ and an $x_{n_3} \in H$ with $\|x_{n_3}\| \leqslant 1$ but $\|(P_{n_3} K - K)x_{n_3}\| > \epsilon$.

$\dots$

Thus $(x_{n_k})_{k\in\mathbb{N}}$ is bounded and $\|(P_{n_k} K - K)x_{n_k}\| > \epsilon$ for all ks.

As $(x_{n_k})_{k\in\mathbb{N}}$ is bounded and K is compact, there exists a subsequence, say $(Kx_{n_{k_\ell}})_{\ell\in\mathbb{N}}$, of $(Kx_{n_k})_{k\in\mathbb{N}}$, that is convergent to y, say. Then we have
$$\epsilon < \|(P_{n_{k_\ell}} K - K)x_{n_{k_\ell}}\| = \|(P_{n_{k_\ell}} - I)y + (P_{n_{k_\ell}} - I)(Kx_{n_{k_\ell}} - y)\|$$
$$\leqslant \underbrace{\|P_{n_{k_\ell}} y - y\|}_{\to 0} + \underbrace{\|P_{n_{k_\ell}} - I\|}_{\leqslant 2} \underbrace{\|Kx_{n_{k_\ell}} - y\|}_{\to 0},$$
a contradiction. This completes the proof of (1).

(2): As K is compact, so is K^*. Thus by (1), $P_n^* K^* = P_n K^* \overset{n\to\infty}{\longrightarrow} K^*$ in $CL(H)$. But $\|KP_n - K\| = \|(KP_n - K)^*\| = \|P_n K^* - K^*\| \overset{n\to\infty}{\longrightarrow} 0$, and so $KP_n \overset{n\to\infty}{\longrightarrow} K$ in $CL(H)$.

(3): Finally,
$$\|P_nKP_n - K\| = \|P_n(KP_n - K) + P_nK - K\|$$
$$\leq \underbrace{\|P_n\|}_{\leq 1} \underbrace{\|KP_n - K\|}_{\to 0} + \underbrace{\|P_nK - K\|}_{\to 0}.$$

This completes the proof $\square$

So, in Theorem 5.6, what is y_0, K_0? We can take $K_0 = P_nKP_n$, where P_n is the orthogonal projection onto span$\{u_1, \cdots, u_n\}$, and $y_0 = P_ny$. We note that $\|y_0 - y\| = \|P_ny - y\|$ is small for large n, and $\|K - P_nKP_n\|$ is small for large n. Thus $\epsilon = \|(K - K_0)(I - K)^{-1}\|$ is small. So if we look at the equation $(I - P_nKP_n)x_0 = P_ny$ instead of $(I - K)x = y$, then $\|x - x_0\|$ can be made as small as we please by taking n large enough. We give a simple toy example.

Example 5.5.

Consider the operator $K = \begin{bmatrix} 0 & \frac{1}{2} & 0 & 0 & \cdots \\ 0 & 0 & \frac{1}{3} & 0 & \cdots \\ 0 & 0 & 0 & \frac{1}{4} & \cdots \\ \vdots & \vdots & \vdots & \vdots & \ddots \end{bmatrix}$ on ℓ^2. For all $\mathbf{x} = (x_n)_{n\in\mathbb{N}} \in \ell^2$,

$$\|K\mathbf{x}\|_2^2 = \sum_{n=1}^\infty \left(\frac{x_{n+1}}{n+1}\right)^2 \leq \frac{1}{4}\sum_{n=1}^\infty x_{n+1}^2 \leq \frac{1}{4}\|\mathbf{x}\|_2^2.$$

So $\|K\| \leq \frac{1}{2}$, and $I - K$ is invertible in $CL(\ell^2)$.

K is Hilbert-Schmidt as $\displaystyle\sum_{n=1}^\infty \|K\mathbf{e}_n\|_2^2 = \sum_{n=2}^\infty \frac{1}{n^2} < \infty$. So K is compact.

Let $\mathbf{y} = \left(\frac{1}{3}, \frac{1}{4}, \frac{1}{5}, \cdots\right) \in \ell^2$. To find approximate solutions of the equation

$$\mathbf{x} - K\mathbf{x} = \mathbf{y},$$

we fix an $n \in \mathbb{N}$, and solve instead $\mathbf{x} - P_nKP_n\mathbf{x} = P_n\mathbf{y}$, that is, the system

$$\begin{bmatrix} 1 & -\frac{1}{2} & & & & \\ & 1 & -\frac{1}{3} & & & \\ & & 1 & \ddots & & \\ & & & \ddots & -\frac{1}{n} & \\ & & & & 1 & \\ \hline & & & & & 1 \\ & & & & & & \ddots \end{bmatrix} \begin{bmatrix} x_1 \\ x_2 \\ x_3 \\ \vdots \\ x_n \\ \hline x_{n+1} \\ \vdots \end{bmatrix} = \begin{bmatrix} \frac{1}{3} \\ \frac{1}{4} \\ \vdots \\ \frac{1}{n+2} \\ \hline 0 \\ \vdots \end{bmatrix}.$$

The approximate solutions for $n = 1, 2, 3, 4, 5$ are given (correct up to four decimal places) by

$$\mathbf{x}^{(1)} = (0.3333, 0, 0, 0, \cdots)$$
$$\mathbf{x}^{(2)} = (0.4583, 0.2500, 0, 0, 0, \cdots)$$
$$\mathbf{x}^{(3)} = (0.4917, 0.3167, 0.2000, 0, 0, 0, \cdots)$$
$$\mathbf{x}^{(4)} = (0.4986, 0.3306, 0.2417, 0.1667, 0, 0, 0, \cdots)$$
$$\mathbf{x}^{(5)} = (0.4998, 0.3329, 0.2488, 0.1952, 0.1428, 0, 0, 0, \cdots),$$

while the exact unique solution to the equation $(I - K)\mathbf{x} = \mathbf{y}$ is given by

$$\mathbf{x} := \left(\frac{1}{2}, \frac{1}{3}, \frac{1}{4}, \cdots\right) \in \ell^2.$$

To 4 decimal places, this is $\mathbf{x} = (0.5, 0.3333, 0.2500, 0.2000, 0.1667, \cdots)$. ◇

5.4 (∗) Spectral Theorem for Compact Operators

In elementary linear algebra, one learns about the Spectral Theorem, which says that every Hermitian matrix $T \in \mathbb{C}^{d \times d}$ is diagonalisable, with a basis of orthonormal eigenvectors $\mathbf{u}_1, \cdots, \mathbf{u}_d \in \mathbb{C}^d$, and corresponding real eigenvalues $\lambda_1 \geq \cdots \geq \lambda_d$, so that

$$T\mathbf{x} = \sum_{n=1}^{d} \lambda_n \langle \mathbf{x}, \mathbf{u}_n \rangle \mathbf{u}_n, \quad \mathbf{x} \in \mathbb{C}^d.$$

Towards seeking a generalisation to the Hilbert space case, we'll now show that while the spectrum of a general self-adjoint operator may be quite complicated, for a *compact* self-adjoint operator, things are quite similar to the finite-dimensional case.

Theorem 5.8. (Spectral Theorem for compact, self-adjoint operators).
Let H be a Hilbert space and $T = T^* \in K(H)$ have infinite rank. Then there exist orthonormal eigenvectors u_n, $n \in \mathbb{N}$, with corresponding eigenvalues λ_n, $n \in \mathbb{N}$, such that $\lim_{n \to \infty} \lambda_n = 0$, and for all $x \in H$,

$$Tx = \sum_{n=1}^{\infty} \lambda_k \langle x, u_n \rangle u_n.$$

We had already seen that the eigenvalues of a self-adjoint operator must be real, and that the eigenvectors corresponding to distinct eigenvalues are orthogonal. It is also clear that for any eigenvalue λ of T, we have that $|\lambda| \leq \|T\|$, since if $v \in H \setminus \{\mathbf{0}\}$ is a corresponding eigenvector, then

$$|\lambda|\|v\| = \|\lambda v\| = \|Tv\| \leq \|T\|\|v\|.$$

Let us make a few more observations which will be used in proving the spectral theorem.

Lemma 5.1. *If $T = T^* \in CL(H)$, then $\|T\| = \sup\limits_{\|x\|=1} |\langle Tx, x \rangle|$.*

Proof. Let $M := \sup\limits_{\|x\|=1} |\langle Tx, x \rangle|$.
If $x \in H$ is such that $\|x\| = 1$, then we have by Cauchy-Schwarz that $|\langle Tx, x \rangle| \leq \|Tx\|\|x\| \leq \|T\|\|x\|\|x\| = \|T\|$. Thus $M \leq \|T\|$.
It remains to show the reverse inequality. For any $x, y \in H$,

$$\langle T(x+y), x+y \rangle = \langle Tx, x \rangle + 2\operatorname{Re}\langle Tx, y \rangle + \langle Ty, y \rangle,$$
$$\langle T(x-y), x-y \rangle = \langle Tx, x \rangle - 2\operatorname{Re}\langle Tx, y \rangle + \langle Ty, y \rangle,$$

and so $4\operatorname{Re}\langle Tx, y \rangle = \langle T(x+y), x+y \rangle - \langle T(x-y), x-y \rangle$. We note that by the definition of M,

$$\langle T(x+y), x+y \rangle \leq |\langle T(x+y), x+y \rangle| \leq M\|x+y\|^2,$$
$$-\langle T(x-y), x-y \rangle \leq |\langle T(x-y), x-y \rangle| \leq M\|x-y\|^2,$$

and so from the above, together with the Parallelogram Law, we obtain

$$4\operatorname{Re}\langle Tx, y \rangle \leq M\|x+y\|^2 + M\|x-y\|^2 = 2M(\|x\|^2 + \|y\|^2).$$

Let $\theta \in \mathbb{R}$ be such that $\langle Tx, y \rangle = |\langle Tx, y \rangle|e^{i\theta}$. Replacing y by $e^{i\theta}y$ yields

$$|\langle Tx, y \rangle| \leq \frac{M}{2}(\|x\|^2 + \|y\|^2).$$

If $Tx = \mathbf{0}$ or $x = \mathbf{0}$, then $\|Tx\| \leq M\|x\|$ is trivially true.
If $Tx \neq \mathbf{0}$ and $x \neq \mathbf{0}$, then with $y := \dfrac{\|x\|}{\|Tx\|}Tx$ in the above, we obtain $\|Tx\|\|x\| \leq M\|x\|^2$ and so $\|Tx\| \leq M\|x\|$. Thus $\|T\| \leq M$. $\square$

Moreover if T is *compact*, then this bound M is achieved, thanks to the following result. Indeed, if x is the unit-norm eigenvector corresponding to the eigenvalue λ whose modulus $|\lambda| = \|T\|$, then

$$|\langle Tx, x \rangle| = |\langle \lambda x, x \rangle| = |\lambda|\|x\|^2 = |\lambda| = \|T\|.$$

Lemma 5.2. *If H is a nontrivial Hilbert space and $T = T^* \in K(H)$, then either $\|T\|$ or $-\|T\|$ is an eigenvalue of T.*

Proof. If $T = \mathbf{0}$, then this is trivial. So let us suppose that T is nonzero. From the previous lemma, it follows that there is a sequence $(x'_n)_{n\in\mathbb{N}}$ of unit norm vectors in H such that $|\langle Tx'_n, x'_n\rangle| \stackrel{n\to\infty}{\longrightarrow} \|T\| \neq 0$. But as $\langle Tx'_n, x'_n\rangle$ is real, we have $\langle Tx'_n, x'_n\rangle$ is either $|\langle Tx'_n, x'_n\rangle|$ or $-|\langle Tx'_n, x'_n\rangle|$. Thus either for infinitely many n, $\langle Tx'_n, x'_n\rangle$ is positive (and then the subsequence $(\langle Tx_n, x_n\rangle)_{n\in\mathbb{N}}$ with these ns converges to $\|T\|$), or for infinitely many n, $\langle Tx'_n, x'_n\rangle$ is negative (and then the subsequence $(\langle Tx_n, x_n\rangle)_{n\in\mathbb{N}}$ with these ns converges to $-\|T\|$). So $\langle Tx_n, x_n\rangle \stackrel{n\to\infty}{\longrightarrow} \lambda$, where λ is either $\|T\|$ or $-\|T\|$. We have

$$\begin{aligned} 0 \leq \|Tx_n - \lambda x_n\|^2 &= \|Tx_n\|^2 - 2\lambda\langle Tx_n, x_n\rangle + \lambda^2 \\ &\leq \|T\|^2 \cdot 1 - 2\lambda\langle Tx_n, x_n\rangle + \lambda^2 \\ &= \lambda^2 \cdot 1 - 2\lambda\langle Tx_n, x_n\rangle + \lambda^2 \stackrel{n\to\infty}{\longrightarrow} 0. \end{aligned}$$

So $Tx_n - \lambda x_n \stackrel{n\to\infty}{\longrightarrow} \mathbf{0}$. As T is compact, there is a subsequence, say $(Tx_{n_k})_{k\in\mathbb{N}}$ of $(Tx_n)_{n\in\mathbb{N}}$ that converges, say, to $y \in H$. Then $(x_{n_k})_{k\in\mathbb{N}}$ converges to y/λ, because $\lambda x_{n_k} = Tx_{n_k} - (Tx_{n_k} - \lambda x_{n_k}) \stackrel{k\to\infty}{\longrightarrow} y - \mathbf{0} = y$. Thanks to the continuity of T, we obtain

$$y = \lim_{k\to\infty} Tx_{n_k} = T\left(\lim_{k\to\infty} x_{n_k}\right) = T\left(\frac{1}{\lambda}y\right) = \frac{1}{\lambda}Ty.$$

Hence $Ty = \lambda y$, and $y \neq \mathbf{0}$ since $\|y\| = \lim_{k\to\infty} \|\lambda x_{n_k}\| = |\lambda| = \|T\| \neq 0$. $\square$

Lemma 5.3. *Let H be a Hilbert space, $T = T^* \in CL(H)$, and Y be a T-invariant closed subspace of H. Then:*
  (1) $Y^\perp$ *is also T-invariant.*
  (2) *The restriction $T|_{Y^\perp} : Y^\perp \to Y^\perp$ of T to the Hilbert space $Y^\perp$ is also self-adjoint.*
  (3) *If T is in addition compact, then $T|_{Y^\perp}$ is also compact.*

Proof.
(1) Let $z \in Y^\perp$. For all $y \in Y$, we have that $Ty \in Y$ (Y is T-invariant!), and so $\langle Tz, y\rangle = \langle z, T^*y\rangle = \langle z, Ty\rangle = 0$. Thus $Tz \in Y^\perp$.
(2) $Y^\perp$, being a closed subspace of a Hilbert space, is itself a Hilbert space. As $Y^\perp$ is T-invariant, the restriction $T|_{Y^\perp} : Y^\perp \to Y^\perp$ is well-defined. Let us denote this restriction by $\widetilde{T}$. For $z_1, z_2 \in Y^\perp$,
$$\begin{aligned}\langle \widetilde{T}z_1, z_2\rangle_{Y^\perp} &= \langle \widetilde{T}z_1, z_2\rangle = \langle Tz_1, z_2\rangle = \langle z_1, T^*z_2\rangle \\ &= \langle z_1, Tz_2\rangle = \langle z_1, \widetilde{T}z_2\rangle = \langle z_1, \widetilde{T}z_2\rangle_{Y^\perp}.\end{aligned}$$
Thus $T|_{Y^\perp}$ is self-adjoint.
(3) Finally, suppose that T is compact. Let $(z_n)_{n\in\mathbb{N}}$ be a bounded sequence

in $Y^\perp$. Then $\|z_n\|_{Y^\perp} = \|z_n\|$. So $(z_n)_{n\in\mathbb{N}}$ is a bounded sequence in H. As T is compact, $(Tz_n)_{n\in\mathbb{N}} = (\widetilde{T}z_n)_{n\in\mathbb{N}}$ has a subsequence $(\widetilde{T}x_{n_k})_{k\in\mathbb{N}}$ that is convergent in H. In particular, $(\widetilde{T}x_{n_k})_{k\in\mathbb{N}}$ is Cauchy in H, and hence also Cauchy in $Y^\perp$ (because $\|\widetilde{T}z_n - \widetilde{T}z_m\| = \|\widetilde{T}z_n - \widetilde{T}z_m\|_{Y^\perp}$). As $Y^\perp$ is complete, it follows that $(\widetilde{T}x_{n_k})_{k\in\mathbb{N}}$ is convergent in $Y^\perp$. So $T|_{Y^\perp}$ is compact. $\square$

Proof. (Of the spectral theorem). Let $H_1 := H$ and $T_1 := T$.
By Lemma 5.2, there exists an eigenvalue λ_1 of T_1 and a corresponding eigenvector u_1 such that $\|u_1\| = 1$ and $|\lambda_1| = \|T_1\|$.

Set $H_2 = (\mathrm{span}\{u_1\})^\perp$. Then H_2 is a closed subspace of H_1, and it is also T-invariant: $TH_2 \subset H_2$. Let $T_2 := T|_{H_2}$. Then T_2 is self-adjoint and compact. There exist an eigenvalue λ_2 of T_2 and a corresponding eigenvector u_2 such that $\|u_2\| = 1$ and $|\lambda_2| = \|T_2\|$. So

$$|\lambda_2| = |\langle T_2 u_2, u_2\rangle| = |\langle Tu_2, u_2\rangle| \leq \|T\| = |\lambda_1|.$$

Clearly, $\{u_1, u_2\}$ are orthonormal, $Tu_1 = \lambda_1 u_1$ and $Tu_2 = \lambda_2 u_2$.

Now let $H_3 := (\mathrm{span}\{u_1, u_2\})^\perp$. Then H_3 is a closed subspace of H, $H_3 \subset H_2$, and as $\mathrm{span}\{u_1, u_2\}$ is T-invariant, we obtain $TH_3 \subset H_3$. Let $T_3 := T|_{H_3}$. Then T_3 is self-adjoint and compact. Thus there exist an eigenvalue λ_3 of T_3 and a corresponding eigenvector $u_3 \in H_3$, such that $\|u_3\| = 1$ and $|\lambda_3| = \|T_3\|$. As $u_3 \in H_3 \subset H_2$, we have that

$$|\lambda_3| = |\langle T_3 u_3, u_3\rangle| = |\langle T_2 u_3, u_3\rangle| \leq \|T_2\| = |\lambda_2|.$$

Continuing in this manner, we get a sequence $\lambda_1, \lambda_2, \lambda_3, \cdots$ of eigenvalues of T, and a corresponding set of eigenvectors $u_1, u_2, u_3, \cdots$, such that

$$|\lambda_{n+1}| = \|T_{n+1}\| \leq \|T_n\| = |\lambda_n|, \quad n = 1, 2, 3, \cdots.$$

The process would stop at some n if $H_n := (\mathrm{span}\{u_1, \cdots, u_{n-1}\})^\perp$ would become $\{\mathbf{0}\}$, but we will now show that thanks to the infinite rank assumption on T, this case is impossible. Suppose, on the contrary, that $H_n = \{\mathbf{0}\}$. For any $x \in H$, we have $x - \sum_{k=1}^{n-1}\langle x, u_k\rangle u_k \in (\mathrm{span}\{u_1, \cdots, u_{n-1}\})^\perp$.
As $H_n = \{\mathbf{0}\}$, we obtain $x = \sum_{k=1}^{n-1}\langle x, u_k\rangle u_k$. Thus $Tx = \sum_{k=1}^{n-1}\langle x, u_k\rangle Tu_k$.
So $\mathrm{ran}\,T$ is spanned by $Tu_1, \cdots, Tu_n$, a contradiction to the assumption that T has infinite rank. Thus one has an infinite sequence of eigenvectors $u_1, u_2, u_3, \cdots$ with eigenvalues $\lambda_1, \lambda_2, \lambda_3, \cdots$.

Let us now show that $(|\lambda_n|)_{n\in\mathbb{N}}$ converges to 0. As it is decreasing, it converges to $\inf_{n\in\mathbb{N}} |\lambda_n| =: \epsilon$. If $\epsilon > 0$, then for $n \neq m$,

$$\|Tu_n - Tu_m\|^2 = \|\lambda_n u_n - \lambda_m u_m\|^2 = \lambda_n^2 + \lambda_m^2 \geq 2\epsilon^2.$$

But this contradicts the fact that T is compact, since the sequence $(Tu_n)_{n\in\mathbb{N}}$ should have some convergent (and hence Cauchy) subsequence. So $(|\lambda_n|)_{n\in\mathbb{N}}$ converges to 0, and hence $(\lambda_n)_{n\in\mathbb{N}}$ also converges to 0.

For all $x \in H$, we have $x - \sum_{k=1}^{n-1} \langle x, u_k\rangle u_k \in H_n$, and so

$$\left\|Tx - \sum_{k=1}^{n-1} \lambda_k \langle x, u_k\rangle u_k\right\| = \left\|T\left(x - \sum_{k=1}^{n-1} \langle x, u_k\rangle u_k\right)\right\|$$

$$= \left\|T_n\left(x - \sum_{k=1}^{n-1} \langle x, u_k\rangle u_k\right)\right\|$$

$$\leq \|T_n\| \left\|x - \sum_{k=1}^{n-1} \langle x, u_k\rangle u_k\right\|$$

$$\leq |\lambda_n|\|x\| \xrightarrow{n\to\infty} 0.$$

The last inequality above follows from Bessel's Inequality (Exercise 4.15, page 172). Hence $Tx = \sum_{k=1}^{\infty} \lambda_k \langle x, u_k\rangle u_k$ for all $x \in H$. $\square$

Exercise 5.6.
Let H be an infinite dimensional Hilbert space, and $T = T^* \in K(H)$ be one-to-one. Show that the eigenvectors of T form an orthonormal basis for H.

Exercise 5.7. Let H be a Hilbert space. Suppose that $T = T^* \in K(H)$ has infinite rank, and is *positive*, that is, $\langle Tx, x\rangle \geq 0$ for all $x \in H$.
Prove that T has a square root, that is, an operator $\sqrt{T} \in CL(H)$ such that $(\sqrt{T})^2 = T$.

Chapter 6

A glimpse of distribution theory

In this last chapter of the book, we will see a generalisation of functions

$$f : \mathbb{R}^d \to \mathbb{C}$$

and their ordinary calculus to "distributions" or "generalised functions". These distributions will be "continuous linear functionals on the vector space of test functions $\mathcal{D}(\mathbb{R}^d)$":

$$\underset{\text{distribution}}{T} : \underset{\text{test functions}}{\mathcal{D}(\mathbb{R}^d)} \to \mathbb{C}.$$

Why study distributions? We list three main reasons:

(1) To mathematically model the situation when one has an impulsive force (imagine a blow to an object which changes its momentum, but the force itself is supposed to act "impulsively", that is the time interval when the force is applied is 0!). Similar situations arise in other instances in mathematics and the applied sciences.
(2) To develop a calculus which captures more general situations than the classical case. For example, what is the derivative of $|x|$ at $x = 0$?
It will turn out that this is also useful to talk about weaker notions of solutions of Partial Differential Equations (PDEs).
(3) To extend the Fourier transform theory to functions that may not be absolutely integrable. For example, what[1] is the Fourier transform of the constant function 1?

It turns out that the theory of distributions solves all of these three problems in one go. This seems like a miracle, and naturally there is a price to pay. The price is that everything classical is now replaced by a weaker notion.

---
[1] Although the classical Fourier transform does not exist, it can be shown that in the sense of distributions, the Fourier transform of the constant function 1 is the Dirac delta distribution δ.

Nevertheless this is useful, since it is often sufficient for what one wants to do. An example is that, as opposed to functions on $\mathbb{R}$, which have a well-defined value at every point $x \in \mathbb{R}$, we can no longer talk about the value of a distribution at a point of $\mathbb{R}$.

Let us elaborate on reason (2) above, in the context of PDEs. An example we met earlier in Exercise 3.17, page 143, is that of the wave equation (which describes the motion of a plucked guitar string), and we had checked that for a twice continuously differentiable f

$$u(x,t) = \frac{f(x-t) + f(x+t)}{2} \tag{6.1}$$

gives a solution with the initial condition described by f (and zero initial velocity). However, when we pluck a guitar string, the initial shape needn't be C^2, and in fact, it could have a corner like this:

Nature of course doesn't care about the lack of classical differentiability of our candidate solution to the PDE model, and produces a travelling wave solution, with time snapshots that look like the ones shown in Figure 6.1. We must have at sometime witnessed choppy waves on the surface of the sea on a windy day.

Thus it is desirable to weaken the notion of differentiability, to allow for such (classically) non-differentiable solutions to nevertheless serve as solutions to the wave equation. We will see that such a weaker notion is provided by viewing our function as a more general object, namely a *distribution*, and with its weaker *distributional calculus*, the wave equation will be satisfied!

So with this motivation, we will begin to learn the very basics of the theory of distributions in this chapter, and also see a glimpse of its applications to PDEs. For a firm foundation of the theory of distributions, one needs preliminaries on topological vector spaces[2]. Here we will adopt a "working" approach, in which we will learn rigorous, but (seemingly) ad

---

[2]That is, vector spaces with a topology making the vector space operations of addition and scalar multiplication continuous. Normed- and inner product- spaces are particular examples of topological vector spaces, where the topology is given by a norm, but it turns out that there are weaker ways of specifying a topology, which aren't generated by a norm.

hoc definitions about continuity and convergence, in the spaces related to distributions. We will make a few remarks that will serve as a guide to the reader who wishes to delve into the subject deeper.

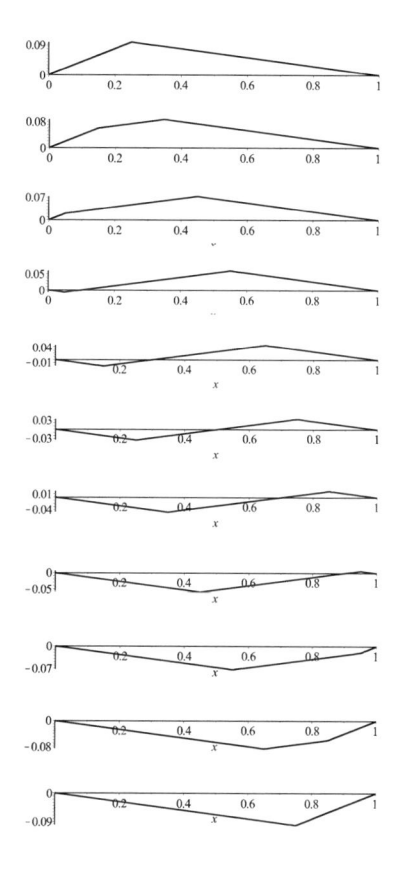

Fig. 6.1 Time snapshots of a plucked guitar string.

We first make a brief historical remark about the story of the development of distributions. The prime example of a distribution, the "delta function" δ_a, was introduced[3] in the 1930s in order to do quantum mechanical computations (as eigenstates of the position operator). However, a firm

---

[3]There were, however, earlier usages of such an object; for example an infinitely tall, unit impulse function was used by Cauchy in the early 19th century. The Dirac delta function as such was introduced as a "convenient notation" by the English physicist Paul Dirac in his book, *The Principles of Quantum Mechanics*, where he called it the "delta function", as a continuous analogue of the discrete Kronecker delta.

mathematical foundation for this and other generalised functions had to wait till the 1950s when the French mathematician Laurent Schwartz introduced the concept of distributions and developed its theory. For this, he was awarded the Fields medal.

6.1 Test functions, distributions, and examples

It will turn out that distributions are functionals on a vector space, namely the vector space of "test functions". Just like a function $f : \mathbb{R}^d \to \mathbb{C}$ can be evaluated at a point $\mathbf{x} \in \mathbb{R}^d$ giving a number $f(\mathbf{x})$, we will see later on that a distribution T on $\mathbb{R}^d$ can be "tested against" a test function φ, giving a number $\langle T, \varphi \rangle$. Let us first begin by describing this vector space $\mathcal{D}(\mathbb{R}^d)$ of test functions.

Definition 6.1. (Test function). A *test function* $\varphi : \mathbb{R}^d \to \mathbb{C}$ is an infinitely differentiable function, for which there exists a compact set, outside which φ vanishes. The set of all test functions is denoted by $\mathcal{D}(\mathbb{R}^d)$. Equipped with pointwise operations, $\mathcal{D}(\mathbb{R}^d)$ is a vector space. Thus:

$\mathcal{D}(\mathbb{R}^d) = $ vector space of test functions

$\phantom{\mathcal{D}(\mathbb{R}^d)} = \{$smooth functions with "compact support"$\}$

$\phantom{\mathcal{D}(\mathbb{R}^d)} = \left\{ \varphi : \mathbb{R}^d \to \mathbb{C} : \begin{array}{l} \varphi \text{ is infinitely many times differentiable, and} \\ \exists K_\varphi \text{ compact in } \mathbb{R}^d, \text{ such that } \varphi \equiv 0 \text{ on } \mathbb{R}^d \backslash K_\varphi \end{array} \right\}.$

Example 6.1. Let $\varphi : \mathbb{R} \to \mathbb{R}$ be given by

$$\varphi(x) = \begin{cases} e^{-\frac{1}{1-x^2}} & \text{if } |x| < 1, \\ 0 & \text{if } |x| \geqslant 1. \end{cases}$$

We claim that φ is a test function, that is, it is an element of $\mathcal{D}(\mathbb{R})$.

It is clear that φ vanishes outside the compact interval $[-1, 1]$.

Moreover, it is also infinitely many times differentiable: Indeed, it can be seen that the function $f : \mathbb{R} \to \mathbb{R}$ given by

$$f(x) = \begin{cases} e^{-\frac{1}{x}} & \text{if } x > 0, \\ 0 & \text{if } x \leqslant 0, \end{cases}$$

is clearly infinitely many times differentiable outside 0, and also

$$\forall n \in \mathbb{N}, \ \lim_{x \to 0} f^{(n)}(x) = 0,$$

A glimpse of distribution theory

showing that f is infinitely many times differentiable everywhere on $\mathbb{R}$. (See Exercise 6.1 where the details are spelt out.)
The graph of f is shown in the following picture.

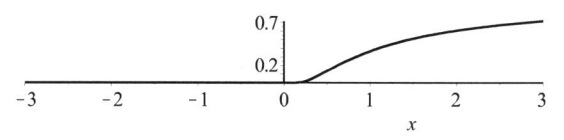

The function φ is just the composition of f with the polynomial $x \mapsto 1-x^2$.

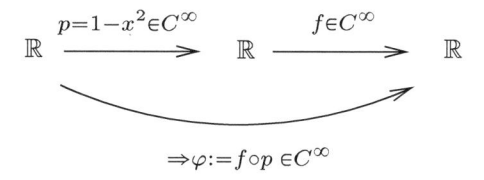

$$\mathbb{R} \xrightarrow{p=1-x^2 \in C^\infty} \mathbb{R} \xrightarrow{f \in C^\infty} \mathbb{R}$$

$$\Rightarrow \varphi := f \circ p \in C^\infty$$

The picture below shows the graph of φ.

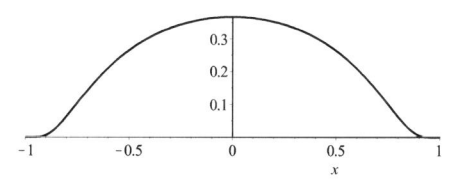

Similarly, we could have composed f with the function
$$\mathbf{x} \mapsto 1 - \|\mathbf{x}\|_2^2 = 1 - (x_1^2 + \cdots + x_d^2) : \mathbb{R}^d \to \mathbb{R}$$
and obtained a function in $\mathcal{D}(\mathbb{R}^d)$ that is C^∞ and is zero outside the closed unit ball $B(\mathbf{0}, 1) := \{\mathbf{x} \in \mathbb{R}^d : \|\mathbf{x}\|_2 \leq 1\}$ in $\mathbb{R}^d$. Note that as $B(\mathbf{0}, 1)$ is closed and bounded in $\mathbb{R}^d$, it is compact. ◇

Based on the above example, it might seem that test functions are rather special, and are few and far between. But let us note that whenever we have a $\varphi \in \mathcal{D}(\mathbb{R}^d)$, then for every $\lambda > 0$ and every $\mathbf{a} \in \mathbb{R}^d$, also the functions
$$\mathbf{x} \mapsto \varphi(\mathbf{x} - \mathbf{a}) \text{ and } \mathbf{x} \mapsto \varphi(\lambda \mathbf{x})$$
belong to $\mathcal{D}(\mathbb{R}^d)$. Moreover, it is easy to see that $\mathcal{D}(\mathbb{R}^d)$ is closed under partial differentiation. (In the following, it will be convenient to introduce the following notation: if $\mathbf{k} = (k_1, \cdots, k_d)$ is a multi-index of nonnegative integers, then

$$D^{\mathbf{k}} := \frac{\partial^{k_1 + \cdots + k_d}}{\partial x_1^{k_1} \cdots \partial x_d^{k_d}}.$$

In this notation, we have: $\varphi \in \mathcal{D}(\mathbb{R}^d) \Rightarrow D^{\mathbf{k}}\varphi \in \mathcal{D}(\mathbb{R}^d)$ for all $\mathbf{k}$.) By taking linear combinations, we see that we thus get a huge abundance of functions in $\mathcal{D}(\mathbb{R}^d)$.

Exercise 6.1. ($*$)(A C^∞ function which is not analytic.)
(1) Suppose that $f : \mathbb{R} \to \mathbb{R}$ is continuous on $\mathbb{R}$, continuously differentiable on $\mathbb{R}_* := \mathbb{R}\setminus\{0\}$, and such that $\lim_{x \to 0} f'(x)$ exists.
Show that f is continuously differentiable on $\mathbb{R}$.
(2) Let $f : \mathbb{R} \to \mathbb{R}$ be $n-1$ times continuously differentiable, n times continuously differentiable on $\mathbb{R}_*$, and such that $\lim_{x \to 0} f^{(n)}(x)$ exists.
Show that f is n times continuously differentiable on $\mathbb{R}$.
(3) Let $f : \mathbb{R} \to \mathbb{R}$ given by $f(x) = \begin{cases} e^{-\frac{1}{x}} & \text{if } x > 0, \\ 0 & \text{if } x \leq 0. \end{cases}$
Show that f is infinitely many times differentiable.
Hint: Using induction on n, show that for $x > 0$, $f^{(n)}(x) = R_n(x)f(x)$, where R_n is a rational function. Conclude that $\lim_{x \to 0} x^{-n} f(x) = 0$.

Exercise 6.2. Solve $\dfrac{\partial u}{\partial x} = 0$ in $\mathcal{D}(\mathbb{R}^2)$.

Exercise 6.3. Show that $\{\Phi' : \Phi \in \mathcal{D}(\mathbb{R})\} = \left\{\varphi \in \mathcal{D}(\mathbb{R}) : \int_{-\infty}^\infty \varphi(x)dx = 0\right\} =: Y$.
Also, show that if $\varphi \in Y$, then the $\Phi \in \mathcal{D}(\mathbb{R})$ such that $\Phi' = \varphi$ is *unique*.

Definition 6.2. (Convergence in $\mathcal{D}(\mathbb{R}^d)$).
We say that a sequence $(\varphi_n)_{n\in\mathbb{N}}$ *converges to* φ *in* $\mathcal{D}(\mathbb{R}^d)$ if
(1) there exists a compact set $K \subset \mathbb{R}^d$ such that
 all the φ_n vanish outside K, and
(2) φ_n converges uniformly[4] to φ, and
 for each multi-index $\mathbf{k}$, $D^{\mathbf{k}}\varphi_n$ converges uniformly to $D^{\mathbf{k}}\varphi$.
We then simply write $\varphi_n \xrightarrow{\mathcal{D}} \varphi$.

Remark 6.1. (Topology on $\mathcal{D}(\mathbb{R}^d)$).
It turns out that $\mathcal{D}(\mathbb{R}^d)$ is a topological space with a certain topology (with a collection of open sets denoted by, say $\mathcal{O}$), allowing one to then talk about convergent sequences in this topology given by $\mathcal{O}$. It can be shown that $(\varphi_n)_{n\in\mathbb{N}}$ converges to φ in the topological space $(\mathcal{D}(\mathbb{R}^d), \mathcal{O})$ if and only if it converges in the above sense. See for example [Bremermann (1965), Appendix 1, pages 37–40].

---
[4]That is, for every $\epsilon > 0$, there exists an $N \in \mathbb{N}$ such that for all $n > N$ and all $\mathbf{x} \in K$, $|\varphi_n(\mathbf{x}) - \varphi(\mathbf{x})| < \epsilon$.

Exercise 6.4. Let $(\varphi_n)_{n\in\mathbb{N}}$ be a sequence in $\mathcal{D}(\mathbb{R})$ such that for each $n \in \mathbb{N}$,
$$\int_{-\infty}^{\infty} \varphi_n(x)dx = 0.$$
From Exercise 6.3, it follows that for each $n \in \mathbb{N}$, there exists a unique $\Phi_n \in \mathcal{D}(\mathbb{R})$, such that $\Phi_n' = \varphi_n$. Suppose moreover that $\varphi_n \xrightarrow{\mathcal{D}} 0$. Prove that $\Phi_n \xrightarrow{\mathcal{D}} 0$.

Now that we have a vector space $\mathcal{D}(\mathbb{R}^d)$, with a topology, one can talk about continuous linear transformations $\mathcal{D}(\mathbb{R}^d)$ from $\mathbb{C}$, and these are called distributions. The precise definition is as follows.

Definition 6.3. (Distribution).
A *distribution* T on $\mathbb{R}^d$ is a map $T : \mathcal{D}(\mathbb{R}^d) \to \mathbb{C}$ such that
  (1) (Linearity) For all $\varphi, \psi \in \mathcal{D}(\mathbb{R}^d)$ and all $\alpha \in \mathbb{C}$,
$$T(\varphi + \psi) = T(\varphi) + T(\psi) \text{ and } T(\alpha \cdot \varphi) = \alpha \cdot T(\varphi).$$
  (2) (Continuity) If $\varphi_n \xrightarrow{\mathcal{D}} \varphi$, then $T(\varphi_n) \to T(\varphi)$.
The set of all distributions is denoted by $\mathcal{D}'(\mathbb{R}^d)$.
With pointwise operations, $\mathcal{D}'(\mathbb{R}^d)$ is a vector space.
We will usually denote $T(\varphi)$ for $\varphi \in \mathcal{D}(\mathbb{R}^d)$, by $\langle T, \varphi \rangle$.

Remark 6.2. It is enough to check the continuity requirement with $\varphi = 0$, since from the linearity of T, it follows that $T(\varphi_n) - T(\varphi) = T(\varphi - \varphi_n)$, and it is clear that $\varphi_n \xrightarrow{\mathcal{D}} \varphi$ if and only if $\varphi_n - \varphi \xrightarrow{\mathcal{D}} 0$.

Remark 6.3. It can be shown that a linear transformation from $\mathcal{D}(\mathbb{R}^d)$ to $\mathbb{C}$ is continuous, with respect to the aforementioned (Remark 6.1) topology on $\mathcal{D}(\mathbb{R}^d)$ given by $\mathcal{O}$, if and only if it is continuous in the above sense (given by item (2) in the above definition). See for example [Rudin (1991), Theorem 6.6 on page 155].

When first encountered, distributions may seem very strange objects indeed, far removed from the world of ordinary functions that we are used to. But now we'll see that practically all functions one meets in practice, can be viewed as distributions. This will enable us later, to apply the distributional calculus we develop, also to functions (by viewing them as distributions), and it will be in *this* sense, that we'll be able check that even with a non-C^2 function f, the u given by (6.1) satisfies the wave equation.

Example 6.2. ($L^1_{\text{loc}}(\mathbb{R}^d)$ functions are distributions.)
Let $f : \mathbb{R}^d \to \mathbb{C}$ be a *locally integrable* function (written $f \in L^1_{\text{loc}}(\mathbb{R}^d)$), that is, for every compact set K,
$$\int_K |f(\mathbf{x})|d\mathbf{x} < \infty.$$

Here $d\mathbf{x}$ stands for the volume element $dx_1 \cdots dx_d$ in $\mathbb{R}^d$.
Then f defines a distribution T_f as follows:

$$\langle T_f, \varphi \rangle := \int_{\mathbb{R}^d} f(\mathbf{x})\varphi(\mathbf{x})d\mathbf{x}, \quad \varphi \in \mathcal{D}(\mathbb{R}^d).$$

The integral exists since φ is bounded and zero outside a compact set, and so we are actually integrating over a compact set.
It is also easy to see that T_f is linear.

Continuity: Let $\varphi_n \xrightarrow{\mathcal{D}} \mathbf{0}$. Then there is a compact set K such that all φ_n, $n \in \mathbb{N}$, vanish outside K. We have

$$|\langle T_f, \varphi_n \rangle| = \left| \int_{\mathbb{R}^d} f(\mathbf{x})\varphi_n(\mathbf{x})d\mathbf{x} \right| \leq \int_K |f(\mathbf{x})||\varphi_n(\mathbf{x})|d\mathbf{x}$$
$$\leq \|\varphi_n\|_\infty \int_K |f(\mathbf{x})|d\mathbf{x} \xrightarrow{n \to \infty} 0,$$

since φ_n converges to $\mathbf{0}$ uniformly on K (the derivatives of φ_n play no role here). Consequently, $T_f \in \mathcal{D}'(\mathbb{R}^d)$.

Distributions T_f, where f is locally integrable, are called *regular distributions*. An example of a function $f \in L^1_{\text{loc}}(\mathbb{R})$ is the Heaviside function, whose graph is displayed in the picture on the left below.

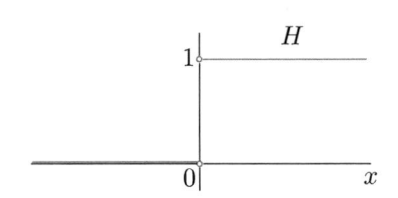

"Why should I refuse a good dinner simply because I don't understand the digestive processes involved?"

The *Heaviside[5] function* H is given by $H(x) = \begin{cases} 1 & \text{if } x > 0, \\ 0 & \text{if } x < 0. \end{cases}$

The value $H(0)$ of the function H at $x = 0$ can be arbitrarily assigned, since for any $\varphi \in \mathcal{D}(\mathbb{R})$, the value of the integral

$$\int_{\mathbb{R}} H(x)\varphi(x)dx = \langle T_H, \varphi \rangle$$

---

[5]Named after Oliver Heaviside (1850–1925), self-taught English physicist, who among other things, developed an "operational calculus" to solve linear differential equations. His methods were not rigorous, and he faced criticism from mathematicians. The Heaviside operational calculus can be justified using distribution theory; see for example [Schwartz (1966), pages 128–130].

won't change, no matter what choice we make for $H(0)$! So the distribution T_H will be the same, irrespective of what number we set $H(0)$ to be. We denote the distribution T_H simply by the symbol H from now on. ◇

Theorem 6.1. *The following are equivalent for $f, g \in L^1_{\text{loc}}(\mathbb{R}^d)$:*
(1) $T_f = T_g$.
(2) *For almost all $\mathbf{x} \in \mathbb{R}^d$, we have $f(\mathbf{x}) = g(\mathbf{x})$.*

In particular, if f and g are continuous, then $T_f = T_g$ if and only if $f = g$.

Theorem 6.1 means that the map $f \mapsto T_f$, from $L^1_{\text{loc}}(\mathbb{R}^d)$ (which is the space of equivalence classes of locally integrable functions that are equal almost everywhere), to $\mathcal{D}'(\mathbb{R}^d)$, is injective. In practice, one identifies (the equivalence class of) $f \in L^1_{\text{loc}}(\mathbb{R}^d)$ with the distribution T_f, and one considers the map $f \mapsto T_f : L^1_{\text{loc}}(\mathbb{R}^d) \to \mathcal{D}'(\mathbb{R}^d)$ as an inclusion: $L^1_{\text{loc}}(\mathbb{R}^d) \subset \mathcal{D}'(\mathbb{R}^d)$. So just like we identify each integer as a rational number, we may think of all locally integrable functions as distributions. Since the map $f \mapsto T_f$ is linear (that is, $T_{f+g} = T_f + T_g$ and $T_{\alpha f} = \alpha T_f$), this identification respects addition and scalar multiplication. As the space $C(\mathbb{R}^d)$ of continuous functions can be considered as a subspace of $L^1_{\text{loc}}(\mathbb{R}^d)$ (two continuous functions on $\mathbb{R}^d$ that are equal almost everywhere are identical), we get the following inclusions:
$$\mathcal{D}(\mathbb{R}^d) \subset C^\infty(\mathbb{R}^d) \subset C^n(\mathbb{R}^d) \subset C(\mathbb{R}^d) \subset L^1_{\text{loc}}(\mathbb{R}^d) \subset \mathcal{D}'(\mathbb{R}^d).$$
We won't include the proof of Theorem 6.1 here. The interested reader is referred to [Schwartz (1966), Theorem 2, page 74]. The following example shows that the inclusion $L^1_{\text{loc}}(\mathbb{R}^d) \subset \mathcal{D}'(\mathbb{R}^d)$ is strict, that is, not all distributions are regular.

Example 6.3. (Dirac delta distribution).
The distribution $\delta \in \mathcal{D}'(\mathbb{R}^d)$ is defined by
$$\langle \delta, \varphi \rangle = \varphi(\mathbf{0}), \quad \varphi \in \mathcal{D}(\mathbb{R}^d).$$
More generally, one defines, for $\mathbf{a} \in \mathbb{R}^d$, a distribution $\delta_\mathbf{a}$ by
$$\langle \delta_\mathbf{a}, \varphi \rangle = \varphi(\mathbf{a}), \quad \varphi \in \mathcal{D}(\mathbb{R}^d).$$
It is evident that $\delta_\mathbf{a}$ is linear and continuous on $\mathcal{D}(\mathbb{R}^d)$, that is, it is a distribution.

The delta distribution is not regular: there is no function $f \in L^1_{\text{loc}}(\mathbb{R}^d)$ such that $\delta_\mathbf{a} = T_f$. Nevertheless, in a huge amount of literature, one encounters a manner of writing that suggests that $\delta_\mathbf{a}$ is a regular distribution. In place of $\langle \delta, \varphi \rangle$, one writes
$$\int_{\mathbb{R}^d} \delta(\mathbf{x})\varphi(\mathbf{x})d\mathbf{x} = \varphi(\mathbf{0}) \text{ or } \int_{\mathbb{R}^d} \delta_\mathbf{a}(\mathbf{x})\varphi(\mathbf{x})d\mathbf{x} = \varphi(\mathbf{a}).$$

One then talks about delta "functions" instead of delta *distributions*. This is of course incorrect (see Exercise 6.5 below), but in some sense useful if one wants to do formal manipulations in order to guess answers, or in order to get physical insights etc.

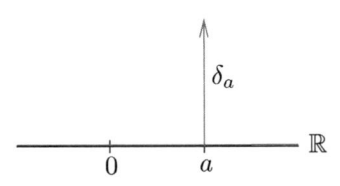

With this fallacious viewpoint, one often depicts the "graph of $\delta_a \in \mathcal{D}'(\mathbb{R})$" as a spike positioned at a, with the intuitive feeling that the "δ_a function is everywhere 0, but is infinity at $x = a$, and has integral over $\mathbb{R}$ equal to 1"! $\diamond$

Exercise 6.5. Show that there is no function $\delta : \mathbb{R} \to \mathbb{R}$ such that for all $a > 0$:
(1) δ is Riemann integrable on $[-a, a]$, and
(2) for every C^∞ function φ vanishing outside $[-a, a]$, $\int_{-a}^{a} \delta(x)\varphi(x)dx = \varphi(0)$.

Example 6.4. $\mathrm{pv}\dfrac{1}{x}$.

Although the function $x \mapsto \frac{1}{x}$ (defined almost everywhere on $\mathbb{R}$) is not locally integrable, nevertheless one can associate a distribution T with this function: for all $\varphi \in \mathcal{D}(\mathbb{R})$,

$$\langle T, \varphi \rangle := \lim_{\epsilon \to 0} \int_{|x|>\epsilon} \frac{\varphi(x)}{x} dx =: \mathrm{pv} \int \frac{\varphi(x)}{x} dx$$

(pv stands for "principal value").

That the limit above exists, and that T defines a distribution can be seen as follows. Suppose that $\varphi = 0$ on $\mathbb{R} \setminus [-a, a]$, where $a > 0$. We have:

$$\varphi(x) = \varphi(0) + \int_0^x \varphi'(y)dy = \varphi(0) + x \int_0^1 \varphi'(tx)dt.$$

Since $\int_{a \geq |x| > \epsilon} \frac{1}{x} dx = 0$, we have

$$\int_{|x|>\epsilon} \frac{\varphi(x)}{x} dx = \int_{a \geq |x| > \epsilon} \frac{\varphi(x)}{x} dx = \varphi(0) \int_{a \geq |x| > \epsilon} \frac{1}{x} dx + \int_{a \geq |x| > \epsilon} dx \int_0^1 \varphi'(tx)dt$$

$$= 0 + \int_{a \geq |x| > \epsilon} dx \int_0^1 \varphi'(tx)dt = \int_{a \geq |x| > \epsilon} dx \int_0^1 \varphi'(tx)dt.$$

Thus $\lim_{\epsilon \to 0} \int_{|x|>\epsilon} \frac{\varphi(x)}{x} dx$ exists, and $\langle T, \varphi \rangle = \int_{-a}^{a} \int_{0}^{1} \varphi'(tx) dt dx$.

Continuity: Let $\varphi_n \xrightarrow{\mathcal{D}} 0$, and suppose that $a > 0$ is such that for all $n \in \mathbb{N}$, $\varphi_n = 0$ outside $[-a, a]$. Then $|\langle T, \varphi_n \rangle| \leq 2a \cdot \sup_{x \in \mathbb{R}} |\varphi_n'(x)| \xrightarrow{n \to \infty} 0$.

This distribution is denoted by $\operatorname{pv} \frac{1}{x}$, so that $\left\langle \operatorname{pv} \frac{1}{x}, \varphi \right\rangle = \operatorname{pv} \int \frac{\varphi(x)}{x} dx$.

(In quantum mechanics (perturbation theory) one encounters
$$\delta^+ := \frac{1}{2}\delta + \frac{1}{2\pi i}\operatorname{pv}\frac{1}{x}, \quad \delta^- := \frac{1}{2}\delta - \frac{1}{2\pi i}\operatorname{pv}\frac{1}{x},$$
having the property $\delta = \delta^+ + \delta^-$.) ◇

Exercise 6.6.
(1) For $\varphi \in \mathcal{D}(\mathbb{R})$, set $\langle T, \varphi \rangle := \sum_{n \in \mathbb{Z}} \varphi(n)$. Show that $T \in \mathcal{D}'(\mathbb{R})$.
(2) Give an example of a sequence of test functions $(\varphi_n)_{n \in \mathbb{N}}$ such that:
  (a) $\varphi_n \xrightarrow{n \to \infty} 0$ uniformly, and for all $k \in \mathbb{N}$, also $\varphi_n^{(k)} \xrightarrow{n \to \infty} 0$ uniformly, but
  (b) $\langle T, \varphi_n \rangle \xrightarrow{n \to \infty} 0$.
  Hint: Consider a sequence of test functions of the type $\varphi_n(x) = \frac{1}{n}\varphi(\frac{x}{n})$ with an appropriate choice of φ.
(3) Does the construction in part (2) contradict our conclusion in (1) that T is a distribution?

6.2 Derivatives in the distributional sense

We will now develop a calculus for distributions.
Let us first consider the case when $d = 1$.

Definition 6.4. (Distributional derivative, $d = 1$). Let $T \in \mathcal{D}'(\mathbb{R})$. Then the *distributional derivative* T' (or $\frac{d}{dx}T$) $\in \mathcal{D}'(\mathbb{R})$ of T is defined by
$$\langle T', \varphi \rangle = -\langle T, \varphi' \rangle, \quad \varphi \in \mathcal{D}(\mathbb{R}).$$

Note that if $\varphi \in \mathcal{D}(\mathbb{R})$, then clearly $\varphi' \in \mathcal{D}(\mathbb{R})$. So the right-hand side above is well defined.
Moreover, the map $\varphi \mapsto -\langle T, \varphi' \rangle$ is linear: for all $\varphi, \psi \in \mathcal{D}(\mathbb{R})$ and $\alpha \in \mathbb{C}$,
$$\langle T', \varphi + \psi \rangle = -\langle T, (\varphi + \psi)' \rangle = -\langle T, \varphi' + \psi' \rangle = -(\langle T, \varphi' \rangle + \langle T, \psi' \rangle)$$
$$= -\langle T, \varphi' \rangle - \langle T, \psi' \rangle = \langle T', \varphi \rangle + \langle T', \psi \rangle,$$
$$\langle T', \alpha\varphi \rangle = -\langle T, (\alpha\varphi)' \rangle = -\langle T, \alpha\varphi' \rangle = -\alpha\langle T, \varphi' \rangle = \alpha\langle T', \varphi \rangle.$$
Continuity of T': If $\varphi_n \xrightarrow{\mathcal{D}} 0$, then also $\varphi_n' \xrightarrow{\mathcal{D}} 0$, and so $-\langle T, \varphi_n' \rangle \to 0$.
Thus $T' \in \mathcal{D}'(\mathbb{R})$.

Q. Is distributional differentiation an extension of classical differentiation?
A. Yes. The result below (Lemma 6.1) shows that our new definition is a sensible generalisation from the classical case: If we have a continuously differentiable function, and if we choose to put on our "distributional glasses", then the distribution we get, by distributionally differentiating the corresponding regular function, is a regular distribution corresponding to the classical derivative.

Lemma 6.1. *If $f \in C^1(\mathbb{R})$, then $(T_f)' = T_{f'}$.*

Proof. Let $\varphi \in \mathcal{D}(\mathbb{R})$ be such that it vanishes outside $[a, b]$. Then using integration by parts,

$$\langle (T_f)', \varphi \rangle = -\langle T_f, \varphi' \rangle = -\int_\mathbb{R} f(x)\varphi'(x)dx = -\int_a^b f(x)\varphi'(x)dx$$

$$= 0 - \left(-\int_a^b f'(x)\varphi(x)dx \right) = \langle T_{f'}, \varphi \rangle.$$

(Here we have used that φ, φ' are zero outside $[a, b]$, and $\varphi(a) = \varphi(b) = 0$.) This completes the proof. $\square$

The above result means that whenever one identifies the function f with the distribution T_f, then the two possible interpretations of the derivative which arise – the classical sense versus the new distributional sense – coincide. In fact, this is the motivation behind our definition of the distributional derivative given in Definition 6.4.

The next example shows that now, endowed with our notion of the distributional derivative, we can differentiate functions which we couldn't earlier (albeit only in the distributional sense).

Example 6.5. ($H' = \delta$).
Let us show that the Dirac distribution is the distributional derivative of regular distribution corresponding to the Heaviside function $H \in L^1_{\text{loc}}(\mathbb{R})$. For any test function $\varphi \in \mathcal{D}(\mathbb{R})$, we know that $\varphi(x) = 0$ for all sufficiently large x, and so

$$\langle H', \varphi \rangle = -\langle H, \varphi' \rangle = -\int_0^\infty \varphi'(x)dx = -\varphi(x)\Big|_0^\infty = \varphi(0) = \langle \delta, \varphi \rangle,$$

and so $H' = \delta$. $\diamond$

Example 6.6. (Dipole).
The derivative δ' of δ, is called the *dipole*, and is given by

$$\langle \delta', \varphi \rangle = -\langle \delta, \varphi' \rangle = -\varphi'(0),$$

for all $\varphi \in \mathcal{D}(\mathbb{R})$. $\diamond$

In Example 6.5, the classical derivative of H for the regions $x < 0$ and $x > 0$ is equal to 0 everywhere, and it is right to think philosophically that the δ appeared owing to the jump in the values of H from 0 to 1 as x went from negative to positive values.

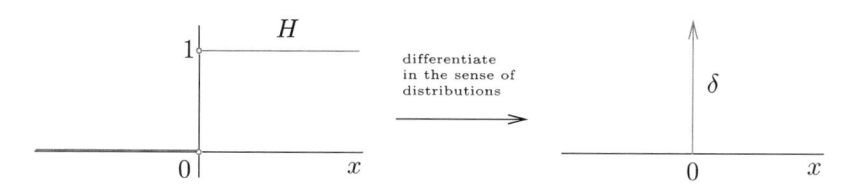

So we can write

$$H' = T_{\mathbf{0}} + 1 \cdot \delta_0,$$

where $T_{\mathbf{0}}$ is the regular distribution corresponding to the zero function $\mathbf{0}$, and the coefficient 1 multiplying δ_0 is the "jump" in the value of H at 0. This is no coincidence, and the observation can be generalised as follows.

Proposition 6.1. (Jump Rule).
Let f be continuously differentiable on $\mathbb{R}$ except at the point $a \in \mathbb{R}$, where the limits $f(a+), f(a-), f'(a+), f'(a-)$ exist.
Then f, f' are locally integrable, and

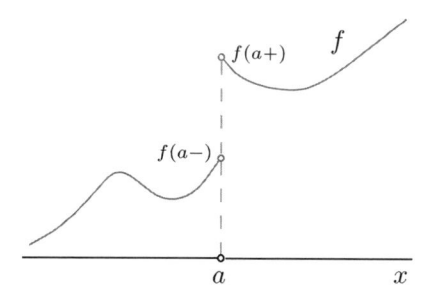

$$(T_f)' = T_{f'} + \bigl(f(a+) - f(a-)\bigr)\delta_a.$$

We think of $f(a+) - f(a-)$ as the jump in f at the point a.
One can formulate this result by saying:

> The derivative of f in the sense of distributions is the classical derivative plus δ_a times the jump in f at a.

Proof. Let $\varphi \in \mathcal{D}(\mathbb{R})$, and suppose that φ is 0 outside $[\alpha, \beta]$, and that $a \in [\alpha, \beta]$. Then

$$\langle (T_f)', \varphi \rangle = -\langle T_f, \varphi' \rangle = -\int_\alpha^\beta f(x)\varphi'(x)dx$$

$$= -\int_\alpha^a f(x)\varphi'(x)dx - \int_a^\beta f(x)\varphi'(x)dx$$

$$= \int_\alpha^a f'(x)\varphi(x)dx - f(a-)\varphi(a) + \int_a^\beta f'(x)\varphi(x)dx + f(a+)\varphi(a)$$

$$= \int_\alpha^\beta f'(x)\varphi(x)dx + \big(f(a+) - f(a-)\big)\varphi(a)$$

$$= \langle T_{f'}, \varphi \rangle + \big(f(a+) - f(a-)\big)\langle \delta_a, \varphi \rangle$$

$$= \langle T_{f'} + \big(f(a+) - f(a-)\big)\delta_a, \varphi \rangle. \qquad \square$$

Example 6.7. ($|x|' = 2H - 1$). In the sense of distributions,

$$\frac{d}{dx}|x| = \begin{cases} 1 & \text{if } x > 0, \\ -1 & \text{if } x < 0 \end{cases} = 2H - 1.$$

Indeed, the jump in $|x|$ at $x = 0$ is $\lim_{x \searrow 0} |x| - \lim_{x \nearrow 0} |x| = 0 - 0 = 0$.

For $x \neq 0$, $|x|$ is differentiable with derivative $\begin{cases} 1 & \text{if } x > 0, \\ -1 & \text{if } x < 0 \end{cases} = 2H(x) - 1$.

Finally, $\lim_{x \searrow 0}(2H(x) - 1) = 1$ and $\lim_{x \nearrow 0}(2H(x) - 1) = -1$.

So by the Jump Rule, $(T_{|x|})' = T_{2H-1}$, or briefly $|x|' = 2H - 1$ in the sense of distributions. $\diamond$

Remark 6.4. This result can be extended to the case when f is continuously differentiable everywhere except for a *finite* number of points a_k, and at these points a_k, the function satisfies the same assumptions as stipulated above. This then leads to

$$(T_f)' = T_{f'} + \sum_k \underbrace{\big(f(a_k+) - f(a_k-)\big)}_{=:\sigma_k}\delta_{a_k}.$$

The proof is analogous.

In fact the result even extends to the case when f has *infinitely* many, but *locally finite*, jump discontinuities: in any compact interval, one finds only finitely many discontinuities. The sum on the right-hand side is the distribution defined by
$$\left\langle \sum_k \sigma_k \delta_{a_k}, \varphi \right\rangle := \sum_k \sigma_k \varphi(a_k), \quad \varphi \in \mathcal{D}(\mathbb{R}),$$
where, for a given test function φ, only finitely many terms on the right-hand side are nonzero.

Example 6.8. $\dfrac{d}{dx}\lfloor x \rfloor = \sum_{n \in \mathbb{Z}} \delta_n$. See the pictures below.

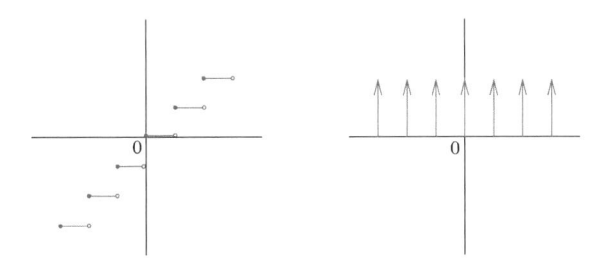

Here $\lfloor \cdot \rfloor$ is the greatest integer function, that is, for $x \in \mathbb{R}$, $\lfloor x \rfloor$ is the greatest integer less than or equal to x. ◇

Exercise 6.7.
Show that $\dfrac{d}{dx} H(x) \cos x = -H(x) \sin x + \delta$ and $\dfrac{d}{dx} H(x) \sin x = H(x) \cos x$.

One can define higher order distributional derivatives by iteratively setting $T^{(n)} := (T^{n-1})'$ for $n \geq 2$.

Exercise 6.8. (Fundamental Solution to the 1D Laplace equation).
Show that the equation $\dfrac{d^2}{dx^2} E = \delta$ is satisfied by $E := \dfrac{1}{2}|x|$.

Exercise 6.9. (Zero distributional derivative implies constancy.)
The aim of this exercise is to show that if $T \in \mathcal{D}'(\mathbb{R})$, and $T' = \mathbf{0}$, then there exists a constant c such that $T = T_c$, the regular distribution associated with the constant function taking value c everywhere on $\mathbb{R}$. To prove this result, we will proceed as follows.
(1) Let V be a complex vector space. If $\ell, L \in L(V, \mathbb{C})$ are such that $\ker \ell \subset \ker L$, then there exists a constant $c \in \mathbb{C}$ such that $L = c\ell$.
   Hint: If $v_0 \in V$ is such that $\ell(v_0) \neq 0$, then show that every vector $v \in V$ can be decomposed as $v = c_v v_0 + w$ for some $c_v \in \mathbb{C}$ and some $w \in \ker \ell$.
(2) Prove, using part (1) and Exercise 6.3, page 232, that if the derivative of the distribution T is zero, then T must be constant.

Exercise 6.10. Show that if $T \in \mathcal{D}'(\mathbb{R})$, then there exists an $S \in \mathcal{D}'(\mathbb{R})$ such that $S' = T$. Moreover, show that such an S is unique up to an additive constant. *Hint:* Fix any $\varphi_0 \in \mathcal{D}(\mathbb{R}) \setminus \{0\}$ which is nonnegative everywhere. For $\psi \in \mathcal{D}(\mathbb{R})$,
$$\varphi := \psi - \Big(\int_{-\infty}^{\infty} \psi(x)dx \Big/ \int_{-\infty}^{\infty} \varphi_0(x)dx \Big) \varphi_0$$
belongs to the subspace Y of $\mathcal{D}(\mathbb{R})$ from Exercise 6.3, page 232. Hence there exists a unique $\Phi \in \mathcal{D}(\mathbb{R})$ such that $\Phi' = \varphi$. Set $\langle S, \psi \rangle = -\langle T, \Phi \rangle$.

Exercise 6.11. Show that for all $n \in \mathbb{N}$, $\delta^{(n)} \neq \mathbf{0}$.

Exercise 6.12. Show that $\{\delta, \delta', \cdots, \delta^{(n)}, \cdots\}$ is linearly independent in $\mathcal{D}'(\mathbb{R})$.

When $d > 1$, the definition of the distributional derivative is analogous.

Definition 6.5. (Distributional partial derivatives).
For $T \in \mathcal{D}'(\mathbb{R}^d)$, the *ith-partial derivative* $\dfrac{\partial T}{\partial x_i}$ of T, $1 \leq i \leq d$, is defined by
$$\Big\langle \frac{\partial T}{\partial x_i}, \varphi \Big\rangle = -\Big\langle T, \frac{\partial \varphi}{\partial x_i} \Big\rangle.$$

Exercise 6.13. Show that for all $T \in \mathcal{D}'(\mathbb{R}^d)$ and all i, j, $\dfrac{\partial^2 T}{\partial x_i \partial x_j} = \dfrac{\partial^2 T}{\partial x_j \partial x_i}$.

Exercise 6.14. The Heaviside function in two variables, $H : \mathbb{R}^2 \to \mathbb{R}$, is defined by
$$H(x,y) := H(x) \otimes H(y) := \begin{cases} 1 \text{ if } x \geq 0 \text{ and } y \geq 0, \\ 0 \text{ if } x < 0 \text{ or } y < 0. \end{cases}$$
(That is, H is the indicator function $1_{[0,\infty)^2}$ of the "first quadrant".)
Show that $\dfrac{\partial^2 H}{\partial x \partial y} = \delta_{\mathbf{0}} = \delta_{(0,0)} =: \delta_0(x) \otimes \delta_0(y)$.

Exercise 6.15. (Fundamental solution of the Laplacian on $\mathbb{R}^2$).
Verify that $\Delta\Big(\dfrac{1}{2\pi} \log r\Big) = \delta_{(0,0)}$, where $\Delta = \dfrac{\partial^2}{\partial x^2} + \dfrac{\partial^2}{\partial y^2}$ and $r = \sqrt{x^2 + y^2}$.

Remark 6.5. (Sobolev spaces).
There exist Hilbert spaces analogous to the spaces $C^n[a,b]$ defined by:
$$H^n(a,b) := \{\mathbf{f} \in L^2(a,b) : \mathbf{f}', \cdots, \mathbf{f}^{(n)} \in L^2(a,b)\}$$
equipped with the norm $\|\cdot\|_n$ defined by:
$$\|\mathbf{f}\|_n^2 = \sum_{k=1}^{n} \|\mathbf{f}^{(k)}\|_2^2, \quad \mathbf{f} \in H^{(n)}(a,b),$$
where $\|\cdot\|_2$ denotes the L^2 norm. This norm is induced by the inner product
$$\langle \mathbf{f}, \mathbf{g} \rangle_n = \sum_{k=1}^{n} \langle \mathbf{f}^{(k)}, \mathbf{g}^{(k)} \rangle_{L^2}, \quad \mathbf{f}, \mathbf{g} \in H^n(a,b).$$

For example: $H^1(a,b) = \{\mathbf{f} \in L^2(a,b) : \mathbf{f}' \in L^2(a,b)\}$ and

$$\langle \mathbf{f}, \mathbf{g} \rangle_1 = \int_a^b \mathbf{f}(x)(\mathbf{g}(x))^* dx + \int_a^b \mathbf{f}'(x)(\mathbf{g}'(x))^* dx, \quad \mathbf{f}, \mathbf{g} \in H^1(a,b).$$

Since $\mathbf{f}$ merely belongs to $L^2(a,b)$, we are aware that $\mathbf{f}^{(k)}$ may not have any meaning in general, and so one ought to make the definition precise. This can be done with the theory of distributions. The space $H^n(a,b)$ is the space of functions $\mathbf{f} \in L^2(a,b)$ such that $\mathbf{f}', \cdots, \mathbf{f}^{(n)}$, defined in the sense of distributions, belong to $L^2(a,b)$. Then it can be showed that the space $H^n(a,b)$ is a Hilbert space. The Sobolev spaces $H^n(a,b)$ are named after the Russian Mathematician Sergei Sobolev.

6.3 Weak solutions

A function satisfying a PDE in the sense of distributions will be called a *weak solution* to a PDE.

Example 6.9. Let $u(x,t) := H(t)e^x$. Then u is locally integrable. We'll show that u is a weak solution of the PDE $\dfrac{\partial u}{\partial x} = u$.
We have for all $\varphi \in \mathcal{D}(\mathbb{R}^2)$ that

$$\left\langle \left(\frac{\partial}{\partial x} - 1\right)u, \varphi \right\rangle = -\left\langle u, \frac{\partial \varphi}{\partial x} \right\rangle - \langle u, \varphi \rangle$$

$$= -\iint_{\mathbb{R}^2} \left(H(t)e^x \frac{\partial \varphi}{\partial x}(x,t) + H(t)e^x \varphi(x,t) \right) dx dt$$

$$= -\int_0^\infty \int_{-\infty}^\infty \left(e^x \frac{\partial \varphi}{\partial x}(x,t) + e^x \varphi(x,t) \right) dx dt$$

$$= -\int_0^\infty \int_{-\infty}^\infty \frac{\partial}{\partial x}(e^x \varphi) dx dt$$

$$= -\int_0^\infty \left(e^x \varphi(x,t) \Big|_{-\infty}^\infty \right) dt$$

$$= -\int_0^\infty 0 dt = 0,$$

and so $\left(\dfrac{\partial}{\partial x} - 1\right)u = \mathbf{0}$ in the sense of distributions.
The following picture shows the graph of this u.

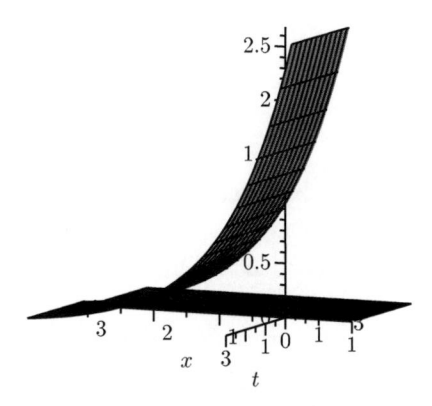

Far from being a classical solution (for which we would want $u \in C^1$) to the PDE, we see that u isn't even continuous! Nevertheless we accept it as a solution to PDE, since it satisfies the PDE, albeit in the weak sense. $\diamond$

Remark 6.6. (Other notions of weak solutions). Besides distributional solutions, there are other notions too of weak solutions in PDE theory, for example "viscosity solutions" (which are natural in certain contexts, such as the Hamilton-Jacobi equation for optimal control).

Remark 6.7. (Why are weak solutions important?) Weak solutions are important, since as mentioned before, many initial/boundary value problems for PDEs encountered in real world may not possess sufficiently smooth solutions, but only weak solutions, which should not be dismissed. We will see two examples below.

Another reason is "theoretical": even when there exist classical solutions, it might be easier to find/show the existence of distributional solutions *first*, and then show later that the solution is in fact sufficiently smooth (and such results are called "regularity results").

Weak solution to the transport equation

The transport equation is given by
$$\begin{cases} \dfrac{\partial u}{\partial t} - c\dfrac{\partial u}{\partial x} = 0, \\ u(x,0) = f(x), \end{cases}$$
where f is the initial condition, and $c \in \mathbb{R}$ is a constant.

This equation arises for instance when one models fluid flow[6].
It is easy to check that if $f \in C^1$, then a solution is given by
$$u(x,t) = f(x+ct).$$
Indeed, $u(x,0) = f(x+c\cdot 0) = f(x+0) = f(x)$, and
$$\frac{\partial u}{\partial t} - c\frac{\partial u}{\partial x} = f'(x+ct)\cdot c - c\cdot f'(x+ct)\cdot 1 = 0.$$
But now, we'll show that even when $f \in L^1_{\text{loc}}(\mathbb{R})$, the same formula, namely $u(x,t) := f(x+ct)$, still gives a solution u to the transport equation. The only change is that it will be a *weak* solution, that is, the PDE will be satisfied in the "distributional sense". To see this, let $\varphi \in \mathcal{D}(\mathbb{R}^2)$. Then:
$$\left\langle \frac{\partial u}{\partial t} - c\frac{\partial u}{\partial x}, \varphi \right\rangle = \left\langle u, -\frac{\partial \varphi}{\partial t} + c\frac{\partial \varphi}{\partial x} \right\rangle$$
$$= \iint_{\mathbb{R}^2} f(x+ct)\left(c\frac{\partial \varphi}{\partial x}(x,t) - \frac{\partial \varphi}{\partial t}(x,t)\right)dxdt. \quad (6.2)$$
Hence to prove our claim, it must be shown that the above integral is zero. To do this, we will make the following change of variables[7]:

$$\boxed{\begin{aligned}\xi &= x+ct\\ \eta &= t\end{aligned}} \quad\longleftrightarrow\quad \boxed{\begin{aligned}x &= \xi - c\eta,\\ t &= \eta.\end{aligned}}$$

Recall that for a double integral, one has the following "change of variables" formula under the change of variables given by the map $(\xi,\eta) \overset{\Psi}{\mapsto} (x,t)$:
$$\iint_{\mathbb{R}^2} F(x,y)dxdy = \iint_{\mathbb{R}^2} (F\circ \Psi)(\xi,\eta)\cdot |J(\xi,\eta)|d\xi d\eta,$$
where $J(\xi,\eta) = \det\begin{bmatrix}\frac{\partial x}{\partial \xi} & \frac{\partial x}{\partial \eta}\\ \frac{\partial t}{\partial \xi} & \frac{\partial t}{\partial \eta}\end{bmatrix}.$

In our case, the derivative of the map $(\xi,\eta) \overset{\Psi}{\mapsto} (x,t)$ is
$$\begin{bmatrix}\frac{\partial x}{\partial \xi} & \frac{\partial x}{\partial \eta}\\ \frac{\partial t}{\partial \xi} & \frac{\partial t}{\partial \eta}\end{bmatrix} = \begin{bmatrix}1 & -c\\ 0 & 1\end{bmatrix},$$
whose determinant is 1.

---
[6]See for example [Pinchover and Rubinstein (2005), page 8].

[7]A motivation for this particular change of variables comes from hindsight, see equation (6.3) on page 246: the aim is to view the integrand in (6.2) as a derivative in one of the variables so that one can apply the fundamental theorem of calculus.

We have

$$(F \circ \Psi)(\xi, \eta) = F(\Psi(\xi, \eta)) = F(\xi - c\eta, \eta)$$
$$= f((\xi - c\eta) + c \cdot \eta)\left(c\frac{\partial \varphi}{\partial x}(\xi - c\eta, \eta) - \frac{\partial \varphi}{\partial t}(\xi - c\eta, \eta)\right)$$
$$= f(\xi)\left(-\frac{\partial(\xi - c\eta)}{\partial \eta}\frac{\partial \varphi}{\partial x}(\xi - c\eta, \eta) - \frac{\partial \eta}{\partial \eta}\frac{\partial \varphi}{\partial t}(\xi - c\eta, \eta)\right)$$
$$= -f(\xi)\left(\frac{\partial \varphi}{\partial x}(\xi - c\eta, \eta)\frac{\partial(\xi - c\eta)}{\partial \eta} + \frac{\partial \varphi}{\partial t}(\xi - c\eta, \eta)\frac{\partial \eta}{\partial \eta}\right)$$
$$= -f(\xi)\frac{\partial}{\partial \eta}\left(\varphi(\xi - c\eta, \eta)\right)$$
$$= -f(\xi)\frac{\partial(\varphi \circ \Psi)}{\partial \eta}(\xi, \eta).$$

Thus

$$\left\langle \frac{\partial u}{\partial t} - c\frac{\partial u}{\partial x}, \varphi \right\rangle = \iint_{\mathbb{R}^2} f(x + ct)\left(c\frac{\partial \varphi}{\partial x}(x, t) - \frac{\partial \varphi}{\partial t}(x, t)\right) dx dt$$
$$= \iint_{\mathbb{R}^2} -f(\xi)\left(\frac{\partial(\varphi \circ \Psi)}{\partial \eta}(\xi, \eta)\right) \cdot |1| d\xi d\eta \quad (6.3)$$
$$= -\int_{-\infty}^{\infty} f(\xi)\left(\int_{-\infty}^{\infty} \frac{\partial(\varphi \circ \Psi)}{\partial \eta}(\xi, \eta) d\eta\right) d\xi$$
$$= -\int_{-\infty}^{\infty} f(\xi)\left((\varphi \circ \Psi)(\xi, \eta)\Big|_{\eta=-\infty}^{\eta=+\infty}\right) d\xi$$
$$= -\int_{-\infty}^{\infty} f(\xi)\left(\varphi(\xi - c\eta, \eta)\Big|_{\eta=-\infty}^{\eta=+\infty}\right) d\xi$$
$$= -\int_{-\infty}^{\infty} f(\xi) \cdot 0 d\xi$$
$$= 0, \quad (6.4)$$

where we have used the Fundamental Theorem of Calculus to simplify the inner integral, and used the fact that φ has compact support to obtain

$$(\varphi \circ \Psi)(\xi, \eta)\Big|_{\eta=-\infty}^{\eta=+\infty} = \lim_{\eta \to \infty}(\varphi \circ \Psi)(\xi, \eta) - \lim_{\eta \to -\infty}(\varphi \circ \Psi)(\xi, \eta)$$
$$= \lim_{\eta \to \infty}\varphi(\xi - c\eta, \eta) - \lim_{\eta \to -\infty}\varphi(\xi - c\eta, \eta)$$
$$= 0 - 0 = 0.$$

See the following picture.

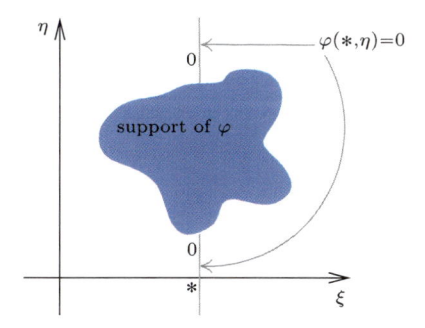

This shows that $u(x,t) := f(x+ct)$ is indeed a weak solution to the transport equation.

Weak solution to the wave equation

Recall that if $f \in C^2(\mathbb{R})$, then

$$u(x,t) := \frac{f(x+t) + f(x-t)}{2} \tag{6.5}$$

is a classical solution to

$$\frac{\partial^2 u}{\partial t^2} - \frac{\partial^2 u}{\partial x^2} = 0$$

with the initial condition $u(x,0) = f(x)$ and with zero initial speed $u_t(x,0) = 0$.

Let us now show that even when f is merely locally integrable, u given by (6.5) satisfies the wave equation, but in the sense of distributions. In order to do this, we will use our result from the previous section, where we considered the transport equation.

Let $f \in L^1_{\text{loc}}(\mathbb{R})$. Putting $c = 1$, we have seen that u_+ given by

$$u_+(x,t) := f(x+t),$$

satisfies, in the sense of distributions, the transport equation

$$\frac{\partial u}{\partial t} - \frac{\partial u}{\partial x} = 0.$$

Similarly, putting $c = -1$, we also see that u_- given by

$$u_-(x,t) := f(x-t)$$

is a weak solution to

$$\frac{\partial u}{\partial t} + \frac{\partial u}{\partial x} = 0.$$

For any distribution $T \in \mathcal{D}'(\mathbb{R}^2)$, $\dfrac{\partial^2 T}{\partial x \partial t} = \dfrac{\partial^2 T}{\partial t \partial x}$ (Exercise 6.13, p. 242). So
$$\left(\frac{\partial}{\partial t} - \frac{\partial}{\partial x}\right)\left(\frac{\partial}{\partial t} + \frac{\partial}{\partial x}\right)T = \left(\frac{\partial^2}{\partial t^2} - \frac{\partial^2}{\partial x^2}\right)T = \left(\frac{\partial}{\partial t} + \frac{\partial}{\partial x}\right)\left(\frac{\partial}{\partial t} - \frac{\partial}{\partial x}\right)T.$$
Using this observation, we will find a weak solution to the wave equation too. Let u be given by (6.5), and $\varphi \in \mathcal{D}(\mathbb{R}^2)$. Then
$$\begin{aligned}\left\langle \frac{\partial^2 u}{\partial t^2} - \frac{\partial^2 u}{\partial x^2}, \varphi \right\rangle &= \frac{1}{2}\left\langle \left(\frac{\partial^2}{\partial t^2} - \frac{\partial^2}{\partial x^2}\right)f(x+t), \varphi \right\rangle \\ &\quad + \frac{1}{2}\left\langle \left(\frac{\partial^2}{\partial t^2} - \frac{\partial^2}{\partial x^2}\right)f(x-t), \varphi \right\rangle \\ &= \frac{1}{2}\left\langle \left(\frac{\partial}{\partial t} + \frac{\partial}{\partial x}\right)\left(\frac{\partial}{\partial t} - \frac{\partial}{\partial x}\right)f(x+t), \varphi \right\rangle \\ &\quad + \frac{1}{2}\left\langle \left(\frac{\partial}{\partial t} - \frac{\partial}{\partial x}\right)\left(\frac{\partial}{\partial t} + \frac{\partial}{\partial x}\right)f(x-t), \varphi \right\rangle \\ &= -\frac{1}{2}\left\langle \left(\frac{\partial}{\partial t} - \frac{\partial}{\partial x}\right)f(x+t), \left(\frac{\partial}{\partial t} + \frac{\partial}{\partial x}\right)\varphi \right\rangle \\ &\quad - \frac{1}{2}\left\langle \left(\frac{\partial}{\partial t} + \frac{\partial}{\partial x}\right)f(x-t), \left(\frac{\partial}{\partial t} - \frac{\partial}{\partial x}\right)\varphi \right\rangle \\ &\stackrel{(*)}{=} -0 - 0 = 0.\end{aligned}$$

To see the equality $(*)$ in the last line, we note that since $\varphi \in \mathcal{D}(\mathbb{R}^2)$, also
$$\left(\frac{\partial}{\partial t} + \frac{\partial}{\partial x}\right)\varphi \in \mathcal{D}(\mathbb{R}^2) \text{ and } \left(\frac{\partial}{\partial t} - \frac{\partial}{\partial x}\right)\varphi \in \mathcal{D}(\mathbb{R}^2).$$
But as
$$\left\langle \left(\frac{\partial}{\partial t} - \frac{\partial}{\partial x}\right)f(x+t), \underset{\text{function}}{\underset{\text{test}}{\text{any}}} \right\rangle = 0 = \left\langle \left(\frac{\partial}{\partial t} + \frac{\partial}{\partial x}\right)f(x-t), \underset{\text{function}}{\underset{\text{test}}{\text{any}}} \right\rangle,$$
we see that the equality $(*)$ above holds.

Exercise 6.16. (Weak solution exists, but no classical solution). Show that $u(x) = \begin{cases} c & \text{if } x < 0, \\ x + c & \text{if } x > 0, \end{cases}$ is a weak solution of the ODE $u' = H$, where H is the Heaviside function.

6.4 Multiplication by C^∞ functions

In general, it is not possible to define the product of two distributions. For example, the product of two locally integrable functions is not in general locally integrable. ($f := 1/\sqrt{|x|}$ is locally integrable, but $f^2 = 1/|x|$ isn't!)

So the product of two regular distributions in general may not define a distribution.

However, one *can* define the product of a function $\alpha \in C^\infty(\mathbb{R}^d)$ with a distribution $T \in \mathcal{D}'(\mathbb{R}^d)$ as follows.

Definition 6.6. (Multiplication of a distribution by a smooth function). Let $\alpha \in C^\infty(\mathbb{R}^d)$ and $T \in \mathcal{D}'(\mathbb{R}^d)$. Then $\alpha T \in \mathcal{D}'(\mathbb{R}^d)$ is defined by

$$\langle \alpha T, \varphi \rangle = \langle T, \alpha \varphi \rangle, \quad \varphi \in \mathcal{D}(\mathbb{R}^d).$$

Note that if $\varphi \in \mathcal{D}(\mathbb{R}^d)$, then it is in particular in $C^\infty(\mathbb{R}^d)$, and so it is clear that $\alpha\varphi$ is infinitely many times differentiable. Moreover, as φ vanishes outside a compact set, so does $\alpha\varphi$. Hence $\alpha\varphi \in \mathcal{D}(\mathbb{R}^d)$, and the right-hand side makes sense. It is also easy to see that the map

$$\varphi \mapsto \langle T, \alpha\varphi \rangle : \mathcal{D}(\mathbb{R}^d) \to \mathbb{R}$$

is linear, thanks to the linearity of T. Finally, the continuity of αT can be established by using the multivariable *Leibniz Rule* for differentiating the product of two functions, which we recall here first:

Leibniz Rule: For a multi-index $\mathbf{n} = (n_1, \cdots, n_d)$ of nonnegative integers $n_1, \cdots, n_d$, define its

- *order* $|\mathbf{n}|$ by $n_1 + \cdots + n_d$, and
- *factorial* by $\mathbf{n}! = n_1! \cdots n_d!$.

Then the (multivariable) *Leibniz Rule* states that for every multi-index $\mathbf{n} := (n_1, \cdots, n_d)$, and functions $f, g \in C^\infty(\mathbb{R}^d)$,

$$D^{\mathbf{n}}(fg) = \sum_{\substack{\text{multi-indices} \\ \mathbf{k}=(k_1,\cdots,k_d): \\ k_i \leq n_i,\, i=1,\cdots,d}} \binom{\mathbf{n}}{\mathbf{k}} (D^{\mathbf{k}} f)(D^{\mathbf{n}-\mathbf{k}} g),$$

where $D^{\mathbf{n}} = \dfrac{\partial^{|\mathbf{n}|}}{\partial x_1^{n_1} \cdots \partial x_d^{n_d}}$, and $\binom{\mathbf{n}}{\mathbf{k}} := \dfrac{\mathbf{n}!}{\mathbf{k}!(\mathbf{n}-\mathbf{k})!}$.

(We will omit the cumbersome, although straightforward, proof of the Leibniz Rule, which proceeds by induction on the order $|\mathbf{n}|$ of $D^{\mathbf{n}}$, and by using the one variable mth derivative formula for the product of two functions.)

Using the Leibniz Rule, it can be seen that if $\varphi_n \xrightarrow{\mathcal{D}} \mathbf{0}$, and if $\alpha \in C^\infty(\mathbb{R}^d)$, then also $\alpha\varphi_n \xrightarrow{\mathcal{D}} \mathbf{0}$. Consequently, $\alpha T \in \mathcal{D}'(\mathbb{R}^d)$.

This product of distributions with smooth functions extends the usual pointwise product of a function with a smooth function.

Proposition 6.2. *If $f \in L^1_{\mathrm{loc}}(\mathbb{R}^d)$ and $\alpha \in C^\infty(\mathbb{R}^d)$, then $\alpha T_f = T_{\alpha f}$.*

Proof. α is bounded on every compact set, and so it follows that αf is locally integrable. For $\varphi \in \mathcal{D}(\mathbb{R}^d)$, we have

$$\langle \alpha T_f, \varphi \rangle = \langle T_f, \alpha\varphi \rangle = \int_{\mathbb{R}^d} f(\mathbf{x})\alpha(\mathbf{x})\varphi(\mathbf{x})d\mathbf{x} = \langle T_{\alpha f}, \varphi \rangle.$$

This completes the proof. $\square$

The above result means that, whenever we identify as usual the elements of $L^1_{\text{loc}}(\mathbb{R}^d)$ with distributions, then the two a priori different manners of forming the product with α lead to the same result.

Example 6.10. One can think of the distribution $H(x)\cos x$ as the product of the C^∞ function $\cos x$ with the distribution $H(x)$. $\Diamond$

Proposition 6.3. *The following calculation rules hold.*
For $T, T_1, T_2 \in \mathcal{D}'(\mathbb{R}^d)$, $\alpha_1, \alpha_2, \alpha, \beta \in C^\infty(\mathbb{R}^d)$, we have

(1) $\alpha(T_1 + T_2) = \alpha T_1 + \alpha T_2$
(2) $(\alpha_1 + \alpha_2)T = \alpha_1 T + \alpha_2 T$
(3) $(\alpha\beta)T = \alpha(\beta T)$
(4) $\mathbf{1}T = T$. (Here $\mathbf{1} \in C^\infty(\mathbb{R}^d)$ is the constant function $\mathbb{R}^d \ni \mathbf{x} \mapsto 1$.)

(Thus $\mathcal{D}'(\mathbb{R}^d)$ is a $C^\infty(\mathbb{R}^d)$-module[8].)

Proof. All of these follow from the definition of multiplication of distributions by C^∞ functions. For example, to check (3), note that for all φ in $\mathcal{D}(\mathbb{R}^d)$, $\langle (\alpha\beta)T, \varphi \rangle = \langle T, (\alpha\beta)\varphi \rangle = \langle T, \beta(\alpha\varphi) \rangle = \langle \beta T, \alpha\varphi \rangle = \langle (\alpha(\beta T)), \varphi \rangle$, proving the claim. $\square$

The product rule for differentiation is valid in the same manner as for functions.

Theorem 6.2. (Product Rule). *For $T \in \mathcal{D}'(\mathbb{R}^d)$ and $\alpha \in C^\infty(\mathbb{R}^d)$,*

$$(d = 1): \qquad (\alpha T)' = \alpha' T + \alpha T'$$

$$(d > 1): \qquad \frac{\partial}{\partial x_i}(\alpha T) = \left(\frac{\partial \alpha}{\partial x_i}\right)T + \alpha\left(\frac{\partial T}{\partial x_i}\right).$$

---

[8] A *module* is just like a vector space, except that the underlying field is replaced by a ring. For the module $\mathcal{D}'(\mathbb{R}^d)$ we consider, the underlying ring is $C^\infty(\mathbb{R}^d)$, with pointwise addition and multiplication. We note that not every nonzero element in $C^\infty(\mathbb{R}^d)$ has a multiplicative inverse; for example a function which is zero on a strict subset of $\mathbb{R}^d$ such as $\mathbf{x} \mapsto x_1$. This is the only thing which is missing for a ring from the list of satisfied field axioms.

Proof. When $d = 1$ and $\varphi \in \mathcal{D}(\mathbb{R})$, we have $(\alpha\varphi)' = \alpha'\varphi + \alpha\varphi'$, and so

$$\begin{aligned}\langle (\alpha T)', \varphi \rangle &= -\langle \alpha T, \varphi' \rangle = -\langle T, \alpha\varphi' \rangle \\ &= -\langle T, (\alpha\varphi)' \rangle + \langle T, \alpha'\varphi \rangle \\ &= \langle T', \alpha\varphi \rangle + \langle \alpha' T, \varphi \rangle \\ &= \langle \alpha T', \varphi \rangle + \langle \alpha' T, \varphi \rangle \\ &= \langle \alpha T' + \alpha' T, \varphi \rangle.\end{aligned}$$

The proof is analogous when $d > 1$. $\square$

Theorem 6.3. *If* $\mathbf{a} \in \mathbb{R}^d$ *and* $\alpha \in C^\infty(\mathbb{R}^d)$, *then* $\alpha\delta_\mathbf{a} = \alpha(\mathbf{a})\delta_\mathbf{a}$.

Proof. For $\varphi \in \mathcal{D}(\mathbb{R}^d)$, we have
$$\langle \alpha\delta_\mathbf{a}, \varphi \rangle = \langle \delta_\mathbf{a}, \alpha\varphi \rangle = (\alpha\varphi)(\mathbf{a}) = \alpha(\mathbf{a})\varphi(\mathbf{a}) = \alpha(\mathbf{a})\langle \delta_\mathbf{a}, \varphi \rangle = \langle \alpha(\mathbf{a})\delta_\mathbf{a}, \varphi \rangle. \quad \square$$

Example 6.11. (δ_a, $a \in \mathbb{R}$, are eigenvectors of the position operator).
Let us recall Exercise 2.35, page 103, where we showed that the position operator $Q : D_Q (\subset L^2(\mathbb{R})) \to L^2(\mathbb{R})$ given by $(Q\mathbf{f})(x) = x\mathbf{f}(x)$, $x \in \mathbb{R}$, $\mathbf{f} \in D_Q$, has empty point spectrum, and so it has no eigenvectors.

But we can "extend" the operator Q to act not just on functions on $\mathbb{R}$, but also distributions:

$$QT := xT, \quad T \in \mathcal{D}'(\mathbb{R}).$$

Then Q is a linear transformation from $\mathcal{D}'(\mathbb{R})$ to itself.

The result in Theorem 6.3 above shows that, for all $a \in \mathbb{R}$,

$$Q\delta_a = x\delta_a = a\delta_a,$$

and so $\delta_a \in \mathcal{D}'(\mathbb{R})$, serves as an eigenvector, with corresponding eigenvalue $a \in \mathbb{R}$, of the position operator $Q \in L(\mathcal{D}'(\mathbb{R}))$. (The physicist Paul Dirac used this in 1926 for Quantum Mechanical computations.) $\diamond$

Example 6.12. We have $x\delta = 0$, $(\cos x)\delta = \delta$, $(\sin x)\delta = 0$. $\diamond$

Exercise 6.17. Redo Exercise 6.7, page 241, using the Product Rule.

Exercise 6.18. (Fundamental Solutions).
Show the following, where $\lambda \in \mathbb{R}$, $n \in \mathbb{N}$, $\omega \in \mathbb{R}\setminus\{0\}$:

(1) $\left(\dfrac{d}{dx} - \lambda\right)\left(H(x)e^{\lambda x}\right) = \delta$

(2) $\dfrac{d^n}{dx^n}\left(H(x)\dfrac{x^{n-1}}{(n-1)!}\right) = \delta$

(3) $\left(\dfrac{d^2}{dx^2} + \omega^2\right)\left(H(x)\dfrac{\sin(\omega x)}{\omega}\right) = \delta$.

Exercise 6.19. Show that if $\alpha \in C^\infty(\mathbb{R})$, then $\alpha \delta' = \alpha(0)\delta' - \alpha'(0)\delta$. Conclude that $x\delta' = -\delta$.

Exercise 6.20. For $T \in \mathcal{D}'(\mathbb{R})$, define $\left[x, \dfrac{d}{dx}\right]T := x\dfrac{dT}{dx} - \dfrac{d}{dx}(xT)$.
Show that for all $T \in \mathcal{D}'(\mathbb{R})$, $\left[x, \dfrac{d}{dx}\right]T = -T$.
(Thus the *commutant* of x, $\dfrac{d}{dx}$, namely $\left[x, \dfrac{d}{dx}\right]$, is -1.)

Exercise 6.21. Show that $u(x,y) := e^{-3yx}H(y)$ is a weak solution of
$$\begin{cases} \dfrac{\partial u}{\partial x} + 3yu = 0, & (x > 0, \ y \in \mathbb{R}) \\ u(0,y) = H(y) & (y \in \mathbb{R}). \end{cases}$$

Exercise 6.22. Show that on $\mathcal{D}'(\mathbb{R})$ it is impossible to define an associative and commutative product such that for $\alpha \in C^\infty(\mathbb{R})$ and $T \in \mathcal{D}'(\mathbb{R})$, it agrees with Definition 6.6. *Hint:* Consider the product of δ, x and pv $\frac{1}{x}$.

Exercise 6.23.
(1) Let T be a distributional solution to the differential equation $\left(\dfrac{d}{dx} - \lambda\right)T = 0$.
Show that T is a classical solution: $T = ce^{\lambda x}$. *Hint:* Differentiate $e^{-\lambda x}T$.
(2) (Hypoellipticity[9] of $\dfrac{d}{dx} - \lambda$.)
Let $f \in C^\infty(\mathbb{R})$, and $T \in \mathcal{D}'(\mathbb{R})$ be a solution to
$$\left(\dfrac{d}{dx} - \lambda\right)T = f. \qquad (*)$$
Show that T is equal to a classical solution, and that $T = F + ce^{\lambda x}$, where F is a classical (namely C^∞) solution of $(*)$.
(3) Consider an ordinary differential operator with constant coefficients:
$$D = \sum_{k=0}^{n} a_k \left(\dfrac{d}{dx}\right)^k.$$
Let $f \in C^\infty(\mathbb{R})$ and let T be a distributional solution of $DT = f$.
Show that T is a classical solution, namely $T \in C^\infty$.
Hint: If λ is a root of the polynomial $P(\xi) = a_0 + a\xi + \cdots + a_n\xi^n$, then D can be written as the product $(\frac{d}{dx} - \lambda)D_1$, where D_1 is a differential operator of order $n-1$.
(4) Let E_* be a *fundamental solution* of the differential operator D, that is, let $DE_* = \delta$. What can one say about the set of all fundamental solutions of D?

Exercise 6.24.
Show that the distributional solutions T to $xT = 0$ are scalar multiples of $\delta = \delta_0$. *Hint:* Show that $\ker \delta = \{x\varphi : \varphi \in \mathcal{D}(\mathbb{R})\}$ and use Exercise 6.9(1), page 241.

---
[9] D is hypoelliptic if $u \in \mathcal{D}'$ and $Du \in C^\infty$ implies $u \in C^\infty$.

6.5 Fourier transform of (tempered) distributions

We make a few final parting remarks for this chapter, which are somewhat sketchy, but aim to give a glimpse of what lies ahead. One would like to extend the classical Fourier transform theory to distributions. From our previous definitions (for example that of differentiation of a distribution), we know that the philosophy is, to transpose the stuff we want to do to a distribution, to an appropriate related thing on the test function, so that the new definition matches with the classical one. Continuing in this spirit, we would like to define the Fourier transform of "nice" distributions $T \in \mathcal{D}'(\mathbb{R})$ in such a manner, so that if $T = T_f$, with $f \in L^1(\mathbb{R})$ (say[10]), then one has $\widehat{T_f} = T_{\hat{f}}$. Proceeding formally, we ought to have for test functions φ that

$$\langle \widehat{T_f}, \varphi \rangle = \langle T_{\hat{f}}, \varphi \rangle = \int_{-\infty}^{\infty} \varphi(x) \hat{f}(x) dx$$

$$= \int_{-\infty}^{\infty} \varphi(x) \int_{-\infty}^{\infty} f(\xi) e^{-2\pi i \xi x} d\xi dx$$

$$= \int_{-\infty}^{\infty} f(\xi) \int_{-\infty}^{\infty} \varphi(x) e^{-2\pi i \xi x} dx d\xi$$

$$= \langle T_f, \hat{\varphi} \rangle.$$

So motivated by this, one could hope to define the Fourier transform of a distribution T by setting

$$\langle \hat{T}, \varphi \rangle = \langle T, \hat{\varphi} \rangle,$$

where $\hat{\varphi}$ is the classical Fourier transform of $\varphi \in \mathcal{D}(\mathbb{R})$, defined by

$$\hat{\varphi}(\xi) = \int_{-\infty}^{\infty} \varphi(x) e^{-2\pi i \xi x} dx.$$

But the above calculation is all wrong! Indeed, for a test function $\varphi \in \mathcal{D}(\mathbb{R})$, the Fourier transform $\hat{\varphi}$ may not have compact support[11], and so $\hat{\varphi}$ does *not* belong to $\mathcal{D}(\mathbb{R})$ (unless $\varphi = \mathbf{0}$). In light of this problem (that the Fourier transform of test functions are no longer test functions), it makes sense to work with a bigger class of test functions that *are* closed under Fourier transformation, and then work with only those distributions that are well-behaved with this larger class of test functions (and then these distributions

---

[10] We start with such functions, since we know that for absolutely integrable functions f, the classical Fourier transform $\hat{f}$ is a well-defined function, which is bounded and continuous on $\mathbb{R}$.

[11] In fact, it can be shown that $\hat{\varphi}$ belongs to $\mathcal{D}$ if and only if $\varphi = \mathbf{0}$.

will be deemed to be "Fourier transform-able"). With this little motivation, we will consider the Schwartz class $\mathcal{S}(\mathbb{R})$ of test functions, defined below. Although this story can be developed in $\mathbb{R}^d$ with $d \geq 1$ in general, we will just work with $d = 1$ here for simplicity.

Definition 6.7. (The Schwartz space $\mathcal{S}(\mathbb{R})$ of test functions).
The *Schwartz space $\mathcal{S}(\mathbb{R})$ of test functions* is the set of all functions $\varphi : \mathbb{R} \to \mathbb{C}$ such that:
    (a) φ is infinitely many times differentiable, and
    (b) for all nonnegative integers ℓ, m, $\sup\limits_{x \in \mathbb{R}} |x^\ell \varphi^{(m)}| < +\infty$.

With pointwise operations, $\mathcal{S}(\mathbb{R})$ is a vector space.
$\mathcal{S}(\mathbb{R})$ is closed under differentiation, and multiplication by polynomials.
It is also immediate that $\mathcal{D}(\mathbb{R}) \subset \mathcal{S}(\mathbb{R})$.
An example of a function in $\mathcal{S}(\mathbb{R}) \setminus \mathcal{D}(\mathbb{R})$ is e^{-x^2}.

Exercise 6.25. Show that $e^{-x^2} \in \mathcal{S}(\mathbb{R})$.

Definition 6.8. (Convergence in $\mathcal{S}(\mathbb{R})$).
A sequence $(\varphi_n)_n$ is said to *converge to $\mathbf{0}$ in $\mathcal{S}(\mathbb{R})$*, written $\varphi_n \xrightarrow{\mathcal{S}} \mathbf{0}$, if for all nonnegative integers ℓ, m, we have $\sup\limits_{x \in \mathbb{R}} |x^\ell \varphi_n^{(m)}(x)| \xrightarrow{n \to \infty} 0$.

Exercise 6.26.
Show that if $(\varphi_n)_{n \in \mathbb{N}}$ is a sequence of test functions in $\mathcal{D}(\mathbb{R})$ such that $\varphi_n \xrightarrow{\mathcal{D}} \mathbf{0}$, then we have $\varphi_n \xrightarrow{\mathcal{S}} \mathbf{0}$.

The following result can be shown, but it will take us a bit far afield, and so we skip its somewhat technical proof.

Proposition 6.4.
$\varphi \mapsto \widehat{\varphi} : \mathcal{S}(\mathbb{R}) \to \mathcal{S}(\mathbb{R})$ *is a (linear and) continuous map, that is,*
    (1) *If $\varphi \in \mathcal{S}(\mathbb{R})$, then $\widehat{\varphi} \in \mathcal{S}(\mathbb{R})$.*
    (2) *If $(\varphi_n)_{n \in \mathbb{N}}$ is a sequence in $\mathcal{S}(\mathbb{R})$ such that $\varphi_n \xrightarrow{\mathcal{S}} \varphi$ as $n \to \infty$, then $\widehat{\varphi}_n \xrightarrow{\mathcal{S}} \widehat{\varphi}$.*

Definition 6.9. (Tempered distribution).
A *tempered distribution T on $\mathbb{R}$* is a map $T : \mathcal{S}(\mathbb{R}) \to \mathbb{C}$ such that
    (1) T is linear, and
    (2) if $(\varphi_n)_{n \in \mathbb{N}}$ is a sequence in $\mathcal{S}(\mathbb{R})$ such that $\varphi_n \xrightarrow{\mathcal{S}} \mathbf{0}$ as $n \to \infty$, then $\langle T, \varphi_n \rangle \to 0$.

The vector space of all tempered distributions (with pointwise operations) is denoted by $\mathcal{S}'(\mathbb{R})$.

Also, since $\mathcal{D}(\mathbb{R}) \subset \mathcal{S}(\mathbb{R})$, and since the inclusion is continuous in the sense of Exercise 6.26, it follows that $\mathcal{S}'(\mathbb{R}) \subset \mathcal{D}'(\mathbb{R})$.

However, it can be shown that the inclusion $\mathcal{S}'(\mathbb{R}) \subset \mathcal{D}'(\mathbb{R})$ is strict, as shown below:

Example 6.13. ($e^{x^2} \in \mathcal{D}'(\mathbb{R}) \setminus \mathcal{S}'(\mathbb{R})$).

e^{x^2}, being continuous, is locally integrable, and hence $T_{e^{x^2}} \in \mathcal{D}'(\mathbb{R})$.

But e^{x^2} does not define a tempered distribution, since, for example, its action on the test function $e^{-x^2} \in \mathcal{S}(\mathbb{R})$, is not finite:

$$\langle T_{e^{x^2}}, e^{-x^2} \rangle = \int_{-\infty}^{\infty} e^{x^2} e^{-x^2} dx = \int_{-\infty}^{\infty} 1 dx \not< \infty.$$

Thus $T_{e^{x^2}} \notin \mathcal{S}'(\mathbb{R})$. ◇

Example 6.14. ($L^1(\mathbb{R}) \subset \mathcal{S}'(\mathbb{R})$).

Let $f \in L^1$, that is, $\|f\|_1 := \int_{\mathbb{R}} |f(x)| dx < \infty$.

We claim that the regular distribution T_f is tempered, that is, $T \in \mathcal{S}'(\mathbb{R})$.

For $\varphi \in \mathcal{S}(\mathbb{R})$, $|\langle T_f, \varphi \rangle| = \left| \int_{\mathbb{R}} f(x)\varphi(x) dx \right| \leq \int_{\mathbb{R}} |f(x)||\varphi(x)| dx \leq \|f\|_1 \|\varphi\|_\infty$.

From here it follows that if $(\varphi_n)_{n \in \mathbb{N}}$ is a sequence in $\mathcal{S}(\mathbb{R})$ such that $\varphi_n \xrightarrow{\mathcal{S}} \mathbf{0}$ as $n \to \infty$, then $\langle T, \varphi_n \rangle \to 0$. Hence $T_f \in \mathcal{S}'(\mathbb{R})$. ◇

Example 6.15. ($\delta \in \mathcal{S}'(\mathbb{R})$).

The map $\varphi \xmapsto{\delta} \varphi(0) : \mathcal{S}(\mathbb{R}) \to \mathbb{C}$ is clearly linear.

It is also continuous, since if $\varphi_n \xrightarrow{\mathcal{S}} \mathbf{0}$, then, in particular, $\varphi_n(0) \to 0$. ◇

Exercise 6.27. ($L^\infty(\mathbb{R}) \subset \mathcal{S}'(\mathbb{R})$).
Show that if f is a bounded function on $\mathbb{R}$, then $T_f \in \mathcal{S}'(\mathbb{R})$.

Derivative of tempered distributions.

If $T \in \mathcal{S}'(\mathbb{R})$, then we define $T' : \mathcal{S}'(\mathbb{R}) \to \mathbb{C}$ by

$$\langle T', \varphi \rangle = -\langle T, \varphi' \rangle, \quad \varphi \in \mathcal{S}(\mathbb{R}).$$

It is easy to see that $T' \in \mathcal{S}'(\mathbb{R})$.

Multiplication of tempered distributions by polynomials.

Recall that elements of $\mathcal{D}'(\mathbb{R})$ could be multiplied by C^∞ functions. This luxury is not available for tempered distributions: just think of multiplying the constant function $\mathbf{1} \in L^\infty(\mathbb{R}) \subset \mathcal{S}'(\mathbb{R})$ by $e^{x^2} \in C^\infty(\mathbb{R})$: their product is not tempered, since by Example 6.13, we know that $e^{x^2} \notin \mathcal{S}'(\mathbb{R})$!

But while elements in $\mathcal{S}'(\mathbb{R})$ can't be multiplied by *general* $\alpha \in C^\infty(\mathbb{R})$, they can nevertheless be multiplied with *polynomials* as follows:
For $T \in \mathcal{S}'(\mathbb{R})$, we define $xT : \mathcal{S}(\mathbb{R}) \to \mathbb{C}$ by

$$\langle xT, \varphi \rangle = \langle T, x\varphi \rangle, \quad \varphi \in \mathcal{S}(\mathbb{R}).$$

Then it is easy to see that $xT \in \mathcal{S}'(\mathbb{R})$.

Fourier transformation of tempered distributions.

Definition 6.10. (Fourier transform of tempered distributions).
If $T \in \mathcal{S}'(\mathbb{R})$, then its *Fourier transform* is the tempered distribution $\widehat{T} \in \mathcal{S}'(\mathbb{R})$, defined by $\langle \widehat{T}, \varphi \rangle = \langle T, \widehat{\varphi} \rangle$, for all $\varphi \in \mathcal{S}(\mathbb{R})$.

Using Proposition 6.4, we can see that $\varphi \mapsto \langle T, \widehat{\varphi} \rangle : \mathcal{S}(\mathbb{R}) \to \mathbb{C}$ defines a tempered distribution.

Exercise 6.28. Show that if $f \in L^1(\mathbb{R})$, then $\widehat{T_f} = T_{\widehat{f}}$.

Example 6.16. (Fourier transform of the Dirac δ).
For $\varphi \in \mathcal{S}(\mathbb{R})$, we have that

$$\langle \widehat{\delta}, \varphi \rangle = \langle \delta, \widehat{\varphi} \rangle = \widehat{\varphi}(0) = \int_{-\infty}^{\infty} \varphi(x) e^{-2\pi i 0 x} dx = \int_{-\infty}^{\infty} \varphi(x) \cdot 1 \, dx = \langle \mathbf{1}, \varphi \rangle.$$

So the Fourier transform $\widehat{\delta}$ of the tempered distribution δ is the regular tempered distribution corresponding to the constant function $\mathbf{1}$. $\diamond$

Much of the classical Fourier transform theory, can be extended appropriately for the class of tempered distributions. For example, for tempered distributions $T \in \mathcal{S}'(\mathbb{R})$, we have

$$\widehat{(T')} = 2\pi i \xi \widehat{T}.$$

This plays an important role in linear differential equation theory: by taking Fourier transforms, the analytic operation of differentiation is converted into the algebraic operation of multiplication by the Fourier transform variable. Another instance is "convolution theorem": we know that if $f, g \in L^1(\mathbb{R})$, then multiplication on the Fourier transform side corresponds to convolution:

$$\widehat{f * g} = \widehat{f} \cdot \widehat{g}.$$

Under certain technical conditions[12] on distributions $T, S \in \mathcal{D}'(\mathbb{R})$, one can define their *convolution* $T * S \in \mathcal{D}'(\mathbb{R})$. Then a distributional analogue of the convolution theorem is often available: for example, for $T \in \mathcal{S}'(\mathbb{R})$, and for a "compactly supported" $S \in \mathcal{D}'(\mathbb{R})$, one has[13]
$$\widehat{T * S} = \widehat{T} \cdot \widehat{S}.$$
These results allow one to rigorously justify some of the formal calculations met in engineering. It also gives rise to some important auxiliary concepts, useful in the theory of PDEs. One such notion, related to convolution of distributions, is the concept of a fundamental solution for a linear PDE.

Definition 6.11. (Fundamental Solution).
Given a linear partial differential operator with constant coefficients,
$$D = \sum_{|\mathbf{k}| \leqslant K} a_{\mathbf{k}} D^{\mathbf{k}},$$
a *fundamental solution* is a distribution $E \in \mathcal{D}'(\mathbb{R}^d)$ such that
$$DE = \delta.$$
In the above, for a multi-index $\mathbf{k} = (k_1, \cdots, k_d)$, $|\mathbf{k}| := k_1 + \cdots + k_d$.

Fundamental solutions are useful, since they allow one to solve the inhomogeneous equation
$$Du = g.$$
For suitable g, it can be shown that $u := E * g$ does the job:
$$Du = D(E * g) = (DE) * g = \delta * g = g.$$
There is also a deep result[14] saying that every nonzero operator D with constant coefficients has a fundamental solution in $\mathcal{D}'(\mathbb{R}^d)$. Fundamental solutions with appropriate boundary conditions specific to a PDE problem are sometimes referred to as *Green's functions*.

Example 6.17. It follows from Exercise 6.8, page 241, that a fundamental solution for the one-dimensional Laplacian operator
$$\frac{d^2}{dx^2}$$
is $E_{\mathrm{p}} := |x|/2$. In fact if we add to E_{p} any solution to the homogeneous equation $u'' = 0$, then it will also be a fundamental solution. So the functions $ax + b + |x|/2$, with arbitrary constants a and b, are all fundamental solutions of the one-dimensional Laplacian operator. ◇

---
[12]See for example, [Hörmander (1990), Chapter IV].
[13]See for example, [Hörmander (1990), Theorem 7.1.15, page 166].
[14]Due to Malgrange and Ehrenpreiss.

Notes

Some parts of this chapter are inspired by the lectures on Distribution Theory given by Professor Erik G.F. Thomas at the University of Groningen [Thomas (1996)].

Solutions

Solution to Exercise 0.1, page ix

We have $f(\mathbf{x}_1) = \int_0^T \left(aQ\dfrac{t}{T} + bQ\dfrac{1}{T}\right) Q\dfrac{1}{T} dt = \dfrac{aQ^2}{2} + \dfrac{bQ^2}{T}$.

On the other hand, $f(\mathbf{x}_2) = \int_0^T \left(aQ\dfrac{t^2}{T^2} + bQ\dfrac{2t}{T^2}\right) Q\dfrac{2t}{T^2} dt = \dfrac{aQ^2}{2} + \dfrac{4}{3}\dfrac{bQ^2}{T}$.

Clearly $f(\mathbf{x}_2) > f(\mathbf{x}_1)$, and so the mining operation $\mathbf{x}_1$ is preferred to $\mathbf{x}_2$ because it incurs a lower cost.

Solutions to the exercises from Chapter 1

Solution to Exercise 1.1, page 7

True. Indeed we have:
(V1) For all $x, y, z > 0$, $x \boldsymbol{+} (y \boldsymbol{+} z) = x \boldsymbol{+} (yz) = x(yz) = (xy)z = (xy) \boldsymbol{+} z = (x \boldsymbol{+} y) \boldsymbol{+} z$.
(V2) For all $x > 0$, $x \boldsymbol{+} 1 = x1 = x = 1x = 1 \boldsymbol{+} x$.
(So 1 serves as the zero vector in this vector space!)
(V3) If $x > 0$, then $1/x > 0$ too, and $x \boldsymbol{+} (1/x) = x(1/x) = 1 = (1/x)x = (1/x) \boldsymbol{+} x$.
(Thus $1/x$ acts as the inverse of x with respect to the operation $\boldsymbol{+}$.)
(V4) For all $x, y > 0$, $x \boldsymbol{+} y = xy = yx = y \boldsymbol{+} x$.
(V5) For all $x > 0$, $1 \cdot x = x^1 = x$.
(V6) For all $x > 0$ and all $\alpha, \beta \in \mathbb{R}$, $\alpha \cdot (\beta \cdot x) = \alpha \cdot x^\beta = (x^\beta)^\alpha = x^{\alpha\beta} = (\alpha\beta) \cdot x$.
(V7) For all $x > 0$ and all $\alpha, \beta \in \mathbb{R}$, $(\alpha + \beta) \cdot x = x^{\alpha+\beta} = x^\alpha x^\beta = x^\alpha \boldsymbol{+} x^\beta = \alpha \cdot x \boldsymbol{+} \beta \cdot x$.
(V8) For all $x, y > 0$, $\alpha \in \mathbb{R}$, $\alpha \cdot (x \boldsymbol{+} y) = \alpha \cdot (xy) = (xy)^\alpha = x^\alpha y^\alpha = x^\alpha \boldsymbol{+} y^\alpha = \alpha \cdot x \boldsymbol{+} \alpha \cdot y$.

We remark that V is isomorphic to the one dimensional vector space $\mathbb{R}$ (with the usual operations): indeed, it can be checked that the maps $\log : V \to \mathbb{R}$ and $\exp : \mathbb{R} \to V$ are linear transformations, and are inverses of each other.

Solution to Exercise 1.2, page 7

We prove this by contradiction. Suppose that $C[0,1]$ has dimension d. Consider functions $\mathbf{x}_n(t) = t^n$, $t \in [0,1]$, $n = 1, \cdots, d$. Since polynomials are continuous, we have $\mathbf{x}_n \in C[0,1]$ for all $n = 1, \cdots, d$.

First we prove that $\mathbf{x}_n$, $n = 1, \cdots, d$, are linearly independent in $C[0,1]$. Suppose not. Then there exist $\alpha_n \in \mathbb{R}$, $n = 1, \cdots, d$, not all zeros, such that $\alpha_1 \cdot \mathbf{x}_1 + \cdots + \alpha_d \cdot \mathbf{x}_d = \mathbf{0}$. Let $m \in \{1, \cdots, d\}$ be the smallest index such that $\alpha_m \neq 0$. Then for all $t \in [0,1]$, $\alpha_m t^m + \cdots + \alpha_d t^d = 0$. In particular, for all $t \in (0,1]$, we have $\alpha_m + \alpha_{m+1} t + \cdots + \alpha_d t^{d-m} = 0$.
Thus for all $n \in \mathbb{N}$ we have $\alpha_m + \dfrac{\alpha_{m+1}}{n} + \cdots + \dfrac{\alpha_d}{n^{d-m}} = 0$.
Passing the limit as $n \to \infty$, we obtain $\alpha_m = 0$, a contradiction. So the functions $\mathbf{x}_n$, $n = 1, \cdots, d$, are linearly independent in $C[0,1]$.

Next, we get the contradiction to $C[0,1]$ having dimension d. Since any independent set of cardinality d in a d-dimensional vector space is a basis for this vector space, $\{\mathbf{x}_n : n = 1, \cdots, d\}$ is a basis for $C[0,1]$. Since the constant function $\mathbf{1}$ (taking value 1 everywhere on $[0,1]$) belongs to $C[0,1]$, there exist $\beta_n \in \mathbb{R}$, $n = 1, \cdots, d$, such that $\mathbf{1} = \beta_1 \cdot \mathbf{x}_1 + \cdots + \beta_d \cdot \mathbf{x}_d$. In particular, putting $t = 0$, we obtain the contradiction that $1 = 0$: $1 = \mathbf{1}(0) = (\beta_1 \cdot \mathbf{x}_1 + \cdots + \beta_d \cdot \mathbf{x}_d)(0) = 0$.

Solution to Exercise 1.3, page 7

("If" part.) Suppose that $y_a = y_b = 0$. Then we have:
(S1) If $\mathbf{x}_1, \mathbf{x}_2 \in S$, then $\mathbf{x}_1 + \mathbf{x}_2 \in S$. As $\mathbf{x}_1, \mathbf{x}_2 \in C^1[a,b]$, also $\mathbf{x}_1 + \mathbf{x}_2 \in C^1[a,b]$. Moreover, $\mathbf{x}_1(a) + \mathbf{x}_2(a) = 0 + 0 = 0 = y_a$ and $\mathbf{x}_1(b) + \mathbf{x}_2(b) = 0 + 0 = 0 = y_b$.
(S2) If $\mathbf{x} \in S$ and $\alpha \in \mathbb{R}$, then $\alpha \cdot \mathbf{x} \in S$. Indeed, as $\mathbf{x} \in C^1[a,b]$, and $\alpha \in \mathbb{R}$, we have $\alpha \cdot \mathbf{x} \in C^1[a,b]$, and $(\alpha \cdot \mathbf{x})(a) = \alpha 0 = 0 = y_a$, $(\alpha \cdot \mathbf{x})(b) = \alpha 0 = 0 = y_b$.
(S3) $\mathbf{0} \in S$, since $\mathbf{0} \in C^1[a,b]$ and $\mathbf{0}(a) = 0 = y_a = y_b = \mathbf{0}(b)$.

Hence, S is a subspace of a vector space $C^1[a,b]$.

("Only if" part.) Suppose that S is a subspace of $C^1[a,b]$. Let $\mathbf{x} \in S$. Then $2 \cdot \mathbf{x} \in S$. Therefore, $(2 \cdot \mathbf{x})(a) = y_a$, and so $y_a = (2 \cdot \mathbf{x})(a) = 2\mathbf{x}(a) = 2y_a$. Thus $y_a = 0$. Moreover, $(2 \cdot \mathbf{x})(b) = y_b$, and so $y_b = (2 \cdot \mathbf{x})(b) = 2\mathbf{x}(b) = 2y_b$. Hence also $y_b = 0$.

Solution to Exercise 1.4, page 10

We have $\|t\|_\infty = \max\limits_{t \in [0,1]} |t| = \max\limits_{t \in [0,1]} t = 1$,

$\|-t\|_\infty = \max\limits_{t \in [0,1]} |-t| = \max\limits_{t \in [0,1]} t = 1$,

$\|t^n\|_\infty = \max\limits_{t \in [0,1]} |t^n| = \max\limits_{t \in [0,1]} t^n = 1$, and

$\|\sin(2\pi n t)\|_\infty = \max\limits_{t \in [0,1]} |\sin(2\pi n t)| = 1$.

Solution to Exercise 1.5, page 14

From the triangle inequality, we have that $\|x\| = \|y + x - y\| \leq \|y\| + \|x - y\|$, for all $x, y \in X$. So for all $x, y \in X$, $\|x\| - \|y\| \leq \|x - y\|$.

Interchanging x, y, we get $\|y\| - \|x\| \leq \|y - x\| = \|(-1)(x - y)\| = |-1|\|x - y\| = \|x - y\|$. So for all $x, y \in X$, $-(\|x\| - \|y\|) \leq \|x - y\|$.

Combining the results from the first two paragraphs, we obtain $\big|\|x\| - \|y\|\big| \leq \|x - y\|$ for all $x, y \in X$.

Solution to Exercise 1.6, page 14

No, since for example (N2) fails if we take $x = 1$ and $\alpha = 2$:
$$\|2 \cdot 1\| = \|2\| = 2^2 = 4 \neq 2 = 2(1) = |2|(1^2) = |2|\|1\|.$$

Solution to Exercise 1.7, page 15

We verify that (N1), (N2), (N3) are satisfied by $\|\cdot\|_Y$:
(N1) For all $y \in Y$, $\|y\|_Y = \|y\|_X \geq 0$.
If $y \in Y$ and $\|y\|_Y = 0$, then $\|y\|_X = 0$, and so $y = \mathbf{0} \in X$.
But $\mathbf{0} \in Y$, and so $y = \mathbf{0} \in Y$.

(N2) If $y \in Y$ and $\alpha \in \mathbb{R}$, then $\alpha \cdot y \in Y$ and $\|\alpha \cdot y\|_Y = \|\alpha \cdot y\|_X = |\alpha|\|y\|_X = |\alpha|\|y\|_Y$.
(N3) If $y_1, y_2 \in Y$, then $y_1 + y_2 \in Y$.
Also, $\|y_1 + y_2\|_Y = \|y_1 + y_2\|_X \leqslant \|y_1\|_X + \|y_2\|_X = \|y_1\|_Y + \|y_2\|_Y$.

Solution to Exercise 1.8, page 15

(1) We first consider the case $1 \leqslant p < \infty$, and then $p = \infty$. Let $1 \leqslant p < \infty$.

(N1) If $\mathbf{x} = (x_1, \cdots, x_d) \in \mathbb{R}^d$, then $\|x\|_p = \Big(\sum_{n=1}^{d} |x_n|^p\Big)^{1/p} \geqslant 0$.

If $\mathbf{x} \in \mathbb{R}^d$ and $\|\mathbf{x}\|_p = 0$, then $\|\mathbf{x}\|_p^p = 0$, that is, $\sum_{n=1}^{d} |x_n|^p = 0$.

So $|x_n| = 0$ for $1 \leqslant n \leqslant d$, that is, $\mathbf{x} = \mathbf{0}$.

(N2) Let $\mathbf{x} = (x_1, \cdots, x_d) \in \mathbb{R}^d$, and $\alpha \in \mathbb{R}$.

Then $\|\alpha \cdot \mathbf{x}\|_p = \Big(\sum_{n=1}^{d} |\alpha x_n|^p\Big)^{1/p} = \Big(|\alpha|^p \sum_{n=1}^{d} |x_n|^p\Big)^{1/p} = |\alpha|\|\mathbf{x}\|_p$.

(N3) Let $\mathbf{x} = (x_1, \cdots, x_d) \in \mathbb{R}^d$ and $\mathbf{y} = (y_1, \cdots, y_d) \in \mathbb{R}^d$.

If $p = 1$, then we have $|x_n + y_n| \leqslant |x_n| + |y_n|$ for $1 \leqslant n \leqslant d$.
By adding these, $\|\mathbf{x} + \mathbf{y}\|_1 \leqslant \|\mathbf{x}\|_1 + \|\mathbf{y}\|_1$, establishing (N3) for $p = 1$.
Now consider the case $1 < p < \infty$.
If $\mathbf{x} + \mathbf{y} = \mathbf{0}$, then $\|\mathbf{x} + \mathbf{y}\|_p = \|\mathbf{0}\|_p = 0 \leqslant \|\mathbf{x}\|_p + \|\mathbf{y}\|_p$ trivially.
So we assume that $\mathbf{x} + \mathbf{y} \neq \mathbf{0}$. By Hölder's Inequality, we have

$$\sum_{n=1}^{d} |x_n||x_n + y_n|^{p-1} \leqslant \Big(\sum_{n=1}^{d} |x_n|^p\Big)^{1/p} \Big(\sum_{n=1}^{d} |x_n + y_n|^{q(p-1)}\Big)^{1/q}$$
$$= \|\mathbf{x}\|_p \|\mathbf{x} + \mathbf{y}\|_p^{p/q},$$

where we used $q(p - 1) = p$ in order to obtain the last equality.

Similarly, $\sum_{n=1}^{d} |y_n||x_n + y_n|^{p-1} \leqslant \|\mathbf{y}\|_p \|\mathbf{x} + \mathbf{y}\|_p^{p/q}$. Consequently,

$$\|\mathbf{x} + \mathbf{y}\|_p^p = \sum_{n=1}^{d} |x_n + y_n|^p = \sum_{n=1}^{d} |x_n + y_n||x_n + y_n|^{p-1}$$
$$\leqslant \sum_{n=1}^{d} \big(|x_n| + |y_n|\big)|x_n + y_n|^{p-1}$$
$$\leqslant \|\mathbf{x}\|_p \|\mathbf{x} + \mathbf{y}\|_p^{p/q} + \|\mathbf{y}\|_p \|\mathbf{x} + \mathbf{y}\|_p^{p/q}$$
$$= \big(\|\mathbf{x}\|_p + \|\mathbf{y}\|_p\big)\|\mathbf{x} + \mathbf{y}\|_p^{p/q}.$$

Dividing throughout by $\|\mathbf{x} + \mathbf{y}\|_p^{p/q} > 0$, we obtain $\|\mathbf{x} + \mathbf{y}\|_p \leqslant \|\mathbf{x}\|_p + \|\mathbf{y}\|_p$.
This completes the proof that $(\mathbb{R}^d, \|\cdot\|_p)$ is a normed space for $1 \leqslant p < \infty$.

Now we consider the case $p = \infty$.

(N1) If $\mathbf{x} = (x_1, \cdots, x_d) \in \mathbb{R}^d$, then $\|\mathbf{x}\|_\infty = \max\{|x_1|, \cdots, |x_d|\} \geq 0$.
If $\mathbf{x} \in \mathbb{R}^d$ and $\|\mathbf{x}\|_\infty = 0$, then $\max\{|x_1|, \cdots, |x_d|\} = 0$, and so $|x_n| = 0$ for $1 \leq n \leq d$, that is, $\mathbf{x} = \mathbf{0}$.

(N2) Let $\mathbf{x} = (x_1, \cdots, x_d) \in \mathbb{R}^d$, and $\alpha \in \mathbb{R}$.
Then $\|\alpha \cdot \mathbf{x}\|_\infty = \max\{|\alpha x_1|, \cdots, |\alpha x_d|\} = |\alpha| \max\{|x_1|, \cdots, |x_d|\} = |\alpha| \|\mathbf{x}\|_\infty$.

(N3) Let $\mathbf{x} = (x_1, \cdots, x_d) \in \mathbb{R}^d$ and $\mathbf{y} = (y_1, \cdots, y_d) \in \mathbb{R}^d$.
We have $|x_n + y_n| \leq |x_n| + |y_n| \leq \|x\|_\infty + \|y\|_\infty$ for $1 \leq n \leq d$.
So it follows that $\|\mathbf{x} + \mathbf{y}\|_\infty \leq \|\mathbf{x}\|_\infty + \|\mathbf{y}\|_\infty$, establishing (N3) for $p = \infty$.

(2) See the following pictures.

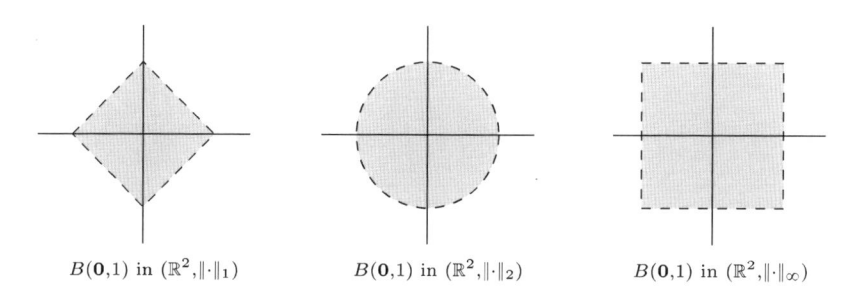

$B(\mathbf{0},1)$ in $(\mathbb{R}^2, \|\cdot\|_1)$ $B(\mathbf{0},1)$ in $(\mathbb{R}^2, \|\cdot\|_2)$ $B(\mathbf{0},1)$ in $(\mathbb{R}^2, \|\cdot\|_\infty)$

(3) We have for $\mathbf{x} = (a,b) \in \mathbb{R}^2$ that

$$(\max\{|a|, |b|\})^p \leq |a|^p + |b|^p \leq (\max\{|a|, |b|\})^p + (\max\{|a|, |b|\})^p$$
$$= 2(\max\{|a|, |b|\})^p.$$

So $\|\mathbf{x}\|_\infty = \max\{|a|, |b|\} \leq \sqrt[p]{|a|^p + |b|^p} = \|\mathbf{x}\|_p \leq \sqrt[p]{2} \max\{|a|, |b|\} = \sqrt[p]{2} \|\mathbf{x}\|_\infty$.
We have $\lim\limits_{p \to \infty} \sqrt[p]{2} = 1$. (As $2^{1/p} \geq 1$, $h_p := 2^{1/p} - 1 \geq 0$. We have

$$2 = (1 + h_p)^p = 1 + ph_p + \text{nongnegative terms} \geq 1 + ph_p,$$

giving $1/p \geq h_p \geq 0$ for all p, and so $h_p \to 0$ as $p \to \infty$.) So it follows by the Sandwich Theorem[1] that $\lim\limits_{p \to \infty} \|\mathbf{x}\|_p = \|\mathbf{x}\|_\infty$.

The balls $B_p(\mathbf{0}, 1)$ grow to $B_\infty(\mathbf{0}, 1)$ as p increases.

Solution to Exercise 1.9, page 16

(1) If $x, y \in \overline{B(\mathbf{0},1)}$, then for all $\alpha \in (0,1)$, $(1-\alpha) \cdot x + \alpha \cdot y \in \overline{B(\mathbf{0},1)}$ too, since

$$\|(1-\alpha)x + \alpha y\| \leq \|(1-\alpha)x\| + \|\alpha y\| = |1-\alpha|\|x\| + |\alpha|\|y\| = (1-\alpha)\|x\| + \alpha\|y\|$$
$$\leq (1-\alpha) \cdot 1 + \alpha \cdot 1 = 1.$$

---
[1] See for example [Sasane (2015), §2.4].

(2) See the following picture.

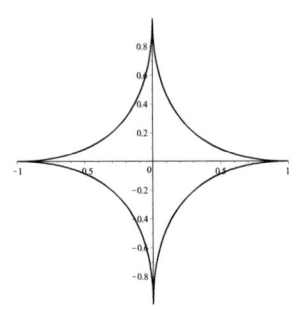

(3) $\overline{B(\mathbf{0},1)}$ is not convex: taking $\mathbf{x} = (1,0)$, $\mathbf{y} = (0,1)$ and $\alpha = 1/2$, we obtain
$$\|(1-\alpha)\mathbf{x} + \alpha\mathbf{y}\|_{1/2} = \left\|\left(\frac{1}{2},\frac{1}{2}\right)\right\|_{1/2} = 2 > 1, \text{ and so } (1-\alpha)\mathbf{x} + \alpha\mathbf{y} \notin \overline{B(\mathbf{0},1)}.$$

Solution to Exercise 1.10, page 16

We'll verify that (N1), (N2), (N3) hold.

(N1) If $\mathbf{x} \in C[a,b]$, then $|\mathbf{x}(t)| \geq 0$ for all $t \in [0,1]$, and so $\|\mathbf{x}\|_1 = \int_a^b |\mathbf{x}(t)|dt \geq 0$.

Let $\mathbf{x} \in C[a,b]$ be such that $\|\mathbf{x}\|_1 = 0$. If $\mathbf{x}(t) = 0$ for all $t \in (a,b)$, then by the continuity of $\mathbf{x}$ on $[a,b]$, it follows that $\mathbf{x}(t) = 0$ for all $t \in [a,b]$ too, and we are done! So suppose that it is not the case that for all $t \in (a,b)$, $\mathbf{x}(t) = 0$. Then there exists a $t_0 \in (a,b)$ such that $\mathbf{x}(t_0) \neq 0$. As $\mathbf{x}$ is continuous at t_0, there exists a $\delta > 0$ small enough so that $a < t_0 - \delta$, $t_0 + \delta < b$, and such that for all $t \in [a,b]$ such that $t_0 - \delta < t < t_0 + \delta$, $|\mathbf{x}(t) - \mathbf{x}(t_0)| < |\mathbf{x}(t_0)|/2$. Then for $t_0 - \delta < t < t_0 + \delta$, we have, using the "reverse" Triangle Inequality from Exercise 1.5, page 14, that
$$|\mathbf{x}(t)| = |\mathbf{x}(t) - \mathbf{x}(t_0) + \mathbf{x}(t_0)|$$
$$\geq |\mathbf{x}(t_0)| - |\mathbf{x}(t) - \mathbf{x}(t_0)| = |\mathbf{x}(t_0)| - |\mathbf{x}(t_0)|/2 = |\mathbf{x}(t_0)|/2 > 0.$$
So $0 = \|\mathbf{x}\|_1 = \int_a^b |\mathbf{x}(t)|dt \geq \int_{t_0-\delta}^{t_0+\delta} |\mathbf{x}(t)|dt \geq 2\delta \cdot |\mathbf{x}(t_0)|/2 = \delta \cdot |\mathbf{x}(t_0)| > 0$.

This is a contradiction. Hence $\mathbf{x} = \mathbf{0}$.

(N2) For $\mathbf{x} \in C[a,b]$, $\alpha \in \mathbb{R}$, $\|\alpha \cdot \mathbf{x}\|_1 = \int_a^b |\alpha \mathbf{x}(t)|dt = |\alpha| \int_a^b |\mathbf{x}(t)|dt = |\alpha|\|\mathbf{x}\|_1$.

(N3) Let $\mathbf{x}, \mathbf{y} \in C[a,b]$. Then
$$\|\mathbf{x}+\mathbf{y}\|_1 = \int_a^b |(\mathbf{x}+\mathbf{y})(t)|dt = \int_a^b |\mathbf{x}(t)+\mathbf{y}(t)|dt$$
$$\leq \int_a^b (|\mathbf{x}(t)| + |\mathbf{y}(t)|)dt = \int_a^b |\mathbf{x}(t)|dt + \int_a^b |\mathbf{y}(t)|dt = \|\mathbf{x}\|_1 + \|\mathbf{y}\|_1.$$

Solution to Exercise 1.11, page 17

(N1) For $\mathbf{x} \in C^n[a,b]$, clearly $\|\mathbf{x}\|_{n,\infty} = \|\mathbf{x}\|_\infty + \cdots + \|\mathbf{x}^{(n)}\|_\infty \geq 0 + \cdots + 0 = 0$.
If $\mathbf{x} \in C^n[a,b]$ is such that $\|\mathbf{x}\|_{n,\infty} = 0$, then $\|\mathbf{x}\|_\infty + \cdots + \|\mathbf{x}^{(n)}\|_\infty = 0$, and since each term in this sum is nonnegative, we have $\|\mathbf{x}\|_\infty = 0$, and so $\mathbf{x} = \mathbf{0}$.

(N2) Let $\mathbf{x} \in C^n[a,b]$ and $\alpha \in \mathbb{R}$. Then

$$\begin{aligned}\|\alpha \cdot \mathbf{x}\|_{n,\infty} &= \|\alpha \cdot \mathbf{x}\|_\infty + \|(\alpha \cdot \mathbf{x})'\|_\infty + \cdots + \|(\alpha \cdot \mathbf{x})^{(n)}\|_\infty \\ &= \|\alpha \cdot \mathbf{x}\|_\infty + \|\alpha \cdot (\mathbf{x}')\|_\infty + \cdots + \|\alpha \cdot (\mathbf{x}^{(n)})\|_\infty \\ &= |\alpha|\|\mathbf{x}\|_\infty + |\alpha|\|\mathbf{x}'\|_\infty + \cdots + |\alpha|\|\mathbf{x}^{(n)}\|_\infty \\ &= |\alpha|(\|\mathbf{x}\|_\infty + \|\mathbf{x}'\|_\infty + \cdots + \|\mathbf{x}^{(n)}\|_\infty) = |\alpha|\|\mathbf{x}\|_{n,\infty}.\end{aligned}$$

(N3) Let $\mathbf{x}, \mathbf{y} \in C^n[a,b]$. For all $0 \leq k \leq n$, $\|\mathbf{x}^{(k)} + \mathbf{y}^{(k)}\|_\infty \leq \|\mathbf{x}^{(k)}\|_\infty + \|\mathbf{y}^{(k)}\|_\infty$, by the Triangle Inequality for $\|\cdot\|_\infty$. Consequently,

$$\begin{aligned}\|\mathbf{x} + \mathbf{y}\|_{n,\infty} &= \|\mathbf{x} + \mathbf{y}\|_\infty + \cdots + \|(\mathbf{x} + \mathbf{y})^{(n)}\|_\infty \\ &= \|\mathbf{x} + \mathbf{y}\|_\infty + \cdots + \|\mathbf{x}^{(n)} + \mathbf{y}^{(n)}\|_\infty \\ &\leq \|\mathbf{x}\|_\infty + \|\mathbf{y}\|_\infty + \cdots + \|\mathbf{x}^{(n)}\|_\infty + \|\mathbf{y}^{(n)}\|_\infty \\ &= \|\mathbf{x}\|_{n,\infty} + \|\mathbf{y}\|_{n,\infty}.\end{aligned}$$

Solution to Exercise 1.12, page 17

(1) Let $k_1, k_2, m_1, m_2, n_1, n_2 \in \mathbb{Z}$, $p \nmid m_1, m_2, n_1, n_2$ and $r = p^{k_1}\dfrac{m_1}{n_1} = p^{k_2}\dfrac{m_2}{n_2}$.
If $k_1 > k_2$, then $p^{k_1-k_2}m_1n_2 = m_2n_1$, which implies that $p \mid m_2n_1$, and as p is prime, this would mean $p \mid m_1$ or $p \mid n_1$, a contradiction. Hence $k_1 \leq k_2$. Similarly, we also obtain $k_2 \leq k_1$.
Thus $k_1 = k_2$. Consequently, $\dfrac{1}{p^{k_1}} = \dfrac{1}{p^{k_2}}$, and so $|\cdot|_p$ is well-defined.

(2) If $0 \neq r \in \mathbb{Q}$, then we can express r as $p^k\dfrac{m}{n}$, with $k, m, n \in \mathbb{Z}$, and $p \nmid m, n$.
We see that $|r|_p = \dfrac{1}{p^k} > 0$. If $r = 0$, then $|r|_p = |0|_p = 0$ by definition.
Thus $|r|_p \geq 0$ for all $r \in \mathbb{R}$. Also if $r \neq 0$, then $|r|_p > 0$. Hence $|r|_p = 0$ implies that $r = 0$.

(3) The claim is obvious if $r_1 = 0$ or $r_2 = 0$. Suppose that $r_1 \neq 0$ and $r_2 \neq 0$.
Let $r_1 = p^{k_1}\dfrac{m_1}{n_1}$ and $r_2 = p^{k_2}\dfrac{m_2}{n_2}$, $k_1, k_2, m_1, m_2, n_1, n_2 \in \mathbb{Z}$, $p \nmid m_1, m_2, n_1, n_2$.
So $r_1 r_2 = p^{k_1+k_2}\dfrac{m_1 m_2}{n_1 n_2}$. As $p \nmid m_1$, $p \nmid m_2$, and p is prime, we have $p \nmid m_1 m_2$.
Similarly $p \nmid n_1 n_2$. Thus $|r_1 r_2|_p = \dfrac{1}{p^{k_1+k_2}} = \dfrac{1}{p^{k_1}} \cdot \dfrac{1}{p^{k_2}} = |r_1|_p |r_2|_p$.

(4) The inequality is trivially true if $r_1 = 0$ or $r_2 = 0$ or if $r_1 + r_2 = 0$.
Assume $r_1 \neq 0$, $r_2 \neq 0$, and $r_1 + r_2 \neq 0$.

Let $r_1 = p^{k_1} \dfrac{m_1}{n_1}$, $r_2 = p^{k_2} \dfrac{m_2}{n_2}$, with $k_1, k_2, m_1, m_2, n_1, n_2 \in \mathbb{Z}$, $p \nmid m_1, m_2, n_1, n_2$.
We have
$$\begin{aligned}
r_1 + r_2 &= p^{k_1} \frac{m_1}{n_1} + p^{k_2} \frac{m_2}{n_2} \\
&= \frac{p^{k_1} m_1 n_2 + p^{k_2} n_1 m_2}{n_1 n_2} \\
&= p^{\min\{k_1, k_2\}} \cdot \frac{p^{k_1 - \min\{k_1, k_2\}} m_1 n_2 + p^{k_2 - \min\{k_1, k_2\}} n_1 m_2}{n_1 n_2} \\
&= p^{\min\{k_1, k_2\}} \cdot \frac{\widetilde{m}}{n_1 n_2},
\end{aligned}$$
where $\widetilde{m} := p^{k_1 - \min\{k_1, k_2\}} m_1 n_2 + p^{k_2 - \min\{k_1, k_2\}} n_1 m_2$ ($\neq 0$, since $r_1 + r_2 \neq 0$). By the Fundamental Theorem of Arithmetic, there exists a unique integer $\widetilde{k} \geqslant 0$ and an integer m such that $\widetilde{m} = p^{\widetilde{k}} m$ and $p \nmid m$. Clearly $p \nmid n_1 n_2$.
Hence $r_1 + r_2 = p^{\widetilde{k} + \min\{k_1, k_2\}} \dfrac{m}{n_1 n_2}$, with $p \nmid m, n_1 n_2$.
So $|r_1 + r_2|_p = \dfrac{1}{p^{\widetilde{k} + \min\{k_1, k_2\}}}$
$$\leqslant \frac{1}{p^{\min\{k_1, k_2\}}} = \max\left\{\frac{1}{p^{k_1}}, \frac{1}{p^{k_2}}\right\} = \max\{|r_1|_p, |r_2|_p\}.$$
This yields the Triangle Inequality:
$$|r_1 + r_2|_p \leqslant \max\{|r_1|_p, |r_2|_p\} \leqslant |r_1|_p + |r_2|_p.$$

Solution to Exercise 1.13, page 17

(N1) Clearly $\|\mathbf{M}\|_\infty = \max\limits_{\substack{1 \leqslant i \leqslant m \\ 1 \leqslant j \leqslant n}} |m_{ij}| \geqslant 0$ for all $\mathbf{M} = [m_{ij}] \in \mathbb{R}^{m \times n}$.

If $\|\mathbf{M}\|_\infty = 0$, then $|m_{ij}| = 0$ for all $1 \leqslant i \leqslant m$, $1 \leqslant j \leqslant n$, that is, $\mathbf{M} = [m_{ij}] = \mathbf{0}$, the zero matrix.

(N2) For $\mathbf{M} = [m_{ij}] \in \mathbb{R}^{m \times n}$ and $\alpha \in \mathbb{R}$, we have
$$\|\alpha \cdot \mathbf{M}\|_\infty = \max_{\substack{1 \leqslant i \leqslant m \\ 1 \leqslant j \leqslant n}} |\alpha m_{ij}| = \max_{\substack{1 \leqslant i \leqslant m \\ 1 \leqslant j \leqslant n}} |\alpha||m_{ij}| = |\alpha| \max_{\substack{1 \leqslant i \leqslant m \\ 1 \leqslant j \leqslant n}} |m_{ij}| = |\alpha| \|\mathbf{M}\|_\infty.$$

(N3) For $\mathbf{P} = [p_{ij}], \mathbf{Q} = [q_{ij}] \in \mathbb{R}^{m \times n}$, $|p_{ij} + q_{ij}| \leqslant |p_{ij}| + |q_{ij}| \leqslant \|\mathbf{P}\|_\infty + \|\mathbf{Q}\|_\infty$. As this holds for all i, j, $\|\mathbf{P} + \mathbf{Q}\|_\infty = \max\limits_{\substack{1 \leqslant i \leqslant m \\ 1 \leqslant j \leqslant n}} |p_{ij} + q_{ij}| \leqslant \|\mathbf{P}\|_\infty + \|\mathbf{Q}\|_\infty$.

Solution to Exercise 1.14, page 19

Consider the open ball $B(x, r) = \{y \in X : \|x - y\| < r\}$ in X. If $y \in B(x, r)$, then $\|x - y\| < r$. Define $r' = r - \|x - y\| > 0$. We claim that $B(y, r') \subset B(x, r)$. Let $z \in B(y, r')$. Then $\|z - y\| < r' = r - \|x - y\|$ and so $\|x - z\| \leqslant \|x - y\| + \|y - z\| < r$. Hence $z \in B(x, r)$. The following picture illustrates this.

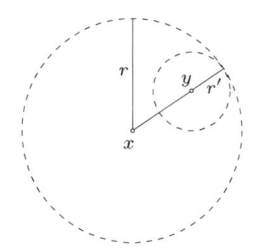

Solution to Exercise 1.15, page 19

The point $\mathbf{c} := \left(\frac{1}{2}, 0\right) \in I$, but for each $r > 0$, the point $\mathbf{y} = \left(\frac{1}{2}, \frac{r}{2}\right)$ belongs to the ball $B(\mathbf{c}, r)$, but not to I, since $\|\mathbf{y} - \mathbf{c}\|_2 = \sqrt{\left(\frac{1}{2} - \frac{1}{2}\right)^2 + \left(\frac{r}{2} - 0\right)^2} = \frac{r}{2} < r$, but $\frac{r}{2} \neq 0$. See the following picture.

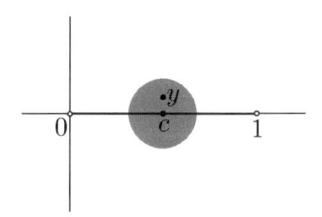

Solution to Exercise 1.16, page 19

Using the following picture, it can be seen that the collections $\mathcal{O}_1$, $\mathcal{O}_2$, $\mathcal{O}_\infty$ of open sets in the normed spaces $(\mathbb{R}^2, \|\cdot\|_1)$, $(\mathbb{R}^2, \|\cdot\|_2)$, $(\mathbb{R}^2, \|\cdot\|_\infty)$, respectively, coincide.

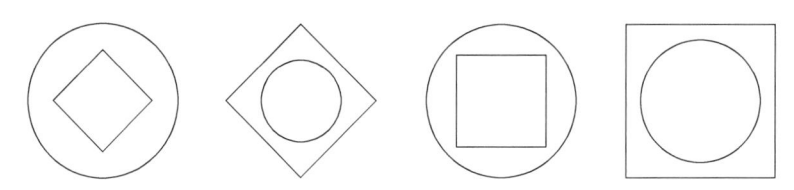

Solution to Exercise 1.17, page 20

If F_i, $i \in I$, is a family of closed sets, then $X \backslash F_i$, $i \in I$, is a family of open sets. Hence $\bigcup_{i \in I} X \backslash F_i = X \backslash \bigcap_{i \in I} F_i$ is open. So $\bigcap_{i \in I} F_i = X \backslash \left(X \backslash \bigcap_{i \in I} F_i\right)$ is closed.

If $F_1, \cdots, F_n$ are closed, then $X\backslash F_1, \cdots, X\backslash F_n$ are open, and so the intersection $\bigcap_{i=1}^{n}(X\backslash F_i) = X\backslash \bigcup_{i=1}^{n} F_i$ of these finitely many open sets is open as well.

Thus $\bigcup_{i=1}^{n} F_i = X\backslash \left(X\backslash \bigcup_{i=1}^{n} F_i\right)$ is closed.

For showing that the finiteness condition cannot be dropped, we'll consider the normed space $X = \mathbb{R}$, and simply rework Example 1.15, page 20, by taking complements.

We know that $F_n := \mathbb{R}\backslash(-1/n, 1/n)$, $n \in \mathbb{N}$, is closed and the union of these, $\bigcup_{n=1}^{\infty} F_n = \bigcup_{n=1}^{\infty} \mathbb{R}\backslash(-1/n, 1/n) = \mathbb{R}\backslash \bigcap_{n=1}^{\infty} (-1/n, 1/n) = \mathbb{R}\backslash\{0\}$ which is not closed, since if it were, its complement $\mathbb{R}\backslash(\mathbb{R}\backslash\{0\}) = \{0\}$ would be open, which is false.

Solution to Exercise 1.18, page 20

Consider the closed ball $\overline{B(x,r)} = \{y \in X : \|x - y\| \leq r\}$ in X. To show that $\overline{B(x,r)}$ is closed, we'll show its complement, $U := \{y \in X : \|x - y\| > r\}$, is open. If $y \in U$, then $\|x - y\| > r$. Define $r' = \|x - y\| - r > 0$. We claim that $B(y, r') \subset U$. Let $z \in B(y, r')$. Then $\|z - y\| < r' = \|x - y\| - r$ and so $\|x - z\| \geq \|x - y\| - \|y - z\| > \|x - y\| - (\|x - y\| - r) = r$. Hence $z \in U$.

Solution to Exercise 1.19, page 20

(1) False.

For example, in the normed space $\mathbb{R}$, consider the set $[0, 1)$. Then $[0, 1)$ is not open, since every open ball B with centre 0 contains at least one negative real number, and so B has points not belonging to $[0, 1)$.

On the other hand, this set $[0, 1)$ is not closed either, as its complement is $C := (-\infty, 0) \cup [1, \infty)$, which is not open, since every open ball B' with centre 1 contains at least one positive real number strictly less than one, and so B' contains points that do not belong to C.

(2) False. $\mathbb{R}$ is open in $\mathbb{R}$, and it is also closed.

(3) True. $\varnothing$ and X are both open and closed in any normed space X.

(4) True. $[0, 1)$ is neither open nor closed in $\mathbb{R}$.

(5) False.

$0 \in \mathbb{Q}$, but every open ball centred at 0 contains irrational numbers; just consider $\sqrt{2}/n$, with a sufficiently large n.

(6) False.

Consider the sequence $(a_n)_{n \in \mathbb{N}}$ given by $a_1 = \dfrac{3}{2}$, and for $n > 1$, $a_{n+1} = \dfrac{4 + 3a_n}{3 + 2a_n}$.

Then it can be shown, using induction on n, that $(a_n)_{n \in \mathbb{N}}$ is bounded below by $\sqrt{2}$, and that $(a_n)_{n \in \mathbb{N}}$ is monotone decreasing. (Example 1.19, page 31.)

So $(a_n)_{n \in \mathbb{N}}$ is convergent with a limit L satisfying $L = \dfrac{4 + 3L}{3 + 2L}$, and so $L^2 = 2$.

As L must be positive (the sequence is bounded below by $\sqrt{2}$), it follows that $L = \sqrt{2}$. So every ball with centre $\sqrt{2}$ and a positive radius contains elements from $\mathbb{Q}$ (terms a_n for large n), showing that $\mathbb{R}\backslash\mathbb{Q}$ is not open, and hence $\mathbb{Q}$ is not closed.

(Alternately, let $c \in \mathbb{R}$ have the decimal expansion $c = 0.101001000100001\cdots$. The number c is irrational because[2] it has a nonterminating and nonrepeating decimal expansion. The sequence of rational numbers obtained by truncation, namely 0.1, 0.101, 0.101001, 0.1010010001, 0.101001000100001, $\cdots$ converges with limit c, and so every ball with centre c and a positive radius contains elements from $\mathbb{Q}$, showing again that $\mathbb{R} \setminus \mathbb{Q}$ is not open, and hence $\mathbb{Q}$ is not closed.)

(7) True. $\mathbb{R}\backslash\mathbb{Z} = \bigcup\limits_{n\in\mathbb{Z}} (n, n+1)$. As each $(n, n+1)$ is open, so is their union. Hence $\mathbb{Z} = \mathbb{R}\backslash(\mathbb{R}\backslash\mathbb{Z})$ is closed.

Solution to Exercise 1.20, page 21

We have already seen in Exercise 1.14, page 19, that the interior of $\mathbb{S}$, namely the open ball $B(\mathbf{0}, 1) = \{\mathbf{x} \in X : \|\mathbf{x}\| < 1\}$ is open. Also, it follows from Exercise 1.18, page 20, that the exterior of the closed ball $\overline{B(\mathbf{0}, 1)}$, namely the set $U = \{\mathbf{x} \in X : \|\mathbf{x}\| > 1\}$ is open as well. Thus, the complement of $\mathbb{S}$, being the union of the two open sets $B(\mathbf{0}, 1)$ and U, is open. Consequently, $\mathbb{S}$ is closed.

Solution to Exercise 1.21, page 21

If $X = \{\mathbf{0}\}$, then $\{\mathbf{0}\}$ is clearly closed, since $X\backslash\{\mathbf{0}\} = \emptyset$ is open.

Now suppose that $X \neq \{\mathbf{0}\}$, and let $x \in X$. We want to show that $U := X\backslash\{x\}$ is open. Let $y \in U := X\backslash\{x\}$, and set $r := \|x - y\| > 0$. We claim that the open ball $B(y, r)$ is contained in U. If $z \in B(y, r)$, then $\|y - z\| < r$, and so $\|z - x\| \geq \|x - y\| - \|y - z\| \geq r - \|y - z\| > r - r = 0$. Hence $z \neq x$, and so $z \in X\backslash\{x\} = U$. Consequently U is open, and so $\{x\} = X\backslash U$ is closed.

If F is empty, then it is closed.

If F is not empty, then $F = \{x_1, \cdots, x_n\} = \bigcup\limits_{i=1}^{n} \{x_i\}$, for some $x_1, \cdots, x_n \in X$. As F is the finite union of the closed sets $\{x_1\}, \cdots, \{x_n\}$, F is closed too.

Solution to Exercise 1.22, page 21

Let $x, y \in \mathbb{R}$ and $x < y$. By the Archimedean property of $\mathbb{R}$, there is a positive integer n such that $n > 1/(y - x)$, that is $n(y - x) > 1$. Also, there are positive integers m_1, m_2 such that $m_1 > nx$ and $m_2 > -nx$, so that $-m_2 < nx < m_1$. Thus we have $nx \in [-m_2, -m_2+1) \cup [-m_2+1, -m_2+2) \cup \cdots \cup [m_1-1, m_1)$. Hence

---
[2]See for example [Sasane (2015), Chapter 6].

there is an integer m such that $m-1 \leqslant nx < m$. We have $nx < m \leqslant 1+nx < ny$, and so dividing by n, we have $x < q := m/n < y$. Consequently, between any two real numbers, there is a rational number.

Let $x \in \mathbb{R}$ and let $\epsilon > 0$. Then there is a rational number y such that $x - \epsilon < y < x + \epsilon$, that is, $|x - y| < \epsilon$. Hence $\mathbb{Q}$ is dense in $\mathbb{R}$.

Solution to Exercise 1.23, page 21

Let $x \in \mathbb{R}$ and let $\epsilon > 0$. If $x \in \mathbb{R}\backslash\mathbb{Q}$, then taking $y = x$, we have $|x - y| = 0 < \epsilon$. If on the other hand, $x \in \mathbb{Q}$, then let $n \in \mathbb{N}$ be such that $n > \sqrt{2}/\epsilon$ so that with $y := x + \sqrt{2}/n$, we have $y \in \mathbb{R}\backslash\mathbb{Q}$, and $|x - y| = \sqrt{2}/n < \epsilon$. So $\mathbb{R}\backslash\mathbb{Q}$ is dense in $\mathbb{R}$.

Solution to Exercise 1.24, page 21

Let $\mathbf{x} = (x_n)_{n \in \mathbb{N}} \in \ell^2$, and $\epsilon > 0$. Let $N \in \mathbb{N}$ be such that $\displaystyle\sum_{n=N+1}^{\infty} |x_n|^2 < \epsilon^2$.

Then $\mathbf{y} := (x_1, \cdots, x_N, 0, \cdots) \in c_{00}$, and $\|\mathbf{x} - \mathbf{y}\|_2^2 = \displaystyle\sum_{n=N+1}^{\infty} |x_n|^2 < \epsilon^2$.

Thus $\|\mathbf{x} - \mathbf{y}\|_2 < \epsilon$. Consequently, c_{00} is dense in ℓ^2.

Solution to Exercise 1.25, page 21

Consider the set D of all finitely supported sequences with rational terms. Then D is a countable set since it is a countable union of countable sets. We now show that D is dense in ℓ^1. Let $\mathbf{x} := (x_n)_{n \in \mathbb{N}} \in \ell^1$ and let $r > 0$.

Let $N \in \mathbb{N}$ be large enough so that $\displaystyle\sum_{n=N+1}^{\infty} |x_n| < \frac{r}{2}$.

As $\mathbb{Q}$ is dense in $\mathbb{R}$, there exist $q_1, \cdots, q_N \in \mathbb{Q}$ such that $\displaystyle\sum_{n=1}^{N} |x_n - q_n| < \frac{r}{2}$.

With $\mathbf{x}' := (q_1, \cdots, q_N, 0, \cdots) \in D$, $\|\mathbf{x} - \mathbf{x}'\| = \displaystyle\sum_{n=1}^{N} |x_n - q_n| + \sum_{n=N+1}^{\infty} |x_n| < r$.

Solution to Exercise 1.26, page 22

By the Binomial Theorem, we have

$$\bigl(t + (1-s)\bigr)^n = \sum_{k=0}^{n} \binom{n}{k} t^k (1-s)^{n-k}. \tag{7.1}$$

Putting $s = t$, we get $1 = \big(t + (1-t)\big)^n = \sum_{k=0}^{n} \binom{n}{k} t^k (1-t)^{n-k} = \sum_{k=0}^{n} \mathbf{p}_{n,k}(t)$.

Keeping s fixed, and differentiating (7.1) with respect to t yields

$$\sum_{k=0}^{n} \binom{n}{k} k t^{k-1} (1-s)^{n-k} = n(t+1-s)^{n-1}.$$

Multiplying throughout by t gives

$$\sum_{k=0}^{n} k \binom{n}{k} t^k (1-s)^{n-k} = nt(t+1-s)^{n-1}. \tag{7.2}$$

With $s = t$, $\sum_{k=0}^{n} k\mathbf{p}_{n,k}(t) = \sum_{k=0}^{n} k\binom{n}{k} t^k (1-t)^{n-k} = nt(t+1-t)^{n-1} = nt1^{n-1} = nt$.

Differentiating (7.2) with respect to t yields

$$\sum_{k=0}^{n} \binom{n}{k} k^2 t^{k-1} (1-s)^{n-k} = n(t+1-s)^{n-1} + nt(n-1)(t+1-s)^{n-2}.$$

Multiplying throughout by t yields

$$\sum_{k=0}^{n} \binom{n}{k} k^2 t^k (1-s)^{n-k} = nt(t+1-s)^{n-1} + n(n-1)t^2(t+1-s)^{n-2}.$$

Setting $s = t$ now gives

$$\sum_{k=0}^{n} k^2 \mathbf{p}_{n,k}(t) = nt(t+1-t)^{n-1} + n(n-1)t^2(t+1-t)^{n-2} = nt + n(n-1)t^2.$$

Hence

$$\sum_{k=0}^{n} (k-nt)^2 \mathbf{p}_{k,n}(t) = \sum_{k=0}^{n} (k^2 - 2ntk + n^2t^2)\mathbf{p}_{k,n}(t)$$

$$= \sum_{k=0}^{n} k^2 \mathbf{p}_{n,k}(t) - 2nt \sum_{k=0}^{n} k\mathbf{p}_{k,n}(t) + n^2t^2 \sum_{k=0}^{n} \mathbf{p}_{k,n}(t)$$

$$= nt + n(n-1)t^2 - 2nt(nt) + n^2t^2 \cdot 1 = nt(1-t).$$

Solution to Exercise 1.27, page 33

(1) We check that the relation $\sim$ is reflexive, symmetric and transitive.

(ER1) (Reflexivity) If $\|\cdot\|$ is a norm on X, then for all $x \in X$, we have that $1 \cdot \|x\| = \|x\| = 1 \cdot \|x\|$, and so $\|\cdot\| \sim \|\cdot\|$.

(ER2) (Symmetry) If $\|\cdot\|_a \sim \|\cdot\|_b$, then there exist positive m, M such that for all $x \in X$, $m\|x\|_b \leq \|x\|_a \leq M\|x\|_b$. A rearrangement of this gives $(1/M)\|x\|_a \leq \|x\|_b \leq (1/m)\|x\|_a$, $x \in X$, and so $\|\cdot\|_2 \sim \|\cdot\|_1$.

(ER3) (Transitivity) If $\|\cdot\|_a \sim \|\cdot\|_b$ and $\|\cdot\|_b \sim \|\cdot\|_c$, then there exist positive constants $M_{ab}, M_{bc}, m_{ab}, m_{bc}$ such that for all $x \in X$, we have that $m_{ab}\|x\|_b \leq \|x\|_a \leq M_{ab}\|x\|_b$ and $m_{bc}\|x\|_c \leq \|x\|_b \leq M_{bc}\|x\|_c$. Thus $m_{ab}m_{bc}\|x\|_c \leq m_{ab}\|x\|_b \leq \|x\|_a \leq M_{ab}\|x\|_b \leq M_{ab}M_{bc}\|x\|_c$, and so $\|\cdot\|_a \sim \|\cdot\|_c$.

(2) Suppose that $\|\cdot\|_a \sim \|\cdot\|_b$. Because $\sim$ is an equivalence relation, it is enough to just prove that if U is open in $(X, \|\cdot\|_b)$, then U is open in $(X, \|\cdot\|_a)$ too, and similarly, if $(x_n)_{n\in\mathbb{N}}$ is Cauchy (respectively) convergent in $(X, \|\cdot\|_b)$, then it is Cauchy (respectively convergent) in $(X, \|\cdot\|_a)$ as well. Let $m, M > 0$ be such that for all $x \in X$, $m\|x\|_b \leq \|x\|_a \leq M\|x\|_b$.

Let U be open in $(X, \|\cdot\|_b)$, and $x \in U$. Then as U is open in $(X, \|\cdot\|_b)$, there exists an $r > 0$ such that $B_b(x, r) := \{y \in X : \|y - x\|_b < r\} \subset U$. But if $y \in X$ satisfies $\|y - x\|_a < mr$, then $\|y - x\|_b \leq (1/m)\|y - x\|_a < (1/m)mr = r$, and so $y \in B_b(x, r) \subset U$. Hence $B_a(x, mr) := \{y \in X : \|y - x\|_a < mr\} \subset U$. So it follows that U is open in $(X, \|\cdot\|_a)$ too.

Now suppose that $(x_n)_{n\in\mathbb{N}}$ is a Cauchy sequence in $(X, \|\cdot\|_b)$. Let $\epsilon > 0$. Then there exists an $N \in \mathbb{N}$ such that for all $n > N$, $\|x_n - x_m\|_b < \epsilon/M$. Hence for all $n > N$, $\|x_n - x_m\|_a \leq M\|x_n - x_m\|_b < M \cdot (\epsilon/M) = \epsilon$.
Consequently, $(x_n)_{n\in\mathbb{N}}$ is a Cauchy sequence in $(X, \|\cdot\|_a)$ as well.

If $(x_n)_{n\in\mathbb{N}}$ is a convergent sequence in $(X, \|\cdot\|_b)$ with limit L, then for $\epsilon > 0$, there exists an $N \in \mathbb{N}$ such that for all $n > N$, $\|x_n - L\|_b < \epsilon/M$. Thus for all $n > N$, $\|x_n - L\|_a \leq M\|x_n - L\|_b < M \cdot (\epsilon/M) = \epsilon$. So $(x_n)_{n\in\mathbb{N}}$ is convergent with limit L in $(X, \|\cdot\|_a)$ too.

Solution to Exercise 1.28, page 42

(1) Let $L > 0$ be such that for all $x, y \in \mathbb{R}$, $|f(x) - f(y)| = |\sqrt{|x|} - \sqrt{|y|}| \leq L|x-y|$. Then in particular, with $x = \dfrac{1}{n^2}$, $n \in \mathbb{N}$, and $y = 0$, we obtain $\dfrac{1}{n} \leq \dfrac{L}{n^2}$.

Thus $n \leq L$ for all $n \in \mathbb{N}$, which is absurd. So f is not Lipschitz.

(2) $\mathbf{x}_1(0) = 0$ and $\mathbf{x}_2(0) = 0^2/4 = 0$, and so $\mathbf{x}_1, \mathbf{x}_2$ satisfy the initial condition. For all $t \geq 0$, $\mathbf{x}_1'(t) = 0 = \sqrt{|0|} = \sqrt{|\mathbf{x}_1(t)|}$, $\mathbf{x}_2'(t) = \dfrac{2t}{4} = \dfrac{t}{2} = \sqrt{\left|\dfrac{t^2}{4}\right|} = \sqrt{|\mathbf{x}_2(t)|}$.

So $\mathbf{x}_1, \mathbf{x}_2$ are both solutions to the given Initial Value Problem.

Solution to Exercise 1.29, page 43

Let F be closed, and $(x_n)_{n \in \mathbb{N}}$ be a sequence in F which converges to x. Suppose that $x \notin F$. Since F is closed, there is an open ball $B(x,r) := \{\tilde{x} \in X : \|x-\tilde{x}\| < r\}$ with $r > 0$, which is contained in $X \backslash F$. But with $\epsilon := r > 0$, there exists an $N \in \mathbb{N}$ such that for all $n > N$, $\|x_n - x\| < r$. In particular, $\|x_{N+1} - x\| < r$, so that $F \ni x_{N+1} \in B(x,r) \subset X \backslash F$, a contradiction. Hence $x \in F$.

Now suppose that for every sequence $(x_n)_{n \in \mathbb{N}}$ in F, convergent in X with a limit $x \in X$, we have that the limit $x \in F$. We want to show that $X \backslash F$ is open. Suppose it isn't. Then[3] $\neg [\forall x \in X \backslash F, \ \exists r > 0 \text{ such that } B(x,r) \subset X \backslash F]$. In other words, $\exists x \in X \backslash F$ such that $\forall r > 0$, $B(x,r) \cap F \neq \emptyset$. So with $r = 1/n$, $n \in \mathbb{N}$, we can find an $x_n \in B(x,r) \cap F$. Then we obtain a sequence $(x_n)_{n \in \mathbb{N}}$ in F satisfying $\|x_n - x\| < 1/n$ for all $n \in \mathbb{N}$. Thus $(x_n)_{n \in \mathbb{N}}$ converges to x. But $x \notin F$, contradicting the hypothesis. Hence $X \backslash F$ is open, that is, F is closed.

Solution to Exercise 1.30, page 43

Let $(\mathbf{x}_n)_{n \in \mathbb{N}}$ in c_{00} be given by $\mathbf{x}_n = \left(1, \dfrac{1}{2}, \dfrac{1}{3}, \cdots, \dfrac{1}{n}, 0, \cdots\right)$, $n \in \mathbb{N}$.
Then with $\mathbf{x} := \left(1, \dfrac{1}{2}, \dfrac{1}{3}, \cdots\right) \in \ell^2 \backslash c_{00}$, we have

$$\|\mathbf{x}_n - \mathbf{x}\|_2^2 = \sum_{k=n+1}^{\infty} \frac{1}{k^2} < \sum_{k=n+1}^{\infty} \frac{1}{k(k-1)} = \sum_{k=n+1}^{\infty} \left(\frac{1}{k-1} - \frac{1}{k}\right)$$
$$= \left(\frac{1}{n} - \frac{1}{n+1}\right) + \left(\frac{1}{n+1} - \frac{1}{n+2}\right) + \cdots = \frac{1}{n} \xrightarrow{n \to \infty} 0,$$

showing that c_{00} is not closed.

Solution to Exercise 1.31, page 43

(1) Suppose that F is a closed set containing S. Let L be a limit point of S. Then there exists a sequence $(x_n)_{n \in \mathbb{N}}$ in $S \backslash \{L\}$ which converges to L. As each $x_n \in S \backslash \{L\} \subset S \subset F$, and since F is closed, it follows that $L \in F$. So all the limit points of S belong to F. Hence $\overline{S} \subset F$.

$\overline{S}$ is closed. Suppose that $(x_n)_{n \in \mathbb{N}}$ is a sequence in $\overline{S}$ that converges to L. We would like to prove that $L \in \overline{S}$. If $L \in S$, then $L \in \overline{S}$, and we are done. So suppose that $L \notin S$. Now for each n, we define the new term x'_n as follows:

1° If $x_n \in S$, then $x'_n := x_n$.

2° If $x_n \notin S$, then x_n must be a limit point of S, and so $B(x_n, 1/n)$ must contain some element, say x'_n, of S.

Hence we have $\|x'_n - L\| = \begin{cases} \|x_n - L\| & \text{if } x_n \in S, \\ < \dfrac{1}{n} + \|x_n - L\| & \text{if } x_n \notin S. \end{cases}$

---
[3] The symbol $\neg$ stands for "negation". It is read as: "It is not the case that $\cdots$".

Thus $(x'_n)_{n\in\mathbb{N}}$ is a sequence in $S\setminus\{L\}$ which converges to L, and so L is a limit point of S, that is, $L \in \overline{S}$. Consequently $\overline{S}$ is closed.

(2) We first note that if $y \in \overline{Y}$, then there exists a $(y_n)_{n\in\mathbb{N}}$ in Y that converges to y. Indeed, this is obvious if y is a limit point of Y, and if $y \in Y$, then we may just take $(y_n)_{n\in\mathbb{N}}$ as the constant sequence with all terms equal to y. We have:

(S1) Let $x, y \in \overline{Y}$. Let $(x_n)_{n\in\mathbb{N}}$, $(y_n)_{n\in\mathbb{N}}$ be sequences in Y that converge to x, y, respectively. Then $x_n + y_n \in Y \subset \overline{Y}$ for each $n \in \mathbb{N}$, and $(x_n + y_n)_{n\in\mathbb{N}}$ converges to $x + y$. But as $\overline{Y}$ is closed, it follows that $x + y \in \overline{Y}$ too.

(S2) Let $\alpha \in \mathbb{K}$, $y \in \overline{Y}$. Let $(y_n)_{n\in\mathbb{N}}$ be a sequence in Y that converges to y. Then $\alpha \cdot y_n \in Y \subset \overline{Y}$ for each $n \in \mathbb{N}$, and $(\alpha \cdot y_n)_{n\in\mathbb{N}}$ converges to $\alpha \cdot y$. But as $\overline{Y}$ is closed, it follows that $\alpha \cdot y \in \overline{Y}$ too.

(S3) $\mathbf{0} \in Y \subset \overline{Y}$.

Hence $\overline{Y}$ is a closed subspace.

(3) The proof is similar to part (2). Let $x, y \in \overline{C}$. Then there exist sequences $(x_n)_{n\in\mathbb{N}}$ and $(y_n)_{n\in\mathbb{N}}$ in C that converge to x, y, respectively. If $\alpha \in (0, 1)$, then $(1 - \alpha)x + \alpha y = (1 - \alpha) \lim_{n\to\infty} x_n + \alpha \lim_{n\to\infty} y_n = \lim_{n\to\infty} ((1-\alpha)x_n + \alpha y_n)$. As $(1 - \alpha)x_n + \alpha y_n \in C \subset \overline{C}$ for all $n \in \mathbb{N}$, and since $\overline{C}$ is closed, it follows that $(1 - \alpha)x + \alpha y \in \overline{C}$ too.

(4) Suppose that D is dense in X. Let $x \in X\setminus D$. If $n \in \mathbb{N}$, then the ball $B(x, 1/n)$ must contain an element $d_n \in D$. The sequence $(d_n)_{n\in\mathbb{N}}$ converges to x because $\|x - d_n\| < 1/n$, $n \in \mathbb{N}$. Hence x is a limit point of D, that is, $x \in \overline{D}$. So $X\setminus D \subset \overline{D}$. Also $D \subset \overline{D}$. Thus $X = D \cup (X\setminus D) \subset \overline{D} \subset X$, and so $X = \overline{D}$. Now suppose that $X = \overline{D}$. If $x \in X\setminus D = \overline{D}\setminus D$, then x is a limit point of D, and so there is a sequence $(d_n)_{n\in\mathbb{N}}$ in D that converges to x. Thus given an $\epsilon > 0$, there is an N such that $\|x - d_N\| < \epsilon$, that is, $d_N \in D \bigcap B(x, \epsilon)$. On the other hand, if $x \in D$, and $\epsilon > 0$, then $x \in B(x, \epsilon) \bigcap D$. Hence $\overline{D}$ is dense in X.

Solution to Exercise 1.32, page 43

Let $(x_n)_{n\in\mathbb{N}} \in \ell^1$. Then $\sum_{n=1}^{\infty} |x_n| < \infty$, and so $\lim_{n\to\infty} |x_n| = 0$. Thus there exists an $N \in \mathbb{N}$ such that $|x_n| \leqslant 1$ for all $n \geqslant N$. For all $n \geqslant N$, $|x_n|^2 = |x_n| \cdot |x_n| \leqslant |x_n| \cdot 1 = |x_n|$. By the Comparison Test[4], $\sum_{n=1}^{\infty} |x_n|^2 < \infty$. Hence $(x_n)_{n\in\mathbb{N}} \in \ell^2$.

$\left(1, \dfrac{1}{2}, \dfrac{1}{3}, \cdots\right) \in \ell^2\setminus\ell^1$, as $\sum_{n=1}^{\infty} \dfrac{1}{n^2} < \infty$, while the Harmonic Series $\sum_{n=1}^{\infty} \dfrac{1}{n}$ diverges.

$(\ell^1, \|\cdot\|_2)$ is not a Banach space: Let us suppose, on the contrary, that it is a

---
[4]See for example [Sasane (2015), page 311].

Banach space, and we will arrive at a contradiction by showing a Cauchy sequence which is not convergent in $(\ell^1, \|\cdot\|_2)$.

Consider for $n \in \mathbb{N}$, $\mathbf{x}_n := \left(1, \frac{1}{2}, \frac{1}{3}, \cdots \frac{1}{n}, 0, \cdots\right) \in \ell^1 \subset \ell^2$. Then $(\mathbf{x}_n)_{n \in \mathbb{N}}$ converges in ℓ^2 to $\mathbf{x} := \left(1, \frac{1}{2}, \frac{1}{3}, \cdots\right) \in \ell^2$, because $\|\mathbf{x}_n - \mathbf{x}\|_2^2 = \sum_{k=n+1}^{\infty} \frac{1}{k^2} \xrightarrow{n \to \infty} 0$.

So $(\mathbf{x}_n)_\mathbb{N}$ is a Cauchy sequence in $(\ell^2, \|\cdot\|_2)$, and so it is also Cauchy in $(\ell^1, \|\cdot\|_2)$. As we have assumed that $(\ell^1, \|\cdot\|_2)$ is a Banach space, it follows that the Cauchy sequence $(\mathbf{x}_n)_{n \in \mathbb{N}}$ must be convergent to some element $\mathbf{x}' \in \ell^1 \subset \ell^2$. But by the uniqueness of limits (when we consider $(\mathbf{x}_n)_{n \in \mathbb{N}}$ as a sequence in ℓ^2), we must have $\mathbf{x} = \mathbf{x}' \in \ell^1$, which is false, since we know that the Harmonic Series diverges. This contradiction proves that $(\ell^1, \|\cdot\|_2)$ is not a Banach space.

Solution to Exercise 1.33, page 43

Let $(\mathbf{a}_n)_{n \in \mathbb{N}}$ be a Cauchy sequence in c_0. Then this is also a Cauchy sequence in ℓ^∞, and hence convergent to a sequence in ℓ^∞, say $\mathbf{a}$. We'll show that $\mathbf{a} \in c_0$. We write $\mathbf{a}_n = (a_n^{(m)})_{m \in \mathbb{N}}$, and $\mathbf{a} = (a^{(m)})_{m \in \mathbb{N}}$. Let $\epsilon > 0$. Then there exists an $N \in \mathbb{N}$ such that $\|\mathbf{a}_N - \mathbf{a}\|_\infty < \epsilon$. In particular, for all $m \in \mathbb{N}$, $|a_N^{(m)} - a^{(m)}| < \epsilon$. But as $\mathbf{a}_N \in c_0$, we can find an M such that for all $m > M$, $|a_N^{(m)}| < \epsilon$. Consequently, for $m > M$, we have from the above that $|a^{(m)}| \leq |a^{(m)} - a_N^{(m)}| + |a_N^{(m)}| < \epsilon + \epsilon = 2\epsilon$. Thus $\mathbf{a} \in c_0$ too.

Solution to Exercise 1.34, page 44

Given $\epsilon > 0$, let $N \in \mathbb{N}$ be large enough so that for all $n > N$, $\|x_n - x\| < \epsilon$. Then for all $n > N$, we have $|\|x_n\| - \|x\|| \leq \|x_n - x\| < \epsilon$, and so it follows that the sequence $(\|x_n\|)_{n \in \mathbb{N}}$ is $\mathbb{R}$ is convergent, with limit $\|x\|$.

Solution to Exercise 1.35, page 44

First consider the case $1 \leq p < \infty$.

(N1) $\|\mathbf{x}\|_p = \sum_{n=1}^{\infty} |x_n|^p \geq 0$ for all $\mathbf{x} = (x_1, x_2, x_3, \cdots) \in \ell^p$.

If $0 = \|\mathbf{x}\|_p = \left(\sum_{n=1}^{\infty} |x_n|^p\right)^{1/p}$, then $|x_n| = 0$ for all n, and so $\mathbf{x} = \mathbf{0}$.

(N2) $\|\alpha \cdot \mathbf{x}\|_p = \left(\sum_{n=1}^{\infty} |\alpha x_n|^p\right)^{1/p} = \left(|\alpha|^p \sum_{n=1}^{\infty} |x_n|^p\right)^{1/p} = |\alpha| \|\mathbf{x}\|_p$, for $\mathbf{x} \in \ell^p$, $\alpha \in \mathbb{K}$.

(N3) Let $\mathbf{x} = (x_1, x_2, \cdots)$ and $\mathbf{y} = (y_1, y_2, \cdots)$ belong to ℓ^p. Let $d \in \mathbb{N}$. By the Triangle Inequality for the $\|\cdot\|_p$-norm on $\mathbb{R}^d$,
$$\left(\sum_{k=1}^{d} |x_k + y_k|^p\right)^{1/p} \leq \left(\sum_{k=1}^{d} |x_k|^p\right)^{1/p} + \left(\sum_{k=1}^{d} |y_k|^p\right)^{1/p}.$$
Passing the limit as d tends to ∞ yields $\|\mathbf{x} + \mathbf{y}\|_p \leq \|\mathbf{x}\|_p + \|\mathbf{y}\|_p$.

Now consider the case $p = \infty$.

(N1) $\|\mathbf{x}\|_\infty = \sup_{n\in\mathbb{N}} |x_n| \geq 0$ for all $\mathbf{x} = (x_1, x_2, x_3, \cdots) \in \ell^\infty$.
If $0 = \|\mathbf{x}\|_\infty = \sup_{n\in\mathbb{N}} |x_n|$, then $|x_n| = 0$ for all n, that is, $\mathbf{x} = \mathbf{0}$.

(N2) $\|\alpha \cdot \mathbf{x}\|_\infty = \sup_{n\in\mathbb{N}} |\alpha x_n| = |\alpha| \sup_{n\in\mathbb{N}} |x_n| = |\alpha| \|\mathbf{x}\|_\infty$, for $\mathbf{x} \in \ell^\infty$, $\alpha \in \mathbb{K}$.

(N3) Let $\mathbf{x} = (x_1, x_2, \cdots)$ and $\mathbf{y} = (y_1, y_2, \cdots)$ belong to ℓ^∞.
Then for all k, $|x_k + y_k| \leq |x_k| + |y_k| \leq \|\mathbf{x}\|_\infty + \|\mathbf{y}\|_\infty$, and so
$$\|\mathbf{x} + \mathbf{y}\|_\infty = \sup_{k\in\mathbb{N}} |x_k + y_k| \leq \|\mathbf{x}\|_\infty + \|\mathbf{y}\|_\infty.$$

Solution to Exercise 1.36, page 44

From Exercise 1.11, page 17, taking $n = 1$, $(C^1[a,b], \|\cdot\|_{1,\infty})$ is a normed space. We show that $(C^1[a,b], \|\cdot\|_{1,\infty})$ is complete. Let $(\mathbf{x}_n)_{n\in\mathbb{N}}$ be a Cauchy sequence in $C^1[a,b]$. Then $\|\mathbf{x}_n - \mathbf{x}_m\|_\infty \leq \|\mathbf{x}_n - \mathbf{x}_m\|_\infty + \|\mathbf{x}'_n - \mathbf{x}'_m\|_\infty = \|\mathbf{x}_n - \mathbf{x}_m\|_{1,\infty}$, and so $(\mathbf{x}_n)_{n\in\mathbb{N}}$ is a Cauchy sequence in $(C[a,b], \|\cdot\|_\infty)$, and hence convergent to, say, $\mathbf{x} \in C[a,b]$. Also, $\|\mathbf{x}'_n - \mathbf{x}'_m\|_\infty \leq \|\mathbf{x}_n - \mathbf{x}_m\|_\infty + \|\mathbf{x}'_n - \mathbf{x}'_m\|_\infty = \|\mathbf{x}_n - \mathbf{x}_m\|_{1,\infty}$, shows that $(\mathbf{x}'_n)_{n\in\mathbb{N}}$ is a Cauchy sequence in $(C[a,b], \|\cdot\|_\infty)$, and hence convergent to, say, $\mathbf{y} \in C[a,b]$. We will now show that $\mathbf{x} \in C^1[a,b]$, and $\mathbf{x}' = \mathbf{y}$. Let $t \in [a,b]$. By the Fundamental Theorem of Calculus, $\mathbf{x}_n(t) - \mathbf{x}_n(a) = \int_a^t \mathbf{x}'_n(\tau)d\tau$, and so

$$\left| \mathbf{x}_n(t) - \mathbf{x}_n(a) - \int_a^t \mathbf{y}(\tau)d\tau \right| = \left| \int_a^t (\mathbf{x}'_n(\tau) - \mathbf{y}(\tau))d\tau \right| \leq \int_a^t |\mathbf{x}'_n(\tau) - \mathbf{y}(\tau)| d\tau$$
$$\leq \|\mathbf{x}'_n - \mathbf{y}\|_\infty (t-a).$$

Passing the limit as n goes to ∞ gives, for all $t \in [a,b]$, $\mathbf{x}(t) = \mathbf{x}(a) + \int_a^t \mathbf{y}(\tau)d\tau$.

By the Fundamental Theorem of Calculus, $\mathbf{x}' = \mathbf{y} \in C[a,b]$. So $\mathbf{x} \in C^1[a,b]$. Finally, we'll show that $(\mathbf{x}_n)_{n\in\mathbb{N}}$ converges to $\mathbf{x}$ in $C^1[a,b]$. Let $\epsilon > 0$, and let N be such that for all $m, n > N$, $\|\mathbf{x}_n - \mathbf{x}_m\|_{1,\infty} < \epsilon$. Then for all $t \in [a,b]$, we have $|\mathbf{x}_n(t) - \mathbf{x}_m(t)| + |\mathbf{x}'_n(t) - \mathbf{x}'_m(t)| \leq \|\mathbf{x}_n - \mathbf{x}_m\|_\infty + \|\mathbf{x}'_n - \mathbf{x}'_m\|_\infty = \|\mathbf{x}_n - \mathbf{x}_m\|_{1,\infty} < \epsilon$. Letting m go to ∞, it follows that for all $n > N$, $|\mathbf{x}_n(t) - \mathbf{x}(t)| + |\mathbf{x}'_n(t) - \mathbf{x}'(t)| \leq \epsilon$. As the choice of $t \in [a,b]$ was arbitrary, it follows that

$$\sup_{t\in[a,b]} |\mathbf{x}_n(t) - \mathbf{x}(t)| + \sup_{t\in[a,b]} |\mathbf{x}'_n(t) - \mathbf{x}'(t)|$$
$$\leq \sup_{t\in[a,b]} (|\mathbf{x}_n(t) - \mathbf{x}(t)| + |\mathbf{x}'_n(t) - \mathbf{x}'(t)|) + \sup_{t\in[a,b]} (|\mathbf{x}_n(t) - \mathbf{x}(t)| + |\mathbf{x}'_n(t) - \mathbf{x}'(t)|) \leq 2\epsilon,$$

that is, $\|\mathbf{x}_n - \mathbf{x}_m\|_{1,\infty} \leq 2\epsilon$.

Solution to Exercise 1.37, page 44

Let $(x_n)_{n\in\mathbb{N}}$ be any Cauchy sequence in X. We construct a subsequence $(x_{n_k})_{k\in\mathbb{N}}$ inductively, possessing the property that if $n > n_k$, then $\|x_n - x_{n_k}\| < 1/2^k$, $k \in \mathbb{N}$.

Choose $n_1 \in \mathbb{N}$ large enough so that if $n,m \geq n_1$, then $\|x_n - x_m\| < 1/2$. Suppose $x_{n_1}, \cdots, x_{n_k}$ have been constructed. Choose $n_{k+1} > n_k$ such that if $n,m \geq n_{k+1}$, then $\|x_n - x_m\| < 1/2^{k+1}$. In particular for $n \geq n_{k+1}$, $\|x_n - x_{n_{k+1}}\| < 1/2^{k+1}$. Now define $u_1 = x_{n_1}$, $u_{k+1} = x_{n_{k+1}} - x_{n_k}$, $k \in \mathbb{N}$.

We have $\displaystyle\sum_{k=1}^{\infty} \|u_k\| \leq \|x_{n_1}\| + \sum_{k=1}^{\infty} \frac{1}{2^k} < \infty$. Thus $\displaystyle\sum_{k=1}^{\infty} u_k$ converges.

But the partial sums of $\displaystyle\sum_{k=1}^{\infty} u_k$ are $\displaystyle\sum_{j=1}^{k} u_j = x_{n_k}$, $k \in \mathbb{N}$.

So $(x_{n_k})_{k \in \mathbb{N}}$ converges in X, to, say $x \in X$. As $(x_{n_k})_{k \in \mathbb{N}}$ is a convergent subsequence of the *Cauchy* sequence $(x_n)_{n \in \mathbb{N}}$, it now follows that $(x_n)_{n \in \mathbb{N}}$ is itself convergent with the same limit x. Indeed, given $\epsilon > 0$, first let N be such that for all $n,m > N$, $\|x_n - x_m\| < \epsilon/2$, and next let $n_K > N$ be such that $\|x_{n_K} - x\| < \epsilon/2$, which yields that for all $n > N$,
$$\|x_n - x\| = \|x_n - x_{n_K} + x_{n_K} - x\| \leq \|x_n - x_{n_K}\| + \|x_{n_K} - x\| < \frac{\epsilon}{2} + \frac{\epsilon}{2} = \epsilon.$$

Solution to Exercise 1.38, page 44

(N1) For $(x,y) \in X \times Y$, $\|(x,y)\| = \max\{\|x\|, \|y\|\} \geq 0$.
If $\|(x,y)\| = 0$, then $0 \leq \|x\| \leq \max\{\|x\|, \|y\|\} = \|(x,y)\| = 0$, and so $\|x\| = 0$, giving $x = \mathbf{0}$. Similarly, $y = \mathbf{0}$ too, and so $(x,y) = \mathbf{0}_{X \times Y}$.

(N2) For $\alpha \in \mathbb{K}$, and $(x,y) \in X \times Y$,
$$\|\alpha(x,y)\| = \|(\alpha x, \alpha y)\| = \max\{\|\alpha x\|, \|\alpha y\|\} = \max\{|\alpha|\|x\|, |\alpha|\|y\|\}$$
$$= |\alpha| \max\{\|x\|, \|y\|\} = |\alpha| \|(x,y)\|.$$

(N3) Let $(x_1, y_1), (x_2, y_2) \in X \times Y$. Then
$$\|x_1 + x_2\| \leq \|x_1\| + \|x_2\| \leq \max\{\|x_1\|, \|y_1\|\} + \max\{\|x_2\|, \|y_2\|\}$$
$$= \|(x_1, y_1)\| + \|(x_2, y_2)\|,$$
$$\|y_1 + y_2\| \leq \|y_1\| + \|y_2\| \leq \max\{\|x_1\|, \|y_1\|\} + \max\{\|x_2\|, \|y_2\|\}$$
$$= \|(x_1, y_1)\| + \|(x_2, y_2)\|,$$

and so $\max\{\|x_1 + x_2\|, \|y_1 + y_2\|\} \leq \|(x_1, y_1)\| + \|(x_2, y_2)\|$. Thus
$$\|(x_1, y_1) + (x_2, y_2)\| = \|(x_1 + x_2, y_1 + y_2)\| = \max\{\|x_1 + x_2\|, \|y_1 + y_2\|\}$$
$$\leq \|(x_1, y_1)\| + \|(x_2, y_2)\|.$$

Hence $(x,y) \mapsto \max\{\|x\|, \|y\|\}$, $(x,y) \in X \times Y$, defines a norm on $X \times Y$.

Let $\big((x_n, y_n)\big)_{n \in \mathbb{N}}$ be Cauchy in $X \times Y$. As $\|x\| \leq \max\{\|x\|, \|y\|\} = \|(x,y)\|$, $(x_n)_{n \in \mathbb{N}}$ is Cauchy in X. As X is Banach, $(x_n)_{n \in \mathbb{N}}$ converges to some $x \in X$. Similarly $(y_n)_{n \in \mathbb{N}}$ converges to a $y \in Y$. Let $\epsilon > 0$. Then there exists an N_x such that for all $n > N_x$, $\|x_n - x\| < \epsilon$, and there is an N_y such that for all $n > N_y$, $\|y_n - y\| < \epsilon$. So with $N := \max\{N_x, N_y\}$, for all $n > N$, we have $\|x_n - x\| < \epsilon$ and $\|y_n - y\| < \epsilon$. Thus $\|(x_n, y_n) - (x,y)\| = \|(x_n - x, y_n - y)\| = \max\{\|x_n - x\|, \|y_n - y\|\} < \epsilon$, showing that $\big((x_n, y_n)\big)_{n \in \mathbb{N}}$ converges to (x,y) in $X \times Y$. So $X \times Y$ is Banach.

Solution to Exercise 1.39, page 50

Since K is compact in $\mathbb{R}^d$, it is closed and bounded. Let $R > 0$ be such that for all $\mathbf{x} \in K$, $\|\mathbf{x}\|_2 \leqslant R$. In particular, for every $\mathbf{x} \in K \cap F$, we have $\|\mathbf{x}\|_2 \leqslant R$. Thus $K \cap F$ is bounded. Also, since both K and F are closed, it follows that even $K \cap F$ is closed. Hence $K \cap F$ is closed and bounded, and so by Theorem 1.10, page 45, we conclude that $K \cap F$ is compact.

Solution to Exercise 1.40, page 50

Clearly $\mathbb{S}^{d-1}$ is bounded. It is also closed, and we prove this below. Let $(\mathbf{x}_n)_{n\in\mathbb{N}}$ be a sequence in $\mathbb{S}^{d-1}$ which converges to $\mathbf{L}$ in $\mathbb{R}^d$. Let $\mathbf{L} = (L_1, \cdots, L_d)$ and $\mathbf{x}_n = \left(x_n^{(1)}, \cdots, x_n^{(d)}\right)$ for $n \in \mathbb{N}$. Then $\lim\limits_{n\to\infty} x_n^{(k)} = L_k$ $(k = 1, \ldots, d)$. Since $\mathbf{x}_n \in \mathbb{S}^{d-1}$ for each $n \in \mathbb{N}$, we have $\|\mathbf{x}_n\|_2^2 = \left(x_1^{(n)}\right)^2 + \cdots + \left(x_d^{(n)}\right)^2 = 1$. Passing the limit as $n \to \infty$, we obtain $\|\mathbf{L}\|_2^2 = L_1^2 + \cdots + L_d^2 = 1$. Hence $\mathbf{L} \in \mathbb{S}^{d-1}$. So $\mathbb{S}^{d-1}$ is closed. As $\mathbb{S}^{d-1}$ is closed and bounded, it follows from Theorem 1.10, page 45, that it is compact.

Solution to Exercise 1.41, page 50

(1) Let $(\mathbf{R}_n)_{n\in\mathbb{N}}$ be a sequence in $O(2)$.

Using $(\mathbf{R}_n)^\top \mathbf{R}_n = \mathbf{I}$, if $\mathbf{R}_n =: \begin{bmatrix} a_n & b_n \\ c_n & d_n \end{bmatrix}$ then $a_n^2 + b_n^2 = 1$ and $c_n^2 + d_n^2 = 1$.

So each of the sequences $(a_n)_{n\in\mathbb{R}}, (b_n)_{n\in\mathbb{R}}, (c_n)_{n\in\mathbb{R}}, (d_n)_{n\in\mathbb{R}}$ is bounded. By successively refining subsequences of these sequences, we can choose a sequence of indices $n_1 < n_2 < n_3 < \cdots$, such that the sequences $(a_{n_k})_{k\in\mathbb{N}}$, $(b_{n_k})_{k\in\mathbb{N}}, (c_{n_k})_{k\in\mathbb{N}}, (d_{n_k})_{k\in\mathbb{N}}$ are convergent, to, say, a, b, c, d, respectively.

Hence $(\mathbf{R}_{n_k})_{k\in\mathbb{N}}$ is convergent with the limit $\mathbf{R} := \begin{bmatrix} a & b \\ c & d \end{bmatrix}$.

From $(\mathbf{R}_n)^\top \mathbf{R}_n = \mathbf{I}$ $(n \in \mathbb{N})$, it follows that also $\mathbf{R}^\top \mathbf{R} = \mathbf{I}$, that is, $\mathbf{R} \in O(2)$.

(2) The hyperbolic rotations $\mathbf{R}(t) := \begin{bmatrix} \cosh t & \sinh t \\ \sinh t & \cosh t \end{bmatrix}$ belong to $O(1,1)$ because

$$(\mathbf{R}(t))^\top \begin{bmatrix} 1 & 0 \\ 0 & -1 \end{bmatrix} \mathbf{R}(t) = \begin{bmatrix} \cosh t & \sinh t \\ \sinh t & \cosh t \end{bmatrix} \begin{bmatrix} 1 & 0 \\ 0 & -1 \end{bmatrix} \begin{bmatrix} \cosh t & \sinh t \\ \sinh t & \cosh t \end{bmatrix}$$

$$= \begin{bmatrix} \cosh t & -\sinh t \\ \sinh t & -\cosh t \end{bmatrix} \begin{bmatrix} \cosh t & \sinh t \\ \sinh t & \cosh t \end{bmatrix}$$

$$= \begin{bmatrix} (\cosh t)^2 - (\sinh t)^2 & 0 \\ 0 & (\sinh t)^2 - (\cosh t)^2 \end{bmatrix}$$

$$= \begin{bmatrix} 1 & 0 \\ 0 & -1 \end{bmatrix}.$$

But $\|\mathbf{R}(t)\|_\infty \geqslant |\cosh(t)| = \cosh t \to \infty$ as $t \to \infty$, showing that $O(1,1)$ is not bounded. Hence $O(1,1)$ can't be compact (as every compact set is necessarily bounded).

Solution to Exercise 1.42, page 51

Let $K := \left\{1, \frac{1}{2}, \frac{1}{3}, \cdots\right\} \cup \{0\}$. Since $K \subset [0,1]$, clearly K is bounded. Moreover, $\mathbb{R}\backslash K = (-\infty, 0) \cup \left(\bigcup_{n=1}^{\infty} \left(\frac{1}{n+1}, \frac{1}{n}\right)\right) \cup (1, \infty)$.

Thus $\mathbb{R}\backslash K$, being the union of open intervals, is open, that is, K is closed. Since K is closed and bounded, it is compact.

Solutions to the exercises from Chapter 2

Solution to Exercise 2.1, page 58

If $\mathbf{1} \in C[0,1]$ denotes the constant function taking value 1 everywhere, then

$$S_1(\mathbf{1}) = \int_0^1 (\mathbf{1}(t))^2 dt = \int_0^1 1 dt = 1,$$

and so $2 \cdot S_1(\mathbf{1}) = 2 \neq 4 = \int_0^1 2^2 dt = \int_0^1 (2\mathbf{1}(t))^2 dt = S_1(2 \cdot \mathbf{1}).$

So (L2) is violated, showing that S_1 is *not* a linear transformation.
On the other hand, S_2 is a linear transformation. For all $\mathbf{x}_1, \mathbf{x}_2 \in C[0,1]$,

$$\begin{aligned}S_2(\mathbf{x}_1 + \mathbf{x}_2) &= \int_0^1 (\mathbf{x}_1 + \mathbf{x}_2)(t^2) dt = \int_0^1 (\mathbf{x}_1(t^2) + \mathbf{x}_2(t^2)) dt \\ &= \int_0^1 \mathbf{x}_1(t^2) dt + \int_0^1 \mathbf{x}_2(t^2) dt = S_2(\mathbf{x}_1) + S_2(\mathbf{x}_2),\end{aligned}$$

and so (L1) holds. Moreover, for all $\alpha \in \mathbb{R}$ and $\mathbf{x} \in C[0,1]$ we have

$$S_2(\alpha \cdot \mathbf{x}) = \int_0^1 (\alpha \cdot \mathbf{x})(t^2) dt = \int_0^1 \alpha \mathbf{x}(t^2) dt = \alpha \int_0^1 \mathbf{x}(t^2) dt = \alpha \cdot S_2(\mathbf{x}),$$

and so (L2) holds as well.

Solution to Exercise 2.2, page 58

(1) Let $\alpha_1, \alpha_2 \in \mathbb{R}$ be such that $\alpha_1 \mathbf{f}_1 + \alpha_2 \mathbf{f}_2 = \mathbf{0}$, that is,

$$\alpha_1 e^{at} \cos(bt) + \alpha_2 e^{at} \sin(bt) = 0 \quad (t \in \mathbb{R}).$$

In particular, with $t = 0$, we obtain $\alpha_1 = 0$. Thus $\alpha_2 e^{at} \sin(bt) = 0$ for all $t \in \mathbb{R}$. With $t = \pi/2b$, we see that $\alpha_2 e^{\frac{a\pi}{2b}} \sin(\pi/2) = 0$, and so $\alpha_2 = 0$. Consequently, $\mathbf{f}_1, \mathbf{f}_2$ are linearly independent.

(2) First of all, D is a well-defined map from $S_{\mathbf{f}_1,\mathbf{f}_2}$ to itself, since
$$\begin{aligned}D(\alpha_1 \mathbf{f}_1 &+ \alpha_2 \mathbf{f}_2) \\ &= D(\alpha_1 e^{at} \cos(bt) + \alpha_2 e^{at} \sin(bt)) \\ &= \alpha_1 a e^{at} \cos(bt) - \alpha_1 e^{at} b \sin(bt) + \alpha_2 a e^{at} \cos(bt) + \alpha_2 e^{at} b \cos(bt) \\ &\in S_{\mathbf{f}_1,\mathbf{f}_2}.\end{aligned}$$
Thus $DS_{\mathbf{f}_1,\mathbf{f}_2} \subset S_{\mathbf{f}_1,\mathbf{f}_2}$.
Furthermore, it is clear that $D(\mathbf{g}_1 + \mathbf{g}_2) = D(\mathbf{g}_1) + D(\mathbf{g}_2)$ for all $\mathbf{g}_1, \mathbf{g}_2 \in C^1(\mathbb{R})$ (and in particular for $\mathbf{g}_1, \mathbf{g}_2 \in S_{\mathbf{f}_1,\mathbf{f}_2} \subset C^1(\mathbb{R})$), and also $D(\alpha \cdot \mathbf{g}) = \alpha \cdot D(\mathbf{g})$ for all $\alpha \in \mathbb{R}$ and all $\mathbf{g} \in C^1(\mathbb{R})$ (and in particular, for all $\mathbf{g} \in S_{\mathbf{f}_1,\mathbf{f}_2}$).
Hence D is a linear transformation from $S_{\mathbf{f}_1,\mathbf{f}_2}$ to itself.

(3) We have $D\mathbf{f}_1 = ae^{at}\cos(bt) - e^{at}b\sin(bt) = a\mathbf{f}_1 - b\mathbf{f}_2$, and
$D\mathbf{f}_2 = ae^{at}\sin(bt) + e^{at}b\cos(bt) = b\mathbf{f}_1 + a\mathbf{f}_2$.

So the matrix of D with respect to the basis $B = (\mathbf{f}_1, \mathbf{f}_2)$ is $[D]_B = \begin{bmatrix} a & b \\ -b & a \end{bmatrix}$.

(4) As $\det[D]_B = a^2 + b^2 \neq 0$, $[D]_B$ is invertible, and $[D]_B^{-1} = \dfrac{1}{a^2+b^2}\begin{bmatrix} a & -b \\ b & a \end{bmatrix}$.

Hence D is invertible, and the inverse $D^{-1} : S_{\mathbf{f}_1,\mathbf{f}_2} \to S_{\mathbf{f}_1,\mathbf{f}_2}$ has the matrix $[D^{-1}]_B$ (with respect to B) given by $[D^{-1}]_B = [D]_B^{-1}$ found above.

(5) We note that $[D^{-1}]_B \begin{bmatrix} 1 \\ 0 \end{bmatrix} = \dfrac{1}{a^2+b^2}\begin{bmatrix} a & -b \\ b & a \end{bmatrix}\begin{bmatrix} 1 \\ 0 \end{bmatrix} = \dfrac{1}{a^2+b^2}\begin{bmatrix} a \\ b \end{bmatrix}$, and so

$$D^{-1}(\mathbf{f}_1) = D^{-1}(e^{at}\cos(bt)) = \dfrac{1}{a^2+b^2}(a\mathbf{f}_1 + b\mathbf{f}_2)$$
$$= \dfrac{1}{a^2+b^2}(ae^{at}\cos(bt) + be^{at}\sin(bt)).$$

By the definition of D, $\dfrac{d}{dt}\left(\dfrac{1}{a^2+b^2}(ae^{at}\cos(bt) + be^{at}\sin(bt))\right) = e^{at}\cos(bt)$.

So $\displaystyle\int e^{at}\cos(bt)dt = \dfrac{1}{a^2+b^2}(ae^{at}\cos(bt) + be^{at}\sin(bt)) + C$; C any constant.

Similarly, as $[D^{-1}]_B\begin{bmatrix} 0 \\ 1 \end{bmatrix} = \dfrac{1}{a^2+b^2}\begin{bmatrix} a & -b \\ b & a \end{bmatrix}\begin{bmatrix} 0 \\ 1 \end{bmatrix} = \dfrac{1}{a^2+b^2}\begin{bmatrix} -b \\ a \end{bmatrix}$, we have

$$D^{-1}(\mathbf{f}_2) = D^{-1}(e^{at}\sin(bt)) = \dfrac{1}{a^2+b^2}(-b\mathbf{f}_1 + a\mathbf{f}_2)$$
$$= \dfrac{1}{a^2+b^2}(-be^{at}\cos(bt) + ae^{at}\sin(bt)),$$

and so $\dfrac{d}{dt}\left(\dfrac{1}{a^2+b^2}(-be^{at}\cos(bt) + ae^{at}\sin(bt))\right) = e^{at}\sin(bt)$.

So $\displaystyle\int e^{at}\sin(bt)dt = \dfrac{1}{a^2+b^2}(-be^{at}\cos(bt) + ae^{at}\sin(bt)) + C$; C any constant.

Solution to Exercise 2.3, page 61

(1) We have $f(\mathbf{0}) = \displaystyle\int_0^1 \sqrt{1+(\mathbf{0}'(t))^2}dt = \int_0^1 \sqrt{1+0^2}dt = \int_0^1 1dt = 1$.

(As expected, the arc length is simply the length of the line segment $[0,1]$.)

(2) We have $\mathbf{x}_n'(t) = -2\pi\sqrt{n}\sin(2\pi nt)$, $t \in [0,1]$, and so

$$f(\mathbf{x}_n) = \int_0^1 \sqrt{1+(\mathbf{x}_n'(t))^2}dt = \int_0^1 \sqrt{1+4\pi^2 n(\sin(2\pi nt))^2}dt$$
$$\geq n\int_{\frac{1}{8n}}^{\frac{3}{8n}} \sqrt{1+4\pi^2 n(\sin(2\pi nt))^2}dt$$
$$\geq n\int_{\frac{1}{8n}}^{\frac{3}{8n}} \sqrt{1+2\pi^2 n}dt = n\cdot\left(\dfrac{3}{8n} - \dfrac{1}{8n}\right)\sqrt{1+2\pi^2 n} = \dfrac{\sqrt{1+2\pi^2 n}}{4}.$$

(3) Suppose that f is continuous at $\mathbf{0}$. Then with $\epsilon := 1 > 0$, there exists a $\delta > 0$ such that whenever $\mathbf{x} \in C^1[0,1]$ and $\|\mathbf{x} - \mathbf{0}\|_\infty < \delta$, we have $|f(\mathbf{x}) - f(\mathbf{0})| < 1$.

We have $\|\mathbf{x}_n - \mathbf{0}\|_\infty = \max\limits_{t \in [0,1]} \dfrac{|\cos(2\pi n t)|}{\sqrt{n}} = \dfrac{1}{\sqrt{n}} < \delta$ for all $n > \dfrac{1}{\delta^2}$.

Hence for such n there must hold that $|f(\mathbf{x}_n) - f(\mathbf{0})| = |f(\mathbf{x}_n) - 1| < 1$.

So for all $n > \dfrac{1}{\delta^2}$, we have $\dfrac{\sqrt{1 + 2\pi^2 n}}{4} \leq |f(\mathbf{x}_n)| \leq |f(\mathbf{x}_n) - 1| + 1 < 1 + 1 = 2$,

which is a contradiction. Hence f is not continuous at $\mathbf{0}$.

Let $\mathbf{x}_0, \mathbf{x} \in C^1[a,b]$. Using the triangle inequality in $(\mathbb{R}^2, \|\cdot\|_2)$, we obtain

$$\left|\sqrt{1^2 + (\mathbf{x}'(t))^2} - \sqrt{1^2 + (\mathbf{x}'_0(t))^2}\right| \leq \sqrt{(1-1)^2 + (\mathbf{x}'(t) - \mathbf{x}'_0(t))^2}$$
$$= |\mathbf{x}'(t) - \mathbf{x}'_0(t)|,$$

and so

$$|f(\mathbf{x}) - f(\mathbf{x}_0)| = \left|\int_0^1 \left(\sqrt{1^2 + (\mathbf{x}'(t))^2} - \sqrt{1^2 + (\mathbf{x}'_0(t))^2}\right) dt\right|$$
$$\leq \int_0^1 \left|\sqrt{1^2 + (\mathbf{x}'(t))^2} - \sqrt{1^2 + (\mathbf{x}'_0(t))^2}\right| dt$$
$$\leq \int_0^1 |\mathbf{x}'(t) - \mathbf{x}'_0(t)| dt \leq \int_0^1 \|\mathbf{x}' - \mathbf{x}'_0\|_\infty dt = \|\mathbf{x}' - \mathbf{x}'_0\|_\infty$$
$$\leq \|\mathbf{x} - \mathbf{x}_0\|_{1,\infty}.$$

Thus given $\epsilon > 0$, if we set $\delta := \epsilon$, then we have for all $\mathbf{x} \in C^1[0,1]$ satisfying $\|\mathbf{x} - \mathbf{x}_0\|_{1,\infty} < \delta$ that $|f(\mathbf{x}) - f(\mathbf{x}_0)| \leq \|\mathbf{x} - \mathbf{x}_0\|_{1,\infty} < \delta = \epsilon$.
So f is continuous at $\mathbf{x}_0$. As the choice of $\mathbf{x}_0$ was arbitrary, f is continuous.

Solution to Exercise 2.4, page 62

Let $x_0 \in X$. Given $\epsilon > 0$, set $\delta := \epsilon$. Then for all $x \in X$ satisfying $\|x - x_0\| < \delta$, we have $\big|\|x\| - \|x_0\|\big| \leq \|x - x_0\| < \delta = \epsilon$. Thus $\|\cdot\|$ is continuous at x_0. As $x_0 \in X$ was arbitrary, it follows that $\|\cdot\|$ is continuous on X.

Solution to Exercise 2.5, page 62

$f^{-1}(\{-1,1\}) = \{n\pi : n \in \mathbb{Z}\}$, $f^{-1}(\{1\}) = \{2n\pi : n \in \mathbb{Z}\}$, $f^{-1}([-1,1]) = \mathbb{R}$, and
$$f^{-1}\left(\left(-\frac{1}{2}, \frac{1}{2}\right)\right) = \bigcup_{n \in \mathbb{Z}} \left((2n+1)\frac{\pi}{2} - \frac{\pi}{6}, (2n+1)\frac{\pi}{2} + \frac{\pi}{6}\right) = \bigcup_{n \in \mathbb{Z}} \left(n\pi + \frac{\pi}{3}, n\pi + \frac{2\pi}{3}\right).$$

Solution to Exercise 2.6, page 62

Since cos is periodic with period 2π (that is, $f(x) = f(x + 2\pi)$ for all $x \in \mathbb{R}$), we have $f(\mathbb{R}) = f([0, 2\pi]) = f([\delta, \delta + 2\pi]) = [-1, 1]$.

Solution to Exercise 2.7, page 64

("If" part) Suppose that for every closed F in Y, $f^{-1}(F)$ is closed in X.
Now let V be open in Y. Then $Y\backslash V$ is closed in Y.
Thus $f^{-1}(Y\backslash V) = f^{-1}(Y)\backslash f^{-1}(V) = X\backslash f^{-1}(V)$ is closed in X.
Hence $f^{-1}(V) = X\backslash(X\backslash f^{-1}(V))$ is open in X.
So for every open V in Y, $f^{-1}(V)$ is open in X.
By Theorem 2.1, page 63, f is continuous on X.

("Only if" part) Suppose that f is continuous.
Let F be closed in Y, that is, $Y\backslash F$ is open in Y.
Hence $f^{-1}(Y\backslash F) = f^{-1}(Y)\backslash f^{-1}(F) = X\backslash f^{-1}(F)$ is open in X.
Consequently, we have that $f^{-1}(F)$ is closed in X.

Solution to Exercise 2.8, page 64

If $x \in (g \circ f)^{-1}(W)$, then $(g \circ f)(x) \in W$, that is, $g(f(x)) \in W$. So $f(x) \in g^{-1}(W)$, that is, $x \in f^{-1}(g^{-1}(W))$. Thus $(g \circ f)^{-1}(W) \subset f^{-1}(g^{-1}(W))$.
If $x \in f^{-1}(g^{-1}(W))$, then $f(x) \in g^{-1}(W)$, that is, $(g \circ f)(x) = g(f(x)) \in W$. Hence $x \in (g \circ f)^{-1}(W)$. So we have $f^{-1}(g^{-1}(W)) \subset (g \circ f)^{-1}(W)$.
Consequently, $(g \circ f)^{-1}(W) = f^{-1}(g^{-1}(W))$.

Solution to Exercise 2.9, page 64

(1) True.
Since $(-\infty, 1)$ is open in $\mathbb{R}$ and $f : X \to \mathbb{R}$ is continuous, it follows that $\{x \in X : f(x) < 1\} = f^{-1}(-\infty, 1)$ is open in X by Theorem 2.1, page 63.

(2) True.
Because $(1, \infty)$ is open in $\mathbb{R}$, and $f : X \to \mathbb{R}$ is continuous, it follows by Theorem 2.1, page 63, that $\{x \in X : f(x) > 1\} = f^{-1}(1, \infty)$ is open in X.

(3) False.
Take for example $X = \mathbb{R}$ with the usual Euclidean norm, and consider the continuous function $f(x) = x$ for all $x \in \mathbb{R}$. Then $\{x \in X : f(x) = 1\} = \{1\}$, which is not open in $\mathbb{R}$.

(4) True.
$(-\infty, 1]$ is closed in $\mathbb{R}$ because its complement is $(1, \infty)$, which is open in $\mathbb{R}$. As $f : X \to \mathbb{R}$ is continuous, $\{x \in X : f(x) \leqslant 1\} = f^{-1}(-\infty, 1]$ is closed in X by Corollary 2.1, page 64.

(5) True.
Since $\{1\}$ is closed in $\mathbb{R}$ and since $f : X \to \mathbb{R}$ is continuous, it follows by Corollary 2.1, page 64, that $\{x \in X : f(x) = 1\} = f^{-1}\{1\}$ is closed in X.

(6) True.
Each of the sets $f^{-1}\{1\}$ and $f^{-1}\{2\}$ are closed, and so their finite union, namely $\{x \in X : f(x) = 1 \text{ or } 2\}$ is closed as well.

(7) False.
Take for example $X = \mathbb{R}$ with the usual Euclidean norm, and consider the continuous function $f(x) = 1$ $(x \in \mathbb{R})$. Then $\{x \in X : f(x) = 1\} = \mathbb{R}$, which is not bounded, and hence can't be compact.

Solution to Exercise 2.10, page 65

For all $x \in X$, we have $f(2x) = -f(x)$, and so
$$f(x) = -f\left(\frac{1}{2}x\right) = (-1)^2 f\left(\frac{1}{4}x\right) = \cdots = (-1)^n f\left(\frac{1}{2^n}x\right)$$
Since the sequence $\left(\frac{1}{2^n}x\right)_{n \in \mathbb{N}}$ converges to $\mathbf{0}$, it follows that
$$f(\mathbf{0}) = f\left(\lim_{n \to \infty} \frac{1}{2^n}x\right) = \lim_{n \to \infty} f\left(\frac{1}{2^n}x\right) = \lim_{n \to \infty} (-1)^n f(x).$$
So we obtain that $((-1)^n f(x))_{n \in \mathbb{N}}$ is convergent with limit $f(\mathbf{0})$. Thus the subsequence $(f(x))_{n \in \mathbb{N}} = ((-1)^{2n} f(x))_{n \in \mathbb{N}}$ of $((-1)^n f(x))_{n \in \mathbb{N}}$ is also convergent with limit $f(\mathbf{0})$. Hence $f(x) = f(\mathbf{0})$ for all $x \in X$. As $f(\mathbf{0}) = f(2 \cdot \mathbf{0}) = -f(\mathbf{0})$, it follows that $f(\mathbf{0}) = \mathbf{0}$. Hence $f(x) = \mathbf{0}$ for all $x \in X$. So if f is continuous and it satisfies the given identity then it must be the constant function $x \mapsto \mathbf{0} : X \to Y$.

Conversely, the constant function $x \mapsto \mathbf{0} : X \to Y$ is indeed continuous and also $f(2x) + f(x) = \mathbf{0} + \mathbf{0} = \mathbf{0}$ for all $x \in X$.

Solution to Exercise 2.11, page 65

The determinant of $\mathbf{M} = [m_{ij}]$ is given by the sum of expressions of the type
$$\pm m_{1p(1)} m_{2p(2)} m_{3p(3)} \cdots m_{np(n)},$$
where $p : \{1, 2, 3, \cdots, n\} \to \{1, 2, 3, \cdots, n\}$ is a permutation. Since each of the maps $\mathbf{M} \mapsto m_{1p(1)} m_{2p(2)} m_{3p(3)} \cdots m_{np(n)}$ is easily seen to be continuous using the characterisation of continuous functions provided by Theorem 2.3, page 64, it follows that their linear combination is also continuous.

$\{0\}$ is closed in $\mathbb{R}$, and so its inverse image $\det^{-1}\{0\} = \{\mathbf{M} \in \mathbb{R}^{n \times n} : \det \mathbf{M} = 0\}$ under the continuous map det is also closed. Thus its complement, namely the set $\{\mathbf{M} \in \mathbb{R}^{n \times n} : \det \mathbf{M} \neq 0\}$, is open. But this is precisely the set of invertible matrices, since $\mathbf{M} \in \mathbb{R}^{n \times n}$ is invertible if and only if $\det \mathbf{M} \neq 0$.

Solution to Exercise 2.12, page 73

We'd seen in Exercise 1.21, page 21, that a singleton set in any normed space is closed. So $\{\mathbf{0}\}$ is closed in $\mathbb{R}^m$. As the linear transformation $T_A : \mathbb{R}^n \to \mathbb{R}^m$ is continuous, its inverse image under T_A, $T_A^{-1}(\{\mathbf{0}\}) = \{\mathbf{x} \in \mathbb{R}^n : A\mathbf{x} = \mathbf{0}\} = \ker A$, is closed in $\mathbb{R}^n$.

Solution to Exercise 2.13, page 73

Let V be a subspace of $\mathbb{R}^n$, and let $\{\mathbf{v}_1, \cdots, \mathbf{v}_k\}$ be a basis for V. Extend this to a basis $\{\mathbf{v}_1, \cdots, \mathbf{v}_k, \mathbf{v}_{k+1}, \cdots, \mathbf{v}_n\}$ for $\mathbb{R}^n$. By using the Gram-Schmidt orthogonalisation procedure, we can find an orthonormal[5] set of vectors $\{\mathbf{u}_1, \cdots, \mathbf{u}_n\}$ such that for each $k \in \{1, \cdots, n\}$, the span of the vectors $\mathbf{v}_1, \cdots, \mathbf{v}_k$ coincides with the span of $\mathbf{u}_1, \cdots, \mathbf{u}_k$. Now define $A \in \mathbb{R}^{(n-k) \times n}$ as follows:

$$A = \begin{bmatrix} \mathbf{u}_{k+1}^\top \\ \vdots \\ \mathbf{u}_n^\top \end{bmatrix}.$$

It is clear from the orthonormality of the $\mathbf{u}_j$s that $A\mathbf{u}_1 = \cdots = A\mathbf{u}_k = \mathbf{0}$, and so it follows that also any linear combination of $\mathbf{u}_1, \cdots, \mathbf{u}_k$ lies in the kernel of A. In other words, $V \subset \ker A$.

On the other hand, if $\mathbf{x} = \alpha_1 \mathbf{u}_1 + \cdots + \alpha_n \mathbf{u}_n$, where $\alpha_1, \cdots, \alpha_n$ are scalars and if $A\mathbf{x} = \mathbf{0}$, then it follows that

$$\mathbf{0} = A\mathbf{x} = \begin{bmatrix} \mathbf{u}_{k+1}^\top(\alpha_1 \mathbf{u}_1 + \cdots + \alpha_n \mathbf{u}_n) \\ \vdots \\ \mathbf{u}_n^\top(\alpha_1 \mathbf{u}_1 + \cdots + \alpha_n \mathbf{u}_n) \end{bmatrix} = \begin{bmatrix} \alpha_{k+1} \\ \vdots \\ \alpha_n \end{bmatrix}.$$

So $\mathbf{x} = \alpha_1 \mathbf{u}_1 + \cdots + \alpha_k \mathbf{u}_k \in V$. Hence $\ker A \subset V$.

Consequently $V = \ker A$, and by the result of the previous exercise, it now follows that V is closed.

Solution to Exercise 2.14, page 73

(1) The linearity of T follows immediately from the properties of the Riemann integral. Continuity follows from the straightforward estimate

$$|T\mathbf{x}| = \left| \int_a^b \mathbf{x}(t) dt \right| \leq \int_a^b |\mathbf{x}(t)| dt \leq \int_a^b \|\mathbf{x}\|_\infty dt = (b-a)\|\mathbf{x}\|_\infty.$$

(2) The partial sums $\mathbf{s}_n$ of the series converge to $\mathbf{f}$. Thus, since the continuous map T preserves convergent sequences, it follows that

$$\int_a^b \mathbf{f}(t) dt = T\mathbf{f} = \lim_{n \to \infty} T\mathbf{s}_n = \lim_{n \to \infty} \int_a^b \mathbf{s}_n(t) dt$$
$$= \lim_{n \to \infty} \sum_{k=1}^n \int_a^b \mathbf{f}_k(t) dt$$
$$= \sum_{k=1}^\infty \int_a^b \mathbf{f}_k(t) dt.$$

---

[5] $\mathbf{u}_i^\top \mathbf{u}_j = 0$, unless $i = i$, in which case $\mathbf{u}_i^\top \mathbf{u}_i = 1$. Here $\cdot^\top$ denotes transpose.

Solution to Exercise 2.15, page 73

We have for all $t \in \mathbb{R}$ that

$$|(\mathbf{f} * \mathbf{g})(t)| = \left| \int_{-\infty}^{\infty} \mathbf{f}(t-\tau)\mathbf{g}(\tau)d\tau \right| \leq \int_{-\infty}^{\infty} |\mathbf{f}(t-\tau)||\mathbf{g}(\tau)|d\tau$$

$$\leq \int_{-\infty}^{\infty} |\mathbf{f}(t-\tau)|\|\mathbf{g}\|_\infty d\tau = \|\mathbf{g}\|_\infty \int_{-\infty}^{\infty} |\mathbf{f}(t-\tau)|d\tau$$

$$= \|\mathbf{g}\|_\infty \int_{-\infty}^{\infty} |\mathbf{f}(\sigma)|d\sigma \text{ (using the substitution } \sigma = t - \tau)$$

$$= \|\mathbf{g}\|_\infty \|\mathbf{f}\|_1 < \infty.$$

Thus $\|\mathbf{f} * \mathbf{g}\|_\infty \leq \|\mathbf{g}\|_\infty \|\mathbf{f}\|_1$ for all $\mathbf{g} \in L^\infty(\mathbb{R})$. So $\mathbf{f}*$ is well-defined. Linearity is easy to see. From the above estimate, it follows that the linear transformation $f*$ is continuous as well.

Solution to Exercise 2.16, page 73

Consider the reflection map $\mathbf{f} \stackrel{R}{\mapsto} \check{\mathbf{f}} : L^2(\mathbb{R}) \to L^2(\mathbb{R})$. Then it is straightforward to check that $R \in L(L^2(\mathbb{R}))$, and moreover it is continuous since $\|\mathbf{f}\|_2 = \|\check{\mathbf{f}}\|_2$ for all $\mathbf{f} \in L^2(\mathbb{R})$. Clearly $Y = \ker(I - R)$, and so, being the inverse image of the closed set $\{\mathbf{0}\}$ under the continuous map $I - R$, it follows that Y is closed.

Solution to Exercise 2.17, page 76

For $(a_n)_{n\in\mathbb{N}}$, $\|\Lambda(a_n)_{n\in\mathbb{N}}\|_2^2 = \sum_{n=1}^{\infty} |\lambda_n a_n|^2 = \sum_{n=1}^{\infty} |\lambda_n|^2 |a_n|^2$

$$\leq \sum_{n=1}^{\infty} \left(\sup_{n\in\mathbb{N}} |\lambda_n|\right)^2 |a_n|^2 = \left(\sup_{n\in\mathbb{N}} |\lambda_n|\right)^2 \|(a_n)_{n\in\mathbb{N}}\|_2^2,$$

and so $\Lambda \in CL(\ell^2)$ and $\|\Lambda\| \leq \sup_{n\in\mathbb{N}} |\lambda_n|$.

Moreover, for $\ell^2 \ni \mathbf{e}_n := (0, \cdots, 0, 1, 0, \cdots) \in \ell^2$ (sequence with all terms equal to 0 and nth term equal to 1), we have

$$\|\Lambda\| = \|\Lambda\| \cdot 1 = \|\Lambda\|\|\mathbf{e}_n\|_2 \geq \|\Lambda\mathbf{e}_n\|_2 = \|(0, \cdots, 0, \lambda_n \cdot 1, 0, \cdots)\|_2 = |\lambda_n|,$$

for all n, and so $\|\Lambda\|$ is an upper bound for $\{|\lambda_n| : n \in \mathbb{N}\}$. Hence $\|\Lambda\| \geq \sup_{n\in\mathbb{N}} |\lambda_n|$.
From the above, it now follows that $\|\Lambda\| = \sup_{n\in\mathbb{N}} |\lambda_n|$.

If $\lambda_n = 1 - \dfrac{1}{n}$, $n \in \mathbb{N}$, then $\|\Lambda\| = \sup_{n\in\mathbb{N}} \left(1 - \dfrac{1}{n}\right) = 1$.

Suppose that $\mathbf{x} = (a_n)_{n\in\mathbb{N}} \in \ell^2$ is such that $\|\mathbf{x}\|_2 \leq 1$ and $\|\Lambda\mathbf{x}\|_2 = \|\Lambda\| = 1$. If $0 = a_2 = a_3 = \cdots$, then $\Lambda\mathbf{x} = \mathbf{0}$, and this contradicts the fact that $\|\Lambda\mathbf{x}\|_2 = 1$. So at least one of the terms $a_2, a_3, \cdots$ must be nonzero.

But then $1 \geqslant \|\mathbf{x}\|_2^2 = |a_1|^2 + |a_2|^2 + |a_3|^2 + \cdots \geqslant |a_2^2| + |a_3|^2 + \cdots$
$$> \frac{1}{4}|a_2|^2 + \frac{1}{9}|a_3|^2 + \cdots = \|\Lambda \mathbf{x}\|_2^2 = 1,$$
a contradiction. So the operator norm is not attained for this particular Λ.

Solution to Exercise 2.18, page 76

Let $\mathbf{x} = (x_n)_{n\in\mathbb{N}} \in \ell^p$, and let $\epsilon > 0$.

Then there exists an N such that $\sum_{k=N+1}^{\infty} |x_k|^p < \epsilon^p$. Let $\mathbf{s}_n := \sum_{k=1}^{n} x_k \mathbf{e}_k$.

Then for $n > N$, $\mathbf{x} - \mathbf{s}_n = (0, \cdots, 0, x_{n+1}, x_{n+2}, x_{n+3}, \cdots)$.

So $\|\mathbf{x} - \mathbf{s}_n\|_p^p = \sum_{k=n+1}^{\infty} |x_k|^p \leqslant \sum_{k=N+1}^{\infty} |x_k|^p < \epsilon^p$, giving $\|\mathbf{x} - \mathbf{s}_n\|_p < \epsilon$.

So $(\mathbf{s}_n)_{n\in\mathbb{N}}$ converges in ℓ^p to $\mathbf{x}$, that is, $\mathbf{x} = \sum_{n=1}^{\infty} x_n \mathbf{e}_n$.

The map $\mathbf{x} = (x_1, x_2, x_3, \cdots) \stackrel{\varphi_n}{\mapsto} x_n : \ell^p \to \mathbb{K}$ is easily seen to be linear.

It's continuous as for all $\mathbf{x} \in \ell^p$, $|\varphi_n(\mathbf{x})| = |x_n| = (|x_n|^p)^{1/p} \leqslant \Big(\sum_{k=1}^{\infty} |x_k|^p\Big)^{1/p} = \|\mathbf{x}\|_p$.

If $\mathbf{x} = \sum_{k=1}^{\infty} \xi_k \mathbf{e}_k = \sum_{k=1}^{\infty} \xi'_k \mathbf{e}_k$, where the ξ_is and ξ'_is are scalars, then applying φ_n,

$$\xi_n = \sum_{k=1}^{\infty} \xi_k \varphi_n(\mathbf{e}_k) = \varphi_n\Big(\sum_{k=1}^{\infty} \xi_k \mathbf{e}_k\Big) = \varphi_n(\mathbf{x}) = \varphi_n\Big(\sum_{k=1}^{\infty} \xi'_k \mathbf{e}_k\Big) = \sum_{k=1}^{\infty} \xi'_k \varphi_n(\mathbf{e}_k) = \xi'_n.$$

As the choice of n was arbitrary, $\xi_n = \xi'_n$ for all n.

Solution to Exercise 2.19, page 77

(1) Let $\mathbf{x} = (x_n)_{n\in\mathbb{N}} \in \ell^\infty$. Then for all $n \in \mathbb{N}$, $|x_n| \leqslant \|\mathbf{x}\|_\infty$.

Thus $\Big|\dfrac{x_1 + \cdots + x_n}{n}\Big| \leqslant \dfrac{|x_1| + \cdots + |x_n|}{n} \leqslant \dfrac{n\|\mathbf{x}\|_\infty}{n} = \|\mathbf{x}\|_\infty$.

Consequently $A\mathbf{x} \in \ell^\infty$. So A is a well-defined map.

The linearity is easy to check.

Also, we see that for all $\mathbf{x} \in \ell^\infty$ that $\|A\mathbf{x}\|_\infty = \sup_{n\in\mathbb{N}} \Big|\dfrac{x_1 + \cdots + x_n}{n}\Big| \leqslant \|\mathbf{x}\|_\infty$.

So $A \in CL(\ell^\infty)$, and $\|A\| \leqslant 1$. Also, with $\mathbf{1} := (1, 1, 1, \cdots) \in \ell^\infty$, we have
$$\|A\| = \|A\| \cdot 1 = \|A\|\|\mathbf{1}\|_\infty \geqslant \|A\mathbf{1}\|_\infty = \|(1, 1, 1, \cdots)\|_\infty = 1.$$
Consequently, $\|A\| = 1$.

(2) Let $\mathbf{x} = (x_n)_{n\in\mathbb{N}} \in c$, and let its limit be denoted by L.
We'll show that $A\mathbf{x} \in c$ as well.

We will prove that $A\mathbf{x}$ is convergent with the same limit L! (Intuitively, this makes sense since for large n, all x_ns look alike, $\approx L$, and the average of these is approximately L, since the first few terms do not "contribute much" if we take a large collection to take an average.)

Let $\epsilon > 0$. Then there exists an $N_1 \in \mathbb{N}$ such that for all $n > N_1$, $|x_n - L| < \epsilon/2$. Since $(x_n)_{n \in \mathbb{N}}$ is convergent, it is bounded, and so there exists an $M > 0$ such that for all $n \in \mathbb{N}$, $|a_n| \leq M$.

Choose $N \in \mathbb{N}$ such that $N > \max\left\{N_1, \dfrac{N_1(M+|L|)}{\epsilon/2}\right\}$.

(This ghastly choice of N is arrived at by working backwards. Since we wish to make $\left|\dfrac{x_1 + \cdots + x_n}{n} - L\right|$ less than ϵ for $n > N$, we manipulate this, as shown in the chain of inequalities below, and then choose N large enough to achieve this.)

So $N > N_1$ and $\dfrac{N_1(M+|L|)}{N} < \dfrac{\epsilon}{2}$. Then for all $n > N$, we have:

$$\left|\frac{x_1 + \cdots + x_{N_1} + x_{N_1+1} + \cdots + x_n}{n} - L\right| = \left|\frac{x_1 + \cdots + x_{N_1} + x_{N_1+1} + \cdots + x_n - nL}{n}\right|$$

$$= \frac{|x_1 + \cdots + x_{N_1} + x_{N_1+1} + \cdots + x_n - nL|}{n}$$

$$\leq \frac{|x_1 - L| + \cdots + |x_{N_1} - L| + |x_{N_1+1} - L| + \cdots + |x_n - L|}{n}$$

$$\leq \frac{(|x_1| + |L| + \cdots + |x_{N_1}| + |L|) + \epsilon/2 + \cdots + \epsilon/2}{n}$$

$$\leq \frac{N_1(M+|L|) + (n - N_1)(\epsilon/2)}{n}$$

$$\leq \frac{N_1(M+|L|)}{n} + \left(1 - \frac{N_1}{n}\right) \cdot \frac{\epsilon}{2}$$

$$< \frac{N_1(M+|L|)}{N} + 1 \cdot \frac{\epsilon}{2} < \frac{\epsilon}{2} + \frac{\epsilon}{2} = \epsilon.$$

So $\left(\dfrac{x_1 + \cdots + x_n}{n}\right)_{n \in \mathbb{N}}$ is a convergent sequence with limit L.

Hence $A\mathbf{x} \in c$. Consequently $Ac \subset c$, and c is an invariant subspace of A.

Solution to Exercise 2.20, page 85

(If part:) Since $\inf_{n \in \mathbb{N}} |\lambda_n| > 0$, we have $|\lambda_k| \geq \inf_{n \in \mathbb{N}} |\lambda_n| > 0$, and so $\lambda_k \neq 0$ for all k.

Moreover, $\sup_{n \in \mathbb{N}} \left|\dfrac{1}{\lambda_k}\right| = \sup_{n \in \mathbb{N}} \dfrac{1}{|\lambda_k|} = \dfrac{1}{\inf_{n \in \mathbb{N}} |\lambda_n|} < \infty$, and so $V : \ell^2 \to \ell^2$ given by

$$V(a_1, a_2, a_3, \cdots) = \left(\frac{a_1}{\lambda_1}, \frac{a_2}{\lambda_2}, \frac{a_3}{\lambda_3}, \cdots\right), \quad (a_n)_{n \in \mathbb{N}} \in \ell^2,$$

belongs to $CL(\ell^2)$. Moreover for all $(a_n)_{n \in \mathbb{N}}$ we have

$$V\Lambda(a_n)_{n \in \mathbb{N}} = V(\lambda_n a_n)_{n \in \mathbb{N}} = \left((\lambda_n a_n)/\lambda_n\right)_{n \in \mathbb{N}} = (a_n)_{n \in \mathbb{N}} = I(a_n)_{n \in \mathbb{N}}$$
$$= \left(\lambda_n (a_n/\lambda_n)\right)_{n \in \mathbb{N}} = \Lambda(a_n/\lambda_n)_{n \in \mathbb{N}} = \Lambda V(a_n)_{n \in \mathbb{N}},$$

and so $V\Lambda = I = \Lambda V$. Hence Λ is invertible in $CL(\ell^2)$, with $\Lambda^{-1} = V$.
(Only if part:) Let Λ be invertible in $CL(\ell^2)$. Then there exists a $\Lambda^{-1} \in CL(\ell^2)$ such that $\Lambda^{-1}\Lambda = I = \Lambda\Lambda^{-1}$. So $\|\mathbf{x}\|_2 = \|\Lambda^{-1}\Lambda\mathbf{x}\|_2 \leqslant \|\Lambda^{-1}\|\|\Lambda\mathbf{x}\|_2$, for all $\mathbf{x} \in \ell^2$.
Hence $\|\Lambda\mathbf{x}\|_2 \geqslant \dfrac{1}{\|\Lambda^{-1}\|}\|\mathbf{x}\|_2$ for all $\mathbf{x} \in \ell^2$. So with $\mathbf{x} := \mathbf{e}_k$ (kth term 1, others 0),

$$|\lambda_k| = \|\Lambda\mathbf{e}_k\|_2 \geqslant \frac{1}{\|\Lambda^{-1}\|}\|\mathbf{e}_k\|_2 = \frac{1}{\|\Lambda^{-1}\|} \text{ for all } k \in \mathbb{N}.$$

Thus $\inf\limits_{k \in \mathbb{N}} |\lambda_k| \geqslant \dfrac{1}{\|\Lambda^{-1}\|} > 0$.

Solution to Exercise 2.21, page 86

We have
$(I + BA)(I - B(I + AB)^{-1}A)$
$= I - B(I + AB)^{-1}A + BA - BAB(I + AB)^{-1}A$
$= I - B(I + AB)^{-1}A + BA - B(I + AB - I)(I + AB)^{-1}A$
$= I - \cancel{B(I+AB)^{-1}A} + BA - B(I + AB)(I + AB)^{-1}A + \cancel{B(I+AB)^{-1}A}$
$= I + BA - BIA = I.$

Similarly,
$(I - B(I + AB)^{-1}A)(I + BA) = I + BA - B(I + AB)^{-1}A - B(I + AB)^{-1}ABA$
$= I + BA - B(I + AB)^{-1}A - B(I + AB)^{-1}(I + AB - I)A$
$= I + \cancel{BA} - \cancel{B(I+AB)^{-1}A} - \cancel{BA} + B(I + AB)^{-1}A = I.$

Solution to Exercise 2.22, page 86

(1) If there exist matrices A, B such that $AB - BA = I$, then
$$n = \operatorname{tr}(I) = \operatorname{tr}(AB - BA) = \operatorname{tr}(AB) - \operatorname{tr}(BA) = 0,$$
a contradiction.

(2) If $n = 1$, then $AB^n - B^nA = AB - BA = I = 1 \cdot B^0 = nB^{n-1}$.
If for some $n \in \mathbb{N}$, we have $AB^n - B^nA = nB^{n-1}$, then
$AB^{n+1} - B^{n+1}A = AB^n \cdot B - B^n \cdot BA$
$= (nB^{n-1} + B^nA)B - B^n \cdot BA$
$= nB^n + B^n \cdot AB - B^n \cdot BA = nB^n + B^n(AB - BA)$
$= nB^n + B^nI = nB^n + B^n = (n+1)B^n = (n+1)B^{(n+1)-1},$

and so the result follows by induction.

Suppose that $AB - BA = I$. Then for all $n \in \mathbb{N}$, $AB^n - B^nA = nB^{n-1}$.
Taking operator norm on both sides yields
$$n\|B^{n-1}\| = \|AB^{n-1}B - BB^{n-1}A\| \leqslant 2\|A\|\|B\|\|B^{n-1}\|. \tag{7.3}$$

We claim that $B^{n-1} \neq \mathbf{0}$ for all $n \in \mathbb{N}$. Indeed, if $n = 1$, then $B^0 := I \neq \mathbf{0}$.
If $B^{n-1} \neq \mathbf{0}$ for some $n \in \mathbb{N}$, then $B^n = \mathbf{0}$ gives the contradiction that
$$\mathbf{0} = A\mathbf{0} - \mathbf{0}A = AB^n - B^nA = nB^{n-1} \neq \mathbf{0},$$

and so we must have $B^n \neq \mathbf{0}$ too. By induction, our claim is proved. Thus in (7.3), we may cancel $\|B^{n-1}\| > 0$ on both sides of the inequality, obtaining $n \leq 2\|A\|\|B\|$ for all $n \in \mathbb{N}$, which is absurd. Consequently, our original assumption that $AB - BA = I$ must be false.

(3) If $\Psi \in C^{\infty}(\mathbb{R})$, then
$$(AB - BA)\Psi = A(B\Psi) - B(A\Psi) = (x\Psi)' - x(\Psi')$$
$$= x\Psi' + \Psi - x\Psi' = \Psi = I\Psi,$$

and so $AB - BA = I$.

Solution to Exercise 2.23, page 87

(1) For $\mathbf{x} = (x_1, x_2) \in \mathbb{R}^2$, we have, using the Cauchy-Schwarz inequality, that
$$\|K\mathbf{x}\|_2^2 = \left(\frac{1}{2}x_1 + \frac{1}{3}x_2\right)^2 + \left(\frac{1}{3}x_1 + \frac{1}{4}x_2\right)^2$$
$$\leq \left(\frac{1}{4} + \frac{1}{9}\right)(x_1^2 + x_2^2) + \left(\frac{1}{9} + \frac{1}{16}\right)(x_1^2 + x_2^2)$$
$$= \left(\frac{1}{4} + \frac{1}{9} + \frac{1}{9} + \frac{1}{16}\right)\|\mathbf{x}\|_2^2.$$

So $\|K\| \leq \sqrt{\frac{1}{4} + \frac{1}{9} + \frac{1}{9} + \frac{1}{16}} \leq \sqrt{\frac{1+16+16+9}{9(16)}} < \sqrt{\frac{4(16)}{9(16)}} = \sqrt{\frac{4}{9}} = \frac{2}{3} < 1$.

By the Neumann Series Theorem, $(I - K)^{-1}$ exists in $CL(\mathbb{R}^2)$.
So there is a unique solution $\mathbf{x} \in \mathbb{R}^2$ to $(I-K)\mathbf{x} = \mathbf{y}$, given by $\mathbf{x} = (I-K)^{-1}\mathbf{y}$.

(2) We have $I - K = \begin{bmatrix} 1/2 & -1/3 \\ -1/3 & 3/4 \end{bmatrix}$, and so $(I - K)^{-1} = \frac{72}{19}\begin{bmatrix} 3/4 & 1/3 \\ 1/3 & 1/2 \end{bmatrix}$.

Thus $\mathbf{x} = (I-K)^{-1}\mathbf{y} = \frac{72}{19}\begin{bmatrix} 3/4 & 1/3 \\ 1/3 & 1/2 \end{bmatrix}\begin{bmatrix} 1 \\ 2 \end{bmatrix} = \frac{72}{19}\begin{bmatrix} 17/12 \\ 4/3 \end{bmatrix}$.

(3) A computer program yielded the following numerical values:

n	approximate solution $\mathbf{x}_n = (I + K + K^2 + \cdots + K^n)\mathbf{y}$	relative error (%) $\|\mathbf{x} - \mathbf{x}_n\|_2/\|\mathbf{x}\|_2$
2	(3.0278, 3.4306)	38.63
3	(3.6574, 3.8669)	28.24
5	(4.4541, 4.4190)	15.09
10	(5.1776, 4.9204)	3.15
15	(5.3286, 5.0250)	0.66
20	(5.3601, 5.0469)	0.14
25	(5.3667, 5.0514)	0.03
30	(5.3681, 5.0524)	0.01

Solution to Exercise 2.24, page 88

If $n = 1$, then $(I - A)P_1 = (I - A)(I + A)(I + A^2) = I - A^4 = I - A^{2^{1+1}}$.

If the claim is true for some $k \in \mathbb{N}$, then
$$(I - A)P_{k+1} = (I - A)P_k(I + A^{2^{k+1}}) = (I - A^{2^{k+1}})(I + A^{2^{k+1}})$$
$$= I - (A^{2^{k+1}})^2 = I - A^{2^{k+1} \cdot 2} = I - A^{2^{k+2}} = I - A^{2^{(k+1)+1}}.$$
So the claim follows by induction for all $n \in \mathbb{N}$.

$(I - A^{2^{n+1}})_{n \in \mathbb{N}}$ converges to I in $\mathcal{L}(X)$ since $\|A\| < 1$ and
$$\|I - A^{2^{n+1}} - I\| = \|-A^{2^{n+1}}\| = \|A^{2^{n+1}}\| \leqslant \|A\|^{2^{n+1}} \xrightarrow{n \to \infty} 0.$$
Also, since $\|A\| < 1$, $I - A$ is invertible in $CL(X)$. We have
$$\|(I - A)^{-1}(I - A^{2^{n+1}}) - (I - A)^{-1}\| \leqslant \|(I - A)^{-1}\| \|(I - A^{2^{n+1}}) - I\| \xrightarrow{n \to \infty} 0,$$
and so $\left((I - A)^{-1}(I - A^{2^{n+1}})\right)_{n \in \mathbb{N}} = \left((I - A)^{-1}(I - A)P_n\right)_{n \in \mathbb{N}} = (P_n)_{n \in \mathbb{N}}$ is convergent with limit $(I - A)^{-1}$.

Solution to Exercise 2.25, page 88

(1) Let $T_0 \in GL(X)$. Then $T_0^{-1} \in CL(X)$, and also $r := \|T_0^{-1}\| \neq 0$.

If $T \in B\left(T_0, \dfrac{1}{2\|T_0^{-1}\|}\right)$, then $\|T - T_0\| < \dfrac{1}{2\|T_0^{-1}\|}$, and in particular,
$$\|(T - T_0)T_0^{-1}\| \leqslant \|T - T_0\| \|T_0^{-1}\| < \frac{1}{2\|T_0^{-1}\|} \|T_0^{-1}\| = \frac{1}{2} < 1,$$
and so by the Neumann Series Theorem, $I + (T - T_0)T_0^{-1}$ belongs to $GL(X)$. But as $T_0 \in GL(X)$ too, it now follows that
$$T = T_0 + (T - T_0) = T_0\left(I + (T - T_0)T_0^{-1}\right) \in GL(X).$$
This completes the proof that $GL(X)$ is an open subset of $CL(X)$.

(2) Let $T_0 \in GL(X)$ and $\epsilon > 0$. Set $\delta := \min\left\{\dfrac{\epsilon}{2\|T_0\|^2}, \dfrac{1}{2\|T_0^{-1}\|}\right\}$.

Let $T \in CL(X)$ be such that $\|T - T_0\| < \delta$.

Then in particular $\|T - T_0\| < \dfrac{1}{2\|T_0^{-1}\|}$ and so by part (1), $T \in GL(X)$, with
$$T^{-1} = \left(I + (T - T_0)T_0^{-1}\right)^{-1} T_0^{-1}.$$
Moreover, we have
$$T^{-1} - T_0^{-1} = T^{-1}(T_0 - T)T_0^{-1} = \left(I + (T - T_0)T_0^{-1}\right)^{-1} T_0^{-1}(T_0 - T)T_0^{-1}.$$
Thus using the estimate from the Neumann Series Theorem,
$$\|T^{-1} - T_0^{-1}\| \leqslant \frac{1}{1 - \|(T - T_0)T_0^{-1}\|} \|T_0^{-1}\|^2 \|T_0 - T\|$$
$$\leqslant \frac{\|T_0^{-1}\|^2 \|T_0 - T\|}{1 - \|(T - T_0)\| \|T_0^{-1}\|} < \frac{\|T_0^{-1}\|^2 \delta}{1 - \delta \|T_0^{-1}\|}$$
$$< \frac{\|T_0^{-1}\|^2 \frac{\epsilon}{2\|T_0\|^2}}{1 - \frac{1}{2\|T_0^{-1}\|} \|T_0^{-1}\|} = \epsilon.$$

Solution to Exercise 2.26, page 92

$A^2 = B^2 = 0$, and so A, B are nilpotent.

Hence $e^A = I + A = \begin{bmatrix} 1 & 1 \\ 0 & 1 \end{bmatrix}$, and $e^B = I + B = \begin{bmatrix} 1 & 0 \\ -1 & 1 \end{bmatrix}$.

We note that $AB = \begin{bmatrix} -1 & 0 \\ 0 & 0 \end{bmatrix} \neq \begin{bmatrix} 0 & 0 \\ 0 & -1 \end{bmatrix} = BA$.

Also, $e^A e^B = \begin{bmatrix} 1 & 1 \\ 0 & 1 \end{bmatrix} \begin{bmatrix} 1 & 0 \\ -1 & 1 \end{bmatrix} = \begin{bmatrix} 0 & 1 \\ -1 & 1 \end{bmatrix}$.

We have $A + B = \begin{bmatrix} 0 & 1 \\ -1 & 0 \end{bmatrix}$, and $(A+B)^2 = \begin{bmatrix} -1 & 0 \\ 0 & -1 \end{bmatrix} = -I$. Thus

$$e^{A+B} = I + \frac{A+B}{1!} + \frac{(A+B)^2}{2!} + \frac{(A+B)^3}{3!} + \frac{(A+B)^4}{4!} + \cdots$$

$$= I + \frac{A+B}{1!} - \frac{I}{2!} - \frac{A+B}{3!} + \frac{I}{4!} + \cdots$$

$$= \Big(1 - \frac{1}{2!} + \frac{1}{4!} - + \cdots\Big)I + \Big(\frac{1}{1!} - \frac{1}{3!} + \frac{1}{5!} - + \cdots\Big)(A+B)$$

$$= (\cos 1)I + (\sin 1)(A+B),$$

and so $e^{A+B} = \begin{bmatrix} \cos 1 & \sin 1 \\ -\sin 1 & \cos 1 \end{bmatrix} \neq \begin{bmatrix} 0 & 1 \\ -1 & 1 \end{bmatrix} = e^A e^B$.

Solution to Exercise 2.27, page 94

Suppose that the Banach space has an infinite countable Hamel basis $\{x_1, x_2, x_3, \cdots\}$. We can ensure that for all $n \in \mathbb{N}$, we have $\|x_n\| = 1$. Let $F_n := \mathrm{span}\{x_1, x_2, \cdots, x_n\}$. Then each F_n is a finite dimensional normed space (with the induced norm from X), and so it is a Banach space. It follows that F_n is a closed subspace of X. By the Baire Lemma, there is an $n \in \mathbb{N}$ such that F_n contains an open set U, and in particular, an open ball $B(x, 2r)$ for some $r > 0$. The vector $y := rx_{n+1} + x$ belongs to $B(x, 2r)$ since $\|y - x\| = \|rx_{n+1}\| = r < 2r$. Since $y, x \in B(x, 2r) \subset F_n$, and as F_n is a subspace, we conclude that $(y - x)/r \in F_n$ too, that is, $x_{n+1} \in F_n = \mathrm{span}\{x_1, \cdots, x_n\}$, a contradiction.

Solution to Exercise 2.28, page 96

In light of the Open Mapping Theorem, such a function must necessarily be nonlinear. If the function is constant on an open interval I, then the image $f(I)$ will be a singleton, which is not closed. The following function does the job:

$$f(x) = \begin{cases} x + 1 & \text{if } x \leqslant -1, \\ 0 & \text{if } -1 < x < 1, \\ x - 1 & \text{if } x \geqslant 1. \end{cases}$$

If $I := (-1, 1)$, then $f(I) = \{0\}$, which is not open. f is surjective and continuous, and its graph is depicted in the following picture.

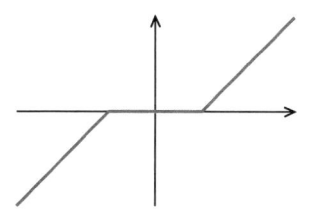

Solution to Exercise 2.29, page 96

From Exercise 1.38, page 44, $X \times Y$ is a Banach space. Since $\mathcal{G}(T)$ is a closed subspace of the Banach space $X \times Y$, it is a Banach space too. Let us now consider the map $p : \mathcal{G}(T) \to X$ defined by $p(x, Tx) = x$ for $x \in X$. Then p is a linear transformation:

$$p\big((x_1, Tx_1) + (x_2, Tx_2)\big) = p\big(x_1 + x_2, T(x_1 + x_2)\big) = x_1 + x_2$$
$$= p(x_1, Tx_1) + p(x_2, Tx_2),$$
$$p\big(\alpha(x, Tx)\big) = p\big(\alpha x, T(\alpha x)\big) = \alpha x = \alpha p(x, Tx),$$

for $\alpha \in \mathbb{K}$, $x, x_1, x_2 \in X$. Moreover, p continuous because
$$\|p(x, Tx)\| = \|x\| \leq \max\{\|x\|, \|Tx\|\} = \|(x, Tx)\|, \quad x \in X.$$
p is also injective since if $p(x, Tx) = \mathbf{0}$, then $x = \mathbf{0}$.
Furthermore, if $x \in X$, then $x = p(x, Tx)$, showing that p is surjective too.
Thus, $p \in CL\big(\mathcal{G}(T), X\big)$ is bijective, and hence invertible in $CL\big(\mathcal{G}(T), X\big)$, with inverse $p^{-1} \in CL\big(X, \mathcal{G}(T)\big)$. Hence for all $x \in X$,
$$\|Tx\| \leq \max\{\|x\|, \|Tx\|\} = \|(x, Tx)\| = \|p^{-1}x\| \leq \|p^{-1}\|\|x\|,$$
showing that $T \in CL(X, Y)$.

Solution to Exercise 2.30, page 102

We have

$$(\lambda I - T)^{-1} - (\mu I - T)^{-1} = (\lambda I - T)^{-1}\big(I - (\lambda I - T)(\mu I - T)^{-1}\big)$$
$$= (\lambda I - T)^{-1}\big((\mu I - T) - (\lambda I - T)\big)(\mu I - T)^{-1}$$
$$= (\lambda I - T)^{-1}(\mu I - \lambda I)(\mu I - T)^{-1}$$
$$= (\lambda I - T)^{-1}(\mu - \lambda)I(\mu I - T)^{-1}$$
$$= (\mu - \lambda)(\lambda I - T)^{-1}(\mu I - T)^{-1}.$$

Solution to Exercise 2.31, page 102

(1) We know that $\sigma(T) \subset \{\lambda \in \mathbb{C} : |\lambda| \leq \|T\|\}$, and so $\|T\|$ is an upper bound for $\{|\lambda| : \lambda \in \sigma(T)\}$. Thus $r_\sigma(T) = \sup_{\lambda \in \sigma(T)} |\lambda| \leq \|T\|$.

(2) We have $\sigma(T_A) = \{\text{eigenvalues of } A\} = \{1\}$, and so $r_\sigma(T_A) = 1$.

On the other hand, with $\mathbf{x}_1 := \left(\frac{1}{\sqrt{2}}, \frac{1}{\sqrt{2}}\right) \in \mathbb{R}^2$, we have $\|\mathbf{x}_1\|_2 = 1$, and so

$$\|T_A\| = \sup_{\|\mathbf{x}\| \leqslant 1} \|T\mathbf{x}\|_2 \geqslant \|T\mathbf{x}_1\|_2 = \sqrt{\left(\frac{1}{\sqrt{2}} + \frac{1}{\sqrt{2}}\right)^2 + \left(\frac{1}{\sqrt{2}}\right)^2} > 1 = r_\sigma(T_A).$$

Solution to Exercise 2.32, page 103

Suppose that $\lambda^2 \notin \sigma(T^2)$. Then $\lambda^2 \in \rho(T^2)$, that is, $\lambda^2 I - T^2$ is invertible in $CL(X)$. From the identity $(\lambda^2 I - T^2) = (\lambda I - T)(\lambda I + T) = (\lambda I + T)(\lambda I - T)$, we then obtain $(\lambda I - T) \underbrace{(\lambda I + T)(\lambda^2 I - T^2)^{-1}}_{=:P} = I = \underbrace{(\lambda^2 I - T^2)^{-1}(\lambda I + T)}_{=:Q}(\lambda I - T)$.
But then $Q = QI = Q(\lambda I - T)P = IP = P$, and so $P = Q \in CL(X)$ is the inverse of $\lambda I - T$, a contradiction to the fact that $\lambda \in \sigma(T)$.

Solution to Exercise 2.33, page 103

If $\mathbf{e}_n \in \ell^2$ denotes the sequence with the nth term equal to 1, and all others equal to 0, then $\Lambda \mathbf{e}_n = \lambda_n \mathbf{e}_n$, and so each λ_n is an eigenvalue of Λ with eigenvector $\mathbf{e}_n \neq \mathbf{0}$. Thus $\{\lambda_n : n \in \mathbb{N}\} \subset \sigma_p(\Lambda)$.
Next we will show that $\sigma(\Lambda) \subset \{\lambda_n : n \in \mathbb{N}\} \bigcup \{0\}$. To this end, suppose that $\mu \notin \{\lambda_n : n \in \mathbb{N}\} \bigcup \{0\}$. Then we claim that $\mu I - \Lambda$ is invertible in $CL(\ell^2)$. By a previous exercise, we know that in order to show the invertibility of

$$\mu I - \Lambda = \begin{bmatrix} \mu - \lambda_1 & & & \\ & \mu - \lambda_2 & & \\ & & \mu - \lambda_3 & \\ & & & \ddots \end{bmatrix},$$

it is enough to show that $|\mu - \lambda_n|$ is bounded away from 0. To see this, note that since $\lambda_n \stackrel{n \to \infty}{\longrightarrow} 0$, there is an N large enough such that $|\lambda_n| < |\mu|/2$ for all $n > N$, and so

$$|\mu - \lambda_n| \geqslant |\mu| - |\lambda_n| > |\mu| - \frac{|\mu|}{2} = \frac{|\mu|}{2} > 0.$$

But also $|\mu - \lambda_1|, \cdots, |\mu - \lambda_N|$ are all positive, so that we do have

$$\inf_{n \in \mathbb{N}} |\mu - \lambda_n| \geqslant \min\left\{|\mu - \lambda_1|, \cdots, |\mu - \lambda_N|, \frac{|\mu|}{2}\right\} > 0.$$

Hence $\mu I - \Lambda \in CL(\ell^2)$ is invertible in $CL(\ell^2)$, that is, $\mu \in \rho(\Lambda)$.
Thus $\sigma(\Lambda) \subset \{\lambda_n : n \in \mathbb{N}\} \bigcup \{0\}$.
But the spectrum $\sigma(\Lambda)$ is closed, and since it contains $\sigma_p(\Lambda) \supset \{\lambda_n : n \in \mathbb{N}\}$, it must contain the limit of $(\lambda_n)_{n \in \mathbb{N}}$, which is $\{0\}$.
So we also obtain $\{\lambda_n : n \in \mathbb{N}\} \bigcup \{0\} \subset \sigma_p(\Lambda) \bigcup \{0\} \subset \sigma(\Lambda)$.
Thus $\sigma(\Lambda) = \{\lambda_n : n \in \mathbb{N}\} \bigcup \{0\}$.
Consequently, $\{\lambda_n : n \in \mathbb{N}\} \subset \sigma_p(\Lambda) \subset \{\lambda_n : n \in \mathbb{N}\} \bigcup \{0\} = \sigma(\Lambda)$.

Solution to Exercise 2.34, page 103

(1) Suppose that $\lambda \in \sigma_{\mathrm{ap}}(T)$. Then there exists a sequence $(x_n)_{n\in\mathbb{C}}$ of vectors in X such that $\|x_n\| = 1$ for all $n \in \mathbb{N}$, and $Tx_n - \lambda x_n \xrightarrow{n\to\infty} \mathbf{0}$ in X.
We will just prove that $\lambda \notin \rho(T)$, and so by definition it will follow that then $\lambda \in \sigma(T)$. Suppose, on the contrary, that $\lambda \in \rho(T)$. Then $T - \lambda I$ is invertible in $CL(X)$. Thus
$$1 = \|x_n\| = \|(T - \lambda I)^{-1}(T - \lambda I)x_n\|$$
$$\leq \|(T - \lambda I)^{-1}\|\|Tx_n - \lambda x_n\| \xrightarrow{n\to\infty} \|(T - \lambda I)^{-1}\|\|\mathbf{0}\| = 0,$$
a contradiction. Consequently, $\lambda \notin \rho(T)$, that is, $\lambda \in \sigma(T)$.

(2) For $k \in \mathbb{N}$, let $\mathbf{e}_k$ denote the sequence in ℓ^2 whose kth term is 1 and all other terms are zeros. Then $\|\mathbf{e}_k\|_2 = 1$, and $\Lambda \mathbf{e}_k = \lambda_k \mathbf{e}_k$, so that
$$\left\|\Lambda \mathbf{e}_k - \left(\lim_{n\to\infty} \lambda_n\right)\mathbf{e}_k\right\|_2 = \left\|\left(\lambda_k - \lim_{n\to\infty}\lambda_n\right)\mathbf{e}_k\right\|_2 = \left|\lambda_k - \lim_{n\to\infty}\lambda_n\right|\|\mathbf{e}_k\|_2$$
$$= \left|\lambda_k - \lim_{n\to\infty}\lambda_n\right| \xrightarrow{k\to\infty} 0,$$
that is, $\Lambda \mathbf{e}_k - \left(\lim_{n\to\infty}\lambda_n\right)\mathbf{e}_k \xrightarrow{k\to\infty} \mathbf{0}$ in ℓ^2. Consequently, $\lim_{n\to\infty}\lambda_n \in \sigma_{\mathrm{ap}}(\Lambda)$.

Solution to Exercise 2.35, page 103

Let $\lambda \in \mathbb{C}$ and $\Psi \in D_Q$ be such that $x\Psi(x) = \lambda\Psi(x)$ for almost all $x \in \mathbb{R}$, that is, $(x - \lambda)\Psi(x) = 0$ for almost all $x \in \mathbb{R}$. Now $x - \lambda \neq 0$ for all $x \in \mathbb{R}\setminus\{\lambda\}$. Hence for almost all $x \in \mathbb{R}$, $\Psi(x) = 0$, that is, $\Psi = \mathbf{0}$ in $L^2(\mathbb{R})$. Consequently, λ can't be an eigenvalue of Q, and so $\sigma_p(Q) = \emptyset$.

Solution to Exercise 2.36, page 105

For simplicity we'll assume $\mathbb{K} = \mathbb{R}$. If $\mathbf{a} = (a_n)_{n\in\mathbb{N}} \in \ell^1$, then define the functional $\varphi_{\mathbf{a}} \in CL(c_0, \mathbb{R}) = (c_0)'$ by
$$\varphi_{\mathbf{a}}(\mathbf{b}) = \sum_{n=1}^{\infty} a_n b_n, \quad \mathbf{b} = (b_n)_{n\in\mathbb{N}} \in c_0.$$
Then $\mathbf{a} \mapsto \varphi_{\mathbf{a}} : \ell^1 \to (c_0)'$ is an injective linear transformation, and it is also continuous because $|\varphi_{\mathbf{a}}(\mathbf{b})| \leq \|\mathbf{b}\|_{\infty}\|\mathbf{a}\|_1$ for all $\mathbf{b} \in c_0$, and $\|\varphi_{\mathbf{a}}\| \leq \|\mathbf{a}\|_1$. To see the surjectivity of this map, we need to show that given $\varphi \in (c_0)'$, there exists an $\mathbf{a} \in \ell^1$ such that $\varphi = \varphi_{\mathbf{a}}$. Let $\mathbf{e}_n \in c_0$ being the sequence with nth term 1 and all others 0. Set $\mathbf{a} = (\varphi(\mathbf{e}_1), \varphi(\mathbf{e}_2), \varphi(\mathbf{e}_3), \cdots)$. We'll show that $\mathbf{a} \in \ell^1$, and that $\varphi = \varphi_{\mathbf{a}}$.

Define the scalars α_n, $n \in \mathbb{N}$, by $\alpha_n = \begin{cases} \dfrac{\varphi(\mathbf{e}_n)}{|\varphi(\mathbf{e}_n)|} & \text{if } \varphi(\mathbf{e}_n) \neq 0, \\ 0 & \text{if } \varphi(\mathbf{e}_n) \neq 0. \end{cases}$

Then for all n we have $\alpha_n \varphi(\mathbf{e}_n) = |\varphi(\mathbf{e}_n)|$.

We have $\|(\alpha_1, \cdots, \alpha_n, 0, \cdots)\|_\infty \leq 1$, and so
$$\begin{aligned}|\varphi(\mathbf{e}_1)| + \cdots + |\varphi(\mathbf{e}_n)| &= \big||\varphi(\mathbf{e}_1)| + \cdots + |\varphi(\mathbf{e}_n)|\big|\\ &= |\varphi(\alpha_1\mathbf{e}_1 + \cdots + \alpha\mathbf{e}_n)|\\ &\leq \|\varphi\|\|(\alpha_1, \cdots, \alpha_n, 0, \cdots)\|_\infty \leq \|\varphi\| \cdot 1,\end{aligned}$$
for all $n \in \mathbb{N}$. Hence $\mathbf{a} \in \ell^1$.

Finally, we need to show $\varphi = \varphi_\mathbf{a}$. Let $\mathbf{b} = (b_n)_{n \in \mathbb{N}} \in c_0$ and $\epsilon > 0$. Then there exists an N such that for all $n > N$, $|b_n| < \epsilon$. Set $\mathbf{b}_\epsilon = (b_1, \cdots, b_N, 0, \cdots) \in c_0$. Then $\|\mathbf{b} - \mathbf{b}_\epsilon\|_\infty = \|(0, \cdots, 0, b_{N+1}, \cdots)\|_\infty \leq \epsilon$. Moreover, we have that
$$\varphi(\mathbf{b}_\epsilon) = b_1\varphi(\mathbf{e}_1) + \cdots + b_N\varphi(\mathbf{e}_N) = b_1a_1 + \cdots + b_Na_N = \varphi_\mathbf{a}(\mathbf{b}_\epsilon).$$
Hence
$$\begin{aligned}|\varphi(\mathbf{b}) - \varphi_\mathbf{a}(\mathbf{b})| &= |\varphi(\mathbf{b}_\epsilon) - \varphi_\mathbf{a}(\mathbf{b}_\epsilon) + \varphi(\mathbf{b} - \mathbf{b}_\epsilon) - \varphi_\mathbf{a}(\mathbf{b} - \mathbf{b}_\epsilon)|\\ &\leq |\varphi(\mathbf{b} - \mathbf{b}_\epsilon)| + |\varphi_\mathbf{a}(\mathbf{b} - \mathbf{b}_\epsilon)| \leq (\|\varphi\| + \|\mathbf{a}\|_1)\epsilon.\end{aligned}$$
As the choice of $\epsilon > 0$ was arbitrary, it follows that $\varphi(\mathbf{b}) = \varphi_\mathbf{a}(\mathbf{b})$ for all $\mathbf{b} \in c_0$, that is, $\varphi = \varphi_\mathbf{a}$.

Solution to Exercise 2.37, page 105

(1) $BV[a,b]$ is a vector space: We prove that $BV[a,b]$ is a subspace of the vector space $\mathbb{R}^{[a,b]}$ of all real valued functions on $[a,b]$ with pointwise operations.

(S1) The zero function $\mathbf{0}$ belongs to $BV[a,b]$.
Indeed, for any partition P, $\sum_{k=0}^{n} |\boldsymbol{\mu}(t_k) - \boldsymbol{\mu}(t_{k-1})| = 0$, and so $\mathrm{var}(\mathbf{0}) = 0 < \infty$.

(S2) Let $\boldsymbol{\mu}_1, \boldsymbol{\mu}_2 \in BV[a,b]$. Then we have
$$\begin{aligned}\mathrm{var}(\boldsymbol{\mu}_1 + \boldsymbol{\mu}_2) &= \sup_{P \in \mathcal{P}} \sum_{k=1}^{n} |(\boldsymbol{\mu}_1 + \boldsymbol{\mu}_2)(t_k) - (\boldsymbol{\mu}_1 + \boldsymbol{\mu}_2)(t_{k-1})|\\ &= \sup_{P \in \mathcal{P}} \sum_{k=1}^{n} |\boldsymbol{\mu}_1(t_k) - \boldsymbol{\mu}_1(t_{k-1}) + \boldsymbol{\mu}_2(t_k) - \boldsymbol{\mu}_2(t_{k-1})|\\ &\leq \sup_{P \in \mathcal{P}} \Big(\sum_{k=1}^{n} |\boldsymbol{\mu}_1(t_k) - \boldsymbol{\mu}_1(t_{k-1})| + \sum_{k=1}^{n} |\boldsymbol{\mu}_2(t_k) - \boldsymbol{\mu}_2(t_{k-1})|\Big)\\ &\leq \sup_{P \in \mathcal{P}} \sum_{k=1}^{n} |\boldsymbol{\mu}_1(t_k) - \boldsymbol{\mu}_1(t_{k-1})| + \sup_{P \in \mathcal{P}} \sum_{k=1}^{n} |\boldsymbol{\mu}_2(t_k) - \boldsymbol{\mu}_2(t_{k-1})|\\ &= \mathrm{var}(\boldsymbol{\mu}_1) + \mathrm{var}(\boldsymbol{\mu}_2) < \infty,\end{aligned}$$
and so $\boldsymbol{\mu}_1 + \boldsymbol{\mu}_2 \in BV[a,b]$.

(S3) Let $\alpha \in \mathbb{R}$ and $\boldsymbol{\mu} \in BV[a,b]$. Then
$$\begin{aligned}\mathrm{var}(\alpha\boldsymbol{\mu}) &= \sup_{P \in \mathcal{P}} \sum_{k=1}^{n} |(\alpha\boldsymbol{\mu})(t_k) - (\alpha\boldsymbol{\mu})(t_{k-1})| = \sup_{P \in \mathcal{P}} |\alpha| \sum_{k=1}^{n} |\boldsymbol{\mu}(t_k) - \boldsymbol{\mu}(t_{k-1})|\\ &= |\alpha| \sup_{P \in \mathcal{P}} \sum_{k=1}^{n} |\boldsymbol{\mu}(t_k) - \boldsymbol{\mu}(t_{k-1})| = |\alpha|\mathrm{var}(\boldsymbol{\mu}),\end{aligned}$$
and so $\alpha\boldsymbol{\mu} \in BV[a,b]$.

(2) We show that $\boldsymbol{\mu} \mapsto \|\boldsymbol{\mu}\|$ defines a norm on $BV[a,b]$.

(N1) If $\boldsymbol{\mu} \in BV[a,b]$, then $\|\boldsymbol{\mu}\| = |\boldsymbol{\mu}(a)| + \mathrm{var}(\boldsymbol{\mu}) \geq 0$.

Let $\boldsymbol{\mu} \in BV[a,b]$ be such that $\|\boldsymbol{\mu}\| = 0$. Then $\mathrm{var}(\boldsymbol{\mu}) = 0$, and $|\boldsymbol{\mu}(a)| = 0$. Hence $\boldsymbol{\mu}(a) = 0$. Suppose that $\boldsymbol{\mu} \neq \mathbf{0}$. Then there exists a $c \in [a,b]$ such that $\boldsymbol{\mu}(c) \neq 0$. Clearly $c \neq a$, since $\boldsymbol{\mu}(a) = 0$. Now consider the partition
$$\begin{cases} a = x_0 < x_1 = c < x_2 = b & \text{if } c < b, \\ a = x_0 < x_1 = c = b & \text{if } c = b. \end{cases}$$
Then $\mathrm{var}(\boldsymbol{\mu}) = \sup_{P \in \mathcal{P}} \sum_{k=1}^{n} |\boldsymbol{\mu}(t_k) - \boldsymbol{\mu}(t_{k-1})|$
$$\geq \begin{cases} |\boldsymbol{\mu}(c) - \boldsymbol{\mu}(a)| + |\boldsymbol{\mu}(b) - \boldsymbol{\mu}(c)| & \text{if } c < b, \\ |\boldsymbol{\mu}(c) - \boldsymbol{\mu}(a)| & \text{if } c = b \end{cases}$$
$$= \begin{cases} |\boldsymbol{\mu}(c)| + |\boldsymbol{\mu}(b) - \boldsymbol{\mu}(c)| & \text{if } c < b, \\ |\boldsymbol{\mu}(c)| & \text{if } c = b \end{cases}$$
$$\geq |\boldsymbol{\mu}(c)| > 0,$$
a contradiction. Hence $\boldsymbol{\mu} = \mathbf{0}$.

(N2) Let $\alpha \in \mathbb{R}$ and $\boldsymbol{\mu} \in BV[a,b]$. Then $\alpha\boldsymbol{\mu} \in BV[a,b]$, and we have seen earlier that $\mathrm{var}(\alpha\boldsymbol{\mu}) = |\alpha|\mathrm{var}(\boldsymbol{\mu})$. Hence
$$\|\alpha\boldsymbol{\mu}\| = |(\alpha\boldsymbol{\mu})(a)| + \mathrm{var}(\alpha\boldsymbol{\mu}) = |\alpha||\boldsymbol{\mu}(a)| + |\alpha|\mathrm{var}(\boldsymbol{\mu})$$
$$= |\alpha|(|\boldsymbol{\mu}(a)| + \mathrm{var}(\boldsymbol{\mu})) = |\alpha|\|\boldsymbol{\mu}\|.$$

(N3) Let $\boldsymbol{\mu}_1, \boldsymbol{\mu}_2 \in BV[a,b]$. Then $\boldsymbol{\mu}_1 + \boldsymbol{\mu}_2 \in BV[a,b]$, and we've seen above that $\mathrm{var}(\boldsymbol{\mu}_1 + \boldsymbol{\mu}_2) \leq \mathrm{var}(\boldsymbol{\mu}_1) + \mathrm{var}(\boldsymbol{\mu}_2)$. Thus
$$\|\boldsymbol{\mu}_1 + \boldsymbol{\mu}_2\| = |(\boldsymbol{\mu}_1 + \boldsymbol{\mu}_2)(a)| + \mathrm{var}(\boldsymbol{\mu}_1 + \boldsymbol{\mu}_2)$$
$$\leq |\boldsymbol{\mu}_1(a)| + |\boldsymbol{\mu}_2(a)| + \mathrm{var}(\boldsymbol{\mu}_1) + \mathrm{var}(\boldsymbol{\mu}_2) = \|\boldsymbol{\mu}_1\| + \|\boldsymbol{\mu}_2\|.$$
Consequently $BV[a,b]$ is a normed space with the norm $\|\cdot\|$.

(3) Let $\mathbf{x} \in C[a,b]$ and $\boldsymbol{\mu} \in BV[a,b]$. Given $\epsilon > 0$, let $\delta > 0$ be such that for every partition P satisfying $\delta_P < \delta$, we have
$$\left| \int_a^b \mathbf{x}(t) d\boldsymbol{\mu} - S(P) \right| < \epsilon.$$
Then
$$\left| \int_a^b \mathbf{x}(t) d\boldsymbol{\mu} \right| < \epsilon + |S(P)| = \epsilon + \left| \sum_{k=1}^n \mathbf{x}(t_k)(\boldsymbol{\mu}(t_k) - \boldsymbol{\mu}(t_{k-1})) \right|$$
$$\leq \epsilon + \sum_{k=1}^n |\mathbf{x}(t_k)||\boldsymbol{\mu}(t_k) - \boldsymbol{\mu}(t_{k-1})|$$
$$\leq \epsilon + \sum_{k=1}^n \|\mathbf{x}\|_\infty |\boldsymbol{\mu}(t_k) - \boldsymbol{\mu}(t_{k-1})|$$
$$= \epsilon + \|\mathbf{x}\|_\infty \sum_{k=1}^n |\boldsymbol{\mu}(t_k) - \boldsymbol{\mu}(t_{k-1})| = \epsilon + \|\mathbf{x}\|_\infty \mathrm{var}(\boldsymbol{\mu}).$$
As the choice of $\epsilon > 0$ was arbitrary, it follows that $\left| \int_a^b \mathbf{x}(t) d\boldsymbol{\mu} \right| \leq \|\mathbf{x}\|_\infty \mathrm{var}(\boldsymbol{\mu})$.

(4) For all $\mathbf{x} \in C[a,b]$, $|\varphi_\mu \mathbf{x}| \leq \|\mathbf{x}\|_\infty \mathrm{var}(\mu)$.
From the linearity of the Riemann-Stieltjes integral, it follows that φ_μ is a linear transformation from $C[a,b]$ to $\mathbb{R}$. From the above estimate, we also see that φ_μ is continuous. Consequently $\varphi_\mu \in CL(C[a,b],\mathbb{R}) = (C[a,b])'$. Moreover $\|\varphi_\mu\| \leq \mathrm{var}(\mu)$.

(5) We will show that $(\mathbf{x} \mapsto \mathbf{x}(a)) = \varphi_\mu$, where $\mu(t) = \begin{cases} -1 & \text{if } t = a, \\ 0 & \text{if } t \in (a,b]. \end{cases}$

First of all, $\mu \in BV[a,b]$, since $\mathrm{var}(\mu) = 1 < \infty$. Let $\mathbf{x} \in C[a,b]$, and $\epsilon > 0$. Let $\delta > 0$ be such that for all t such that $t - a < \delta$, we have $|\mathbf{x}(t) - \mathbf{x}(a)| < \epsilon$. Then for all partitions P with $\delta_P < \delta$, we have

$$\left|\mathbf{x}(a) - \sum_{k=1}^n \mathbf{x}(t_k)\big(\mu(t_k) - \mu(t_{k-1})\big)\right| = \big|\mathbf{x}(a) - \mathbf{x}(t_1)\big(0 - (-1)\big) - 0 - \cdots - 0\big|$$

$$= |\mathbf{x}(a) - \mathbf{x}(t_1)| < \epsilon,$$

where the last inequality follows from the fact that $|a - t_1| \leq \delta_P < \delta$.

So $\int_a^b \mathbf{x}(t)d\mu = \mathbf{x}(a)$. ($\mu$ is not unique: for any $c \in \mathbb{R}$, $\mu + c$ also works!)

Solution to Exercise 2.38, page 109

On the one dimensional subspace $Y := \mathrm{span}\{x_*\} \subset X$, we have a continuous linear map $\varphi: Y \to \mathbb{C}$. (Simply define $\varphi(\alpha x_*) = \alpha$, then $|\varphi(\alpha x_*)| = |\alpha| = \|\alpha x_*\|/\|x_*\|$, and so $\|\varphi\| = 1/\|x_*\| < \infty$.) By the Hahn-Banach Theorem, there exists an extension $\varphi_* \in CL(X,\mathbb{C})$ of φ, and so $\varphi_*(x_*) = \varphi(x_*) = 1 \neq 0$. (Alternatively, one could just use Corollary 2.7, page 109, with $x = x_*$ and $y = \mathbf{0}$: there exists a functional $\varphi_* \in CL(X,\mathbb{C})$ such that $\varphi_*(x_*) \neq \varphi_*(\mathbf{0}) = 0$.)

Solution to Exercise 2.39, page 115

Consider the collection P of all linearly independent subsets $S \subset X$. Consider the partial order which is simply set inclusion $\subset$. Then every chain in P has an upper bound, as explained below.

If C is a chain in P, then $U := \bigcup_{S \in C} S$ is an upper bound of C.

We just need to show the linear independence of this set U. To this end, let $v_1, \cdots, v_n$ be any set of vectors from U for which there exist scalars $\alpha_1, \cdots, \alpha_n$ in $\mathbb{F}$ such that $\alpha_1 v_1 + \cdots + \alpha_n v_n = \mathbf{0}$. Let the sets $S_1, \cdots, S_n \in C$ be such that $v_1 \in S_1, \cdots, v_n \in S_n$. As C is a chain, we can arrange the finitely many S_ks in "ascending order", and there exists a $k_* \in \{1, \cdots, n\}$ such that $S_1, \cdots, S_n \subset S_{k_*}$. Then $v_1, \cdots, v_n \in S_{k_*}$. But by the linear independence of S_{k_*}, we conclude that $\alpha_1 = \cdots = \alpha_n = 0$. Thus U is linearly independent, showing that every chain in P has an upper bound.

By Zorn's Lemma, P has a maximal element B. We claim that $\mathrm{span}\, B = X$. For if not, then there exists an $x \in X \backslash \mathrm{span}\, B$. We will show $B' := B \cup \{x\}$ is

linearly independent. Suppose that $\alpha_1, \cdots, \alpha_n, \alpha \in \mathbb{K}$ and $v_1, \cdots, v_n \in B$ are such that $\alpha x + \alpha_1 v_1 + \cdots + \alpha_n v_n = \mathbf{0}$. First we note that $\alpha = 0$, since otherwise
$$x = \left(-\frac{\alpha_1}{\alpha}\right)v_1 + \cdots + \left(-\frac{\alpha_n}{\alpha}\right)v_1 \in \operatorname{span} B,$$
which is false. As $\alpha = 0$, the equality $\alpha x + \alpha_1 v_1 + \cdots + \alpha_n v_n = \mathbf{0}$ now becomes $\alpha_1 v_1 + \cdots + \alpha_n v_n = \mathbf{0}$. But by the independence of the set B, we conclude that $\alpha_1 = \cdots = \alpha_n = 0$ too. Hence B' is linearly independent, and so B' belongs to P. As $B' = B \cup \{x\} \supsetneq B$, we obtain a contradiction (to the maximality of B). Consequently, span $B = X$, and as $B \in P$, B is also linearly independent.

Solution to Exercise 2.40, page 115

Let $B = \{v_i : i \in I\}$. Every $x \in X$ has a unique decomposition
$$x = \alpha_1 x_{i_1} + \cdots + \alpha_n x_{i_n},$$
for some finite number of indices $i_1, \cdots, i_n \in I$ and scalars $\alpha_1, \cdots, \alpha_n$ in $\mathbb{F}$. Define $F(x) = \alpha_1 f(v_{i_1}) + \cdots + \alpha_n f(v_{i_n})$. It is clear that $F(v_i) = f(v_i)$, $i \in I$. Let us check that $F : X \to Y$ is linear.

(L1) Given $x_1, x_2 \in X$, there exist scalars $\alpha_1, \cdots, \alpha_n$ and $\beta_1, \cdots, \beta_n$ (possibly several of them equal to zero) and indices $i_1, \cdots, i_n \in I$, such that
$$x_1 = \alpha_1 v_{i_1} + \cdots + \alpha_n v_{i_n} \text{ and } x_2 = \beta_1 v_{i_1} + \cdots + \beta_n v_{i_n}.$$
Then $x_1 + x_2 = (\alpha_1 + \beta_1)v_{i_1} + \cdots + (\alpha_n + \beta_n)v_{i_n}$. So
$$\begin{aligned}F(x_1 + x_2) &= (\alpha_1 + \beta_1)f(v_{i_1}) + \cdots + (\alpha_n + \beta_n)f(v_{i_n})\\ &= \alpha_1 f(v_{i_1}) + \cdots + \alpha_n f(v_{i_n}) + \beta_1 f(v_{i_1}) + \cdots + \beta_n f(v_{i_n})\\ &= F(x_1) + F(x_2).\end{aligned}$$

(L2) Let $\alpha \in \mathbb{F}$. Given $x \in X$, there exist $\beta_1, \cdots, \beta_n \in \mathbb{F}$ and $i_1, \cdots, i_n \in I$, such that $x = \beta_1 v_{i_1} + \cdots + \beta_n v_{i_n}$. Then $\alpha x = (\alpha \beta_1)v_{i_1} + \cdots + (\alpha \beta_n)v_{i_n}$. So $F(\alpha x) = (\alpha \beta_1)f(v_{i_1}) + \cdots + (\alpha \beta_n)f(v_{i_n})$
$$= \alpha\big(\beta_1 f(v_{i_1}) + \cdots + \beta_n f(v_{i_n})\big) = \alpha F(x).$$

Solution to Exercise 2.41, page 115

Let B be a Hamel basis for X. As X is infinite dimensional, B is an infinite set. Let $\{v_n : n \in \mathbb{N}\}$ be a countable subset of B. Let $y_* \in Y$ be any nonzero vector. Let $f : B \to Y$ be defined by $f(v) = \begin{cases} n\|v_n\|y_* & \text{if } v = v_n,\ n \in \mathbb{N}, \\ 0 & \text{if for all } n \in \mathbb{N}, v \neq v_n. \end{cases}$
By the previous exercise, this f extends to a linear transformation F from X to Y. We claim that $F \notin CL(X, Y)$. Suppose that it does. Then there exists an $M > 0$ such that for all $x \in X$, $\|F(x)\| \leqslant M\|x\|$. But if we put $x = v_n$, $n \in \mathbb{N}$, this yields $n\|v_n\|\|y_*\| = \|f(v_n)\| = \|F(v_n)\| \leqslant M\|v_n\|$, and so for all $n \in \mathbb{N}$, $n \leqslant M/\|y_*\|$, which is absurd. Thus F is a linear transformation from X to Y, but is not continuous.

Solution to Exercise 2.42, page 115

If $\mathbb{R}$ were finite dimensional, say d-dimensional over $\mathbb{Q}$, then there would exist a one-to-one correspondence between $\mathbb{R}$ and $\mathbb{Q}^d$. But $\mathbb{Q}^d$ is countable, while $\mathbb{R}$ isn't, a contradiction. So $\mathbb{R}$ is an infinite dimensional vector space over $\mathbb{Q}$.

Suppose that $\mathbb{R}$ has a countable basis $B = \{v_n : n \in \mathbb{N}\}$ over $\mathbb{Q}$.

We will define an injective map $f : \mathbb{R} \to \bigcup_{n=1}^{\infty} \mathbb{Q}^n$, yielding a contradiction.

Set $f(0) := 0 \in \mathbb{Q}^1$. If $x \neq 0$, then x has a decomposition $x = q_1 v_1 + \cdots + q_n v_n$, where $q_1, \cdots, q_n \in \mathbb{Q}$ and $q_n \neq 0$. In this case, set $f(x) = (q_1, \cdots, q_n) \in \mathbb{Q}^n$. It can be seen that if $f(x) = f(y)$, for some $x, y \in \mathbb{R}$, then $x = y$. So f is injective.

As $\bigcup_{n=1}^{\infty} \mathbb{Q}^n$ is countable, follows that $\mathbb{R}$ is countable too, a contradiction.

Hence B can't be countable.

Solution to Exercise 2.43, page 115

The set $\mathbb{R}$ is an infinite dimensional vector space over $\mathbb{Q}$. Let $\{v_i : i \in I\}$ be a Hamel basis for this vector space. Fix any $i_* \in I$.

We define a function $f : B \to \mathbb{R}$ on the basis elements: $f(v_i) = \begin{cases} 1 & \text{if } i = i_*, \\ 0 & \text{if } i \neq i_*. \end{cases}$

Let F be an extension of f from B to $\mathbb{R}$, as provided by Exercise 2.40, page 115. Then F is linear, and in particular, additive. So $F(x+y) = F(x) + F(y)$ for all $x, y \in \mathbb{R}$.

We now show that F is not continuous on $\mathbb{R}$: for otherwise, for any $v_i \neq v_{i_*}$, if $(q_n)_{n \in \mathbb{N}}$ is a sequence in $\mathbb{Q}$ converging to the real number v_i/v_{i_*} ($v_{i_*} \neq 0$ since it is a basis vector), then we would have

$$0 = f(v_i) = F(v_i) = F\left(\frac{v_i}{v_{i_*}} \cdot v_{i_*}\right) = F\left(\lim_{n \to \infty} q_n \cdot v_{i_*}\right)$$

$$= \lim_{n \to \infty} F(q_n \cdot v_{i_*}) \quad \text{(continuity of } F\text{)}$$

$$= \lim_{n \to \infty} q_n F(v_{i_*}) \quad \text{(linearity of } F \text{ for rational scalars)}$$

$$= \frac{v_i}{v_{i_*}} f(v_{i_*}) = \frac{v_i}{v_{i_*}} \neq 0,$$

a contradiction!

Solution to Exercise 2.44, page 116

(1) By the Algebra of Limits, the map l is linear.

Let $(x_n)_{n \in \mathbb{N}} \in c$. For all $n \in \mathbb{N}$, $|x_n| \leq \|(x_n)_{n \in \mathbb{N}}\|_\infty$.

Passing the limit as $n \to \infty$, $|l(x_n)_{n \in \mathbb{N}}| = \left|\lim_{n \to \infty} x_n\right| = \lim_{n \to \infty} |x_n| \leq \|(x_n)_{n \in \mathbb{N}}\|_\infty$.

Thus $l \in CL(c, \mathbb{K})$.

(2) Y is a subspace of ℓ^∞. Indeed we have:

(S1) Clearly $(0)_{n\in\mathbb{N}} \in Y$, since $\lim_{n\to\infty} \dfrac{x_1 + \cdots + x_n}{n} = \lim_{n\to\infty} \dfrac{0}{n} = 0$.

(S2) Let $(x_n)_{n\in\mathbb{N}}, (y_n)_{n\in\mathbb{N}} \in Y$.

Then $\lim_{n\to\infty} \dfrac{x_1 + \cdots + x_n}{n}$ and $\lim_{n\to\infty} \dfrac{y_1 + \cdots + y_n}{n}$ exist.

As $\dfrac{(x_1 + y_1) + \cdots + (x_n + y_n)}{n} = \dfrac{x_1 + \cdots + x_n}{n} + \dfrac{y_1 + \cdots + y_n}{n}$,

we conclude that $\lim_{n\to\infty} \dfrac{(x_1 + y_1) + \cdots + (x_n + y_n)}{n}$ exists as well.

Thus $(x_n)_{n\in\mathbb{N}} + (y_n)_{n\in\mathbb{N}} \in Y$ too.

(S3) Let $(x_n)_{n\in\mathbb{N}} \in Y$ and $\alpha \in \mathbb{K}$. Then $\lim_{n\to\infty} \dfrac{x_1 + \cdots + x_n}{n}$ exists.

As $\dfrac{\alpha x_1 + \cdots + \alpha x_n}{n} = \alpha \cdot \dfrac{x_1 + \cdots + x_n}{n}$, it follows that

$\lim_{n\to\infty} \dfrac{\alpha x_1 + \cdots + \alpha x_n}{n}$ exists, and so $\alpha \cdot (x_n)_{n\in\mathbb{N}} \in Y$.

Consequently, Y is a subspace of ℓ^∞.

(3) For all $\mathbf{x} \in \ell^\infty$, $\mathbf{x} - S\mathbf{x} \in Y$: Let $\mathbf{x} = (x_n)_{n\in\mathbb{N}} \in \ell^\infty$. Then we have
$$\mathbf{x} - S\mathbf{x} = (x_1, x_2, x_3, \cdots) - (x_2, x_3, x_4, \cdots)$$
$$= (x_1 - x_2, x_2 - x_3, x_3 - x_4, \cdots).$$

We have $\dfrac{(x_1 - x_2) + (x_2 - x_3) + \cdots + (x_n - x_{n+1})}{n} = \dfrac{x_1 - x_{n+1}}{n}$.

As $\mathbf{x} \in \ell^\infty$, it follows that $\lim_{n\to\infty} \dfrac{x_1 - x_{n+1}}{n} = 0$, and so $\mathbf{x} - S\mathbf{x} \in Y$.

(4) If $\mathbf{x} = (x_n)_{n\in\mathbb{N}} \in c$, then $A\mathbf{x} \in c$, where A denotes the averaging operator (Exercise 2.19, page 77).

Hence $\lim_{n\to\infty} \dfrac{x_1 + \cdots + x_n}{n}$ exists, and so $\mathbf{x} \in Y$. Consequently, $c \subset Y$.

(5) Define $L_0 : Y \to \mathbb{K}$ by $L_0(x_n)_{n\in\mathbb{N}} = \lim_{n\to\infty} \dfrac{x_1 + \cdots + x_n}{n}$, for $(x_n)_{n\in\mathbb{N}} \in Y$.

Then it is easy to check that $L_0 : Y \to \mathbb{K}$ is a linear transformation.

Moreover, if $\mathbf{x} \in Y$, then $|L_0\mathbf{x}| = \left|\lim_{n\to\infty} \dfrac{x_1 + \cdots + x_n}{n}\right| = \lim_{n\to\infty} \left|\dfrac{x_1 + \cdots + x_n}{n}\right|$.

But $\left|\dfrac{x_1 + \cdots + x_n}{n}\right| \leqslant \dfrac{|x_1| + \cdots + |x_n|}{n} \leqslant \dfrac{n\|\mathbf{x}\|_\infty}{n} = \|\mathbf{x}\|_\infty$.

Hence $|L_0\mathbf{x}| \leqslant \|\mathbf{x}\|_\infty$. Consequently, $L_0 \in CL(Y, \mathbb{K})$.

We had seen that if $\mathbf{x} \in c$, then $A\mathbf{x} \in c$, and that $l(A\mathbf{x}) = l(\mathbf{x})$.
Hence for all $\mathbf{x} \in c$, $L_0(\mathbf{x}) = l(A\mathbf{x}) = l(\mathbf{x})$, that is, $L_0|_c = l$.
Using the Hahn-Banach Theorem, there exists an $L \in CL(\ell^\infty, \mathbb{K})$ such that $L|_Y = L_0$ (and $\|L\| = \|L_0\|$).
In particular, if $\mathbf{x} \in c$, then $\mathbf{x} \in Y$ and so $L\mathbf{x} = L_0\mathbf{x} = l\mathbf{x}$. Thus $L|_c = l$.
Also, if $\mathbf{x} = (x_n)_{n\in\mathbb{N}} \in \ell^\infty$, then $\mathbf{x} - S\mathbf{x} \in Y$.

Hence $L(\mathbf{x} - S\mathbf{x}) = L_0(\mathbf{x} - S\mathbf{x}) = \lim\limits_{n\to\infty} \dfrac{x_1 - x_{n+1}}{n} = 0$.

Thus $L\mathbf{x} = LS\mathbf{x}$ for all $\mathbf{x} \in \ell^\infty$, that is, $L = LS$.

(6) We have $((-1)^n)_{n\in\mathbb{N}} + S((-1)^n)_{n\in\mathbb{N}} = ((-1)^n)_{n\in\mathbb{N}} + ((-1)^{n+1})_{n\in\mathbb{N}} = (0)_{n\in\mathbb{N}}$.

So $0 = L((0)_{n\in\mathbb{N}}) = L\big(((-1)^n)_{n\in\mathbb{N}} + S((-1)^n)_{n\in\mathbb{N}}\big)$
$= L((-1)^n)_{n\in\mathbb{N}} + LS((-1)^n)_{n\in\mathbb{N}} = L((-1)^n)_{n\in\mathbb{N}} + L((-1)^n)_{n\in\mathbb{N}}$
$= 2L((-1)^n)_{n\in\mathbb{N}}.$

Consequently, $L((-1)^n)_{n\in\mathbb{N}} = 0$.

Solutions to the exercises from Chapter 3

Solution to Exercise 3.1, page 124

f is a continuous linear transformation. Thus it follows that $f'(\mathbf{x}_0) = f$ for all $\mathbf{x}_0$, and in particular also for $\mathbf{x}_0 = \mathbf{0}$.

Solution to Exercise 3.2, page 125

Suppose that $f'(x_0) = L \in CL(X, Y)$. Let $M > 0$ be such that $\|Lh\| \leq M\|h\|$, for all $h \in X$. Let $\epsilon > 0$. Then there exists a $\delta_1 > 0$ such that whenever $x \in X$ satisfies $0 < \|x - x_0\| < \delta_1$, we have
$$\frac{\|f(x) - f(x_0) - L(x - x_0)\|}{\|x - x_0\|} < \epsilon.$$
So if $x \in X$ satisfies $\|x - x_0\| < \delta_1$, then $\|f(x) - f(x_0) - L(x - x_0)\| \leq \epsilon\|x - x_0\|$. Let $\delta := \min\left\{\delta_1, \dfrac{\epsilon}{\epsilon + M}\right\}$. Then for all $x \in X$ satisfying $\|x - x_0\| < \delta$, we have
$$\|f(x) - f(x_0)\| \leq \|f(x) - f(x_0) - L(x - x_0)\| + \|L(x - x_0)\|$$
$$\leq \epsilon\|x - x_0\| + M\|x - x_0\| = (\epsilon + M)\|x - x_0\|$$
$$< (\epsilon + M) \cdot \frac{\epsilon}{\epsilon + M} = \epsilon.$$
Hence f is continuous at x_0.

Solution to Exercise 3.3, page 125

(Rough work: We have for $\mathbf{x} \in C^1[0, 1]$ that
$$f(\mathbf{x}) - f(\mathbf{x}_0) = (\mathbf{x}'(1))^2 - (\mathbf{x}_0'(1))^2 = (\mathbf{x}'(1) + \mathbf{x}_0'(1))(\mathbf{x} - \mathbf{x}_0)'(1)$$
$$\approx 2\mathbf{x}_0'(1)(\mathbf{x} - \mathbf{x}_0)'(1) = L(\mathbf{x} - \mathbf{x}_0),$$
where $L : C^1[0, 1] \to \mathbb{R}$ is the map given by $L\mathbf{h} = 2\mathbf{x}_0'(1)\mathbf{h}'(1)$, $\mathbf{h} \in C^1[0, 1]$. So we make the guess that $f'(\mathbf{x}_0) = L$.)
Let us first check that L is a continuous linear transformation. L is linear because:
(L1) For all $\mathbf{h}_1, \mathbf{h}_2 \in C^1[0, 1]$, we have
$$L(\mathbf{h}_1 + \mathbf{h}_2) = 2\mathbf{x}_0'(1)(\mathbf{h}_1 + \mathbf{h}_2)'(1) = 2\mathbf{x}_0'(1)\mathbf{h}_1'(1) + 2\mathbf{x}_0'(1)\mathbf{h}_2'(1) = L(\mathbf{h}_1) + L(\mathbf{h}_2).$$
(L2) For all $\mathbf{h} \in C^1[0, 1]$ and $\alpha \in \mathbb{R}$, we have
$$L(\alpha \cdot \mathbf{h}) = 2\mathbf{x}_0'(1)(\alpha \cdot \mathbf{h})'(1) = 2\mathbf{x}_0'(1)\alpha \mathbf{h}'(1) = \alpha L(\mathbf{h}).$$
Also, L is continuous since for all $\mathbf{h} \in C^1[0, 1]$, we have
$$|L(\mathbf{h})| = |2\mathbf{x}_0'(1)\mathbf{h}'(1)| = 2|\mathbf{x}_0'(1)||\mathbf{h}'(1)| \leq 2|\mathbf{x}_0'(1)|\|\mathbf{h}'\|_\infty \leq 2|\mathbf{x}_0'(1)|\|\mathbf{h}\|_{1,\infty}.$$
So L is a continuous linear transformation. Moreover, for all $\mathbf{x} \in C^1[0, 1]$,
$$f(\mathbf{x}) - f(\mathbf{x}_0) - L(\mathbf{x} - \mathbf{x}_0) = (\mathbf{x}'(1))^2 - (\mathbf{x}_0'(1))^2 - 2\mathbf{x}_0'(1)(\mathbf{x} - \mathbf{x}_0)'(1) = ((\mathbf{x} - \mathbf{x}_0)'(1))^2$$
so that $|f(\mathbf{x}) - f(\mathbf{x}_0) - L(\mathbf{x} - \mathbf{x}_0)| = |(\mathbf{x} - \mathbf{x}_0)'(1)|^2 \leq \|(\mathbf{x} - \mathbf{x}_0)'\|_\infty^2 \leq \|\mathbf{x} - \mathbf{x}_0\|_{1,\infty}^2.$

Given $\epsilon > 0$, set $\delta = \epsilon$. Then if $\mathbf{x} \in C^1[0,1]$ satisfies $0 < \|\mathbf{x} - \mathbf{x}_0\|_{1,\infty} < \delta$, we have
$$\frac{|f(\mathbf{x}) - f(\mathbf{x}_0) - L(\mathbf{x} - \mathbf{x}_0)|}{\|\mathbf{x} - \mathbf{x}_0\|_{1,\infty}} \leqslant \frac{\|\mathbf{x} - \mathbf{x}_0\|_{1,\infty}^2}{\|\mathbf{x} - \mathbf{x}_0\|_{1,\infty}} = \|\mathbf{x} - \mathbf{x}_0\|_{1,\infty} < \delta = \epsilon.$$

Solution to Exercise 3.4, page 125

Given $\epsilon > 0$, let $\epsilon' > 0$ be such that $\epsilon'\|x_2 - x_1\| < \epsilon$. Let $\delta' > 0$ such that whenever $0 < \|x - \gamma(t_0)\| < \delta'$, we have
$$\frac{|f(x) - f(\gamma(t_0)) - f'(\gamma(t_0))(x - \gamma(t_0))|}{\|x - \gamma(t_0)\|} < \epsilon'.$$
Let $\delta > 0$ be such that $\delta\|x_2 - x_1\| < \delta'$. For all $t \in \mathbb{R}$ satisfying $0 < |t - t_0| < \delta$,
$$\gamma(t) - \gamma(t_0) = (1-t)x_1 + tx_2 - (1-t_0)x_1 - t_0 x_2 = (t_0 - t)x_1 + (t - t_0)x_2$$
$$= (t - t_0)(x_2 - x_1),$$
and so $\|\gamma(t) - \gamma(t_0)\| = |t - t_0|\|x_2 - x_1\| \leqslant \delta\|x_2 - x_1\| < \delta'$. Thus for all $t \in \mathbb{R}$ satisfying $0 < |t - t_0| < \delta$, we have
$$\left|\frac{f(\gamma(t)) - f(\gamma(t_0))}{t - t_0} - f'(\gamma(t_0))(x_2 - x_1)\right|$$
$$= \frac{|f(\gamma(t)) - f(\gamma(t_0)) - f'(\gamma(t_0))((t - t_0)(x_2 - x_1))|}{|t - t_0|\|x_2 - x_1\|}\|x_2 - x_1\|$$
$$= \frac{|f(\gamma(t)) - f(\gamma(t_0)) - f'(\gamma(t_0))(\gamma(t) - \gamma(t_0))|}{\|\gamma(t) - \gamma(t_0)\|}\|x_2 - x_1\| < \epsilon'\|x_2 - x_1\| < \epsilon.$$
Thus $f \circ \gamma$ is differentiable at t_0 and $\dfrac{d}{dt}(f \circ \gamma)(t_0) = f'(\gamma(t_0))(x_2 - x_1)$.

Let $x_1, x_2 \in X$ be such that $g(x_1) \neq g(x_2)$. With γ the same as above, we have for *all* $t \in \mathbb{R}$ that
$$\frac{d}{dt}(g \circ \gamma)(t) = g'(\gamma(t))(x_2 - x_1) = \mathbf{0}(x_2 - x_1) = 0.$$
So $g \circ \gamma$ is constant. Thus $(g \circ \gamma)(1) = g(x_2) = g(x_1) = (g \circ \gamma)(0)$, a contradiction. Consequently, g is constant.

Solution to Exercise 3.5, page 128

Suppose that $f'(\mathbf{x}_0) = \mathbf{0}$. Then for every $\mathbf{h} \in C[a,b]$, $2\displaystyle\int_a^b \mathbf{x}_0(t)\mathbf{h}(t)dt = 0$.

In particular, setting $\mathbf{h} = \mathbf{x}_0$, we have $\displaystyle\int_a^b (\mathbf{x}_0(t))^2 dt = 0$, giving $\mathbf{x}_0 = \mathbf{0} \in C[a,b]$.

Vice versa, if $\mathbf{x}_0 = \mathbf{0}$, then
$$(f'(\mathbf{x}_0))(\mathbf{h}) = 2\int_a^b \mathbf{x}_0(t)\mathbf{h}(t)dt = 2\int_a^b \mathbf{0}(t)\mathbf{h}(t)dt = 2\int_a^b 0 \cdot \mathbf{h}(t)dt = 2\int_a^b \mathbf{0}dt = 0$$
for all $\mathbf{h} \in C[a,b]$, that is, $f'(\mathbf{0}) = \mathbf{0}$.

Consequently, $f'(\mathbf{x}_0) = \mathbf{0}$ if and only if $\mathbf{x}_0 = \mathbf{0}$.

So we see that if $\mathbf{x}_*$ is a minimiser, then $f'(\mathbf{x}_*) = \mathbf{0}$, and so from the above $\mathbf{x}_* = \mathbf{0}$. We remark that $\mathbf{0}$ is easily seen to be the minimiser because
$$f(\mathbf{x}) = \int_a^b \underbrace{(\mathbf{x}(t))^2}_{\geq 0} \, dt \geq 0 = f(\mathbf{0}) \text{ for all } \mathbf{x} \in C[a,b].$$

Solution to Exercise 3.6, page 129

If $\mathbf{x}_1, \mathbf{x}_2 \in S$, $\alpha \in (0,1)$, then $\mathbf{x}_1, \mathbf{x}_2 \in C^1[a,b]$. So $(1-\alpha)\mathbf{x}_1 + \alpha\mathbf{x}_2 \in C^1[a,b]$. Moreover, as $\mathbf{x}_1(a) = \mathbf{x}_2(a) = y_a$ and $\mathbf{x}_1(b) = \mathbf{x}_2(b) = y_b$, we also have that
$$\big((1-\alpha)\mathbf{x}_1 + \alpha\mathbf{x}_2\big)(a) = (1-\alpha)\mathbf{x}_1(a) + \alpha\mathbf{x}_2(a) = (1-\alpha)y_a + \alpha y_a = y_a,$$
$$\big((1-\alpha)\mathbf{x}_1 + \alpha\mathbf{x}_2\big)(b) = (1-\alpha)\mathbf{x}_1(b) + \alpha\mathbf{x}_2(b) = (1-\alpha)y_b + \alpha y_b = y_b.$$
Thus $(1-\alpha)\mathbf{x}_1 + \alpha\mathbf{x}_2 \in S$. Consequently, S is convex.

Solution to Exercise 3.7, page 129

For $x_1, x_2 \in X$ and $\alpha \in (0,1)$ we have by the triangle inequality that
$$\|(1-\alpha)x_1 + \alpha x_2\| \leq \|(1-\alpha)x_1\| + \|\alpha x_2\| = |1-\alpha|\|x_1\| + |\alpha|\|x_2\|$$
$$= (1-\alpha)\|x_1\| + \alpha\|x_2\|.$$
Thus $\|\cdot\|$ is convex.

Solution to Exercise 3.8, page 129

(If part:) Let $x_1, x_2 \in C$ and $\alpha \in (0,1)$. Then we have that $(x_1, f(x_1)) \in U(f)$ and $(x_2, f(x_2)) \in U(f)$. Since $U(f)$ is convex,
$$(1-\alpha) \cdot (x_1, f(x_1)) + \alpha \cdot (x_2, f(x_2))$$
$$= (\underbrace{(1-\alpha) \cdot x_1 + \alpha \cdot x_2}_{=:x \in C}, \underbrace{(1-\alpha)f(x_1) + \alpha f(x_2)}_{=:y}) \in U(f).$$
Consequently, $(1-\alpha)f(x_1) + \alpha f(x_2) = y \geq f(x) = f\big((1-\alpha) \cdot x_1 + \alpha \cdot x_2\big)$.
Hence f is convex.

(Only if part:) Let $(x_1, y_1), (x_2, y_2) \in U(f)$ and $\alpha \in (0,1)$. Then we know that $y_1 \geq f(x_1)$ and $y_2 \geq f(x_2)$ and so
$$(1-\alpha)y_1 + \alpha y_2 \geq (1-\alpha)f(x_1) + \alpha f(x_2) \geq f\big((1-\alpha) \cdot x_1 + \alpha \cdot x_2\big).$$
Consequently, $\big((1-\alpha) \cdot x_1 + \alpha \cdot x_2, (1-\alpha)y_1 + \alpha y_2\big) \in U(f)$, that is,
$$(1-\alpha) \cdot (x_1, y_1) + \alpha \cdot (x_2, y_2) \in U(f).$$
So $U(f)$ is convex.

Solution to Exercise 3.9, page 129

We prove this using induction on n. The result is trivially true when $n = 1$, and in fact we have equality in this case. Suppose the inequality has been established for some $n \in \mathbb{N}$. If $x_1, \cdots, x_n, x_{n+1}$ are $n+1$ vectors, and $\alpha := \frac{1}{n+1} \in (0,1)$, then

$$f\left(\frac{1}{n+1}(x_1 + \cdots + x_n + x_{n+1})\right)$$
$$= f\left(\frac{n}{n+1} \cdot \frac{1}{n}(x_1 + \cdots + x_n) + \frac{1}{n+1} \cdot x_{n+1}\right)$$
$$= f\left(\left(1 - \frac{1}{n+1}\right) \cdot \frac{1}{n}(x_1 + \cdots + x_n) + \frac{1}{n+1}x_{n+1}\right)$$
$$= f\left((1-\alpha) \cdot \frac{1}{n}(x_1 + \cdots + x_n) + \alpha \cdot x_{n+1}\right)$$
$$\leq (1-\alpha) \cdot f\left(\frac{1}{n}(x_1 + \cdots + x_n)\right) + \alpha \cdot f(x_{n+1})$$
$$\leq (1-\alpha) \cdot \frac{f(x_1) + \cdots + f(x_n)}{n} + \alpha \cdot f(x_{n+1})$$
$$= \frac{\not{n}}{n+1} \cdot \frac{f(x_1) + \cdots + f(x_n)}{\not{n}} + \frac{1}{n+1} \cdot f(x_{n+1})$$
$$= \frac{f(x_1) + \cdots + f(x_n) + f(x_{n+1})}{n+1},$$

and so the claim follows for all n.

Solution to Exercise 3.10, page 130

We have for all $x \in \mathbb{R}$
$$f'(x) = \frac{2x}{2\sqrt{1+x^2}} = \frac{x}{\sqrt{1+x^2}},$$
$$f''(x) = \frac{1}{\sqrt{1+x^2}} + x\left(-\frac{1}{2}\right) \cdot \frac{2x}{\sqrt{(1+x^2)^3}} = \frac{1+x^2-x^2}{\sqrt{(1+x^2)^3}} = \frac{1}{\sqrt{(1+x^2)^3}} > 0.$$

Thus f is convex.

(Alternately, one could note that $(x,y) \mapsto \|(x,y)\|_2 = \sqrt{x^2+y^2}$ is a norm on $\mathbb{R}^2$, and so it is convex. Now fixing $y = 1$, and keeping x variable, we get convexity of $x \mapsto \sqrt{1+x^2}$.)

Solution to Exercise 3.11, page 132

For $\mathbf{x}_1, \mathbf{x}_2 \in C^1[0,1]$ and $\alpha \in (0,1)$, we have, using the convexity of function $\xi \mapsto \sqrt{1+\xi^2} : \mathbb{R} \to \mathbb{R}$ (Exercise 3.10, page 130), that

$$f((1-\alpha)\mathbf{x}_1 + \alpha\mathbf{x}_2) = \int_0^1 \sqrt{1 + ((1-\alpha)\mathbf{x}_1'(t) + \alpha\mathbf{x}_2'(t))^2}\,dt$$
$$\leq \int_0^1 \left((1-\alpha)\sqrt{1 + (\mathbf{x}_1'(t))^2} + \alpha\sqrt{1 + (\mathbf{x}_2'(t))^2}\right)dt$$
$$= (1-\alpha)\int_0^1 \sqrt{1 + (\mathbf{x}_1'(t))^2}\,dt + \alpha\int_0^1 \sqrt{1 + (\mathbf{x}_2'(t))^2}\,dt$$
$$= (1-\alpha)f(\mathbf{x}_1) + \alpha f(\mathbf{x}_2).$$

Solution to Exercise 3.12, page 133

(If:) Suppose that $\mathbf{x}_0(t) = \mathbf{0}$ for all $t \in [0,1]$. Then we have that for all $\mathbf{h} \in C[0,1]$,
$$(f'(\mathbf{x}_0))(\mathbf{h}) = 2\int_0^1 \mathbf{x}_0(t)\mathbf{h}(t)dt = 2\int_0^1 \mathbf{0} \cdot \mathbf{h}(t)dt = 2\int_0^1 0 dt = 2 \cdot 0 = 0,$$
and so $f'(\mathbf{x}_0) = \mathbf{0}$.

(Only if:) Now suppose that $f'(\mathbf{x}_0) = \mathbf{0}$. Thus for every $\mathbf{h} \in C[0,1]$, we have
$$(f'(\mathbf{x}_0))(\mathbf{h}) = 2\int_0^1 \mathbf{x}_0(t)\mathbf{h}(t)dt = 0.$$
In particular, taking $\mathbf{h} := \mathbf{x}_0 \in C[0,1]$, we obtain $2\int_0^1 \mathbf{x}_0(t)\mathbf{x}_0(t)dt = 0.$

So $\int_0^1 (\mathbf{x}_0(t))^2 dt = 0$. As $\mathbf{x}_0$ is continuous on $[0,1]$, it follows that $\mathbf{x}_0 = \mathbf{0}$.
By the necessary condition for $\mathbf{x}_0$ to be a minimiser, we have that $f'(\mathbf{x}_0) = \mathbf{0}$ and so $\mathbf{x}_0$ must be the zero function $\mathbf{0}$ on $[0,1]$. Furthermore, as f is convex and $f'(\mathbf{0}) = \mathbf{0}$, it follows that the zero function is a minimiser. Consequently, there exists a unique solution to the optimisation problem, namely the zero function $\mathbf{0} \in C[0,1]$. The conclusion is also obvious from the fact that for all $\mathbf{x} \in C[0,1]$,
$$f(\mathbf{x}) = \int_0^1 \underbrace{(\mathbf{x}(t))^2}_{\geqslant 0} dt \geqslant \int_0^1 0 dt = 0 = f(\mathbf{0}).$$

Solution to Exercise 3.13, page 141

We have $F(\xi, \eta, \tau) = \sqrt{1+\eta^2}$. Then $\dfrac{\partial F}{\partial \xi} = 0$ and $\dfrac{\partial F}{\partial \eta} = \dfrac{\eta}{\sqrt{1+\eta^2}}$.

The Euler-Lagrange equation is $0 - \dfrac{d}{dt}\left(\dfrac{\mathbf{x}'_*(t)}{\sqrt{1+(\mathbf{x}'_*(t))^2}}\right) = 0$, $t \in [a,b]$.

Upon integrating, we obtain $\dfrac{\mathbf{x}'_*(t)}{\sqrt{1+(\mathbf{x}'_*(t))^2}} = C$ on $[a,b]$ for some constant C.

Thus $(\mathbf{x}'_*(t))^2 = \dfrac{C^2}{1-C^2} =: A$, for all $t \in [a,b]$.

So $A \geqslant 0$, and $\mathbf{x}'_*(t) = \pm\sqrt{A}$ for each $t \in [a,b]$. As $\mathbf{x}'_*$ is continuous, we can conclude that $\mathbf{x}'_*$ must be either everywhere equal to $\sqrt{A}$, or everywhere equal to $-\sqrt{A}$. In either case, $\mathbf{x}'_*$ is constant, and so $\mathbf{x}_*$ is given by $\mathbf{x}_*(t) = \alpha t + \beta$, $t \in [a,b]$. Since $\mathbf{x}_*(a) = x_a$ and $\mathbf{x}_*(b) = x_b$, we have
$$\alpha = \frac{x_b - x_a}{b-a} \text{ and } \beta = \frac{bx_a - ax_b}{b-a},$$
and $\mathbf{x}_*(t) = \dfrac{x_b - x_a}{b-a}t + \dfrac{bx_a - ax_b}{b-a}$, for all $t \in [a,b]$.

That this $\mathbf{x}_* \in S$ is indeed a minimiser can be concluded by noticing that the map $\mathbf{x} \mapsto L(\gamma_\mathbf{x}) : S \to \mathbb{R}$ is convex, thanks to the convexity of $\eta \overset{\varphi}{\mapsto} \sqrt{1+\eta^2} : \mathbb{R} \to \mathbb{R}$;

$$\varphi''(\eta) = \frac{1}{(1+\eta^2)^{3/2}} \geq 0, \text{ for all } \eta \in \mathbb{R} \text{ (Exercise 3.10, page 130)}.$$

(The fact that $\mathbf{x}_*$ is a minimiser, is of course expected geometrically, since the straight line is the curve of shortest length between two points in the Euclidean plane.)

Solution to Exercise 3.14, page 141

We have $L_0(\gamma_\mathbf{x}) := \displaystyle\int_a^b \|\gamma'_\mathbf{x}(t)\|_0 dt = \int_a^b \sqrt{0\cdot (\mathbf{x}'(t))^2 + 1}\, dt = \int_a^b 1\, dt = b - a.$

Solution to Exercise 3.15, page 141

With $F(\xi,\eta,\tau) := \sqrt{1-\eta^2}$, we have
$$L(\gamma_\mathbf{x}) = \int_a^b \|\gamma'_\mathbf{x}(t)\|_{-1} dt \int_a^b \sqrt{|1-(\mathbf{x}'(t))^2|}\, dt$$
$$\stackrel{|\mathbf{x}'(t)|<1}{=} \int_a^b \sqrt{1-(\mathbf{x}'(t))^2}\, dt = \int_a^b F(\mathbf{x}(t),\mathbf{x}'(t),t)\, dt.$$

Then $\dfrac{\partial F}{\partial \xi} = 0$ and $\dfrac{\partial F}{\partial \eta} = \dfrac{-\eta}{\sqrt{1-\eta^2}}$.

The Euler-Lagrange equation is $0 - \dfrac{d}{dt}\left(\dfrac{-\mathbf{x}'_*(t)}{\sqrt{1+(\mathbf{x}'_*(t))^2}}\right) = 0,\ t \in [a,b].$

Upon integrating, we obtain $\dfrac{\mathbf{x}'_*(t)}{\sqrt{1-(\mathbf{x}'_*(t))^2}} = C$ on $[a,b]$ for some constant C.

Thus $\left(\mathbf{x}'_*(t)\right)^2 = \dfrac{C^2}{1+C^2} =: A,\ t \in [a,b].$

So $A \geq 0$, and $\mathbf{x}'_*(t) = \pm\sqrt{A}$ for each $t \in [a,b]$. As $\mathbf{x}'_*$ is continuous, we can conclude that $\mathbf{x}'_*$ must be either everywhere equal to $\sqrt{A}$, or everywhere equal to $-\sqrt{A}$. In either case, $\mathbf{x}'_*$ is constant, and so $\mathbf{x}_*$ is given by $\mathbf{x}_*(t) = \alpha t + \beta$, $t \in [a,b]$. Since $\mathbf{x}_*(a) = x_a$ and $\mathbf{x}_*(b) = x_b$, we have
$$\alpha = \frac{x_b - x_a}{b-a} \text{ and } \beta = \frac{bx_a - ax_b}{b-a},$$
and $\mathbf{x}_*(t) = \dfrac{x_b - x_a}{b-a} t + \dfrac{bx_a - ax_b}{b-a}$, for all $t \in [a,b]$.

We will now show that this $\mathbf{x}_*$ is a maximiser of $\mathbf{x} \mapsto L(\gamma_\mathbf{x}) : S \to \mathbb{R}$, that is, it is a minimiser of $\mathbf{x} \mapsto -L(\gamma_\mathbf{x})$. Note that the map $\eta \stackrel{\varphi}{\mapsto} -\sqrt{1-\eta^2} : (-1,1) \to \mathbb{R}$ is convex because
$$\varphi''(\eta) = \frac{1}{(1-\eta^2)^{3/2}} \geq 0, \text{ for all } \eta \in (-1,1).$$
Hence $\mathbf{x} \mapsto -L(\gamma_\mathbf{x}) : S \to \mathbb{R}$ is convex too, and this proves our claim.

Solution to Exercise 3.16, page 142

We have $F(\xi_1, \xi_2, \eta_1, \eta_2) = \dfrac{m}{2}\eta_1^2 + \dfrac{m}{2}\xi_1^2\eta_2^2 + \dfrac{GMm}{\xi_1}$. Thus

$$\dfrac{\partial F}{\partial \xi_1} = m\xi_1\eta_2^2 - \dfrac{GMm}{\xi_1^2}, \quad \dfrac{\partial F}{\partial \eta_1} = m\eta_1, \quad \dfrac{\partial F}{\partial \xi_2} = 0, \quad \dfrac{\partial F}{\partial \eta_2} = m\xi_1^2\eta_2.$$

So the Euler-Lagrange equations are

$$0 = m\mathbf{r}(t)\big(\varphi'(t)\big)^2 - \dfrac{GMm}{(\mathbf{r}(t))^2} - \dfrac{d}{dt}\big(m\mathbf{r}'(t)\big),$$

$$0 = 0 - \dfrac{d}{dt}\Big(m(\mathbf{r}(t))^2\varphi'(t)\Big),$$

that is,

$$\mathbf{r}''(t) = \mathbf{r}(t)\big(\varphi'(t)\big)^2 - \dfrac{GM}{(\mathbf{r}(t))^2}, \text{ and}$$

$$\dfrac{d}{dt}\Big(m(\mathbf{r}(t))^2\varphi'(t)\Big) = 0.$$

Solution to Exercise 3.17, page 143

(1) With $\mathcal{L}(X_1, X_2, U, V_1, V_2) := \sqrt{1 + V_1^2 + V_2^2}$, we have that

$$I(u) = \iint_\Omega \mathcal{L}(x, y, u, u_x, u_y)\,dx\,dy.$$

We have $\dfrac{\partial \mathcal{L}}{\partial U} = 0$, $\dfrac{\partial \mathcal{L}}{\partial V_1} = \dfrac{V_1}{\sqrt{1 + V_1^2 + V_2^2}}$, $\dfrac{\partial \mathcal{L}}{\partial V_2} = \dfrac{V_2}{\sqrt{1 + V_1^2 + V_2^2}}$.

So the Euler-Lagrange equation is

$$0 - \dfrac{\partial}{\partial x}\Big(\dfrac{u_x}{\sqrt{1 + u_x^2 + u_y^2}}\Big) - \dfrac{\partial}{\partial y}\Big(\dfrac{u_y}{\sqrt{1 + u_x^2 + u_y^2}}\Big) = 0.$$

We have

$$\dfrac{\partial}{\partial x}\Big(\dfrac{u_x}{\sqrt{1 + u_x^2 + u_y^2}}\Big)$$

$$= \dfrac{u_{xx}\sqrt{1 + u_x^2 + u_y^2} - u_x \cdot \big(\dfrac{1}{2\sqrt{1 + u_x^2 + u_y^2}}\big)(2u_x u_{xx} + 2u_y u_{xy})}{1 + u_x^2 + u_y^2}$$

$$= \dfrac{u_{xx}(1 + u_x^2 + u_y^2) - u_x(u_x u_{xx} + u_y u_{xy})}{(1 + u_x^2 + u_y^2)^{3/2}}$$

$$= \dfrac{u_{xx}(1 + u_y^2) - u_x u_y u_{xy}}{(1 + u_x^2 + u_y^2)^{3/2}}.$$

Similarly $\dfrac{\partial}{\partial y}\Big(\dfrac{u_y}{\sqrt{1 + u_x^2 + u_y^2}}\Big) = \dfrac{u_{yy}(1 + u_x^2) - u_x u_y u_{yx}}{(1 + u_x^2 + u_y^2)^{3/2}}.$

Thus the Euler-Lagrange equation becomes (using $u_{xy} = u_{yx}$)

$$u_{xx}(1 + u_y^2) - 2u_x u_y u_{xy} + u_{yy}(1 + u_x^2) = 0.$$

If $u = Ax + By + C$, then $u_{xx} = 0$, $u_{xy} = 0$ and $u_{yy} = 0$, so that all the three summands on the left-hand side of the Euler-Lagrange equation vanish, and so we see that the Euler-Lagrange equation is satisfied.

If $u = \tan^{-1}(y/x)$, then we have
$$u_x = \frac{1}{1+\frac{y^2}{x^2}} \cdot \left(-\frac{y}{x^2}\right) = \frac{-y}{x^2+y^2}, \text{ and } u_y = \frac{1}{1+\frac{y^2}{x^2}} \cdot \frac{1}{x} = \frac{x}{x^2+y^2}.$$

Thus $u_{xx} = \dfrac{2yx}{(x^2+y^2)^2}$, $u_{xy} = u_{yx} = \dfrac{y^2-x^2}{(x^2+y^2)^2}$, and $u_{yy} = \dfrac{-2xy}{(x^2+y^2)^2}$.

Hence
$$u_{xx}(1+u_y^2) - 2u_x u_y u_{xy} + u_{yy}(1+u_x^2)$$
$$= \frac{2yx}{(x^2+y^2)^2}\left(1 + \frac{x^2}{(x^2+y^2)^2}\right)$$
$$+ \frac{2xy}{(x^2+y^2)^2} \cdot \frac{(y^2-x^2)}{(x^2+y^2)^2} - \frac{2xy}{(x^2+y^2)^2}\left(1 + \frac{y^2}{(x^2+y^2)^2}\right)$$
$$= \frac{2xy(x^2-y^2) + 2xy(y^2-x^2)}{(x^2+y^2)^4} = 0.$$

With $s := \sqrt{x^2+y^2}$ and $t = \tan^{-1}(y/x) = u$, we have $\tan t = \dfrac{y}{x}$, and so
$$\frac{1}{(\cos t)^2} = 1 + (\tan t)^2 = \frac{x^2+y^2}{x^2}.$$

Thus $x = \sqrt{x^2+y^2} \cdot \cos t = s \cdot \cos t$. Then
$$y = x \cdot \tan t = (s \cdot \cos t) \cdot \tan t = s \cdot \sin t.$$

Vice versa, if $x = s \cdot \cos t$, $y = s \cdot \sin t$ and $u = t$, then
$$x^2 + y^2 = (s \cdot \cos t)^2 + (s \cdot \sin t)^2 = s^2 \cdot 1 = s^2,$$
and so $s = \sqrt{x^2+y^2}$. Also $\dfrac{y}{x} = \dfrac{s \cdot \cos t}{s \cdot \sin t} = \tan t$, and so $u = \tan^{-1}(y/x) = t$.

Using the Maple command given in the exercise we obtain the following:

(2) If $\mathcal{L}(X_1, X_2, U, V_1, V_2) := \dfrac{V_2^2 - V_1^2}{2}$, then $I(u) = \displaystyle\int_0^T \int_0^1 \mathcal{L}(x, t, u, u_x, u_t) dx dt$.

We have $\dfrac{\partial \mathcal{L}}{\partial U} = 0$, $\dfrac{\partial \mathcal{L}}{\partial V_1} = -V_1$, $\dfrac{\partial \mathcal{L}}{\partial V_2} = V_2$.

So the Euler-Lagrange equation is:
$$0 - \frac{\partial}{\partial x}\left(-\frac{\partial u_*}{\partial x}\right) - \frac{\partial}{\partial t}\left(\frac{\partial u_*}{\partial t}\right) = 0.$$
Thus u_* satisfies the wave equation $\dfrac{\partial^2 u_*}{\partial t^2} - \dfrac{\partial^2 u_*}{\partial x^2} = 0.$

We can check this by direct differentiation that the given u in terms of f satisfies the wave equation. We have
$$\frac{\partial u}{\partial t} = \frac{\partial}{\partial t}\left(\frac{f(x-t)+f(x+t)}{2}\right) = \frac{-f'(x-t)+f'(x+t)}{2}.$$
Differentiating again with respect to t, we obtain
$$\frac{\partial^2 u}{\partial t^2} = \frac{f''(x+t)+f''(x-t)}{2}. \qquad (*)$$
Similarly, by differentiating u with respect to x we obtain
$$\frac{\partial u}{\partial x} = \frac{\partial}{\partial x}\left(\frac{f(x-t)+f(x+t)}{2}\right) = \frac{f'(x-t)+f'_*(x+t)}{2}.$$
Differentiating again with respect to x, we obtain
$$\frac{\partial^2 u}{\partial x^2} = \frac{f''(x+t)+f''(x-t)}{2}. \qquad (**)$$
It follows from $(*)$ and $(**)$ that $\dfrac{\partial^2 u}{\partial t^2} - \dfrac{\partial^2 u}{\partial x^2} = 0.$

Let us check that the boundary conditions are satisfied.

Note that $u(0,t) = \dfrac{f(-t)+f(t)}{2} = 0$ since f is odd.

Now we would like to check $u(1,t) = 0$ too.

Using the oddness and 2-periodicity of f, we have
$$f(1+t) = f(1+t-2) = f(t-1) = f(-(1-t)) = -f(1-t).$$
So $u(1,t) = \dfrac{f(1-t)+f(1+t)}{2} = 0.$

Finally, we can check if the initial conditions is satisfied.

We have $u(x,0) = \dfrac{f(x)+f(x)}{2} = f(x)$ for all x.

Also, from our previous calculation, we have
$$\frac{\partial u}{\partial t}(x,0) = \frac{f'(x+0)-f'(x-0)}{2} = \frac{f'(x)-f'(x)}{2} = 0$$
for all x.

For a fixed t, the graph of $f(\cdot - t)$ is just a shifted version of the graph of f by t units to the right. As t increases, the graph travels to the right, representing a travelling wave, moving to the right with a speed 1. Similarly the graph of $f(\cdot + t)$ with increasing t represents a travelling wave moving to the left with speed 1. The solution of the wave equation is an average of these two travelling waves moving in opposite directions, and the shape of the wave is determined by the initial shape of the string.

Solution to Exercise 3.18, page 153

We have (suppressing the argument (q,p) everywhere)
$$\{F,G\} = \frac{\partial F}{\partial q} \cdot \frac{\partial G}{\partial p} - \frac{\partial G}{\partial q} \cdot \frac{\partial F}{\partial p} = -\Big(\frac{\partial G}{\partial q} \cdot \frac{\partial F}{\partial p} - \frac{\partial F}{\partial q} \cdot \frac{\partial G}{\partial p}\Big) = -\{G,F\}.$$
Also,
$$\begin{aligned}\{\alpha F + \beta G, H\} &= \frac{\partial(\alpha F + \beta G)}{\partial q} \cdot \frac{\partial H}{\partial p} - \frac{\partial H}{\partial q} \cdot \frac{\partial(\alpha F + \beta G)}{\partial p} \\ &= \Big(\alpha\frac{\partial F}{\partial q} + \beta\frac{\partial G}{\partial q}\Big)\frac{\partial H}{\partial p} - \frac{\partial H}{\partial q}\Big(\alpha\frac{\partial F}{\partial p} + \beta\frac{\partial G}{\partial p}\Big) \\ &= \alpha\Big(\frac{\partial F}{\partial q} \cdot \frac{\partial H}{\partial p} - \frac{\partial H}{\partial q} \cdot \frac{\partial F}{\partial p}\Big) + \beta\Big(\frac{\partial G}{\partial q} \cdot \frac{\partial H}{\partial p} - \frac{\partial H}{\partial q} \cdot \frac{\partial G}{\partial p}\Big) \\ &= \alpha\{F,H\} + \beta\{G,H\}.\end{aligned}$$

Finally, we will prove the Jacobi Identity. In order to simplify the notation, we will use subscripts to denote partial derivatives, for example F_p will mean $\frac{\partial F}{\partial p}$. First we note that
$$\begin{aligned}\{F,\{G,H\}\} &= F_q\{G,H\}_p - F_p\{G,H\}_q \\ &= F_q(G_qH_p - G_pH_q)_p - F_p(G_qH_p - G_pH_q)_q \\ &= F_q(G_{qp}H_p + G_qH_{pp} - G_{pp}H_q - G_pH_{qp}) \\ &\quad - F_p(G_{qq}H_p + G_qH_{pq} - G_{pq}H_q - G_pH_{qq}).\end{aligned}$$
Similarly, by making cyclic substitutions $F \to G \to H$ above, we obtain
$$\begin{aligned}\{G,\{H,F\}\} &= G_q(H_{qp}F_p + H_qF_{pp} - H_{pp}F_q - H_pF_{qp}) \\ &\quad - G_p(H_{qq}F_p + H_qF_{pq} - H_{pq}F_q - H_pF_{qq}), \\ \{H,\{F,G\}\} &= H_q(F_{qp}G_p + F_qG_{pp} - F_{pp}G_q - F_pG_{qp}) \\ &\quad - H_p(F_{qq}G_p + F_qG_{pq} - F_{pq}G_q - F_pG_{qq}).\end{aligned}$$
Thanks to the symmetry of the left-hand side of the expression in Jacobi's Identity in F, G, H, it is enough to show that after collecting all the F_q, F_p terms, their overall coefficients are zero.

The overall coefficient of F_q is
$$G_{qp}H_p + G_qH_{pp} - G_{pp}H_q - G_pH_{qp} - G_qH_{pp} + G_pH_{pq} + G_{pp}H_q - G_{pq}H_p.$$
Since $G_{pq} = G_{qp}$ and $H_{pq} = H_{qp}$, we see that the above expression is 0.
The overall coefficient of F_p is
$$-G_{qq}H_p - G_qH_{pq} + G_{pq}H_q + G_pH_{qq} + G_qH_{qp} - G_pH_{qq} - G_{qp}H_q + G_{qq}H_p = 0.$$
This completes the proof of the Jacobi Identity.

Solution to Exercise 3.19, page 153

We have $\{Q,P\} = \dfrac{\partial Q}{\partial q} \cdot \dfrac{\partial P}{\partial p} - \dfrac{\partial P}{\partial q} \cdot \dfrac{\partial Q}{\partial p} = 1 \cdot 1 - 0 \cdot 0 = 1.$

Solutions to the exercises from Chapter 4

Solution to Exercise 4.1, page 162

With $\mathbf{x} := \mathbf{1} = (t \mapsto 1)$, and $\mathbf{y} := (t \mapsto t)$, $2\|\mathbf{x}\|_\infty^2 + 2\|\mathbf{y}\|_\infty^2 = 2 \cdot 1^2 + 2 \cdot 1^2 = 4$, while $\|\mathbf{x} + \mathbf{y}\|_\infty^2 + \|\mathbf{x} - \mathbf{y}\|_\infty^2 = \|1 + t\|_\infty^2 + \|1 - t\|_\infty^2 = 2^2 + 1^2 = 5$. So $\|\cdot\|_\infty$ does not obey the Parallelogram Law, and hence $\|\cdot\|_\infty$ cannot be a norm induced by some inner product on $C[0, 1]$.

Solution to Exercise 4.2, page 162

Let $x, y, z \in X$. Then
$$\|z - x\|^2 = \|z\|^2 - \langle x, z \rangle - \langle z, x \rangle + \|x\|^2,$$
$$\|z - y\|^2 = \|z\|^2 - \langle y, z \rangle - \langle z, y \rangle + \|y\|^2,$$
$$-\frac{1}{2}\|x - y\|^2 = -\frac{1}{2}\|x\|^2 + \frac{1}{2}\langle y, x \rangle + \frac{1}{2}\langle x, y \rangle - \frac{1}{2}\|y\|^2,$$
$$-2\left\|z - \frac{x+y}{2}\right\|^2 = -2\|z\|^2 + 2\left\langle \frac{x+y}{2}, z \right\rangle + 2\left\langle z, \frac{x+y}{2} \right\rangle - 2\left\|\frac{x+y}{2}\right\|^2.$$

Adding these, we obtain
$$\|z - x\|^2 + \|z - y\|^2 - \frac{1}{2}\|x - y\|^2 - 2\left\|z - \frac{1}{2}(x+y)\right\|^2$$
$$= \frac{1}{2}\|x\|^2 + \frac{1}{2}\|y\|^2 + \frac{1}{2}\langle y, x \rangle + \frac{1}{2}\langle x, y \rangle - 2 \cdot \frac{1}{4}\|x + y\|^2$$
$$= \frac{1}{2}\left(\|x\|^2 + \|y\|^2 + \langle y, x \rangle + \langle x, y \rangle - \langle x + y, x + y \rangle\right) = 0.$$

Geometric interpretation in $\mathbb{R}^2$: If x, y, z are the vertices of a triangle ABC, then $\left\|z - \frac{x+y}{2}\right\|$ is the length of the median AD (see the picture).

The Appollonius Identity gives $AB^2 + AC^2 = \frac{1}{2}BC^2 + 2AD^2$.

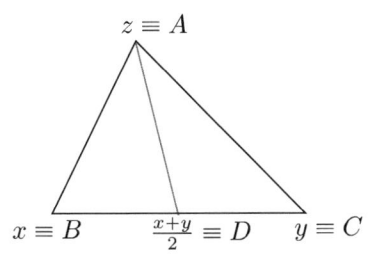

$$z \equiv A$$

$$x \equiv B \qquad \frac{x+y}{2} \equiv D \qquad y \equiv C$$

Solution to Exercise 4.3, page 162

Let $\epsilon > 0$. Let $N_1 \in \mathbb{N}$ be such that for all $n > N_1$, $\|x_n - x\| < \dfrac{\epsilon}{2(\|y\| + 1)}$.

314   A Friendly Approach to Functional Analysis

Let $N_2 \in \mathbb{N}$ be such that for all $n > N_2$, $\|y_n - y\| < \dfrac{\epsilon}{2(M+1)}$, where the number $M := \sup\limits_{n \in \mathbb{N}} \|x_n\| < \infty$ (this exists since $(x_n)_{n \in \mathbb{N}}$, being convergent, is bounded).
Consequently, for all $n > N := \max\{N_1, N_2\}$,

$$\begin{aligned}
|\langle x_n, y_n \rangle - \langle x, y \rangle| &= |\langle x_n, y_n \rangle - \langle x_n, y \rangle + \langle x_n, y \rangle - \langle x, y \rangle| \\
&= |\langle x_n, y_n - y \rangle + \langle x_n - x, y \rangle| \\
&\leq |\langle x_n, y_n - y \rangle| + |\langle x_n - x, y \rangle| \\
&\leq \|x_n\|\|y_n - y\| + \|x_n - x\|\|y\| \\
&\leq M \cdot \frac{\epsilon}{2(M+1)} + \frac{\epsilon}{2(\|y\|+1)}\|y\| \leq \epsilon.
\end{aligned}$$

Hence $(\langle x_n, y_n \rangle)_{n \in \mathbb{N}}$ is convergent in $\mathbb{K}$, with limit $\langle x, y \rangle$.

Solution to Exercise 4.4, page 162

If the ellipse has major and minor axis lengths as $2a$ and $2b$, respectively, then observe that the perimeter is given by

$$P = \int_0^{2\pi} \sqrt{(a\cos t)^2 + (b\sin t)^2}\, dt = \int_0^{2\pi} \sqrt{(a\sin t)^2 + (b\cos t)^2}\, dt,$$

where the last expression is obtained by rotating the ellipse through $90°$, obtaining a new ellipse with the same perimeter.

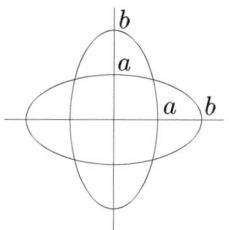

Using Cauchy-Schwarz Inequality we obtain

$$\begin{aligned}
P^2 &= \left(\int_0^{2\pi} \sqrt{(a\cos t)^2 + (b\sin t)^2}\, dt\right)\left(\int_0^{2\pi} \sqrt{(a\sin t)^2 + (b\cos t)^2}\, dt\right) \\
&\geq \left(\int_0^{2\pi} \sqrt[4]{((a\cos t)^2 + (b\sin t)^2)((a\sin t)^2 + (b\cos t)^2)}\, dt\right)^2 \\
&\geq \left(\int_0^{2\pi} \sqrt[4]{((a\cos t)(b\cos t) + (b\sin t)(a\sin t))^2}\, dt\right)^2 \\
&= \left(\int_0^{2\pi} \sqrt{ab}\, dt\right)^2 = \left(2\pi\sqrt{ab}\right)^2.
\end{aligned}$$

Thus $P \geq 2\pi\sqrt{ab}$. Since the areas of the circle and the ellipse are equal, it follows that $\pi r^2 = \pi ab$, where r denotes the radius of the circle. Hence $r = \sqrt{ab}$. So we have $P \geq 2\pi\sqrt{ab} = 2\pi r$, that is, the perimeter P of the ellipse is at least as large as the circumference of the circle.

Solution to Exercise 4.5, page 163

(IP1) If $A \in \mathbb{R}^{m \times n}$, then $\langle A, A \rangle = \text{tr}(A^\top A) = \sum_{i=1}^{n} \sum_{k=1}^{m} a_{ki} a_{ki} = \sum_{i=1}^{n} \sum_{k=1}^{m} a_{ki}^2 \geq 0$.

If $A \in \mathbb{R}^{m \times n}$ and $\langle A, A \rangle = 0$, then $\sum_{i=1}^{n} \sum_{k=1}^{m} a_{ki}^2 = 0$, and so for all $k \in \{1, \cdots, m\}$ and all $i \in \{1, \cdots, n\}$, $a_{ki} = 0$, that is, $A = 0$.

(IP2) For all $A_1, A_2, B \in \mathbb{R}^{m \times n}$,
$$\langle A_1 + A_2, B \rangle = \text{tr}((A_1 + A_2)^\top B) = \text{tr}(A_1^\top B) + \text{tr}(A_2^\top B) = \langle A_1, B \rangle + \langle A_2, B \rangle.$$
For all $A, B \in \mathbb{R}^{m \times n}$ and $\alpha \in \mathbb{R}$,
$$\langle \alpha A, B \rangle = \text{tr}((\alpha A)^\top B) = \alpha \text{tr}(A^\top B) = \alpha \langle A, B \rangle.$$

(IP3) For all $A, B \in \mathbb{R}^{m \times n}$,
$$\langle A, B \rangle = \text{tr}(A^\top B) = \text{tr}((A^\top B)^\top) = \text{tr}(B^\top A^{\top\top}) = \text{tr}(B^\top A) = \langle B, A \rangle.$$

This is a Hilbert space, since finite-dimensional normed spaces are complete.

Solution to Exercise 4.6, page 163

Let $x, y \in X$. Then
$$\begin{aligned} 0 &= \langle T(x+y), x+y \rangle = \langle Tx + Ty, x + y \rangle \\ &= \langle Tx, x \rangle + \langle Tx, y \rangle + \langle Ty, x \rangle + \langle Ty, y \rangle \\ &= 0 + \langle Tx, y \rangle + \langle Ty, x \rangle + 0 \\ &= \langle Tx, y \rangle + \langle Ty, x \rangle. \quad (*) \end{aligned}$$

Also,
$$\begin{aligned} 0 &= \langle T(x + iy), x + iy \rangle = \langle Tx + iTy, x + iy \rangle \\ &= \langle Tx, x \rangle - i\langle Tx, y \rangle + i\langle Ty, x \rangle + i(-i)\langle Ty, y \rangle \\ &= 0 - i\langle Tx, y \rangle + i\langle Ty, x \rangle + 0 \\ &= i(\langle Ty, x \rangle - \langle Tx, y \rangle). \quad (**) \end{aligned}$$

From (*) and (**) it follows that for all $x, y \in X$, $\langle Tx, y \rangle = 0$.
In particular, with $y = Tx$, we get $\langle Tx, Tx \rangle = 0$, that is, $\|Tx\|^2 = 0$.
Hence for all $x \in X$, $Tx = 0$, that is, $T = \mathbf{0}$.

We have $\langle Tx, x \rangle = \left\langle \begin{bmatrix} -x_2 \\ x_1 \end{bmatrix}, \begin{bmatrix} x_1 \\ x_2 \end{bmatrix} \right\rangle = -x_2 x_1 + x_1 x_2 = 0$, for all $x = \begin{bmatrix} x_1 \\ x_2 \end{bmatrix} \in \mathbb{R}^2$.

There is no contradiction to the previous part since the vector space $\mathbb{R}^2$ is a vector space over the *real* scalars.

Solution to Exercise 4.7, page 163

R is an equivalence relation on $\mathcal{C}$:

(ER1) If $\mathbf{x} = (x_n)_{n \in \mathbb{N}} \in \mathcal{C}$, then $\lim_{n \to \infty} \|x_n - x_n\|_X = \lim_{n \to \infty} 0 = 0$, and so $(\mathbf{x}, \mathbf{x}) \in R$.

(ER2) If $\mathbf{x} = (x_n)_{n \in \mathbb{N}}$, $\mathbf{y} = (y_n)_{n \in \mathbb{N}} \in \mathcal{C}$, and $(\mathbf{x}, \mathbf{y}) \in R$, then $\lim_{n \to \infty} \|x_n - y_n\|_X = 0$.
So $\lim_{n \to \infty} \|y_n - x_n\|_X = \lim_{n \to \infty} |-1| \|x_n - y_n\|_X = \lim_{n \to \infty} \|x_n - y_n\|_X = 0$.
Hence $(\mathbf{y}, \mathbf{x}) \in R$.

(ER3) Let $\mathbf{x} = (x_n)_{n\in\mathbb{N}}$, $\mathbf{y} = (y_n)_{n\in\mathbb{N}}$, $\mathbf{z} = (z_n)_{n\in\mathbb{N}}$ be in $\mathcal{C}$, such that $(\mathbf{x},\mathbf{y}) \in R$ and $(\mathbf{y},\mathbf{z}) \in R$. Then $\lim_{n\to\infty} \|x_n - y_n\|_X = 0$ and $\lim_{n\to\infty} \|y_n - z_n\|_X = 0$. As $0 \leq \|x_n - z_n\|_X \leq \|x_n - y_n\|_X + \|y_n - z_n\|_X$, we get $\lim_{n\to\infty} \|x_n - z_n\|_X = 0$. So $(\mathbf{x},\mathbf{z}) \in R$.

Consequently, R is an equivalence relation on $\mathcal{C}$.

$+ : \overline{X} \times \overline{X} \to \overline{X}$ **is well-defined:**

If $[(x_n)_{n\in\mathbb{N}}] = [(x'_n)_{n\in\mathbb{N}}]$ and $[(y_n)_{n\in\mathbb{N}}] = [(y'_n)_{n\in\mathbb{N}}]$, then we wish to show that $[(x_n + y_n)_{n\in\mathbb{N}}] = [(x'_n + y'_n)_{n\in\mathbb{N}}]$. We have that $(x_n + y_n)_{n\in\mathbb{N}} \in \mathcal{C}$, since $(x_n)_{n\in\mathbb{N}}, (y_n)_{n\in\mathbb{N}} \in \mathcal{C}$ and $\|x_n + y_n - (x_m + y_m)\|_X \leq \|x_n - x_m\|_X + \|y_n - y_m\|_X$. Similarly, $(x'_n + y'_n)_{n\in\mathbb{N}} \in \mathcal{C}$.

Furthermore, $0 \leq \|(x_n + y_n) - (x'_n + y'_n)\|_X \leq \|x_n - x'_n\|_X + \|y_n - y'_n\|_X$, and so
$$\lim_{n\to\infty} \|(x_n + y_n) - (x'_n + y'_n)\|_X = 0,$$
that is, $\big((x_n + y_n)_{n\in\mathbb{N}}, (x'_n + y'_n)_{n\in\mathbb{N}}\big) \in R$. So $[(x_n + y_n)_{n\in\mathbb{N}}] = [(x'_n + y'_n)_{n\in\mathbb{N}}]$.

$\cdot : \mathbb{K} \times \overline{X} \to \overline{X}$ **is well-defined:**

Let $\alpha \in \mathbb{K}$ and $[(x_n)_{n\in\mathbb{N}}] = [(x'_n)_{n\in\mathbb{N}}]$. Since $\|\alpha x_n - \alpha x_m\|_X = |\alpha|\|x_n - x_m\|_X$, clearly $(\alpha x_n)_{n\in\mathbb{N}} \in \mathcal{C}$. Similarly, $(\alpha x'_n)_{n\in\mathbb{N}} \in \mathcal{C}$. We have
$$\lim_{n\to\infty} \|\alpha x_n - \alpha x'_n\|_X = \lim_{n\to\infty} |\alpha|\|x_n - x'_n\|_X = |\alpha| \cdot 0 = 0,$$
and so $\big((\alpha x_n)_{n\in\mathbb{N}}, (\alpha x'_n)_{n\in\mathbb{N}}\big) \in R$. So $[(\alpha x_n)_{n\in\mathbb{N}}] = [(\alpha x'_n)_{n\in\mathbb{N}}]$.

$\langle \cdot, \cdot \rangle : \overline{X} \times \overline{X} \to \mathbb{K}$ **is well-defined:**

Since Cauchy sequences are bounded, given $(x_n)_{n\in\mathbb{N}}, (y_n)_{n\in\mathbb{N}}$ in $\mathcal{C}$, we have that $M_{\mathbf{x}} := \sup_{n\in\mathbb{N}} \|x_n\|_X < \infty$ and $M_{\mathbf{y}} := \sup_{n\in\mathbb{N}} \|y_n\|_X < \infty$.

Let N be large enough so that if $m, n > N$, then
$$\|x_n - x_m\|_X < \frac{\epsilon}{2(M_{\mathbf{y}} + 1)}, \quad \|y_n - y_m\|_X < \frac{\epsilon}{2(M_{\mathbf{x}} + 1)}.$$
Thus for $m, n > N$,
$$|\langle x_n, y_n \rangle_X - \langle x_m, y_m \rangle_X| = |\langle x_n, y_n \rangle_X - \langle x_n, y_m \rangle_X + \langle x_n, y_m \rangle_X - \langle x_m, y_m \rangle_X|$$
$$= |\langle x_n, y_n - y_m \rangle_X + \langle x_n - x_m, y_m \rangle_X|$$
$$\leq |\langle x_n, y_n - y_m \rangle_X| + |\langle x_n - x_m, y_m \rangle_X|$$
$$\leq \|x_n\|_X \|y_n - y_m\|_X + \|x_n - x_m\|_X \|y_m\|_X$$
$$\leq M_{\mathbf{x}} \cdot \frac{\epsilon}{2(M_{\mathbf{x}} + 1)} + M_{\mathbf{y}} \cdot \frac{\epsilon}{2(M_{\mathbf{y}} + 1)} < \epsilon.$$
So $\big(\langle x_n, y_n \rangle_X\big)_{n\in\mathbb{N}}$ is a Cauchy sequence in $\mathbb{K}$, and as $\mathbb{K}$ ($= \mathbb{R}$ or $\mathbb{C}$) is complete, it follows that $\lim_{n\to\infty} \langle x_n, y_n \rangle_X$ exists.

Now suppose that $[(x_n)_{n\in\mathbb{N}}] = [(x'_n)_{n\in\mathbb{N}}]$ and $[(y_n)_{n\in\mathbb{N}}] = [(y'_n)_{n\in\mathbb{N}}]$.

Given $\epsilon > 0$, let N be such that for all $n > N$,
$$\|x_n - x'_n\|_X < \frac{\epsilon}{2(M_\mathbf{y} + 1)} \text{ and } \|y_n - y'_n\|_X < \frac{\epsilon}{2(M_{\mathbf{x}'} + 1)},$$
where $M_{\mathbf{x}'} = \sup_{n \in \mathbb{N}} \|x'_n\|_X < \infty$. For $n > N$, we have

$$|\langle x_n, y_n \rangle_X - \langle x'_n, y'_n \rangle_X| = |\langle x_n, y_n \rangle_X - \langle x'_n, y_n \rangle_X + \langle x'_n, y_n \rangle_X - \langle x'_n, y'_n \rangle_X|$$
$$= |\langle x_n - x'_n, y_n \rangle_X + \langle x'_n, y_n - y'_n \rangle_X|$$
$$\leqslant |\langle x_n - x'_n, y_n \rangle_X| + |\langle x'_n, y_n - y'_n \rangle_X|$$
$$\leqslant \|x_n - x'_n\|_X \|y_n\|_X + \|x'_n\|_X \|y_n - y'_n\|_X$$
$$\leqslant \frac{\epsilon}{2(M_\mathbf{y} + 1)} \cdot M_\mathbf{y} + M_{\mathbf{x}'} \cdot \frac{\epsilon}{2(M_{\mathbf{x}'} + 1)} < \epsilon.$$

Passing the limit as $n \to \infty$, we obtain $\left| \lim_{n \to \infty} \langle x_n, y_n \rangle_X - \lim_{n \to \infty} \langle x'_n, y'_n \rangle_X \right| \leqslant \epsilon$.

$\langle \cdot, \cdot \rangle$ defines an inner product on $\overline{X}$:

(IP1) If $\mathbf{x} = [(x_n)_{n \in \mathbb{N}}] \in \overline{X}$, then $\langle \mathbf{x}, \mathbf{x} \rangle_{\overline{X}} = \lim_{n \to \infty} \langle x_n, x_n \rangle_X \geqslant 0$.

Let $\mathbf{x} = [(x_n)_{n \in \mathbb{N}}] \in \overline{X}$ be such that $\langle \mathbf{x}, \mathbf{x} \rangle_{\overline{X}} = 0$.
Then $\lim_{n \to \infty} \langle x_n, x_n \rangle_X = \lim_{n \to \infty} \|x_n\|_X^2 = 0$.
$(\mathbf{0})_{n \in \mathbb{N}} \in \mathcal{C}$ and $\lim_{n \to \infty} \|x_n - \mathbf{0}\|_X = \lim_{n \to \infty} \|x_n\|_X = 0$ (using the above).
Thus $[(x_n)_{n \in \mathbb{N}}] = [(\mathbf{0})_{n \in \mathbb{N}}]$.

(IP2) For all $\mathbf{x}_1, \mathbf{x}_2, \mathbf{y} \in \overline{X}$,
$$\langle \mathbf{x}_1 + \mathbf{x}_2, \mathbf{y} \rangle_{\overline{X}} = \lim_{n \to \infty} \langle x_{1,n} + x_{2,n}, y_n \rangle_X$$
$$= \lim_{n \to \infty} \left(\langle x_{1,n}, y_n \rangle_X + \langle x_{2,n}, y_n \rangle_X \right)$$
$$= \lim_{n \to \infty} \langle x_{1,n}, y_n \rangle_X + \lim_{n \to \infty} \langle x_{2,n}, y_n \rangle_X = \langle \mathbf{x}_1, \mathbf{y} \rangle_{\overline{X}} + \langle \mathbf{x}_2, \mathbf{y} \rangle_{\overline{X}}.$$

For all $\alpha \in \mathbb{K}$ and $\mathbf{x}, \mathbf{y} \in \overline{X}$, we have
$$\langle \alpha \mathbf{x}, \mathbf{y} \rangle_{\overline{X}} = \lim_{n \to \infty} \langle \alpha x_n, y_n \rangle_X = \lim_{n \to \infty} \alpha \langle x_n, y_n \rangle_X$$
$$= \alpha \lim_{n \to \infty} \langle x_n, y_n \rangle_X = \alpha \langle \mathbf{x}, \mathbf{y} \rangle_{\overline{X}}.$$

(IP3) For all $\mathbf{x}, \mathbf{y} \in \overline{X}$, $\langle \mathbf{x}, \mathbf{y} \rangle_{\overline{X}} = \lim_{n \to \infty} \langle x_n, y_n \rangle_X = \lim_{n \to \infty} \langle y_n, x_n \rangle_X^*$
$$= \left(\lim_{n \to \infty} \langle y_n, x_n \rangle_X \right)^* = \langle \mathbf{y}, \mathbf{x} \rangle_{\overline{X}}^*.$$

ι is a linear transformation:

(L1) $\iota(x + y) = [(x + y)_{n \in \mathbb{N}}] = [(x)_{n \in \mathbb{N}}] + [(y)_{n \in \mathbb{N}}] = \iota(x) + \iota(y)$ for all $x, y \in X$.

(L2) $\iota(\alpha x) = [(\alpha x)_{n \in \mathbb{N}}] = \alpha[(x)_{n \in \mathbb{N}}] = \alpha \iota(x)$ for all $\alpha \in \mathbb{K}$ and all $x \in X$.

ι is injective:

If $\iota(x) = [(x)_{n \in \mathbb{N}}] = [(\mathbf{0})_{n \in \mathbb{N}}]$, then $\|x\| = \lim_{n \to \infty} \|x - \mathbf{0}\| = 0$, and so $x = \mathbf{0}$.

ι preserves inner products: For $x, y \in X$, $\langle \iota(x), \iota(y) \rangle_{\overline{X}} = \lim_{n \to \infty} \langle x, y \rangle_X = \langle x, y \rangle_X$.

Solution to Exercise 4.8, page 168

As span$\{v_1\}$ = span$\{x_1\}$ = span$\{u_1\}$, it follows that $v_1 = \alpha_1 u_1$.
Thus $1 = \|v_1\| = |\alpha_1|\|u_1\| = |\alpha_1| \cdot 1 = |\alpha_1|$.
For $n > 1$, $v_n \in \text{span}\{v_1, \cdots, v_n\} = \text{span}\{x_1, \cdots, x_n\} = \text{span}\{u_1, \cdots, u_n\}$.
So there are scalars $\beta_1, \cdots, \beta_{n-1}, \alpha_n$ such that $v_n = \beta_1 u_1 + \cdots + \beta_{n-1} u_{n-1} + \alpha_n u_n$.
We also know that for all $k < n$, $\langle v_n, v_k \rangle = 0$. So it follows that $\langle v_n, v \rangle = 0$ for all $v \in \text{span}\{v_1, \cdots, v_{n-1}\} = \text{span}\{x_1, \cdots, x_{n-1}\} = \text{span}\{u_1, \cdots, u_{n-1}\}$. Thus $\langle v_n, u_k \rangle = 0$ for all $k < n$. This gives $\beta_1 = \cdots = \beta_{n-1} = 0$, and $v_n = \alpha_n u_n$.
Moreover, $1 = \|v_n\| = |\alpha_n|\|u_n\| = |\alpha_n| \cdot 1 = |\alpha_n|$.

Solution to Exercise 4.9, page 171

Let us first note that the derivative of an even monomial t^{2k} is odd, and that of an odd monomial t^{2k+1} is even. From here it follows that the derivative of a polynomial with only even monomials is a polynomial consisting of only odd monomials, while that of a polynomial with only odd monomials is a polynomial with only even monomials.

By the Binomial Theorem, we see that the polynomial $(t^2 - 1)^n$ is the sum of even monomials of the form $c_k t^{2k}$, for suitable scalars c_k, $k = 0, \cdots, n$.

So $\left(\dfrac{d}{dt}\right)^n (t^2 - 1)^n$ will be a polynomial $\mathbf{p}$ with:

(1) only even monomials if n is even,

(2) only odd monomials if n is odd.

In the former case, when n is even, $\mathbf{p}$, being the sum of even functions will be even, while in the latter case, $\mathbf{p}$, being the sum of odd functions, will be odd. Thus $\mathbf{P}_n$ is even when n is even, and odd if n is odd.

If n is odd, then each of the terms $c_k t^{2k-n}$ is an odd polynomial, and hence so is their sum. Consequently, $\mathbf{P}_n$ is odd if n is odd.

We have $\mathbf{P}_n(-1) = (-1)^n \mathbf{P}_n(1) = (-1)^n \cdot 1 = (-1)^n$ for all $n \geqslant 0$.

Solution to Exercise 4.10, page 171

With $\mathbf{y}(t) := (t^2 - 1)^n$, we have $\mathbf{y}'(t) = n(t^2 - 1)^{n-1} \cdot 2t$. So
$$(t^2 - 1)\mathbf{y}'(t) = (t^2 - 1)n(t^2 - 1)^{n-1} 2t = 2nt(t^2 - 1)^n = 2nt\mathbf{y}(t). \quad (*)$$
By differentiating the left-hand side of $(*)$, we obtain
$$\left(\frac{d}{dt}\right)^{n+1}((t^2-1)\mathbf{y}'(t)) = \sum_{k=0}^{n+1}\binom{n+1}{k}\left(\frac{d}{dt}\right)^k (t^2-1) \cdot \left(\frac{d}{dt}\right)^{n+1-k}\mathbf{y}'(t)$$
$$= (t^2 - 1)\left(\frac{d}{dt}\right)^{n+1}\mathbf{y}'(t) + (n+1)2t\left(\frac{d}{dt}\right)^n \mathbf{y}'(t) + \frac{(n+1)n}{2}2\left(\frac{d}{dt}\right)^{n-1}\mathbf{y}'(t),$$
and by differentiating the right-hand side of $(*)$, we have
$$\left(\frac{d}{dt}\right)^{n+1}(2nt\mathbf{y}(t)) = 2nt\left(\frac{d}{dt}\right)^{n+1}\mathbf{y}(t) + (n+1)2n\left(\frac{d}{dt}\right)^n \mathbf{y}(t).$$

Equating the final expressions from the above calculations, we obtain
$$(t^2-1)\frac{d^2}{dt^2}\Big(\frac{d}{dt}\Big)^n(t^2-1)^n + 2t\frac{d}{dt}\Big(\frac{d}{dt}\Big)^n(t^2-1)^n - n(n+1)\Big(\frac{d}{dt}\Big)^n(t^2-1)^n = 0.$$
Multiplying by $\frac{-1}{n!2^n}$, we get $(1-t^2)\mathbf{P}_n''(t) - 2t\mathbf{P}_n'(t) + n(n+1)\mathbf{P}_n(t) = 0.$

Solution to Exercise 4.11, page 171

t^2-1 is zero at ± 1. By Rolle's Theorem, it follows that $(d/dt)(t^2-1)$ is zero at some $t^{(1)} \in (-1,1)$. But we had seen that $(d/dt)(t^2-1)$ is also zero at the end points ± 1. So by Rolle's Theorem applied to the function $(d/dt)(t^2-1)$ on the two intervals $[-1, t^{(1)}]$ and $[t^{(1)}, 1]$, we get the existence of points $t_1^{(2)} \in (-1, t^{(1)})$ and $t_2^{(2)} \in (t^{(1)}, 1)$, where $(d/dt)^2(t^2-1)$ is zero. Proceeding in this manner, we get the existence of points $t_1^{(n)}, \cdots, t_n^{(n)} \in (-1,1)$ where $(d/dt)^n(t^2-1)^n$ vanishes. So $\mathbf{P}_n$ has *at least* n zeros on $(-1,1)$. But $\mathbf{P}_n$ has degree n, and hence it can have *at most* n zeros in $\mathbb{C}$. This shows that all the zeros of $\mathbf{P}_n$ are real, and all of them lie in the open interval $(-1,1)$.

Solution to Exercise 4.12, page 171

The set $\{e_{ij} : 1 \leqslant i \leqslant m, \ 1 \leqslant j \leqslant n\}$, where e_{ij} is the matrix with 1 in the ith row and jth column, and all other entries 0, is a basis for $\mathbb{R}^{m \times n}$. To see that this basis is in fact orthonormal, observe that the map $\iota : \mathbb{R}^{m \times n} \to \mathbb{R}^{mn}$ given by $A = [a_{ij}] \mapsto (a_{11}, \cdots, a_{1n}, a_{21}, \cdots, a_{2n}, \cdots, a_{m1}, \cdots, a_{mn})$ (that is, lay out the rows of A next to each other in one long row), is an isomorphism that preserves inner products:
$$\begin{aligned}\langle A, B\rangle = \operatorname{tr}(A^\top B) &= \sum_{i=1}^n \sum_{k=1}^m a_{ki} b_{ki}\\ &= \langle (a_{11}, \cdots, a_{1m}, a_{21}, \cdots, a_{2n}, \cdots, a_{m1}, \cdots, a_{mn}),\\ &\quad (b_{11}, \cdots, b_{1n}, b_{21}, \cdots, b_{2n}, \cdots, b_{m1}, \cdots, b_{mn})\rangle\\ &= \langle \iota(A), \iota(B)\rangle.\end{aligned}$$
$\{\iota(e_{ij}) : 1 \leqslant i \leqslant m, \ 1 \leqslant j \leqslant n\}$ is orthonormal, and so it follows that the set $\{e_{ij} : 1 \leqslant i \leqslant m, \ 1 \leqslant j \leqslant n\}$ is orthonormal as well.

Solution to Exercise 4.13, page 172

(1) We have $\mathbf{H}_0 = e^{x^2} e^{-x^2} = \mathbf{1}$. For $n \geqslant 0$,
$$\begin{aligned}\mathbf{H}_{n+1}(x) &= (-1)^{n+1} e^{x^2} \Big(\frac{d}{dx}\Big)^{n+1} e^{-x^2} = -e^{x^2} \frac{d}{dx}\Big(e^{-x^2} \cdot (-1)^n e^{x^2} \Big(\frac{d}{dx}\Big)^n e^{-x^2}\Big)\\ &= -e^{x^2} \frac{d}{dx}\big(e^{-x^2} \mathbf{H}_n(x)\big) = -e^{x^2}\big(e^{-x^2}(-2x)\mathbf{H}_n(x) + e^{-x^2}\mathbf{H}_n'(x)\big)\\ &= 2x\mathbf{H}_n(x) - \mathbf{H}_n'(x).\end{aligned}$$

Thus if $\mathbf{H}_n$ is a polynomial, then $2x\mathbf{H}_n$, $\mathbf{H}_n'$ are polynomials too, and so is $\mathbf{H}_{n+1} = 2x\mathbf{H}_n - \mathbf{H}_n'$. Since $\mathbf{H}_0 = 1$ is a nonzero polynomial of degree 0, it follows by induction on n that each $\mathbf{H}_n$, $n \geq 0$ is a polynomial. Moreover, if $\mathbf{H}_n$ has degree d, and its leading term is $c_d x^d$, then $\mathbf{H}_n'$ has degree $d-1$, while $2x\mathbf{H}_n$ has degree $d+1$ with the leading term $2c_d x^{d+1}$. Consequently, the recurrence relation together with $\mathbf{H}_0 = 1$ also reveals that $\mathbf{H}_n$ has the leading term $2^n x^n$, and in particular has degree n.

Using the recursion relation, we get $\mathbf{H}_1 = 2x$, $\mathbf{H}_2 = 4x^2 - 2$, $\mathbf{H}_3 = 8x^3 - 12x$.

(2) Let $m < n$. Then we have
$$\langle \varphi_m, \varphi_n \rangle = (-1)^n \int_{-\infty}^{\infty} \mathbf{H}_m(x) \left(\frac{d}{dx}\right)^n e^{-x^2} dx$$
$$= (-1)^n \left(\mathbf{H}_m(x) \left(\frac{d}{dx}\right)^{n-1} e^{-x^2} \Big|_{-\infty}^{\infty} - \int_{-\infty}^{\infty} \mathbf{H}_m'(x) \left(\frac{d}{dx}\right)^{n-1} e^{-x^2} dx \right).$$

As $(d/dx)^{n-1} e^{-x^2}$ is a sum of terms of the form $c_k x^k e^{-x^2}$, and because $\mathbf{H}_m$ is a polynomial, it follows that the first summand in the right-hand side is 0. So we have $\langle \varphi_m, \varphi_n \rangle = (-1)^{n+1} \int_{-\infty}^{\infty} \mathbf{H}_m'(x) \left(\frac{d}{dx}\right)^{n-1} e^{-x^2} dx$.

We can continue this process of integration by parts, until we arrive at
$$\langle \varphi_m, \varphi_n \rangle = \int_{-\infty}^{\infty} e^{-x^2} \left(\frac{d}{dx}\right)^n \mathbf{H}_m(x) dx.$$

But as $\mathbf{H}_m$ has degree $m < n$, $(d/dx)^n \mathbf{H}_m = 0$, so that $\langle \varphi_m, \varphi_n \rangle = 0$. The case $m > n$ also follows from here, since the inner product is conjugate symmetric. Finally,
$$\|\varphi_n\|_2^2 = \langle \varphi_n, \varphi_n \rangle = (-1)^n \int_{-\infty}^{\infty} \mathbf{H}_n \left(\frac{d}{dx}\right)^n e^{-x^2} dx$$
$$= \int_{-\infty}^{\infty} e^{-x^2} \left(\frac{d}{dx}\right)^n \mathbf{H}_n(x) dx = \int_{-\infty}^{\infty} e^{-x^2} \left(\frac{d}{dx}\right)^n 2^n x^n dx$$
$$= \int_{-\infty}^{\infty} e^{-x^2} 2^n n! dx = 2^n n! \sqrt{\pi}.$$

(The last equality can be justified as follows. With $I := \int_{-\infty}^{\infty} e^{-x^2} dx$, we have
$$I^2 = \left(\int_{-\infty}^{\infty} e^{-x^2} dx \right) \left(\int_{-\infty}^{\infty} e^{-y^2} dy \right) = \int_{-\infty}^{\infty} \int_{-\infty}^{\infty} e^{-(x^2+y^2)} dx dy$$
$$= \int_0^{2\pi} \int_0^{\infty} e^{-r^2} r \, dr \, d\theta = \pi.$$
So $I = \int_{-\infty}^{\infty} e^{-x^2} dx = \sqrt{\pi}$.)

(3) For $n \geq 0$, we have
$$\left(x - \frac{d}{dx}\right) \varphi_n = \left(x - \frac{d}{dx}\right) e^{-x^2/2} \mathbf{H}_n = xe^{-x^2/2} \mathbf{H}_n + xe^{-x^2/2} \mathbf{H}_n - e^{-x^2/2} \mathbf{H}_n'$$
$$= e^{-x^2/2} (2x\mathbf{H}_n - \mathbf{H}_n') = e^{-x^2/2} \mathbf{H}_{n+1} = \varphi_{n+1}.$$

(4) First let us note that if $n \geq 1$, then we have

$$\mathbf{H}'_n(x) = (-1)^n \frac{d}{dx}\left(e^{x^2}\left(\frac{d}{dx}\right)^n e^{-x^2}\right)$$

$$= (-1)^n \left(2xe^{x^2}\left(\frac{d}{dx}\right)^n e^{-x^2} + e^{x^2}\left(\frac{d}{dx}\right)^n(-2xe^{-x^2})\right)$$

$$= (-1)^n \left(2xe^{x^2}\left(\frac{d}{dx}\right)^n e^{-x^2} + e^{x^2}\binom{n}{0}(-2x)\left(\frac{d}{dx}\right)^n e^{-x^2}\right.$$
$$\left. + e^{x^2}\binom{n}{1}(-2)\left(\frac{d}{dx}\right)^{n-1} e^{-x^2}\right)$$

$$= 2(-1)^{n-1} n e^{x^2}\left(\frac{d}{dx}\right)^{n-1} e^{-x^2} = 2n\mathbf{H}_{n-1}(x).$$

Hence for $n \geq 1$,

$$\left(x + \frac{d}{dx}\right)\varphi_n = \left(x + \frac{d}{dx}\right)e^{-x^2/2}\mathbf{H}_n = xe^{-x^2/2}\mathbf{H}_n - xe^{-x^2/2}\mathbf{H}_n + e^{-x^2/2}\mathbf{H}'_n$$

$$= e^{-x^2/2}\mathbf{H}'_n = e^{-x^2/2}2n\mathbf{H}_{n-1} = 2n\varphi_{n-1}.$$

(5) We have for all φ

$$\left(x + \frac{d}{dx}\right)\left(x - \frac{d}{dx}\right)\varphi = x\left(x - \frac{d}{dx}\right)\varphi + \frac{d}{dx}(x\varphi) - \left(\frac{d}{dx}\right)^2\varphi$$

$$= x\left(x - \frac{d}{dx}\right)\varphi + \varphi + x\frac{d}{dx}\varphi - \left(\frac{d}{dx}\right)^2\varphi$$

$$= \left(-\left(\frac{d}{dx}\right)^2 + x^2\right)\varphi + \varphi.$$

Hence for all $n \geq 0$,

$$\left(-\left(\frac{d}{dx}\right)^2 + x^2\right)\varphi_n = \left(x + \frac{d}{dx}\right)\left(x - \frac{d}{dx}\right)\varphi_n - \varphi_n = \left(x + \frac{d}{dx}\right)\varphi_{n+1} - \varphi_n$$

$$= 2(n+1)\varphi_n - \varphi_n = (2n+1)\varphi_n.$$

(6) We have $\psi'_n(x) = \sqrt{a}\varphi'_n(\sqrt{a}x)$ and $\psi''_n(x) = a\varphi''_n(\sqrt{a}x)$.
From the previous part, we have $\left(-(d/dx)^2 + x^2\right)\varphi_n = (2n+1)\varphi_n$, giving

$$\left(-\varphi''(\sqrt{a}x) + (\sqrt{a}x)^2\varphi_n(\sqrt{a}x)\right) = (2n+1)\varphi_n(\sqrt{a}x).$$

We have

$$\left(-\left(\frac{d}{dx}\right)^2 + a^2x^2\right)\psi_n = -a\varphi''_n(\sqrt{a}x) + a^2x^2\varphi_n(\sqrt{a}x)$$

$$= a\left(-\varphi''(\sqrt{a}x) + (\sqrt{a}x)^2\varphi_n(\sqrt{a}x)\right)$$

$$= a(2n+1)\varphi_n(\sqrt{a}x) = a(2n+1)\psi_n(x).$$

In Schrödinger's equation, $a^2 = \frac{\frac{1}{2}m\omega^2}{\frac{\hbar^2}{2m}} = \frac{m^2\omega^2}{\hbar^2}$, and so $\frac{E_n}{\frac{\hbar^2}{2m}} = a(2n+1)$.

So $E_n = a(2n+1)\frac{\hbar^2}{2m} = \sqrt{\frac{m^2\omega^2}{\hbar^2}}(2n+1)\frac{\hbar^2}{2m} = \left(n + \frac{1}{2}\right)\omega\hbar$, for $n \geq 0$.

Solution to Exercise 4.14, page 172

Since $\sum_{n=1}^{\infty} \left\|\frac{u_n}{n}\right\| = \sum_{n=1}^{\infty} \frac{1}{n}$ diverges, $\sum_{n=1}^{\infty} \frac{u_n}{n}$ does not converge absolutely.

If s_n is the nth partial sum of $\sum_{k=1}^{\infty} \frac{u_k}{k}$, then for $n > m$, we have

$$\|s_n - s_m\|^2 = \left\|\sum_{k=m+1}^{n} \frac{u_k}{k}\right\|^2 = \left\langle \sum_{k=m+1}^{n} \frac{u_k}{k}, \sum_{j=m+1}^{n} \frac{u_j}{j} \right\rangle$$

$$= \sum_{k=m+1}^{n} \sum_{j=m+1}^{n} \frac{1}{jk}\langle u_k, u_j \rangle = \sum_{k=m+1}^{n} \frac{1}{k^2},$$

and this can be made as small as we please since $\sum_{n=1}^{\infty} \frac{1}{n^2} < \infty$.

Hence $(s_n)_{n \in \mathbb{N}}$ is Cauchy in H, and since H is a Hilbert space, it converges.

Solution to Exercise 4.15, page 172

For all $N \in \mathbb{N}$, we have

$$0 \leq \left\|x - \sum_{n=1}^{N} \langle x, u_n \rangle u_n\right\|^2$$

$$= \langle x, x \rangle - 2\operatorname{Re}\left\langle x, \sum_{n=1}^{N} \langle x, u_n \rangle u_n \right\rangle + \left\langle \sum_{n=1}^{N} \langle x, u_n \rangle u_n, \sum_{m=1}^{N} \langle x, u_m \rangle u_m \right\rangle$$

$$= \|x\|^2 - 2\operatorname{Re}\sum_{n=1}^{N} (\langle x, u_n \rangle^* \langle x, u_n \rangle) + \sum_{n=1}^{N}\sum_{m=1}^{N} \langle x, u_n \rangle \langle x, u_m \rangle^* \langle u_n, u_m \rangle$$

$$= \|x\|^2 - 2\sum_{n=1}^{N} |\langle x, u_n \rangle|^2 + \sum_{n=1}^{N} |\langle x, u_n \rangle|^2 = \|x\|^2 - \sum_{n=1}^{N} |\langle x, u_n \rangle|^2.$$

Thus $\sum_{n=1}^{N} |\langle x, u_n \rangle|^2 \leq \|x\|^2$, and as N was arbitrary, $\sum_{n=1}^{\infty} |\langle x, u_n \rangle|^2 \leq \|x\|^2$.

Solution to Exercise 4.16, page 173

Let $y \in Y \cap Y^\perp$. As $y \in Y^\perp$, we know that for all $y' \in Y$, $\langle y, y' \rangle = 0$. Taking $y' := y \in Y$, we obtain $0 = \langle y, y' \rangle = \langle y, y \rangle = \|y\|^2$, and so $\|y\| = 0$, giving $y = \mathbf{0}$. So $Y \cap Y^\perp \subset \{\mathbf{0}\}$. Also, since $Y, Y^\perp$ are subspaces, it follows that each contains the zero vector $\mathbf{0}$. So $Y \cap Y^\perp = \{\mathbf{0}\}$.

Solution to Exercise 4.17, page 173

(1) If $y \in Y$, then for each $x \in Y^\perp$, $\langle y, x \rangle = \langle x, y \rangle^* = 0$, and so $y \in (Y^\perp)^\perp$. Thus $Y \subset (Y^\perp)^\perp$.

(2) Let $x \in Z^\perp$. Then $\langle x, z \rangle = 0$ for all $z \in Z$. As $Y \subset Z$, we also have $\langle x, y \rangle = 0$ in particular for all $y \in Y$. Hence $x \in Y^\perp$. This shows that $Z^\perp \subset Y^\perp$.

(3) As $Y \subset \overline{Y}$, it follows from part (2) that $\overline{Y}^\perp \subset Y^\perp$.

Now let $x \in Y^\perp$. Then $\langle x, y \rangle = 0$ for all $y \in Y$.

If $y' \in \overline{Y}$, then there exists a sequence $(y_n)_{n \in \mathbb{N}}$ in Y such that $\lim_{n \to \infty} y_n = y'$.

Thus $\langle x, y' \rangle = \left\langle x, \lim_{n \to \infty} y_n \right\rangle = \lim_{n \to \infty} \langle x, y_n \rangle = \lim_{n \to \infty} 0 = 0$.

Hence $x \in \overline{Y}^\perp$, showing that $Y^\perp \subset \overline{Y}^\perp$ as well.

(4) Suppose that $x \in Y^\perp$.

As Y is dense in X, there is a sequence $(y_n)_{n \in \mathbb{N}}$ in Y converging to x in X.

Thus $\langle x, x \rangle = \lim_{n \to \infty} \langle x, y_n \rangle = \lim_{n \to \infty} 0 = 0$.

(5) Suppose $\mathbf{x} = (x_n)_{n \in \mathbb{N}} \in Y_{\text{even}}^\perp$. Since $\mathbf{e}_{2n} \in Y_{\text{even}}$ for each $\mathbb{N}$, $x_{2n} = \langle \mathbf{x}, \mathbf{e}_{2n} \rangle = 0$. Hence the subspace $Y_{\text{even}}^\perp \subset Y_{\text{odd}}$, where Y_{odd} denotes the subspace of ℓ^2 all sequences whose evenly indexed terms are 0.

Vice versa, if $\mathbf{x} \in Y_{\text{odd}}$, it is clear that for all $\mathbf{y} \in Y_{\text{even}}$, $\langle \mathbf{x}, \mathbf{y} \rangle = 0$. Thus $Y_{\text{odd}} \subset Y_{\text{even}}^\perp$.

Consequently, $Y_{\text{even}}^\perp = Y_{\text{odd}}$.

Similarly, $Y_{\text{odd}}^\perp = Y_{\text{even}}$. And so, $Y_{\text{even}}^{\perp\perp} = Y_{\text{odd}}^\perp = Y_{\text{even}}$.

(6) We know that c_{00} is dense in ℓ^2. (Just truncate the series to the desired accuracy to get a finitely supported approximation!)

So $c_{00}^\perp = \{\mathbf{0}\}$. But then $c_{00}^{\perp\perp} = \{\mathbf{0}\}^\perp = \ell^2 \neq c_{00}$.

Solution to Exercise 4.18, page 176

Let $\mathbf{Y}_1 := \begin{bmatrix} 1 \\ \vdots \\ 1 \end{bmatrix}$, $\mathbf{Y}_2 := \begin{bmatrix} x_1 \\ \vdots \\ x_n \end{bmatrix}$ and $\mathbf{X} := \begin{bmatrix} y_1 \\ \vdots \\ y_n \end{bmatrix}$.

Then $E(m, b) = \sqrt{\sum_{i=1}^n (y_i - mx_i - b)^2} = \|\mathbf{X} - m\mathbf{Y}_2 - b\mathbf{Y}_1\|_2$.

Thus the problem of finding the least square regression line is:

$$\boxed{\text{Find } \mathbf{Y} \in \text{span}\{\mathbf{Y}_1, \mathbf{Y}_2\} \text{ such that } \|\mathbf{X} - \mathbf{Y}\|_2 \text{ is minimised.}}$$

It follows from Theorem 4.5, page 174, that a minimiser $\mathbf{Y}_*$ is given by

$$\mathbf{Y}_* = \langle \mathbf{X}, \mathbf{U}_1 \rangle \mathbf{U}_1 + \langle \mathbf{X}, \mathbf{U}_2 \rangle \mathbf{U}_2,$$

where $\{\mathbf{U}_1, \mathbf{U}_2\}$ is any orthonormal basis for the subspace $Y := \text{span}\{\mathbf{Y}_1, \mathbf{Y}_2\}$ of $\mathbb{R}^n$ with the usual Euclidean inner product. By the Gram-Schmidt Orthonormalisation Procedure, $\mathbf{U}_1 = \dfrac{\mathbf{Y}_1}{\|\mathbf{Y}_1\|_2} = \dfrac{\mathbf{Y}_1}{\sqrt{n}}$, and $\mathbf{U}_2 = \dfrac{\mathbf{Y}_2 - \langle \mathbf{Y}_2, \mathbf{U}_1 \rangle \mathbf{U}_1}{\|\mathbf{Y}_2 - \langle \mathbf{Y}_2, \mathbf{U}_1 \rangle \mathbf{U}_1\|_2}$.

$$\langle \mathbf{Y}_2, \mathbf{U}_1 \rangle \mathbf{U}_1 = \langle \mathbf{Y}_2, \mathbf{Y}_1 \rangle \frac{\mathbf{Y}_1}{n} = \left(\frac{1}{n}\sum_{i=1}^n x_i\right)\mathbf{Y}_1 = \overline{x}\mathbf{Y}_1, \text{ where } \overline{x} := \frac{1}{n}\sum_{i=1}^n x_i.$$

$$\mathbf{Y}_2 - \langle \mathbf{Y}_2, \mathbf{U}_1 \rangle \mathbf{U}_1 = \begin{bmatrix} x_1 - \overline{x} \\ \vdots \\ x_n - \overline{x} \end{bmatrix}, \text{ and } \|\mathbf{Y}_2 - \langle \mathbf{Y}_2, \mathbf{U}_1 \rangle \mathbf{U}_1\|^2 = \sum_{i=1}^n (x_i - \overline{x})^2.$$

$$\langle \mathbf{X}, \mathbf{U}_1 \rangle \mathbf{U}_1 = \langle \mathbf{X}, \mathbf{Y}_1 \rangle \frac{\mathbf{Y}_1}{n} = \left(\frac{1}{n}\sum_{i=1}^n y_i\right)\mathbf{Y}_1 = \overline{y}\mathbf{Y}_1, \text{ where } \overline{y} := \frac{1}{n}\sum_{i=1}^n y_i.$$

$$\langle \mathbf{X}, \mathbf{U}_2 \rangle \mathbf{U}_2 = \left\langle \begin{bmatrix} y_1 \\ \vdots \\ y_n \end{bmatrix}, \begin{bmatrix} x_1 - \overline{x} \\ \vdots \\ x_n - \overline{x} \end{bmatrix} \right\rangle \left(\frac{\mathbf{Y}_2 - \overline{x}\mathbf{Y}_1}{\sum_{i=1}^n (x_i - \overline{x})^2} \right) = \frac{\sum_{i=1}^n x_i y_i - n\overline{x}\overline{y}}{\sum_{i=1}^n (x_i - \overline{x})^2}(\mathbf{Y}_2 - \overline{x}\mathbf{Y}_1).$$

$$\text{So } m = \frac{\sum_{i=1}^n x_i y_i - n\overline{x}\overline{y}}{\sum_{i=1}^n (x_i - \overline{x})^2} = \frac{\sum_{i=1}^n (y_i - \overline{y})(x_i - \overline{x})}{\sum_{i=1}^n (x_i - \overline{x})^2}, \text{ and } b = \overline{y} - \frac{\sum_{i=1}^n (y_i - \overline{y})(x_i - \overline{x})}{\sum_{i=1}^n (x_i - \overline{x})^2}\overline{x}.$$

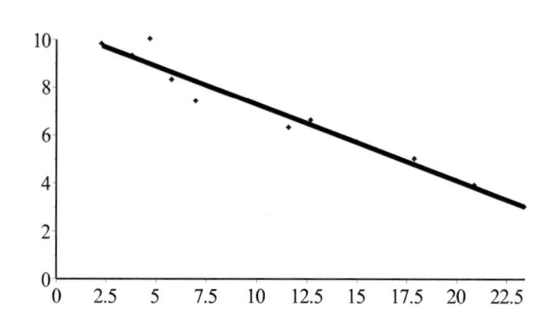

For the given data, using the above formulae, we obtain $m = -0.3184$ million tonnes coal per °C, and $b = 10.4667$ million tonnes of coal. The y-intercept is $b = 10.4667$ million tonnes of coal, and this is the inland energy consumption when the mean temperature is $0°C$ (that is when it is freezing!). The x-intercept is $10.4667/0.3184 = 32.8728$, which is the mean temperature when the inland consumption is 0 (that is, no heating required). The slope is $m = -0.3184$ million tonnes of coal per °C. Thus for each °C drop in temperature, the inland energy consumption increases by 0.3184 million tonnes of coal. Finally, the forecast of the energy consumption for a month with mean temperature $9°C$ is given by $y = mx + b = (-0.3184)(9) + 10.4667 = 7.6011$ million tonnes of coal.

Solution to Exercise 4.19, page 179

Let $C := L^2_+(\mathbb{R})$. Then C is convex. Thus $\overline{C}$ is convex too. We will show that $\mathbf{g}_* := \max\{\mathbf{f}, \mathbf{0}\} \in L^2_+(\mathbb{R}) = C \subset \overline{C}$ satisfies: for all $\mathbf{g} \in \overline{C}$, $\langle \mathbf{f} - \mathbf{g}_*, \mathbf{g} - \mathbf{g}_* \rangle \leq 0$.
We have $\mathbf{f} = \max\{\mathbf{f}, \mathbf{0}\} + \min\{\mathbf{f}, \mathbf{0}\}$. So $\mathbf{f} - \mathbf{g}_* = \min\{\mathbf{f}, \mathbf{0}\}$. Also,
$$\langle \max\{\mathbf{f}, \mathbf{0}\}, \min\{\mathbf{f}, \mathbf{0}\}\rangle = \int_{\mathbb{R}} \max\{\mathbf{f}, \mathbf{0}\}(x) \min\{\mathbf{f}, \mathbf{0}\}(x) dx = 0.$$
Hence we obtain for all $\mathbf{g} \in \overline{C}$ that
$$\langle \mathbf{f} - \mathbf{g}_*, \mathbf{g} - \mathbf{g}_* \rangle = \langle \min\{\mathbf{f}, \mathbf{0}\}, \mathbf{g} \rangle - \langle \min\{\mathbf{f}, \mathbf{0}\}, \max\{\mathbf{f}, \mathbf{0}\}\rangle$$
$$= \int_{\mathbb{R}} \underbrace{\min\{\mathbf{f}, \mathbf{0}\}(x)}_{\leq 0} \underbrace{\mathbf{g}(x)}_{\geq 0} dx - 0 \leq 0.$$
So for all $\mathbf{g} \in \overline{C}$, $\|\mathbf{f} - \mathbf{g}_*\| \leq \|\mathbf{f} - \mathbf{g}\|$. In particular, for all $\mathbf{g} \in L^2_+(\mathbb{R}) = C \subset \overline{C}$, we also have $\|\mathbf{f} - \mathbf{g}_*\| \leq \|\mathbf{f} - \mathbf{g}\|$.

Solution to Exercise 4.20, page 182

We'd seen in Exercise 4.17, page 173, that $Y^\perp = \overline{Y}^\perp$. So $(Y^\perp)^\perp = (\overline{Y}^\perp)^\perp = \overline{Y}$, where the last equality follows from Corollary 4.1, page 182, since $\overline{Y}$ is closed.

Solution to Exercise 4.21, page 182

For all $\mathbf{f} \in L^2(\mathbb{R})$, it is easy to check that $\mathbf{f}_e := (\mathbf{f} + \check{\mathbf{f}})/2$ is even, and $\mathbf{f}_o := (\mathbf{f} - \check{\mathbf{f}})/2$ is odd. Thus for all $\mathbf{g} \in Y$, we have
$$\langle \mathbf{f} - \mathbf{f}_e, \mathbf{g} \rangle = \left\langle \mathbf{f} - \frac{\mathbf{f} + \check{\mathbf{f}}}{2}, \mathbf{g} \right\rangle = \left\langle \frac{\mathbf{f} - \check{\mathbf{f}}}{2}, \mathbf{g} \right\rangle = \int_{-\infty}^{\infty} \underbrace{\mathbf{f}_o(x)}_{\text{odd}} \underbrace{\mathbf{g}(x)}_{\text{even}} dx = 0.$$
Thus, by Theorem 4.7, page 180, $P_Y \mathbf{f} = \mathbf{f}_e$ for all $\mathbf{f} \in L^2(\mathbb{R})$.
By Theorem 4.8, page 180, we have
$$Y^\perp = \ker P_Y = \{\mathbf{f} \in L^2(\mathbb{R}) : P_Y \mathbf{f} = \mathbf{0}\} = \{\mathbf{f} \in L^2(\mathbb{R}) : (\mathbf{f} + \check{\mathbf{f}})/2 = \mathbf{0}\}$$
$$= \{\mathbf{f} \in L^2(\mathbb{R}) : \mathbf{f} = -\check{\mathbf{f}}\} = \{\mathbf{f} \in L^2(\mathbb{R}) : \mathbf{f} \text{ is odd}\}.$$
$P_{Y^\perp} = I - P_Y$, and so for all $\mathbf{f} \in L^2(\mathbb{R})$, $P_{Y^\perp} \mathbf{f} = \mathbf{f} - \dfrac{\mathbf{f} + \check{\mathbf{f}}}{2} = \dfrac{\mathbf{f} - \check{\mathbf{f}}}{2}$.

We have $\mathbf{f} = I\mathbf{f} = P_Y \mathbf{f} + P_{Y^\perp} \mathbf{f} = \underbrace{\dfrac{\mathbf{f} + \check{\mathbf{f}}}{2}}_{\text{even}} + \underbrace{\dfrac{\mathbf{f} - \check{\mathbf{f}}}{2}}_{\text{odd}}$.

Moreover, by Theorem 4.8, this decomposition is unique.

Solution to Exercise 4.22, page 182

$Y = \ker(I - S)$, and so Y is a closed subspace of H.
For all $x \in H$, $S\left(\dfrac{x + Sx}{2}\right) = \dfrac{Sx + S^2 x}{2} = \dfrac{Sx + Ix}{2} = \dfrac{Sx + x}{2} = \dfrac{x + Sx}{2}$.

So $\dfrac{x+Sx}{2} \in Y$ for all $x \in H$. Moreover, for all $y \in Y$, we have
$$\left\langle x - \dfrac{x+Sx}{2}, y \right\rangle = \left\langle \dfrac{x-Sx}{2}, y \right\rangle = \dfrac{\langle x,y \rangle - \langle Sx, y \rangle}{2} = \dfrac{\langle x,y \rangle - \langle x, Sy \rangle}{2}$$
$$= \dfrac{\langle x,y \rangle - \langle x,y \rangle}{2} = 0.$$
Thus, by Theorem 4.7, page 180, $P_Y x = \dfrac{x+Sx}{2}$ for all $x \in H$.

By Theorem 4.8, page 180, we have
$$Y^\perp = \ker P_Y = \{x \in H : P_Y x = \mathbf{0}\} = \left\{x \in H : \dfrac{x+Sx}{2} = \mathbf{0}\right\}$$
$$= \{x \in H : Sx = -x\} = Z.$$
Thus $Z^\perp = (Y^\perp)^\perp = Y$.

$P_{Y^\perp} = I - P_Y$, and so for all $x \in H$, $P_{Y^\perp} x = x - \dfrac{x+Sx}{2} = \dfrac{x-Sx}{2}$.

Solution to Exercise 4.23, page 182

Consider the map $L^2(\mathbb{R}) \ni \mathbf{f} \overset{M}{\mapsto} \mathbf{1}_{\complement A} \cdot \mathbf{f}$, where $\mathbf{1}_{\complement A}$ is the indicator function of $\complement A$.
As $\|M\mathbf{f}\|_2^2 = \|\mathbf{1}_{\complement A} \cdot \mathbf{f}\|_2^2 = \int_{\complement A} |\mathbf{f}(x)|^2 dx \leqslant \int_{\mathbb{R}} |\mathbf{f}(x)|^2 dx = \|\mathbf{f}\|_2^2 < \infty$, $M\mathbf{f} \in L^2(\mathbb{R})$.
It is also easy to see that M is linear. The above inequality then establishes that $M \in CL(L^2(\mathbb{R}))$. We have
$$\ker M = \{\mathbf{f} \in L^2(\mathbb{R}) : M\mathbf{f} = \mathbf{0}\} = \{\mathbf{f} \in L^2(\mathbb{R}) : \mathbf{1}_{\complement A} \cdot \mathbf{f} = \mathbf{0}\}$$
$$= \{\mathbf{f} \in L^2(\mathbb{R}) : \mathbf{f}(x) = 0 \text{ for almost all } x \in \complement A\} = Y_A.$$
Thus Y_A is closed.

For $\mathbf{f} \in L^2(\mathbb{R})$, $\mathbf{1}_A \mathbf{f} \in Y_A$, and moreover, for any $\mathbf{g} \in Y_A$,
$$\langle \mathbf{f} - \mathbf{1}_A \mathbf{f}, \mathbf{g} \rangle = \langle \mathbf{1}_{\complement A} \mathbf{f}, \mathbf{g} \rangle = \int_{\complement A} \mathbf{f}(x) \underbrace{(\mathbf{g}(x))^*}_{\equiv 0 \text{ on } \complement A} dx = 0.$$
Thus $P_A \mathbf{f} = \mathbf{1}_A \mathbf{f}$ for all $\mathbf{f} \in L^2(\mathbb{R})$.

Solution to Exercise 4.24, page 182

Suppose that $D^\perp = \{\mathbf{0}\}$. Then $\overline{D} = (D^\perp)^\perp = \{\mathbf{0}\}^\perp = H$. So D is dense in H.
Now suppose that D is dense in H. Then $\overline{D} = H$. Thus $D^\perp = (\overline{D})^\perp = H^\perp = \{\mathbf{0}\}$.

Solution to Exercise 4.25, page 184

Let $\mathbf{x} \in C[-1,1]$ and $\epsilon > 0$. By Weierstrass's Approximation Theorem (Exercise 1.26, page 22), there is a polynomial $\mathbf{p} \in C[-1,1]$ such that $\|\mathbf{x}-\mathbf{p}\|_\infty < \epsilon/\sqrt{2}$.
Then $\|\mathbf{x}-\mathbf{p}\|_2^2 = \int_{-1}^{1} |\mathbf{x}(t) - \mathbf{p}(t)|^2 dt \leqslant \int_{-1}^{1} \|\mathbf{x}-\mathbf{p}\|_\infty^2 dt < \int_{-1}^{1} \dfrac{\epsilon^2}{2} dt = \epsilon^2.$
Hence $\|\mathbf{x}-\mathbf{p}\|_2 < \epsilon$. Consequently the polynomials are dense in $C[-1,1]$ (with the usual inner product).

Solution to Exercise 4.26, page 185

(L1) For $x, y \in H$, $\iota(x+y) = (\langle x+y, u_n \rangle)_{n \in \mathbb{N}} = (\langle x, u_n \rangle + \langle y, u_n \rangle)_{n \in \mathbb{N}}$
$= (\langle x, u_n \rangle)_{n \in \mathbb{N}} + (\langle y, u_n \rangle)_{n \in \mathbb{N}} = \iota(x) + \iota(y).$

(L2) For $\alpha \in \mathbb{K}$ and $x \in H$, $\iota(\alpha x) = (\langle \alpha x, u_n \rangle)_{n \in \mathbb{N}} = (\alpha \langle x, u_n \rangle)_{n \in \mathbb{N}}$
$= \alpha (\langle x, u_n \rangle)_{n \in \mathbb{N}} = \alpha \iota(x).$

Moreover ι is continuous because $\|\iota(x)\|^2 = \sum_{n=1}^{\infty} |\langle x, u_n \rangle|^2 = \|x\|^2$ for all $x \in H$.

If $x \in H$ is such that $\iota(x) = \mathbf{0}$, then $\|x\| = \|\iota(x)\| = 0$, and so $x = \mathbf{0}$.
Hence ι is injective.

If $(c_n)_{n \in \mathbb{N}} \in \ell^2$, then $x := \sum_{n=1}^{\infty} c_n u_n \in H$, and for all $k \in \mathbb{N}$,

$$\langle x, u_k \rangle = \Big\langle \lim_{N \to \infty} \sum_{n=1}^{N} c_n u_n, u_k \Big\rangle = \lim_{N \to \infty} \Big\langle \sum_{n=1}^{N} c_n u_n, u_k \Big\rangle$$
$$= \lim_{N \to \infty} \sum_{n=1}^{N} c_n \langle u_n, u_k \rangle = \lim_{N \to \infty} c_k = c_k.$$

So $\iota(x) = (c_n)_{n \in \mathbb{N}}$, showing that ι is surjective too.

As $\iota \in CL(H, \ell^2)$ is a bijection, it has a continuous inverse $\iota^{-1} \in CL(\ell^2, H)$ (by Corollary 2.4 on page 96). Moreover, $\|\iota(x)\| = \|x\|$ for all $x \in H$, and so ι is an isometry.

Solution to Exercise 4.27, page 187

Let $\mathbf{x} \in C[0,1]$ be the function $t \mapsto t$.

For $n \neq 0$, we have $\langle \mathbf{x}, \mathbf{T}_n \rangle = \int_0^1 t e^{-2\pi i n t} dt = \dfrac{1}{-2\pi i n}$, using integration by parts.

Also $\langle \mathbf{x}, \mathbf{T}_0 \rangle = 1/2$. By Parseval's Identity,

$$\frac{1}{3} = \int_0^1 t^2 dt = \int_0^1 |\mathbf{x}(t)|^2 dt = \sum_{n \in \mathbb{Z}} |\langle \mathbf{x}, \mathbf{T}_n \rangle|^2 = \frac{1}{4} + \sum_{n \in \mathbb{Z} \setminus \{0\}} \frac{1}{(2\pi n)^2} = \frac{1}{4} + 2 \sum_{n=1}^{\infty} \frac{1}{(2\pi n)^2},$$

which yields $\sum_{n=1}^{\infty} \dfrac{1}{n^2} = \dfrac{\pi^2}{6}$.

Solution to Exercise 4.28, page 187

Let $[(x_n)_{n \in \mathbb{N}}] \in \overline{X}$. Consider the sequence $(x_n)_{n \in \mathbb{N}}$ in X. Since $(x_n)_{n \in \mathbb{N}} \in \mathcal{C}$, given any $\epsilon > 0$, there exists an $N \in \mathbb{N}$ such that for all $m, n > N$, $\|x_n - x_m\| < \epsilon$. Consequently, for all $m > N$, $\|\iota(x_m) - [(x_n)_{n \in \mathbb{N}}]\|_{\overline{X}} = \lim_{n \to \infty} \|x_m - x_n\| \leq \epsilon$.

Hence $\lim_{n \to \infty} \iota(x_n) = [(x_n)_{n \in \mathbb{N}}]$.

Solution to Exercise 4.29, page 187

We have
$$L^2 - 4\pi A = \sum_{n=1}^{\infty} \left(2\pi^2 n^2(a_n^2 + b_n^2 + c_n^2 + d_n^2) - 4\pi^2 n(a_n d_n - b_n c_n)\right)$$
$$= \sum_{n=1}^{\infty} 2\pi^2 n\left(n(a_n^2 + b_n^2 + c_n^2 + d_n^2) - 2a_n d_n + 2b_n c_n\right)$$
$$= 2\pi^2\left((a_1 - d_1)^2 + (c_1 + b_1)^2\right)$$
$$+ 2\pi^2 \sum_{n=2}^{\infty} n\left((a_n - d_n)^2 + (b_n + c_n)^2 + (n-1)(a_n^2 + b_n^2 + c_n^2 + d_n^2)\right)$$
$$\geq 0,$$
with equality if and only if $\boxed{a_1 = d_1,\ c_1 = -b_1,\ a_n = b_n = c_n = d_n = 0 \text{ for all } n \geq 2}$.
Thus the curve enclosing the maximum area is given by
$$\mathbf{x}(s) = a_0 + a_1 \cos\left(\frac{2\pi}{L}s\right) + b_1 \sin\left(\frac{2\pi}{L}s\right)$$
$$\mathbf{y}(s) = c_0 - b_1 \cos\left(\frac{2\pi}{L}s\right) + a_1 \sin\left(\frac{2\pi}{L}s\right),$$
with $L^2 = 4\pi^2(a_1^2 + b_1^2)$.

Let $\alpha \in [0, 2\pi)$ be such that $\cos\alpha = \dfrac{a_1}{\sqrt{a_1^2 + b_1^2}}$ and $\sin\alpha = \dfrac{b_1}{\sqrt{a_1^2 + b_1^2}}$. Then
$$\mathbf{x}_*(s) = a_0 + \sqrt{a_1^2 + b_1^2}\left(\cos\alpha \cos\left(\frac{2\pi}{L}s\right) + \sin\alpha \sin\left(\frac{2\pi}{L}s\right)\right)$$
$$= a_0 + \frac{L}{2\pi}\cos\left(\frac{2\pi}{L}s - \alpha\right)$$
$$\mathbf{y}_*(s) = c_0 + \sqrt{a_1^2 + b_1^2}\left(-\sin\alpha \cos\left(\frac{2\pi}{L}s\right) + \cos\alpha \sin\left(\frac{2\pi}{L}s\right)\right)$$
$$= c_0 + \frac{L}{2\pi}\sin\left(\frac{2\pi}{L}s - \alpha\right).$$
Hence $\left(\mathbf{x}_*(s) - a_0\right)^2 + \left(\mathbf{y}_*(s) - c_0\right)^2 = \left(\dfrac{L}{2\pi}\right)^2$.
Consequently, $s \mapsto (\mathbf{x}_*(s), \mathbf{y}_*(s)) : [0, L] \to \mathbb{R}^2$ is the parametric representation of a circle with centre at $(a_0, c_0) \in \mathbb{R}^2$ and radius equal to $\dfrac{L}{2\pi}$.

Solution to Exercise 4.30, page 188

(1) Call $\mathbf{u}_n$ the nth vector in the list. If $\{\mathbf{u}_n : n \in \mathbb{N}\}$ were an orthonormal basis, then
$$(1, 0, \cdots) =: \mathbf{e}_1 = \sum_{n=1}^{\infty} \langle \mathbf{e}_1, \mathbf{u}_n \rangle \mathbf{u}_n = \langle \mathbf{e}_1, \mathbf{u}_1 \rangle \mathbf{u}_1 = \frac{1}{\sqrt{5}}(1, 2, 0, \cdots),$$
a contradiction. So the given set is *not* an orthonormal basis.

(2) Let us call the evenly indexed vectors as $\mathbf{v}_n$, and the oddly indexed ones as $\mathbf{w}_n$. Then clearly $\langle \mathbf{v}_i, \mathbf{v}_j \rangle = \langle \mathbf{w}_i, \mathbf{w}_j \rangle = \langle \mathbf{v}_i, \mathbf{w}_j \rangle = 0$ whenever $i \neq j$, since there are no overlapping nonzero terms. Also $\langle \mathbf{v}_i, \mathbf{w}_i \rangle = 0$.
Finally $\|\mathbf{v}_i\| = \|\mathbf{w}_i\| = 1$. This shows that the given set B is orthonormal.
In order to show density, we note that $\dfrac{\mathbf{v}_n + \mathbf{w}_n}{\sqrt{2}} = \mathbf{e}_{2n-1}$ and $\dfrac{\mathbf{w}_n - \mathbf{v}_n}{\sqrt{2}} = \mathbf{e}_{2n}$.
Thus $\operatorname{span} B = \operatorname{span}\{\mathbf{e}_n : n \in \mathbb{N}\}$, and the latter is dense in ℓ^2.

Solution to Exercise 4.31, page 188

If X is a real vector space, then let $\mathbb{K}_\mathbb{Q} := \mathbb{Q}$, while if X is a complex vector space, then let $\mathbb{K}_\mathbb{Q} := \mathbb{Q} + i\mathbb{Q}$. Set
$$D := \Big\{ \sum_{n=1}^N c_n u_n : N \in \mathbb{N}, \, c_1, \cdots, c_N \in \mathbb{K}_\mathbb{Q} \Big\}.$$
Then D is countable. Let $x \in X$, and $\epsilon > 0$.
Then there exists an N such that $\Big\| x - \sum_{n=1}^N \langle x, u_n \rangle u_n \Big\| < \dfrac{\epsilon}{2}$.
Let $c_n \in \mathbb{K}_\mathbb{Q}$, $n = 1, \cdots, N$, be such that $\sum_{n=1}^N |\langle x, u_n \rangle - c_n|^2 < \dfrac{\epsilon^2}{4}$.
Then with $y := \sum_{n=1}^N c_n u_n \in B$, we have
$$\|x - y\| \leq \Big\| x - \sum_{n=1}^N \langle x, u_n \rangle u_n \Big\| + \Big\| \sum_{n=1}^N (\langle x, u_n \rangle - c_n) u_n \Big\| < \frac{\epsilon}{2} + \frac{\epsilon}{2} = \epsilon.$$
Thus X is separable.

Solution to Exercise 4.32, page 188

We have for $\lambda \neq \mu$ that
$$\langle e^{i\lambda x}, e^{i\mu x} \rangle = \lim_{T \to \infty} \frac{1}{2T} \int_{-T}^{T} e^{i\lambda x} e^{-i\mu x} dx = \lim_{T \to \infty} \frac{1}{2T} \int_{-T}^{T} e^{i(\lambda-\mu)x} dx$$
$$= \lim_{T \to \infty} \frac{1}{2T} \frac{1}{i(\lambda-\mu)} e^{i(\lambda-\mu)x} \Big|_{-T}^{T} = \lim_{T \to \infty} \frac{1}{2T} \frac{1}{i(\lambda-\mu)} 2i \sin\big((\lambda-\mu)T\big)$$
$$= 0.$$
On the other hand, $\|e^{i\lambda x}\|^2 = 1$. Thus
$$\|e^{i\lambda x} - e^{i\mu x}\|^2 = 1 + 1 - 2\operatorname{Re}\langle e^{i\lambda x}, e^{i\mu x} \rangle = 1 + 1 - 2 \cdot 0 = 2.$$
Hence $\|e^{i\lambda x} - e^{i\lambda \mu}\| = \sqrt{2}$.

Suppose now that $\overline{X}$ is separable, with a dense subset $D = \{d_1, d_2, d_3, \cdots\}$. Then for each $\lambda \in \mathbb{R}$, there exists a $d_\lambda \in D$ such that $\|e^{i\lambda x} - d_\lambda\| < 1/\sqrt{2}$.

This gives us the existence[6] of a map $\lambda \mapsto d_\lambda : \mathbb{R} \to D$.
This map is injective since if $\lambda \neq \mu$, then
$$\sqrt{2} = \|e^{i\lambda x} - e^{i\mu x}\| = \|e^{i\lambda x} - d_\lambda + d_\lambda - d_\mu + e^{i\mu x} - d_\mu\| < \frac{1}{\sqrt{2}} + \|d_\lambda - d_\mu\| + \frac{1}{\sqrt{2}},$$
giving $\|d_\lambda - d_\mu\| > 0$, and in particular $d_\lambda \neq d_\mu$.
But this is absurd, since $\mathbb{R}$ is uncountable, while D is countable!
So $\overline{X}$ is not separable.

Solution to Exercise 4.33, page 189

For $n \in \mathbb{N}$, set $U_n := \left\{ u_i : |\langle x, u_i \rangle|^2 > \frac{\|x\|^2}{n} \right\}$.

If U_n has more than $n - 1$ elements, then for any distinct $u_{i_1}, \cdots, u_{i_n} \in U_n$,
$$\|x\|^2 = n \cdot \frac{\|x\|^2}{n} < |\langle x, u_{i_1} \rangle|^2 + \cdots + |\langle x, u_{i_n} \rangle|^2 \leq \|x\|^2,$$
(where the former inequality is by virtue of the fact that the u_{i_k}'s belong to U_n, and the latter is Bessel's Inequality). So we obtain $\|x\|^2 < \|x\|^2$, which is absurd. Thus U_n has at most $n - 1$ elements. Hence each U_n is finite. But
$$U = \{u_i : \langle x, u_i \rangle \neq 0\} = \{u_i : |\langle x, u_i \rangle| > 0\} = \bigcup_{n \in \mathbb{N}} \left\{ u_i : |\langle x, u_i \rangle|^2 > \frac{\|x\|^2}{n} \right\} = \bigcup_{n \in \mathbb{N}} U_n,$$
and as each U_n is finite, their union U is at most countable.
Consequently, $\langle x, u_i \rangle$ is nonzero for at most a countable number of the u_i's.

Solution to Exercise 4.34, page 190

(1) We have for all $x \in H$ that $|\varphi_y(x)| = |\langle x, y \rangle| \leq \|x\|\|y\|$, and so $\|\varphi_y\| \leq \|y\|$.
If $y = 0$, then $\|\varphi_y\| \leq \|y\| = 0$, and so $\|\varphi_y\| = 0 = \|y\|$.
If $y \neq 0$, then define $z = \dfrac{y}{\|y\|}$, and observe that $\|z\| = 1$, so that
$$\|\varphi_y\| = \sup_{\|x\| \leq 1} |\varphi_y(x)| \geq |\varphi_y(z)| = |\langle z, y \rangle| = \frac{\langle y, y \rangle}{\|y\|} = \|y\|.$$
Hence it follows that $\|\varphi_y\| = \|y\|$.

(2) Let $y \in H \backslash \{0\}$. Then for $x \in H$,
$$\varphi_{iy}(x) = \langle x, iy \rangle = i^* \langle x, y \rangle = -i \langle x, y \rangle = -i\varphi_y(x) = (-i\varphi_y)(x),$$
and so $\varphi_{iy} = -i\varphi_y$.
Also $\|\varphi_y\| = \|y\| \neq 0$, so that $\varphi_y \neq \mathbf{0}$, the zero linear functional.
If the map $\eta \mapsto \varphi_\eta : H \to CL(H, \mathbb{C})$ were linear, then in particular, we would have $\varphi_{iy} = i\varphi_y$, and from the above, we would then get $i\varphi_y = -i\varphi_y$, giving $\varphi_y = \mathbf{0}$, which is absurd.

---

[6] By the Axiom of Choice!

Solution to Exercise 4.35, page 195

We will show that $Y := \operatorname{ran} P = \ker(I - P)$, and since the kernel of the continuous linear transformation $I - P$ is closed, it follows that Y is closed.

That $\operatorname{ran} P = \ker(I - P)$: If $y \in \operatorname{ran} P$, then $y = Px$ for some $x \in H$. Then
$$(I - P)y = y - Py = Px - P(Px) = Px - P^2x = Px - Px = \mathbf{0}.$$
So $y \in \ker(I - P)$. Hence $\operatorname{ran} P \subset \ker(I - P)$.

On the other hand, if $y \in \ker(I - P)$, then $(I - P)y = \mathbf{0}$ and so $y = Py \in \operatorname{ran} P$. Thus $\ker(I - P) \subset \operatorname{ran} P$ as well.

It remains to show that $P = P_Y$. We will use $(\operatorname{ran} P)^{\perp} = \ker(P^*) = \ker P$, where the last equality follows thanks to the self-adjointness of P. Let $x \in H$. Then $x = P_Y x + P_{Y^\perp} x$. But $P_{Y^\perp} x \in Y^{\perp} = \ker P$, and so
$$Px = P(P_Y x + P_{Y^\perp} x) = P(P_Y x) + P(P_{Y^\perp} x) = P(P_Y x) + \mathbf{0} = P(P_Y x).$$
As $P_Y x \in Y = \operatorname{ran} P$, $P_Y x = Px_1$ for some $x_1 \in H$.
Thus $P(P_Y x) = P(Px_1) = P^2 x_1 = Px_1 = P_Y x$. Hence $Px = P(P_Y x) = P_Y x$.

Solution to Exercise 4.36, page 195

We have $T_1^* = \left(\dfrac{T + T^*}{2}\right)^* = \dfrac{T^* + T^{**}}{2} = \dfrac{T^* + T}{2} =: T_1$,

$T_2^* = \left(\dfrac{T - T^*}{2}\right)^* = \dfrac{T^* - T^{**}}{2} = \dfrac{T^* - T}{2} =: -T_2$,

and so T_1 is self-adjoint, while T_2 is skew-adjoint.

Moreover, $T_1 + T_2 = \dfrac{T + T^*}{2} + \dfrac{T - T^*}{2} = T$.

In order to show uniqueness, suppose that T_1', T_2' are self-adjoint and skew-adjoint respectively such that $T = T_1' + T_2'$. Then $T_1 + T_2 = T_1' + T_2'$, and so we obtain $T_1 - T_1' = T_2' - T_2$. As the left-hand side is self-adjoint, and the right-hand side is skew-adjoint, both sides must be zero. (Indeed, if $S := T_1 - T_1' = T_2' - T_2$ is the common value, then $S = S^* = -S$, and so $2S = \mathbf{0}$, that is, $S = \mathbf{0}$.)

Solution to Exercise 4.37, page 195

$\sup\limits_{n \in \mathbb{N}} |\lambda_n^*| = \sup\limits_{n \in \mathbb{N}} |\lambda_n| < \infty$. Define $T : \ell^2 \to \ell^2$ by $T\mathbf{k} = (\lambda_n^* k_n)_{n \in \mathbb{N}}$, $\mathbf{k} = (k_n)_{n \in \mathbb{N}} \in \ell^2$.
Then T is well-defined and $T \in CL(\ell^2)$. We will show that $\Lambda^* = T$.
For all $\mathbf{h} = (h_n)_{n \in \mathbb{N}}$ and $\mathbf{k} = (k_n)_{n \in \mathbb{N}}$ in ℓ^2, we have
$$\langle \Lambda \mathbf{h}, \mathbf{k} \rangle = \langle (\lambda_n h_n)_{n \in \mathbb{N}}, (k_n)_{n \in \mathbb{N}} \rangle = \sum_{n=1}^{\infty} (\lambda_n h_n) k_n^*$$
$$= \sum_{n=1}^{\infty} h_n (\lambda_n^* k_n)^* = \langle (h_n)_{n \in \mathbb{N}}, (\lambda_n^* k_n)_{n \in \mathbb{N}} \rangle = \langle \mathbf{h}, T\mathbf{k} \rangle.$$
Thus $\Lambda^* = T$.

Solution to Exercise 4.38, page 195

We'll show that I^* is given by $(I^*\mathbf{k})(t) = \int_t^1 \mathbf{k}(\tau)d\tau$, $t \in [0,1]$, $\mathbf{k} \in L^2[0,1]$.
$I^* \in CL(L^2[0,1])$ by Example 2.10 (page 70), with $A(t,\tau) = \begin{cases} 0 & \text{if } t \geq \tau, \\ 1 & \text{otherwise} \end{cases}$.
For $\mathbf{h}, \mathbf{k} \in L^2[0,1]$, we have

$$\langle I\mathbf{h}, \mathbf{k}\rangle = \int_0^1 (I\mathbf{h})(t)\mathbf{k}(t)dt = \int_0^1 \Big(\int_0^t \mathbf{h}(\tau)d\tau\Big)\mathbf{k}(t)dt = \int_0^1 \int_0^t \mathbf{h}(\tau)\mathbf{k}(t)d\tau dt$$

$$= \int_0^1 \int_\tau^1 \mathbf{h}(\tau)\mathbf{k}(t)dt d\tau = \int_0^1 \mathbf{h}(\tau)\Big(\int_\tau^1 \mathbf{k}(t)dt\Big)d\tau$$

$$= \Big\langle \mathbf{h}, \int_\cdot^1 \mathbf{k}(t)dt\Big\rangle,$$

and so $I^* \in CL(L^2[0,1])$ is given by $(I^*\mathbf{k})(t) = \int_t^1 \mathbf{k}(\tau)d\tau$, $t \in [0,1]$, $\mathbf{k} \in L^2[0,1]$.

Solution to Exercise 4.39, page 195

$T_A^* = T_{A^*}$, where $A^* = \begin{bmatrix} \cos\theta & \sin\theta \\ -\sin\theta & \cos\theta \end{bmatrix} = \begin{bmatrix} \cos(-\theta) & -\sin(-\theta) \\ \sin(-\theta) & \cos(-\theta) \end{bmatrix}$ $(= A^{-1})$.
Thus T_A^* is *clockwise* rotation through an angle θ in the plane.

Solution to Exercise 4.40, page 196

For $x \in H$, we have $x = \sum_{k=1}^\infty \langle x, u_k\rangle u_k = \underbrace{\sum_{k=1}^n \langle x, u_k\rangle u_k}_{\in Y_n} + \underbrace{\sum_{k=n+1}^\infty \langle x, u_k\rangle u_k}_{=:x' \in Y_n^\perp}.$

(Note that $x' \in Y_n^\perp$ because $\langle x', u_i\rangle = 0$ for all $i = 1, \cdots, n$.)
So $P_n x = \sum_{k=1}^n \langle x, u_k\rangle u_k$ for all $x \in H$. For all $x \in H$, we have

$$\|P_n x - x\|^2 = \Big\|\sum_{k=n+1}^\infty \langle x, u_k\rangle u_k\Big\|^2 = \sum_{k=n+1}^\infty |\langle x, u_k\rangle|^2 \xrightarrow{n\to\infty} 0,$$

since $\sum_{k=1}^\infty |\langle x, u_k\rangle|^2 = \|x\|^2 < \infty$.

Solution to Exercise 4.41, page 196

(1) If $B' = \{u_n' : n \in \mathbb{N}\}$ is another orthonormal basis, then

$$\sum_{n=1}^\infty \|Tu_n\|^2 = \sum_{n=1}^\infty \sum_{m=1}^\infty |\langle Tu_n, u_m'\rangle|^2 = \sum_{n=1}^\infty \sum_{m=1}^\infty |\langle u_n, T^*u_m'\rangle|^2$$

$$= \sum_{m=1}^\infty \sum_{n=1}^\infty |\langle T^*u_m', u_n\rangle|^2 = \sum_{m=1}^\infty \|T^*u_m'\|^2.$$

On the other hand, we also have

$$\sum_{n=1}^{\infty} \|Tu_n'\|^2 = \sum_{n=1}^{\infty}\sum_{m=1}^{\infty} |\langle Tu_n', u_m'\rangle|^2 = \sum_{n=1}^{\infty}\sum_{m=1}^{\infty} |\langle u_n', T^*u_m'\rangle|^2$$

$$= \sum_{m=1}^{\infty}\sum_{n=1}^{\infty} |\langle T^*u_m', u_n'\rangle|^2 = \sum_{m=1}^{\infty} \|T^*u_m'\|^2,$$

and so $\sum_{n=1}^{\infty} \|Tu_n\|^2 = \sum_{m=1}^{\infty} \|T^*u_m'\|^2 = \sum_{n=1}^{\infty} \|Tu_n'\|^2.$

(2) We will verify simultaneously the norm and subspace axioms:

(N1/S3) For all $T \in S_2(H)$ that $\|T\|_{\text{HS}} = \left(\sum_{n=1}^{\infty} \|Tu_n\|^2\right)^{1/2} \geq 0.$

Now let $T \in S_2(H)$ and $\|T\|_{\text{HS}} = 0$. Then $\sum_{n=1}^{\infty} \|Tu_n\|^2 = 0.$
So $Tu_n = \mathbf{0}$ for all n. But then for all $x \in H$, we have

$$Tx = T\left(\sum_{n=1}^{\infty} \langle x, u_n\rangle u_n\right) = \sum_{n=1}^{\infty} \langle x, u_n\rangle Tu_n = \sum_{n=1}^{\infty} \langle x, u_n\rangle \mathbf{0} = \mathbf{0}.$$

Consequently $T = \mathbf{0}.$

Clearly $\mathbf{0} \in S_2(H)$ since $\|\mathbf{0}\|_{\text{HS}} = 0 < \infty.$

(N2/S2) For all $T \in S_2(H)$ and $\alpha \in \mathbb{K}$, we have

$$\|\alpha \cdot T\|_{\text{HS}}^2 = \sum_{n=1}^{\infty} \|(\alpha \cdot T)u_n\|^2 = \sum_{n=1}^{\infty} \|\alpha \cdot (Tu_n)\|^2$$

$$= \sum_{n=1}^{\infty} |\alpha|^2 \|Tu_n\|^2 = |\alpha|^2 \sum_{n=1}^{\infty} \|Tu_n\|^2 = |\alpha|^2 \|T\|_{\text{HS}}^2,$$

and so $\|\alpha \cdot T\|_{\text{HS}} = |\alpha|\|T\|_{\text{HS}}.$

Note that we've also shown for all $T \in S_2(H)$, $\alpha \in \mathbb{K}$, that $\alpha \cdot T \in S_2(H).$

(N3/S1) Finally, if $T_1, T_2 \in S_2(H)$, then we have

$$\|T_1 + T_2\|_{\text{HS}}^2 = \sum_{n=1}^{\infty} \|(T_1 + T_2)u_n\|^2 = \sum_{n=1}^{\infty} \|T_1 u_n + T_2 u_n\|^2$$

$$\leq \sum_{n=1}^{\infty} (\|T_1 u_n\| + \|T_2 u_n\|)^2$$

$$= \|T_1\|_{\text{HS}}^2 + 2\sum_{n=1}^{\infty} \|T_1 u_n\|\|T_2 u_n\| + \|T_2\|_{\text{HS}}^2$$

$$\leq \|T_1\|_{\text{HS}}^2 + 2\left(\sum_{n=1}^{\infty} \|T_1 u_n\|^2\right)^{1/2}\left(\sum_{n=1}^{\infty} \|T_2 u_n\|^2\right)^{1/2} + \|T_2\|_{\text{HS}}^2$$

$$= \|T_1\|_{\text{HS}}^2 + 2\|T_1\|_{\text{HS}}\|T_2\|_{\text{HS}} + \|T_2\|_{\text{HS}}^2 = (\|T_1\|_{\text{HS}} + \|T_2\|_{\text{HS}})^2,$$

and so $\|T_1 + T_2\|_{\text{HS}} \leq \|T_1\|_{\text{HS}} + \|T_2\|_{\text{HS}}.$

Also, this shows that for all $T_1, T_2 \in S_2(H)$, $T_1 + T_2 \in S_2(H).$

(3) We have for all $x \in H$ that
$$\|T^*x\|^2 = \sum_{n=1}^{\infty} |\langle T^*x, u_n\rangle|^2 = \sum_{n=1}^{\infty} |\langle x, Tu_n\rangle|^2 \leq \sum_{n=1}^{\infty} \|x\|^2 \|Tu_n\|^2 = \|x\|^2 \|T\|_{\text{HS}},$$
and so $\|T\| = \|T^*\| \leq \|T\|_{\text{HS}}$.

Solution to Exercise 4.42, page 197

As $CL(H)$ is an algebra, $\Lambda(T) \in CL(H)$. We verify linearity:
(L1) For $T_1, T_2 \in CL(H)$,
$$\Lambda(T_1 + T_2) = A^*(T_1 + T_2) + (T_1 + T_2)A = A^*T_1 + A^*T_2 + T_1A + T_2A$$
$$= (A^*T_1 + T_1A) + (A^*T_2 + T_2A) = \Lambda(T_1) + \Lambda(T_2).$$
(L2) $\Lambda(\alpha T) = A^*(\alpha T) + (\alpha T)A = \alpha(A^*T + TA) = \alpha\Lambda T$, $T \in CL(H)$, $\alpha \in \mathbb{K}$.
Continuity: For $T \in CL(H)$,
$$\|\Lambda(T)\| = \|A^*T + TA\| \leq \|A^*T\| + \|TA\|$$
$$\leq \|A^*\|\|T\| + \|T\|\|A\|$$
$$= (\|A^*\| + \|A\|)\|T\| = 2\|A\|\|T\|,$$
and so $\Lambda \in CL(CL(H))$.
If $T \in CL(H)$ is such that $T = T^*$, then
$$(\Lambda(T))^* = (A^*T + TA)^* = (A^*T)^* + (TA)^* = T^*A^{**} + A^*T^*$$
$$= TA + A^*T = \Lambda(T).$$
So $\Lambda(T)$ is self-adjoint.

Solution to Exercise 4.43, page 197

Let $(T_n)_{n \in \mathbb{N}}$ be a sequence of self-adjoint operators in $CL(H)$ that converges to $T \in CL(H)$. We'd like to show that for all $x, y \in H$, $\langle Tx, y\rangle = \langle x, Ty\rangle$. As we have $\|T_nx - Tx\| \leq \|T_n - T\|\|x\|$, it follows that $(T_nx)_{n \in \mathbb{N}}$ converges to Tx, and similarly, $(T_ny)_{n \in \mathbb{N}}$ converges to Ty. Thus
$$\langle Tx, y\rangle = \left\langle \lim_{n\to\infty} T_nx, y\right\rangle = \lim_{n\to\infty} \langle T_nx, y\rangle = \lim_{n\to\infty} \langle x, T_n^*y\rangle = \lim_{n\to\infty} \langle x, T_ny\rangle$$
$$= \left\langle x, \lim_{n\to\infty} T_ny\right\rangle = \langle x, Ty\rangle.$$

Solution to Exercise 4.44, page 197

Let $\mu \in \rho(T)$. Then there is an $S \in CL(H)$ such that $S(\mu I - T) = I = (\mu I - T)S$. Taking adjoints, we obtain
$$(\mu^*I - T^*)S^* = (S(\mu I - T))^* = I^* = I = I^* = ((\mu I - T)S)^* = S^*(\mu^*I - T^*).$$
Thus $\mu^*I - T^*$ is invertible in $CL(H)$, and so $\mu^* \in \rho(T^*)$.
So we have proved that $\boxed{\mu \in \rho(T) \Rightarrow \mu^* \in \rho(T^*)}$ for all $T \in CL(H)$.

Applying this to T^* instead of T gives:
$$\boxed{\mu^* \in \rho(T^*) \quad \Rightarrow \quad \mu = (\mu^*)^* \in \rho((T^*)^*) = \rho(T).}$$
Consequently for all $T \in CL(H)$, $\mu \in \rho(T)$ if and only if $\mu^* \in \rho(T^*)$.
We had seen that $R = L^*$ and that $\sigma(L) = \{z \in \mathbb{C} : |z| \leq 1\}$.
From the above, we obtain $\sigma(R) = \mathbb{C}\backslash\rho(R) = \mathbb{C}\backslash(\rho(L))^* = \mathbb{C}\backslash\rho(L) = \sigma(L)$.
Consequently the spectrum of R is the same as that of L, namely the closed unit disc $\{z \in \mathbb{C} : |z| \leq 1\}$ in the complex plane.

Solution to Exercise 4.45, page 197

We have for $\lambda \notin \{0, 1\}$,
$$\frac{1}{\lambda}\left(I + \frac{P_Y}{\lambda - 1}\right)(\lambda I - P_Y) = \frac{1}{\lambda}\left(\lambda I - P_Y + \frac{\lambda}{\lambda - 1}P_Y - \frac{P_Y^2}{\lambda - 1}\right)$$
$$= \frac{1}{\lambda}\left(\lambda I - P_Y + \frac{\lambda}{\lambda - 1}P_Y - \frac{P_Y}{\lambda - 1}\right) = I,$$
and similarly $(\lambda I - P_Y)\frac{1}{\lambda}\left(I + \frac{P_Y}{\lambda - 1}\right) = I$.

The previous part shows that $\sigma(P_Y) \subset \{0, 1\}$.
We now show that both 0 and 1 are eigenvalues, so that $\sigma(P_Y) = \sigma_p(P_Y) = \{0, 1\}$.
As Y is a proper subspace, $Y \neq \{\mathbf{0}\}$. So there exist nonzero vectors y in Y, and all of these are eigenvectors of P_Y with eigenvalue 1: $P_Y y = y = 1 \cdot y$.

Also, as Y is a proper subspace, $Y \neq H$.
If $Y^\perp = \{\mathbf{0}\}$, then we have that $Y = (Y^\perp)^\perp = \{\mathbf{0}\}^\perp = H$, a contradiction.
Thus $Y^\perp \neq \{\mathbf{0}\}$. But this means that there exist nonzero vectors x in $Y^\perp$.
All of these are eigenvectors of P_Y with eigenvalue 0, since $P_Y x = \mathbf{0} = 0 \cdot x$.

Solution to Exercise 4.46, page 197

Let $\lambda \in \sigma_p(U)$ with eigenvector $v \neq \mathbf{0}$.
Then $Uv = \lambda v$, and so $|\lambda|^2 \|v\|^2 = \langle \lambda v, \lambda v\rangle = \langle Uv, Uv\rangle = \langle U^*Uv, v\rangle = \langle Iv, v\rangle = \|v\|^2$.
Thus $|\lambda| = 1$, that is, λ lies on the unit circle with centre 0 in the complex plane.

If $v_1, v_2 \in H\backslash\{\mathbf{0}\}$ are eigenvectors of U corresponding to distinct eigenvalues λ_1, λ_2, then we have
$$(\lambda_1 - \lambda_2)\langle v_1, v_2\rangle = \lambda_1 \langle v_1, v_2\rangle - \lambda_2 \langle v_1, v_2\rangle$$
$$= \langle \lambda_1 v_1, v_2\rangle - \lambda_2 \langle v_1, I v_2\rangle$$
$$= \langle Uv_1, v_2\rangle - \lambda_2 \langle v_1, U^*Uv_2\rangle$$
$$= \langle Uv_1, v_2\rangle - \lambda_2 \langle Uv_1, Uv_2\rangle$$
$$= \langle Uv_1, v_2\rangle - \lambda_2 \langle Uv_1, \lambda_2 v_2\rangle$$
$$= \langle Uv_1, v_2\rangle - |\lambda_2|^2 \langle Uv_1, v_2\rangle$$
$$= \langle Uv_1, v_2\rangle - 1\langle Uv_1, v_2\rangle = 0,$$
and so $\langle v_1, v_2\rangle = 0$.

Solution to Exercise 4.47, page 199

The spectrum of T is real, and hence $T + iI$ is invertible in $CL(H)$. Since $(T + iI)(T - iI) = T^2 + I = (T - iI)(T + iI)$, it follows by pre- and post-multiplying with $(T + iI)^{-1}$ that $(T - iI)(T + iI)^{-1} = (T + iI)^{-1}(T - iI) =: U$. Hence we have

$$U^* = \big((T - iI)(T + iI)^{-1}\big)^* = \big((T + iI)^{-1}\big)^*(T - iI)^*$$
$$= \big((T + iI)^*\big)^{-1}(T^* + iI) = (T^* - iI)^{-1}(T^* + iI) = (T - iI)^{-1}(T + iI).$$

So

$$UU^* = (T + iI)^{-1}(T - iI)U^* = (T + iI)^{-1}(T - iI)(T - iI)^{-1}(T + iI) = I,$$
$$U^*U = (T - iI)^{-1}(T + iI)U = (T - iI)^{-1}(T + iI)(T + iI)^{-1}(T - iI) = I.$$

Thus U is unitary. We have

$$I - U = I - (T - iI)(T + iI)^{-1} = \big(T + iI - (T - iI)\big)(T + iI)^{-1} = 2i(T + iI)^{-1}.$$

Hence $I - U$ is invertible in $CL(H)$ with inverse $\dfrac{T + iI}{2i}$. Similarly,

$$I + U = I + (T - iI)(T + iI)^{-1} = (T + iI + T - iI)(T + iI)^{-1} = 2T(T + iI)^{-1}.$$

So $i(I + U)(I - U)^{-1} = i\big(2T(T + iI)^{-1}\big)\dfrac{(T + iI)}{2i} = T.$

Solution to Exercise 4.48, page 199

(1) Suppose that $P_Y \leqslant P_Z$. If $y \in Y$, then

$$\|P_Z y\|^2 + \|P_{Z^\perp} y\|^2 = \langle P_Z y + P_{Z^\perp} y, P_Z y + P_{Z^\perp} y\rangle = \langle y, y\rangle = \langle P_Y y, y\rangle$$
$$\leqslant \langle P_Z y, y\rangle = \langle P_Z y, P_Z y + P_{Z^\perp} y\rangle = \|P_Z y\|^2.$$

So $P_{Z^\perp} y = \mathbf{0}$, giving $y = P_Z y + P_{Z^\perp} y = P_Z y + \mathbf{0} = P_Z y \in Z$. Thus $Y \subset Z$.

(2) Now let $Y \subset Z$ and $x \in H$. We have $P_Z x = P_Y x + (P_Z x - P_Y x)$.
We first show that $P_Z x - P_Y x$ is perpendicular to $P_Y x$.
As $x = P_Y x + P_{Y^\perp} x = P_Z x + P_{Z^\perp} x$, we have $P_Z x - P_Y x = P_{Y^\perp} x - P_{Z^\perp} x$.
So $\langle P_Y x, P_Z x - P_Y x\rangle = \langle P_Y x, P_{Y^\perp} x - P_{Z^\perp} x\rangle = -\langle \underbrace{P_Y x}_{\in Y \subset Z}, P_{Z^\perp} x\rangle = 0.$

Hence

$$\langle P_Z x, x\rangle = \langle P_Z x, P_Z x + P_{Z^\perp} x\rangle = \langle P_Z x, P_Z x\rangle$$
$$= \langle P_Y x + (P_Z x - P_Y x), P_Y x + (P_Z x - P_Y x)\rangle$$
$$= \langle P_Y x, P_Y x\rangle + \langle P_Z x - P_Y x, P_Z x - P_Y x\rangle$$
$$\geqslant \langle P_Y x, P_Y x\rangle = \langle P_Y x, P_Y x + P_{Y^\perp} x\rangle = \langle P_Y x, x\rangle.$$

Consequently, $P_Y \leqslant P_Z$.

Solution to Exercise 4.49, page 204

(1) By the Fundamental Theorem of Calculus, $f(x) - f(0) = \displaystyle\int_0^x f'(x)dx$.

So $\displaystyle\lim_{x \to \infty} f(x) = f(0) + \int_0^\infty f'(x)dx =: L.$

As $f(x) \geq 0$ for all x, we must have that $L \geq 0$. Suppose that $L > 0$. Then there exists an $R > 0$ such that for all $x > R$, $L - f(x) \leq |f(x) - L| < \frac{L}{2}$, and in particular, $f(x) > \frac{L}{2}$ for all $x > R$. Hence for all $x > R$,
$$\int_0^\infty f(t)dt \geq \int_0^x f(t)dt \geq \int_R^x f(t)dt > (x-R)\frac{L}{2} \stackrel{x\to\infty}{\longrightarrow} \infty,$$
which is absurd. Hence $L = 0$.

(2) We apply part (1) with $f(x) := |\Psi(x)|^2$.
We note that $f' = (|\Psi|^2)' = (\Psi\Psi^*)' = \Psi'\Psi^* + \Psi(\Psi')^*$, and so $|f'| \leq 2|\Psi||\Psi'|$.
Thus $\int_0^\infty |f'(x)|dx \leq 2\int_0^\infty |\Psi(x)||\Psi'(x)|dx \leq 2\|\Psi\|_2\|\Psi'\|_2 < \infty$.
So $\int_0^\infty f'(x)dx$ exists. By part (1), $|\Psi(x)|^2 \stackrel{x\to\infty}{\to} 0$, and so $\Psi(x) \stackrel{x\to\infty}{\to} 0$.
To show that $\Psi(x) \stackrel{x\to-\infty}{\to} 0$, we apply the above to $x \mapsto \Psi(-x)$, and note that if $\Psi, \Psi' \in L^2(\mathbb{R})$, then so do $\Psi(-\cdot), (\Psi(-\cdot))' = -\Psi'(-\cdot)$.

Solution to Exercise 4.50, page 205

We have for self-adjoint A, B that
$$[A, B]^* = (AB - BA)^* = B^*A^* - A^*B^* = BA - AB = -(AB - BA) = -[A, B].$$

Solution to Exercise 4.51, page 205

We have
$$[A, [B, C]] = A[B, C] - [B, C]A = A(BC - CB) - (BC - CB)A$$
$$= ABC - ACB - BCA + CBA.$$
Similarly,
$$[B, [C, A]] = BCA - BAC - CAB + ACB, \text{ and}$$
$$[C, [A, B]] = CAB - CBA - ABC + BAC.$$
Hence
$$[A, [B, C]] + [B, [C, A]] + [C, [A, B]] = ABC - ACB - BCA + CBA$$
$$+ BCA - BAC - CAB + ACB$$
$$+ CAB - CBA - ABC + BAC$$
$$= 0.$$

Solution to Exercise 4.52, page 205

If $n = 1$, then $[Q, P] = -[P, Q] = -(-i\hbar I) = i\hbar 1 Q^{1-1}$, and so the claim is true.
If $[Q^n, P] = i\hbar n Q^{n-1}$ for some $n \in \mathbb{N}$, then we have
$$[Q^{n+1}, P] = Q^{n+1}P - PQ^{n+1} = Q(Q^n P) - PQ^{n+1}$$
$$= Q(i\hbar n Q^{n-1} + PQ^n) - PQ^{n+1} = i\hbar n Q^n + QPQ^n - PQQ^n$$
$$= i\hbar n Q^n + (QP - PQ)Q^n = i\hbar n Q^n + (i\hbar)Q^n = i\hbar(n+1)Q^{(n+1)-1},$$
and so the claim follows for all $n \in \mathbb{N}$ by induction.

Solution to Exercise 4.53, page 207

We have in the classical case that
$$\{Q^2, P^2\} = \frac{\partial q^2}{\partial q} \cdot \frac{\partial p^2}{\partial p} - \frac{\partial p^2}{\partial q} \cdot \frac{\partial q^2}{\partial p} = (2q)(2p) - (0)(0) = 4qp = (4QP)(q,p).$$
Thus $\{Q^2, P^2\} = 4QP$.

In the quantum mechanical case, we have, using Exercise 4.52, page 205, that
$$[Q^2, P^2] = Q^2PP - P^2Q^2 = (PQ^2 + i\hbar 2Q)P - P^2Q^2 = 2i\hbar QP + PQ^2P - PPQ^2$$
$$= 2i\hbar QP + P(Q^2P - PQ^2) = 2i\hbar QP + P(i\hbar 2Q)$$
$$= i\hbar 2(QP + PQ).$$

Thus $\dfrac{[Q^2, P^2]}{i\hbar} = 2(QP + PQ) \neq 4QP$ (since otherwise $QP = PQ$, which is false since $[Q, P] = i\hbar I \neq \mathbf{0}$).

QP is not self-adjoint, since if it were self-adjoint, then for all compactly supported Ψ and Φ, we would have
$$\langle \Psi, QP\Phi \rangle = \langle QP\Psi, \Phi \rangle = \langle P\Psi, Q\Phi \rangle = \langle \Psi, PQ\Phi \rangle,$$
which would give $i\hbar \Phi = [Q, P]\Phi = \mathbf{0}$, which is clearly false for nonzero Φ!

On the other hand, for all Ψ and Φ, we have
$$\langle (QP + PQ)\Psi, \Phi \rangle = \langle QP\Psi, \Phi \rangle + \langle PQ\Psi, \Phi \rangle = \langle \Psi, PQ\Phi \rangle + \langle \Psi, QP\Phi \rangle$$
$$= \langle \Psi, (QP + PQ)\Phi \rangle.$$

Solution to Exercise 4.54, page 207

We have
$$\frac{d}{dt}\|\psi(t)\|^2 = \frac{d}{dt}\langle \psi(t), \psi(t) \rangle = \left\langle \frac{d}{dt}\psi(t), \psi(t) \right\rangle + \left\langle \psi(t), \frac{d}{dt}\psi(t) \right\rangle$$
$$= \left\langle \frac{d}{dt}\psi(t), \psi(t) \right\rangle + \left\langle \frac{d}{dt}\psi(t), \psi(t) \right\rangle^* = 2\mathrm{Re}\left\langle \frac{d}{dt}\psi(t), \psi(t) \right\rangle$$
$$= 2\mathrm{Re}\left\langle -\frac{i}{\hbar}H\psi(t), \psi(t) \right\rangle = 2\mathrm{Re}\left(-\frac{i}{\hbar}\underbrace{\langle H\psi(t), \psi(t) \rangle}_{\in \mathbb{R}} \right) = 0,$$
and so $t \mapsto \|\psi(t)\|^2$ is constant, giving $\|\psi(t)\|^2 = \|\psi(0)\|^2 = 1$.

Solution to Exercise 4.55, page 207

As $V \equiv 0$ for $x \in (0, \pi)$, we have $\dfrac{\hbar^2}{2m}X'' + EX = 0$, that is, $X'' + \dfrac{2mE}{\hbar^2}X = 0$.

Depending on the sign of E, the solution is given by
$$X(x) = \begin{cases} A\cos\left(\sqrt{\frac{2mE}{\hbar^2}}x\right) + B\sin\left(\sqrt{\frac{2mE}{\hbar^2}}x\right) & \text{if } E > 0, \\ A + Bx & \text{if } E = 0, \\ A\cosh\left(\sqrt{\frac{2m(-E)}{\hbar^2}}x\right) + B\sinh\left(\sqrt{\frac{2m(-E)}{\hbar^2}}x\right) & \text{if } E < 0. \end{cases}$$

If $E = 0$, then the conditions $X(0) = X(\pi) = 0$ give $A = B = 0$. So $X \equiv 0$.
If $E < 0$, then the conditions $X(0) = X(\pi)$ imply that $A = B = 0$ so that $X \equiv 0$.
So only the case $E > 0$ remains. The condition $X(0) = 0$ gives $A = 0$.
The condition $X(\pi) = 0$ implies $B \sin\left(\sqrt{\dfrac{2mE}{\hbar^2}}\pi\right) = 0$.
As we want nontrivial solutions, we know $B \neq 0$ (otherwise $X \equiv 0$).
So $\sin\left(\underbrace{\sqrt{\dfrac{2mE}{\hbar^2}}\pi}_{>0}\right) = 0$, giving $\sqrt{\dfrac{2mE}{\hbar^2}}\pi \in \pi\mathbb{N}$.

Thus $E = \dfrac{n^2\hbar^2}{2m}$, $n \in \mathbb{N}$ (discrete/"quantised" energy levels!).
We have $|\Psi(x,t)| = |X(x)||T(t)| = |X(x)| \cdot |C| = |C| \cdot |B| \cdot |\sin(nx)|$.
The plots of $|\Psi|^2 = \text{constant} \cdot \big(\sin(nx)\big)^2$ when $n = 1, 2$ are shown below.

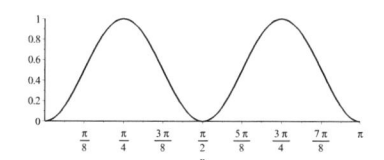

When $n = 1$, the probability is

$$\dfrac{\displaystyle\int_0^{\frac{1}{4}} |\Psi(x,t)|^2 dx}{\displaystyle\int_0^{\pi} |\Psi(x,t)|^2 dx} = \dfrac{\displaystyle\int_0^{\frac{1}{4}} (\sin x)^2 dx}{\displaystyle\int_0^{\pi} (\sin x)^2 dx} = \dfrac{\dfrac{1}{2} \cdot \dfrac{1}{4} - \dfrac{\sin(2x)}{4}\Big|_0^{\frac{1}{4}}}{\dfrac{1}{2} \cdot \pi - \dfrac{\sin(2x)}{4}\Big|_0^{\pi}} = \dfrac{\dfrac{1}{8} - \dfrac{\sin \frac{1}{2}}{4}}{\dfrac{\pi}{2}} \approx 0.0034.$$

When $n = 2$, the probability is

$$\dfrac{\displaystyle\int_0^{\frac{1}{4}} |\Psi(x,t)|^2 dx}{\displaystyle\int_0^{\pi} |\Psi(x,t)|^2 dx} = \dfrac{\displaystyle\int_0^{\frac{1}{4}} (\sin(2x))^2 dx}{\displaystyle\int_0^{\pi} (\sin(2x))^2 dx} = \dfrac{\dfrac{1}{2} \cdot \dfrac{1}{4} - \dfrac{\sin(4x)}{8}\Big|_0^{\frac{1}{4}}}{\dfrac{1}{2} \cdot \pi - \dfrac{\sin(4x)}{8}\Big|_0^{\pi}} = \dfrac{\dfrac{1}{8} - \dfrac{\sin 1}{8}}{\dfrac{\pi}{2}} \approx 0.0126.$$

Solutions to the exercises from Chapter 5

Solution to Exercise 5.1, page 215

(1) T_m is linear:
(L1) For all $x_1, x_2 \in H$,
$$T_m(x_1 + x_2) = \sum_{n=1}^{m} \langle x_1 + x_2, u_n \rangle Tu_n$$
$$= \sum_{n=1}^{m} \langle x_1, u_n \rangle Tu_n + \sum_{n=1}^{m} \langle x_2, u_n \rangle Tu_n = T_m x_1 + T_m x_2.$$
(L2) For all $x \in H$ and $\alpha \in \mathbb{K}$,
$$T_m(\alpha \cdot x) = \sum_{n=1}^{m} \langle \alpha \cdot x, u_n \rangle Tu_n = \sum_{n=1}^{m} \alpha \langle x, u_n \rangle Tu_n = \alpha \cdot (T_m x).$$
So T_m is a linear transformation. Next we prove continuity: for all $x \in H$,
$$\|T_m x\| = \Big\| \sum_{n=1}^{m} \langle x, u_n \rangle Tu_n \Big\|$$
$$\leq \sum_{n=1}^{m} |\langle x, u_n \rangle| \|T\| \underbrace{\|u_n\|}_{=1} = \|T\| \sum_{n=1}^{m} |\langle x, u_n \rangle|$$
$$\leq \|T\| \sum_{n=1}^{m} \|x\| \underbrace{\|u_n\|}_{=1} = m\|T\|\|x\|.$$
Conclusion: $T_m \in CL(H)$.
For $x \in H$ we have
$$\|(T - T_m)x\|^2 = \Big\| T\Big(\sum_{n=1}^{\infty} \langle x, u_n \rangle u_n\Big) - \sum_{n=1}^{m} \langle x, u_n \rangle Tu_n \Big\|^2$$
$$= \Big\| \sum_{n=1}^{\infty} \langle x, u_n \rangle Tu_n - \sum_{n=1}^{m} \langle x, u_n \rangle Tu_n \Big\|^2$$
$$= \Big\| \sum_{n=m+1}^{\infty} \langle x, u_n \rangle Tu_n \Big\|^2$$
$$\leq \Big(\sum_{n=m+1}^{\infty} |\langle x, u_n \rangle| \|Tu_n\| \Big)^2 \leq \|x\|^2 \sum_{n=m+1}^{\infty} \|Tu_n\|^2.$$

(2) As $\sum_{n=1}^{\infty} \|Tu_n\|^2 < \infty$, we have $\|T - T_m\|^2 \leq \sum_{n=m+1}^{\infty} \|Tu_n\|^2 \stackrel{m \to \infty}{\longrightarrow} 0$.
Thus $(T_m)_{m \in \mathbb{N}}$ converges to T in $CL(H)$. Since the range of T_m is contained in the span of $Tu_1, \cdots, Tu_m$, it follows that T_m has finite rank, and so T_m is compact. As T is the limit in $CL(H)$ of a sequence of compact operators, it follows that T is compact.

Solution to Exercise 5.2, page 216

(1) (L1) For $x_1, x_2 \in H$, we have
$$(x_0 \otimes y_0)(x_1 + x_2) = \langle x_1 + x_2, y_0 \rangle x_0 = \big(\langle x_1, y_0 \rangle + \langle x_2, y_0 \rangle\big) x_0$$
$$= \langle x_1, y_0 \rangle x_0 + \langle x_2, y_0 \rangle x_0$$
$$= (x_0 \otimes y_0)(x_1) + (x_0 \otimes y_0)(x_2).$$

(L2) For $\alpha \in \mathbb{K}$ and $x \in H$, we have
$$(x_0 \otimes y_0)(\alpha x) = \langle \alpha x, y_0 \rangle x_0 = \alpha \langle x, y_0 \rangle x_0 = \alpha (x_0 \otimes y_0)(x).$$

Continuity: For $x \in H$, we have
$$\|(x_0 \otimes y_0)(x)\| = \|\langle x, y_0 \rangle x_0\| = |\langle x, y_0 \rangle| \|x_0\|$$
$$\leqslant \|x\| \|y_0\| \|x_0\| \quad \text{(Cauchy-Schwarz)}.$$

So $x_0 \otimes y_0 \in CL(H)$, and $\|x_0 \otimes y_0\| \leqslant \|x_0\| \|y_0\|$.

(2) As $\operatorname{ran}(x_0 \otimes y_0) \subset \operatorname{span}\{x_0\}$, we have that $x_0 \otimes y_0$ has finite rank, and so it is compact.

(3) For all $x \in H$,
$$\big(A(x_0 \otimes y_0)B\big)x = A\big((x_0 \otimes y_0)(Bx)\big) = A\big(\langle Bx, y_0 \rangle x_0\big)$$
$$= \langle Bx, y_0 \rangle Ax_0 = \langle x, B^* y_0 \rangle Ax_0$$
$$= \big((Ax_0) \otimes (B^* y_0)\big)(x).$$

Since this is true for all $x \in H$, we conclude that $A(x_0 \otimes y_0)B = (Ax_0) \otimes (B^* y_0)$.

Solution to Exercise 5.3, page 217

(1) Let $H = \ell^2$, and T be diagonal with 2×2 nilpotent blocks $\begin{bmatrix} 0 & 1 \\ 0 & 0 \end{bmatrix}$.

More explicitly, $T(a_1, a_2, a_3, a_4, a_5, a_6, \cdots) = (a_2, 0, a_4, 0, a_6, 0, \cdots)$, for all $(a_n)_{n \in \mathbb{N}} \in \ell^2$. Thus $T \in CL(\ell^2)$. Also, $T^2 = \mathbf{0}$ is compact. But if we take the bounded sequence $(\mathbf{e}_{2n})_{n \in \mathbb{N}}$, then $(T\mathbf{e}_{2n})_{n \in \mathbb{N}} = (\mathbf{e}_{2n-1})_{n \in \mathbb{N}}$, and this has no convergent subsequence. Hence T is not compact.

(2) Suppose that $(x_n)_{n \in \mathbb{N}}$ is a bounded sequence in H, and $\|x_n\| \leqslant M$ for all n. Since T^2 is compact, $(T^2 x_n)_{n \in \mathbb{N}}$ has a convergent subsequence, say $(T^2 x_{n_k})_{k \in \mathbb{N}}$. We will show that $(T x_{n_k})_{k \in \mathbb{N}}$ is also convergent, by showing that it is Cauchy. We have for j, k that
$$\|T x_{n_j} - T x_{n_k}\|^2 = \langle T(x_{n_j} - x_{n_k}), T(x_{n_j} - x_{n_k}) \rangle$$
$$= \langle x_{n_j} - x_{n_k}, T^* T(x_{n_j} - x_{n_k}) \rangle$$
$$= \langle x_{n_j} - x_{n_k}, T^2(x_{n_j} - x_{n_k}) \rangle$$
$$\leqslant \|x_{n_j} - x_{n_k}\| \|T^2(x_{n_j} - x_{n_k})\| \leqslant 2M \|T^2(x_{n_j} - x_{n_k})\|,$$

and so $(T x_{n_k})_{k \in \mathbb{N}}$ is Cauchy. As H is a Hilbert space, it follows that $(T x_{n_k})_{k \in \mathbb{N}}$ is convergent. Hence T is compact.

Solution to Exercise 5.4, page 217

(1) True.

(2) False.
Neither I nor $-I$ is compact, but their sum is $\mathbf{0}$, which is compact.
(3) True.
(4) False.
See the example in the solution to Exercise 5.3, part (1), page 217.
Alternately, we could take two diagonal operators on ℓ^2 corresponding to the sequences $(1, 0, 1, 0, 1, 0, \cdots)$ and $(0, 1, 0, 1, 0, 1, \cdots)$.

Solution to Exercise 5.5, page 217

If $T \in K(H)$, then as $A^* \in CL(H)$, we have $A^*T \in K(H)$. Also, $TA \in K(H)$ because $T \in K(H)$ and $A \in CL(H)$. Since A^*T and TA are in $K(H)$, also their sum $A^*T + TA \in K(H)$, that is, $\Lambda(T) \in K(H)$. Thus $K(H)$ is Λ-invariant.

Solution to Exercise 5.6, page 226

We have $\ker T = \{\mathbf{0}\}$. So $\overline{\operatorname{ran} T} = (\ker T^*)^\perp = (\ker T)^\perp = \{\mathbf{0}\}^\perp = H$.
So T has infinite rank. Let $x \in H = \overline{\operatorname{ran} T}$, and $\epsilon > 0$. Then there exists a $y \in \operatorname{ran} T$, such that $\|x - y\| < \epsilon/2$.
As $y \in \operatorname{ran} T$, we have $y = Tx'$, for some $x' \in H$, and $y = Tx' = \displaystyle\sum_{n=1}^{\infty} \lambda_n \langle x', u_n \rangle u_n$.
So there exists an N such that with
$$z := \sum_{n=1}^{N} \lambda_n \langle x', u_n \rangle u_n \in \operatorname{span}\{u_n : n \in \mathbb{N}\},$$
we have $\|y - z\| < \epsilon/2$. Consequently, $\|x - z\| \leq \|x - y\| + \|y - z\| < \epsilon/2 + \epsilon/2 = \epsilon$, and so $\operatorname{span}\{u_n : n \in \mathbb{N}\}$ is dense in H. Since $\{u_n : n \in \mathbb{N}\}$ is also an orthonormal set, it follows that it is an orthonormal basis for H.

Solution to Exercise 5.7, page 226

We note that each eigenvalue λ of T is nonnegative because if u is a corresponding unit-norm eigenvector, then $\lambda = \lambda \cdot 1 = \lambda \langle u, u \rangle = \langle Tu, u \rangle \geq 0$. By the spectral theorem, we know that there exists a sequence of orthonormal eigenvectors $u_1, u_2, u_3, \cdots$ of T with corresponding eigenvalues $\lambda_1 \geq \lambda_2 \geq \lambda_3 \geq \cdots \geq 0$.
We will show that for all $x \in H$, $\sqrt{T}x := \displaystyle\sum_{n=1}^{\infty} \sqrt{\lambda_n} \langle x, u_n \rangle u_n$ converges in H.
For $N > M$, we have,
$$\Big\| \sum_{n=1}^{N} \sqrt{\lambda_n} \langle x, u_n \rangle u_n - \sum_{n=1}^{M} \sqrt{\lambda_n} \langle x, u_n \rangle u_n \Big\|^2 = \Big\| \sum_{n=M+1}^{N} \sqrt{\lambda_n} \langle x, u_n \rangle u_n \Big\|^2$$
$$= \sum_{n=M+1}^{N} \lambda_n |\langle x, u_n \rangle|^2 \leq \lambda_{M+1} \sum_{n=M+1}^{N} |\langle x, u_n \rangle|^2 \leq \lambda_{M+1} \|x\|^2 \xrightarrow{M \to \infty} 0.$$

In the above, we have used Bessel's Inequality to get the last inequality.
Hence $\left(\sum_{n=1}^{N}\sqrt{\lambda_n}\langle x, u_n\rangle u_n\right)_{N\in\mathbb{N}}$ is Cauchy in H. As H is a Hilbert space,
$$\sqrt{T}x := \sum_{n=1}^{\infty}\sqrt{\lambda_n}\langle x, u_n\rangle u_n$$
converges in H. Consequently $\sqrt{T}x$ is well-defined for all $x \in H$.
Also, it is easy to see that $\sqrt{T}$ is a linear transformation.

Continuity: For all $N \in \mathbb{N}$,
$$\Big\|\sum_{n=1}^{N}\sqrt{\lambda_n}\langle x, u_n\rangle u_n\Big\|^2 = \sum_{n=1}^{N}\lambda_n|\langle x, u_n\rangle|^2 \leqslant \lambda_1\sum_{n=1}^{N}|\langle x, u_n\rangle|^2 \leqslant \lambda_1\|x\|^2.$$

Passing the limit $N \to \infty$, we obtain $\|\sqrt{T}x\|^2 \leqslant \lambda_1\|x\|^2$, and so $\sqrt{T} \in CL(H)$.
We have for all $x \in H$ that
$$(\sqrt{T})^2 x = \sqrt{T}(\sqrt{T}x) = \sqrt{T}\Big(\sum_{n=1}^{\infty}\sqrt{\lambda_n}\langle x, u_n\rangle u_n\Big)$$
$$= \sum_{n=1}^{\infty}\sqrt{\lambda_n}\langle x, u_n\rangle\sqrt{T}u_n \quad (\text{since } \sqrt{T} \text{ is continuous})$$
$$= \sum_{n=1}^{\infty}\sqrt{\lambda_n}\langle x, u_n\rangle\sqrt{\lambda_n}u_n = \sum_{n=1}^{\infty}\lambda_n\langle x, u_n\rangle u_n = Tx.$$

So $(\sqrt{T})^2 = T$.

Solutions to the exercises from Chapter 6

Solution to Exercise 6.1, page 232

(1) Since $\lim_{x\to 0} f'(x) =: L$ exists, given an $\epsilon > 0$, there exists a $\delta > 0$ such that
$$|f'(x) - L| < \epsilon$$
whenever $0 < |h| < \delta$. Consider the interval $[0, h]$ for some h which satisfies $0 < h < \delta$. Since f is differentiable in $(0, h)$ and continuous on $[0, h]$, it follows from the Mean Value Theorem that
$$\frac{f(h) - f(0)}{h - 0} = f'(\theta h) \text{ where } 0 < \theta < 1.$$
Thus $|\theta h| < \delta$ and so $\left|\dfrac{f(h) - f(0)}{h - 0} - L\right| = |f'(\theta h) - L| < \epsilon$.

So for all $h \in (0, \delta)$, $\left|\dfrac{f(h) - f(0)}{h - 0} - L\right| < \epsilon$.

Applying the Mean Value Theorem on $[-h, 0]$, where $0 < h < \delta$, we also get
$$\left|\frac{f(h) - f(0)}{h - 0} - L\right| < \epsilon$$
for all $h \in (-\delta, 0)$. Consequently, for all h satisfying $0 < |h| < \delta$, we have
$$\left|\frac{f(h) - f(0)}{h - 0} - L\right| < \epsilon,$$
that is, f is differentiable at 0, and $f'(0) = L = \lim_{x\to 0} f'(x)$. $f'(0) = \lim_{x\to 0} f'(x)$ shows that f' is continuous at 0. It was given that f' is also continuous on $\mathbb{R}_*$. So f is continuously differentiable on $\mathbb{R}$.

(2) Applying the result from part (1) above, to the function $f^{(n-1)} : \mathbb{R} \to \mathbb{R}$, we obtain that $f^{(n-1)}$ is continuously differentiable on $\mathbb{R}$, that is, f is n times continuously differentiable on $\mathbb{R}$.

(3) We'll show that for $x > 0$, $f^{(k)}(x) = \dfrac{p_k(x)}{x^{2k}} e^{-1/x}$, where p_k is a polynomial. This holds for $k = 1$: $f(x) = e^{-1/x}$ for $x > 0$, and so $f'(x) = \dfrac{1}{x^2} e^{-1/x}$.

If the claim holds for some k, then
$$f^{(k+1)}(x) = \left(\frac{p_k'(x)x^{2k} - 2kx^{2k-1}p_k(x)}{x^{4k}} + \frac{p_k(x)}{x^{2k+2}}\right)e^{-1/x}$$
$$= \frac{p_k'(x)x^2 - 2kxp_k(x) + p_k(x)}{x^{2k+2}} e^{-1/x} = \frac{p_{k+1}(x)}{x^{2(k+1)}} e^{-1/x},$$
where $p_{k+1}(x) := p_k'(x)x^2 - 2kxp_k(x) + p_k(x)$ is a polynomial.

Now $e^{1/x} e^{-1/x} = 1$, and since we have $e^{1/x} > \dfrac{1}{(2n+1)! x^{2n+1}}$, it follows that $e^{-1/x} < (2n+1)! x^{2n+1}$ for $x > 0$. So $0 < x^{-2n} e^{-1/x} < (2n+1)! x$ for $x > 0$. Thus $\lim_{x\searrow 0} x^{-2n} e^{-1/x} = 0$. Consequently
$$\lim_{x\searrow 0} f^{(n)}(x) = \lim_{x\searrow 0} \frac{p_n(x)}{x^{2n}} e^{-1/x} = \left(\lim_{x\searrow 0} p_n(x)\right)\left(\lim_{x\searrow 0} x^{-2n} e^{-1/x}\right) = p_n(0) \cdot 0 = 0.$$
By the previous part, it follows that $f \in C^\infty(\mathbb{R})$.

Solution to Exercise 6.2, page 232

The equation $\dfrac{\partial u}{\partial x} = 0$ says that u is constant along the lines parallel to the x-axis. So for each fixed y, there is a number C_y such that $u(x, y) = C_y$ for all $x \in \mathbb{R}$. But $u \in \mathcal{D}(\mathbb{R}^2)$ must have compact support, and so it is zero outside a ball $B(\mathbf{0}, R)$ with a large enough radius R. So C_y is forced to be 0 for all y! Hence $u \equiv 0$ is the only solution.

Solution to Exercise 6.3, page 232

It is clear that if $\Phi \in \mathcal{D}(\mathbb{R})$, then $\Phi' \in \mathcal{D}(\mathbb{R})$. Moreover,
$$\int_{-\infty}^{\infty} \Phi'(x)dx = \Phi(x)\Big|_{x=-\infty}^{x=+\infty} = 0 - 0 = 0.$$
So we have $\{\Phi' : \Phi \in \mathcal{D}(\mathbb{R})\} \subset \Big\{\varphi \in \mathcal{D}(\mathbb{R}) : \int_{-\infty}^{\infty} \varphi(x)dx = 0\Big\}$.

Now suppose that $\varphi \in \mathcal{D}(\mathbb{R})$ is such that $\int_{-\infty}^{\infty} \varphi(x)dx = 0$.

Define Φ by $\Phi(x) = \int_{-\infty}^{x} \varphi(\xi)d\xi$ for $x \in \mathbb{R}$. Then $\Phi' = \varphi$, and so $\Phi \in C^{\infty}$.

If $a > 0$ is such that φ is zero outside $[-a, a]$, then we have for $x < -a$ that
$$\Phi(x) = \int_{-\infty}^{x} \varphi(\xi)d\xi = 0.$$
On the other hand, for $x > a$, $\Phi(x) = \int_{-\infty}^{x} \varphi(\xi)d\xi = \int_{-\infty}^{\infty} \varphi(\xi)d\xi = 0$.

So φ also vanishes outside $[-a, a]$, and hence $\Phi \in \mathcal{D}(\mathbb{R})$.

Finally, let $\varphi \in Y$, and suppose that $\Phi_1, \Phi_2 \in \mathcal{D}(\mathbb{R})$ are such that $\Phi_1' = \varphi = \Phi_2'$. Then $(\Phi_1 - \Phi_2)' = 0$, and so $\Phi_1 - \Phi_2 \equiv C$, where C is a constant. But as Φ_1, Φ_2 both have compact supports, it follows that C must be zero. Hence $\Phi_1 = \Phi_2$.

Solution to Exercise 6.4, page 233

From the solution to Exercise 6.3, page 232, we know that the Φ_ns are given by
$$\Phi_n(x) = \int_{-\infty}^{x} \varphi_n(\xi)d\xi, \quad x \in \mathbb{R}.$$
As $\varphi_n \xrightarrow{\mathcal{D}} \mathbf{0}$, there is some $a > 0$ such that all the φ_n vanish outside $[-a, a]$. Then it follows that each Φ_n also vanishes outside $[-a, a]$. Also,
$$|\Phi_n(x)| = \Big|\int_{-\infty}^{x} \varphi_n(\xi)d\xi\Big| \leqslant 2a\|\varphi_n\|_{\infty}.$$
Hence it follows that $(\Phi_n)_{n \in \mathbb{N}}$ converges uniformly to $\mathbf{0}$ as $n \to \infty$. Since $\Phi_n' = \varphi_n$, it follows that $\Phi_n^{(k)} = \varphi_n^{(k-1)}$ for $k \geqslant 1$. Thus for each $k \geqslant 1$, we have that $(\Phi_n^{(k)})_{n \in \mathbb{N}}$ converges uniformly to $\mathbf{0}$ (thanks to the fact that $\varphi_n \xrightarrow{\mathcal{D}} \mathbf{0}$). This completes the proof that $\Phi_n \xrightarrow{\mathcal{D}} \mathbf{0}$.

Solution to Exercise 6.5, page 236

Suppose that such a function δ exists. Let $\varphi(x) := \begin{cases} e^{-\frac{1}{1-x^2}} & \text{if } |x| < 1, \\ 0 & \text{if } |x| \geq 1. \end{cases}$

For $n \in \mathbb{N}$, and let $\varphi_n : \mathbb{R} \to \mathbb{R}$ be defined by $\varphi_n(x) := \varphi(nx)$, $x \in \mathbb{R}$. Then φ_n is smooth, takes values in $[0, 1]$, and vanishes outside $[-1/n, 1/n]$. So we have

$$\frac{1}{e} = \varphi_n(0) = \int_{-1/n}^{1/n} \delta(x)\varphi(nx)dx \leq \int_{-1/n}^{1/n} \left(\sup_{x \in [-1/n, 1/n]} \delta(x)\right) \cdot 1 dx$$

$$\leq \left(\sup_{x \in [-1/n, 1/n]} \delta(x)\right) \frac{2}{n}$$

$$\leq \left(\sup_{x \in [-1, 1]} \delta(x)\right) \frac{2}{n} \xrightarrow{n \to \infty} 0,$$

a contradiction.

Solution to Exercise 6.6, page 237

(1) For all $\varphi \in \mathcal{D}(\mathbb{R})$, there exists an $N \in \mathbb{N}$ such that $\varphi = 0$ on $\mathbb{R}\setminus[-n, n]$.
So the sum in the definition of $\langle T, \varphi \rangle$ is finite: $\sum_{n \in \mathbb{Z}} \varphi(n) = \sum_{n=-N}^{N} \varphi(n)$.
Hence $\langle T, \varphi \rangle$ is well defined for each $\varphi \in \mathcal{D}(\mathbb{R})$. The linearity is obvious.
Now suppose that $\varphi_n \xrightarrow{\mathcal{D}} \mathbf{0}$. Then there exists an $K \in \mathbb{N}$ such that each φ_n, vanishes outside $[-K, K]$. Also, for all $|k| \leq K$, $\varphi_n(k) \xrightarrow{n \to \infty} 0$.
Thus $\langle T, \varphi_n \rangle = \sum_{k \in \mathbb{Z}} \varphi_n(k) = \sum_{k=-K}^{K} \varphi_n(k) \xrightarrow{n \to \infty} 0$, and so $T \in \mathcal{D}'(\mathbb{R})$.

(2) Take any $\varphi \in \mathcal{D}(\mathbb{R})$ that is positive in $(0, 1)$ and zero outside $[0, 1]$.
(From Example 6.1, page 230, there is a $\psi \in \mathcal{D}(\mathbb{R})$ that is positive on $(-1, 1)$ and zero outside $[-1, 1]$. By shifting and scaling, we see that the function φ defined by $\varphi(x) := \psi(2x - 1)$, $x \in \mathbb{R}$, is one such function.)
Now define $\varphi_n \in \mathcal{D}(\mathbb{R})$, $n \in \mathbb{N}$, by $\varphi_n(x) = \frac{1}{n}\varphi\left(\frac{x}{n}\right)$, $x \in \mathbb{R}$.
We have for $k \in \mathbb{N}$ that $\varphi_n^{(k)}(x) = \frac{1}{n^{k+1}}\varphi^{(k)}\left(\frac{x}{n}\right)$, $x \in \mathbb{R}$.
Thus for all $k \geq 0$, $\sup_{x \in \mathbb{R}} |\varphi_n^{(k)}(x) - 0| \leq \frac{1}{n^{k+1}} \sup_{x \in \mathbb{R}} |\varphi^{(k)}(x)| \xrightarrow{n \to \infty} 0$.
Hence for all $k \geq 0$, we have $\varphi_n^{(k)} \xrightarrow{n \to \infty} \mathbf{0}$ uniformly. However, we have

$$\langle T, \varphi_n \rangle = \sum_{k \in \mathbb{Z}} \varphi_n(k) = \sum_{k \in \mathbb{Z}} \frac{1}{n}\varphi\left(\frac{k}{n}\right) = \sum_{k=0}^{n} \frac{1}{n}\varphi\left(\frac{k}{n}\right) \xrightarrow{n \to \infty} \int_0^1 \varphi(x)dx > 0.$$

(3) There is no contradiction to our conclusion from (1) that T is a distribution, since we observe that there is no compact set $K \subset \mathbb{R}$ such that for all $n \in \mathbb{N}$, φ_n is zero outside K: Indeed, $\varphi_{n+1}(n) = \frac{1}{n+1}\varphi\left(\frac{n}{n+1}\right) > 0$, $n \in \mathbb{N}$.

Solution to Exercise 6.7, page 241

The function $x \overset{f}{\mapsto} H(x)\cos x$ is continuously differentiable on $\mathbb{R}\setminus\{0\}$, and has a jump $f(0+) - f(0-) = 1$ at 0. For $x < 0$, $H(x) = 0$ and so $(H(x)\cos x)' = 0$. For $x > 0$, $H(x) = 1$, and so $(H(x)\cos x)' = (\cos x)' = -\sin x$.
Moreover, $\lim\limits_{x \searrow 0} f'(x) = \lim\limits_{x \searrow 0}(-\sin x) = 0$ and $\lim\limits_{x \nearrow 0} f'(x) = \lim\limits_{x \nearrow 0} 0 = 0$.
Consequently, $\dfrac{d}{dx} H(x)\cos x = -H(x)\sin x + 1 \cdot \delta = -H(x)\sin x + \delta$.

The function $x \overset{g}{\mapsto} H(x)\sin x$ is continuously differentiable on $\mathbb{R}\setminus\{0\}$, and has a jump $g(0+) - g(0-) = 0$ at 0. For $x < 0$, $H(x) = 0$ and so $(H(x)\sin x)' = 0$. For $x > 0$, $H(x) = 1$, and so $(H(x)\sin x)' = (\sin x)' = \cos x$.
Moreover, $\lim\limits_{x \searrow 0} g'(x) = \lim\limits_{x \searrow 0} \cos x = 1$ and $\lim\limits_{x \nearrow 0} g'(x) = \lim\limits_{x \nearrow 0} 0 = 0$.
Consequently, $\dfrac{d}{dx} H(x)\sin x = H(x)\cos x + 0 \cdot \delta = H(x)\cos x$.

Solution to Exercise 6.8, page 241

The function $x \overset{f}{\mapsto} |x|/2$ is continuously differentiable on $\mathbb{R}\setminus\{0\}$, and has a jump $f(0+) - f(0-) = 0$ at 0. Moreover, for $x > 0$, $|x|/2 = x/2$, and so we have $(|x|/2)' = (x/2)' = 1/2$ for $x > 0$. On the other hand, for $x < 0$, $|x|/2 = -x/2$, and so we obtain $(|x|/2)' = (-x/2)' = -1/2$ for $x < 0$.
Also, $\lim\limits_{x \searrow 0} f'(x) = \lim\limits_{x \searrow 0} \dfrac{x}{2} = \dfrac{1}{2}$ and $\lim\limits_{x \nearrow 0} f'(x) = \lim\limits_{x \nearrow 0}\left(-\dfrac{x}{2}\right) = -\dfrac{1}{2}$.
Hence $\dfrac{d}{dx}\dfrac{|x|}{2} = T_g + 0 \cdot \delta = T_g$, where $g(x) = \begin{cases} -1/2 \text{ if } x < 0, \\ 1/2 \text{ if } x > 0. \end{cases}$
Again, g is continuously differentiable on $\mathbb{R}\setminus\{0\}$.
g has a jump of $g(0+) - g(0-) = \dfrac{1}{2} - \left(-\dfrac{1}{2}\right) = 1$ at 0.
Also g is constant for $x > 0$ (respectively for $x < 0$), and so $g'(x) = 0$ for $x > 0$ (respectively for $x < 0$).
Also, $\lim\limits_{x \searrow 0} g'(x) = \lim\limits_{x \searrow 0} 0 = 0$ and $\lim\limits_{x \nearrow 0} g'(x) = \lim\limits_{x \nearrow 0} 0 = 0$.
Hence $\dfrac{d^2}{dx^2}\dfrac{|x|}{2} = T_{g'} + 1 \cdot \delta = 0 + \delta = \delta$.

Solution to Exercise 6.9, page 241

(1) Let us first consider the case when $\ell \equiv \mathbf{0}$.
Then $V = \ker \ell \subset \ker L$ implies that $\ker L = V$ too, and so $L = \mathbf{0}$ as well. Thus we may simply take $c = 0$, and then clearly $L = \mathbf{0} = 0\ell$ is valid.

Now let us suppose that $\ell \neq \mathbf{0}$.
Then there is a vector $v_0 \in V$ such that $\ell(v_0) \neq 0$.
This vector v_0 must be nonzero, for otherwise $\ell(v_0) = 0$.
(To show the desired decomposition of an arbitrary vector as $v = c_v v_0 + w$,

with $w \in \ker \ell$, we need to find the appropriate scalar c_v, because then we can set $w := v - c_v v_0$. To find what c_v might work, we apply ℓ on both sides to obtain $\ell(v) = c_v \ell(v_0) + \ell(w) = c_v \ell(v_0) + 0 = c_v \ell(v_0)$.
So it seems that $c_v = \dfrac{\ell(v)}{\ell(v_0)}$ should do the trick!)

Given $v \in V$, we now proceed to show that $w := v - \dfrac{\ell(v)}{\ell(v_0)} v_0 \in \ker \ell$.

We have $\ell(w) = \ell(v) - \dfrac{\ell(v)}{\ell(v_0)} \ell(v_0) = 0$, and so $w \in \ker \ell$.

As $w \in \ker \ell \subset \ker L$, we have $L(w) = 0$, and
$$L(v) = L(c_v v_0 + w) = c_v L(v_0) + L(w) = c_v L(v_0) = \dfrac{\ell(v)}{\ell(v_0)} L(v_0) = \dfrac{L(v_0)}{\ell(v_0)} \ell(v).$$
Hence with $c := \dfrac{L(v_0)}{\ell(v_0)} \in \mathbb{C}$, we have $L = c\ell$.

(2) For $\varphi \in \mathcal{D}(\mathbb{R})$, $0 = \langle \mathbf{0}, \varphi \rangle = \langle T', \varphi \rangle = -\langle T, \varphi' \rangle$. So $\{\varphi' : \varphi \in \mathcal{D}(\mathbb{R})\} \subset \ker T$.
Let $\mathbf{1}$ denote the constant function $\mathbb{R} \ni x \mapsto 1$. By Exercise 6.3, page 232
$$\ker T_{\mathbf{1}} = \left\{ \psi \in \mathcal{D}(\mathbb{R}) : \int_{-\infty}^{\infty} \psi(x)dx = 0 \right\} = \{\varphi' : \varphi \in \mathcal{D}(\mathbb{R})\} \subset \ker T.$$
Finally, by part (1), applied to the vector space $V = \mathcal{D}(\mathbb{R})$, with $L := T$ and $\ell := T_{\mathbf{1}}$, we get the existence of a $c \in \mathbb{C}$ so that $T = cT_{\mathbf{1}} = T_c$.
(Here T_c denotes the regular distribution corresponding to the constant function taking value c everywhere on $\mathbb{R}$.)

Solution to Exercise 6.10, page 242

Fix any $\varphi_0 \in \mathcal{D}(\mathbb{R}) \setminus \{\mathbf{0}\}$ which is nonnegative everywhere. For $\psi \in \mathcal{D}(\mathbb{R})$, set
$$\varphi := \psi - \left(\int_{-\infty}^{\infty} \psi(x)dx \Big/ \int_{-\infty}^{\infty} \varphi_0(x)dx \right) \varphi_0.$$
As ψ and φ_0 belong to $\mathcal{D}(\mathbb{R})$, so does φ. Moreover,
$$\int_{-\infty}^{\infty} \varphi(x)dx = \int_{-\infty}^{\infty} \psi(x)dx - \left(\int_{-\infty}^{\infty} \psi(x)dx \Big/ \int_{-\infty}^{\infty} \varphi_0(x)dx \right) \int_{-\infty}^{\infty} \varphi_0(x)dx = 0.$$
Thus $\varphi \in Y := \left\{ \varphi \in \mathcal{D}(\mathbb{R}) : \int_{-\infty}^{\infty} \varphi(x)dx = 0 \right\}$.

By Exercise 6.3, page 232, there is a unique $\Phi \in \mathcal{D}(\mathbb{R})$ such that $\Phi' = \varphi$.
We define $S : \mathcal{D}(\mathbb{R}) \to \mathbb{C}$ by $\langle S, \psi \rangle = -\langle T, \Phi \rangle$. Let us check that S is linear. Let $\psi_1, \psi_2 \in \mathcal{D}(\mathbb{R})$, and let $\Phi_1, \Phi_2 \in \mathcal{D}(\mathbb{R})$ be such that
$$\Phi_1' = \varphi_1 := \psi_1 - \left(\int_{-\infty}^{\infty} \psi_1(x)dx \Big/ \int_{-\infty}^{\infty} \varphi_0(x)dx \right) \varphi_0,$$
$$\Phi_2' = \varphi_2 := \psi_2 - \left(\int_{-\infty}^{\infty} \psi_2(x)dx \Big/ \int_{-\infty}^{\infty} \varphi_0(x)dx \right) \varphi_0.$$
Then $(\Phi_1 + \Phi_2)' = \psi_1 + \psi_2 - \left(\int_{-\infty}^{\infty} (\psi_1 + \psi_2)(x)dx \Big/ \int_{-\infty}^{\infty} \varphi_0(x)dx \right) \varphi_0$.
So $\langle S, \psi_1 + \psi_2 \rangle = -\langle T, \Phi_1 + \Phi_2 \rangle = -\langle T, \Phi_1 \rangle - \langle T, \Phi_2 \rangle = \langle S, \psi_1 \rangle + \langle S, \psi_2 \rangle$.
Similarly, $\langle S, \alpha\psi \rangle = \alpha \langle S, \psi \rangle$ for all $\psi \in \mathcal{D}(\mathbb{R})$ and all $\alpha \in \mathbb{C}$.

Now we check the continuity of S. Let $(\psi_n)_{n\in\mathbb{N}}$ be a sequence in $\mathcal{D}(\mathbb{R})$ such that $\psi_n \xrightarrow{\mathcal{D}} \mathbf{0}$. Then there exists an $a > 0$ such that all the ψ_n vanish outside $[-a, a]$, and $(\psi_n)_{n\in\mathbb{N}}$ converges uniformly to $\mathbf{0}$ as $n \to \infty$, giving
$$\left|\int_{-\infty}^{\infty} \psi_n(x)dx\right| \leq 2a\|\psi_n\|_\infty \xrightarrow{n\to\infty} 0.$$
Now set $\varphi_n := \psi_n - \left(\int_{-\infty}^{\infty} \psi_n(x)dx \Big/ \int_{-\infty}^{\infty} \varphi_0(x)dx\right)\varphi_0$.
Then there exists a $b > 0$ such that each φ_n vanishes outside $[-b, b]$.
Also, for $k \geq 0$, $\varphi_n^{(k)} = \psi_n^{(k)} - \left(\int_{-\infty}^{\infty} \psi_n(x)dx \Big/ \int_{-\infty}^{\infty} \varphi_0(x)dx\right)\varphi_0^{(k)}$.
So for each $k \geq 0$, $(\varphi_n^{(k)})_{n\in\mathbb{N}}$ converges uniformly to $\mathbf{0}$. Thus $\varphi_n \xrightarrow{\mathcal{D}} \mathbf{0}$.
Let Φ_n be the unique element in $\mathcal{D}(\mathbb{R})$ such that $\Phi_n' = \varphi_n$, $n \in \mathbb{N}$.
From Exercise 6.4, page 233, we can conclude that $\Phi_n \xrightarrow{\mathcal{D}} \mathbf{0}$.
Consequently, $\langle S, \psi_n\rangle = -\langle T, \Phi_n\rangle \to 0$ as $n \to \infty$. Hence $S \in \mathcal{D}'(\mathbb{R})$.
Finally, we'll show that $S' = T$.
If $\Phi \in \mathcal{D}(\mathbb{R})$, then $\varphi := \Phi' - \left(\int_{-\infty}^{\infty} \Phi'(x)dx \Big/ \int_{-\infty}^{\infty} \varphi_0(x)dx\right)\varphi_0 = \Phi' - 0 = \Phi'$.
Thus $\langle S', \Phi\rangle = -\langle S, \Phi'\rangle = -(-\langle T, \Phi\rangle) = \langle T, \Phi\rangle$. So $S' = T$.

Solution to Exercise 6.11, page 242

Let $\varphi \in \mathcal{D}(\mathbb{R})$ be such that $\varphi(0) \neq 0$. (For example we can simply take the test function from Example 6.1, page 230.) Then $x^n\varphi \in \mathcal{D}(\mathbb{R})$ too, and we have
$$\langle \delta^{(n)}, x^n\varphi\rangle = (-1)^n\langle \delta, (x^n\varphi)^{(n)}\rangle = (-1)^n\Big\langle \delta, \sum_{k=0}^{n}\binom{n}{k}\Big(\frac{d^k}{dx^k}x^n\Big)\varphi^{(n-k)}\Big\rangle$$
$$= (-1)^n\binom{n}{n}n!\varphi(0) \neq 0.$$
So $\delta^{(n)} \neq \mathbf{0}$.

Solution to Exercise 6.12, page 242

It is enough to show the linear independence of $\delta, \delta', \cdots, \delta^{(n)}$ for each n. Suppose that there are scalars $c_0, c_1, \cdots, c_n$ such that $c_0\delta + c_1\delta' + \cdots + c_n\delta^{(n)} = \mathbf{0}$. Let $\varphi \in \mathcal{D}(\mathbb{R})$, and for $\lambda > 0$, set $\varphi_\lambda(x) := \varphi(\lambda x)$, for all $x \in \mathbb{R}$. Then
$$0 = \langle \mathbf{0}, \varphi_\lambda\rangle = \langle c_0\delta + c_1\delta' + \cdots + c_n\delta^{(n)}, \varphi_\lambda\rangle$$
$$= c_0\varphi(0) + (-c_1\varphi'(0))\lambda + \cdots + ((-1)^nc_n\varphi^{(n)}(0))\lambda^n.$$
The polynomial $p(z) := c_0\varphi(0) + (-c_1\varphi'(0))z + \cdots + ((-1)^nc_n\varphi^{(n)}(0))z^n$ is zero on $\{\lambda : \lambda > 0\}$, and hence must be identically zero. So $c_0\varphi(0) = \cdots = c^n\varphi^{(n)}(0) = 0$. As the choice of φ was arbitrary, we have that for *all* test functions $\varphi \in \mathcal{D}(\mathbb{R})$,
$$c_0\varphi(0) = \cdots = c^n\varphi^{(n)}(0) = 0. \qquad (*)$$
But if we look at the φ from Example 6.1, page 230, then $\varphi(0) \neq 0$, and also $x^n\varphi$,

$n \in \mathbb{N}$, belongs to $\mathcal{D}(\mathbb{R})$, which moreover satisfies
$$(x^n \varphi)^{(n)}(0) = \sum_{k=0}^{n} \binom{n}{k} \left(\frac{d^k}{dx^k} x^n\right)\bigg|_{x=0} \varphi^{(n-k)}(0) = (-1)^n \binom{n}{n} n! \varphi(0) \neq 0.$$
So using $\varphi, x\varphi, \cdots, x^n \varphi$ as the test functions in $(*)$, we obtain $c_0 = \cdots = c_n = 0$.

Solution to Exercise 6.13, page 242

For any $\varphi \in \mathcal{D}(\mathbb{R}^d)$, we have
$$\left\langle \frac{\partial^2 T}{\partial x_i \partial x_j}, \varphi \right\rangle = -\left\langle \frac{\partial T}{\partial x_j}, \frac{\partial \varphi}{\partial x_i} \right\rangle = -\left(-\left\langle T, \frac{\partial^2 \varphi}{\partial x_j \partial x_i}\right\rangle\right) = \left\langle T, \frac{\partial^2 \varphi}{\partial x_j \partial x_i}\right\rangle$$
$$= \left\langle T, \frac{\partial^2 \varphi}{\partial x_i \partial x_j}\right\rangle \quad (\text{as } \varphi \in \mathcal{D}(\mathbb{R}^d))$$
$$= -\left\langle \frac{\partial T}{\partial x_i}, \frac{\partial \varphi}{\partial x_j}\right\rangle = -\left(-\left\langle \frac{\partial^2 T}{\partial x_j \partial x_i}, \varphi\right\rangle\right) = \left\langle \frac{\partial^2 T}{\partial x_j \partial x_i}, \varphi \right\rangle.$$
So $\dfrac{\partial^2 T}{\partial x_i \partial x_j} = \dfrac{\partial^2 T}{\partial x_j \partial x_i}$.

Solution to Exercise 6.14, page 242

$H \in L^1_{\mathrm{loc}}(\mathbb{R}^2)$, and so it defines a regular distribution on $\mathbb{R}^2$.
For $\varphi \in \mathcal{D}(\mathbb{R}^2)$, with $a > 0$ such that $\varphi \equiv 0$ on $\mathbb{R}^2 \setminus (-a, a)^2$, we have
$$\left\langle \frac{\partial^2 H}{\partial x \partial y}, \varphi \right\rangle = -\left\langle \frac{\partial H}{\partial y}, \frac{\partial \varphi}{\partial x} \right\rangle = \left\langle H, \frac{\partial^2 \varphi}{\partial y \partial x} \right\rangle = \iint_{\mathbb{R}^2} H \frac{\partial^2 \varphi}{\partial y \partial x} dxdy = \iint_{\mathbb{R}^2} H \frac{\partial^2 \varphi}{\partial x \partial y} dxdy$$
$$= \int_0^a \int_0^a \frac{\partial^2 \varphi}{\partial x \partial y} dxdy = \int_0^a \left(\underbrace{\frac{\partial \varphi}{\partial y}(a, y)}_{=0} - \frac{\partial \varphi}{\partial y}(0, y)\right) dy = -\int_0^a \frac{\partial \varphi}{\partial y}(0, y) dy.$$
Thus $\left\langle \dfrac{\partial^2 H}{\partial x \partial y}, \varphi \right\rangle = \varphi(0, 0) - \underbrace{\varphi(0, a)}_{=0} = \varphi(0, 0) = \langle \delta_0, \varphi \rangle$. So $\dfrac{\partial^2 H}{\partial x \partial y} = \delta_0$.

Solution to Exercise 6.15, page 242

If $u : \mathbb{R}^2 \to \mathbb{R}$ is a radial function, say $u(\mathbf{x}) = f(r)$, where $r = \|\mathbf{x}\|_2$, then
$$\Delta u = f''(r) + \frac{f'(r)}{r}.$$
Thus $\Delta \log r = -\dfrac{1}{r^2} + \dfrac{1}{r} \cdot \dfrac{1}{r} = 0$. Since for all $R > 0$ we have
$$\int_0^{2\pi} \int_0^R (\log r) r\, dr\, d\theta = 2\pi \int_0^R r \log r\, dr < \infty,$$
we conclude that $(\mathbf{x} \mapsto \log r) \in L^1_{\mathrm{loc}}(\mathbb{R}^2)$.
For $\varphi \in \mathcal{D}(\mathbb{R}^2)$ which vanishes outside the ball $B(\mathbf{0}, R)$, we have
$$\langle \Delta \log r, \varphi \rangle = \langle \log r, \Delta \varphi \rangle = \int_{\|\mathbf{x}\|_2 < R} (\log r)(\Delta \varphi) d\mathbf{x} = \lim_{\epsilon \to 0} \int_{\epsilon < \|\mathbf{x}\|_2 < R} (\log r)(\Delta \varphi) d\mathbf{x}.$$
$(\log r)(\Delta \varphi)$ is integrable, as $\log r$ is locally integrable, and $\Delta \varphi = 0$ outside a ball.

Let $\epsilon > 0$.

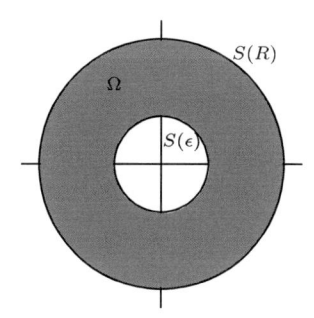

Using Green's formula in the annulus $\Omega := \{\mathbf{x} \in \mathbb{R}^2 : \epsilon < \|\mathbf{x}\|_2 < R\}$ (with the boundary $\partial\Omega$ being the union of the two circles $S(\epsilon) = \{\mathbf{x} : \|\mathbf{x}\|_2 = \epsilon\}$ and $S(R) = \{\mathbf{x} : \|\mathbf{x}\|_2 = R\}$), for the functions $u = \log r$ and $v = \varphi$, we obtain

$$\int_{\epsilon < \|\mathbf{x}\|_2 < R} (\log r)(\Delta\varphi) d\mathbf{x} = \int_{\|\mathbf{x}\|_2 = \epsilon} (\log r) \frac{\partial \varphi}{\partial n} ds - \int_{\|\mathbf{x}\|_2 = \epsilon} \left(\frac{\partial}{\partial n}(\log r)\right) \varphi ds.$$

We'll show below that the first integral on the right-hand side is $O(\epsilon)$, and thus it tends to 0 as $\epsilon \to 0$.

As $\dfrac{\partial \varphi}{\partial n} = \mathbf{n}(\mathbf{x}) \cdot \nabla\varphi(\mathbf{x})$, and $\|\mathbf{n}(\mathbf{x})\|_2 = 1$, the Cauchy-Schwarz Inequality gives

$$\left|\frac{\partial \varphi}{\partial n}(\mathbf{x})\right| \leqslant \|\nabla\varphi(\mathbf{x})\|_2 \cdot 1 \leqslant M,$$

where $M := \sup_{\|\mathbf{x}\|_2 \leqslant R} \|\nabla\varphi(\mathbf{x})\|_2 < \infty$. Finally, $\int_{\|\mathbf{x}\|_2 = \epsilon} ds(\mathbf{x}) = 2\pi\epsilon$, and $2\pi\epsilon \log \epsilon \xrightarrow{\epsilon \to 0} 0$.

Next we will look at the second integral $\int_{\|\mathbf{x}\|_2 = \epsilon} \left(\frac{\partial}{\partial n}(\log r)\right) \varphi ds$.

First, $\dfrac{\partial}{\partial n}(\log r) = -\dfrac{\partial}{\partial r}(\log r) = -\dfrac{1}{r} = -\dfrac{1}{\epsilon}$. Moreover,

$$-\int_{\|\mathbf{x}\|_2 = \epsilon} \left(\frac{\partial}{\partial n}(\log r)\right) \varphi ds = \frac{1}{\epsilon} \int_{\|\mathbf{x}\|_2 = \epsilon} \varphi(\mathbf{x}) ds(\mathbf{x}) = 2\pi \left(\frac{1}{2\pi\epsilon} \int_{\|\mathbf{x}\|_2 = \epsilon} \varphi(\mathbf{x}) ds(\mathbf{x})\right).$$

Given $\eta > 0$,

$$\left|\frac{1}{2\pi\epsilon} \int_{\|\mathbf{x}\|_2 = \epsilon} \varphi(\mathbf{x}) ds(\mathbf{x}) - \varphi(\mathbf{0})\right| = \left|\frac{1}{2\pi\epsilon} \int_{\|\mathbf{x}\|_2 = \epsilon} \varphi(\mathbf{x}) ds(\mathbf{x}) - \frac{1}{2\pi\epsilon} \int_{\|\mathbf{x}\|_2 = \epsilon} \varphi(\mathbf{0}) ds(\mathbf{x})\right|$$

$$\leqslant \frac{1}{2\pi\epsilon} \int_{\|\mathbf{x}\|_2 = \epsilon} |\varphi(\mathbf{x}) - \varphi(\mathbf{0})| ds(\mathbf{x})$$

$$\leqslant \frac{1}{2\pi\epsilon} \eta \cdot 2\pi\epsilon = \eta,$$

where first $\epsilon_0 > 0$ is chosen small enough so that $|\varphi(\mathbf{x}) - \varphi(\mathbf{0})| \leqslant \eta$ if $\|\mathbf{x}\|_2 \leqslant \epsilon_0$, and ϵ satisfies $0 < \epsilon \leqslant \epsilon_0$.

Thus $\lim_{\epsilon \to 0} \int_{\epsilon < \|\mathbf{x}\|_2 < R} (\log r)(\Delta\varphi) d\mathbf{x} = 2\pi\varphi(\mathbf{0}) = \langle 2\pi\delta_{(0,0)}, \varphi \rangle$.

So $\langle \Delta \log r, \varphi \rangle = \langle 2\pi\delta_{(0,0)}, \varphi \rangle$ for all $\varphi \in \mathcal{D}(\mathbb{R}^2)$. Hence $\left(\dfrac{1}{2\pi} \log r\right) = \delta_{(0,0)}$.

Solution to Exercise 6.16, page 248

u is continuous on $\mathbb{R}$, and continuously differentiable on $\mathbb{R}\setminus\{0\}$.
For $x < 0$, we have $u'(x) = 0$. For $x > 0$, $u'(x) = 1$.
Also, $\lim\limits_{x \nearrow 0} u'(x) = 0$ and $\lim\limits_{x \searrow 0} u'(x) = 1$.
Thus by the Jump Rule, $\dfrac{du}{dx} = H + 0 \cdot \delta = H$ in the sense of distributions.
So u is a weak solution of $u' = H$.

Solution to Exercise 6.17, page 251

We view $H(x)\cos x$ as the product of the C^∞ function cos with the regular distribution H. Using the Product Rule, we have

$$\frac{d}{dx}(H(x)\cos x) = (-\sin x)H(x) + (\cos x)\delta = -H(x)\sin x + (\cos 0)\delta$$

$$= -H(x)\sin x + \delta.$$

Similarly,

$$\frac{d}{dx}(H(x)\sin x) = (\cos x)H(x) + (\sin x)\delta = H(x)\cos x + (\sin 0)\delta$$

$$= H(x)\cos x + 0\delta = H(x)\cos x.$$

Solution to Exercise 6.18, page 251

(1) We have
$$\frac{d}{dx}(H(x)e^{\lambda x}) = \lambda e^{\lambda x}H(x) + e^{\lambda x}\delta = \lambda e^{\lambda x}H(x) + e^{\lambda 0}\delta = \lambda e^{\lambda x}H(x) + \delta.$$
Hence $\left(\dfrac{d}{dx} - \lambda\right)H(x)e^{\lambda x} = \delta$.

(2) When $n = 1$, $\dfrac{d}{dx}\left(H(x)\dfrac{x^0}{0!}\right) = \dfrac{d}{dx}H(x) = \delta$.
If the claim is true for some $n \in \mathbb{N}$, then
$$\frac{d^{n+1}}{dx^{n+1}}\left(H(x)\frac{x^n}{n!}\right) = \frac{d^n}{dx^n}\left(\frac{d}{dx}\left(H(x)\frac{x^n}{n!}\right)\right) = \frac{d^n}{dx^n}\left(\frac{nx^{n-1}}{n!}H(x) + \frac{x^n}{n!}\delta\right)$$
$$= \frac{d^n}{dx^n}\left(\frac{x^{n-1}}{(n-1)!}H(x) + 0\delta\right) = \frac{d^n}{dx^n}\left(\frac{x^{n-1}}{(n-1)!}H(x)\right) = \delta.$$
So the claim follows for all $n \in \mathbb{N}$ by induction.

(3) We have
$$\frac{d}{dx}\left(H(x)\frac{\sin(\omega x)}{\omega}\right) = \frac{\omega\cos(\omega x)}{\omega}H(x) + \frac{\sin(\omega x)}{\omega}\delta = (\cos(\omega x))H(x) + 0\delta$$
$$= (\cos(\omega x))H(x).$$

Thus
$$\frac{d^2}{dx^2}\left(H(x)\frac{\sin(\omega x)}{\omega}\right) = \frac{d}{dx}\left(\cos(\omega x)H(x)\right) = -\omega\sin(\omega x)H(x) + \left(\cos(\omega x)\right)\delta$$
$$= -\omega^2 \frac{\sin(\omega x)}{\omega}H(x) + (\cos 0)\delta = -\omega^2 \frac{\sin(\omega x)}{\omega}H(x) + 1\delta$$
$$= -\omega^2 \frac{\sin(\omega x)}{\omega}H(x) + \delta.$$

Consequently, $\left(\dfrac{d^2}{dx^2} + \omega^2\right)\left(H(x)\dfrac{\sin(\omega x)}{\omega}\right) = \delta.$

Solution to Exercise 6.19, page 252

We have for all $\varphi \in \mathcal{D}(\mathbb{R})$ that
$$\langle \alpha\delta', \varphi\rangle = \langle \delta', \alpha\varphi\rangle = -\langle \delta, (\alpha\varphi)'\rangle = -\langle \delta, \alpha'\varphi + \alpha\varphi'\rangle$$
$$= -\alpha'(0)\varphi(0) - \alpha(0)\varphi'(0) = -\alpha'(0)\langle\delta, \varphi\rangle + \alpha(0)\langle \delta', \varphi\rangle$$
$$= \langle \alpha(0)\delta' - \alpha'(0)\delta, \varphi\rangle.$$
So $\alpha\delta' = \alpha(0)\delta' - \alpha'(0)\delta$. In particular, $x\delta' = 0\delta' - 1\delta = -\delta$.

Solution to Exercise 6.20, page 252

For all $\varphi \in \mathcal{D}(\mathbb{R})$, we have that
$$\langle xT' - (xT)', \varphi\rangle = \langle xT', \varphi\rangle - \langle (xT)', \varphi\rangle = \langle T', x\varphi\rangle + \langle xT, \varphi'\rangle$$
$$= -\langle T, (x\varphi)'\rangle + \langle T, x\varphi'\rangle$$
$$= -\langle T, \varphi + x\varphi'\rangle + \langle T, x\varphi'\rangle$$
$$= -\langle T, \varphi\rangle = \langle -T, \varphi\rangle.$$
So $\left[x, \dfrac{d}{dx}\right]T = -T$.

Solution to Exercise 6.21, page 252

With $u := e^{-3yx}H(y)$, we have
$$\frac{\partial u}{\partial x} = \frac{\partial}{\partial x}\left(e^{-3yx}H(y)\right) = e^{-3yx}(-3y)H(y) + e^{-3yx}\frac{\partial}{\partial x}H(y).$$
For all $\varphi \in \mathcal{D}(\mathbb{R}^2)$, we have
$$\left\langle \frac{\partial}{\partial x}H(y), \varphi\right\rangle = -\left\langle H(y), \frac{\partial \varphi}{\partial x}\right\rangle = -\int_{-\infty}^{\infty}\int_{-\infty}^{\infty} H(y)\frac{\partial\varphi}{\partial x}dxdy$$
$$= -\int_0^{\infty}\int_{-\infty}^{\infty}\frac{\partial\varphi}{\partial x}dxdy = -\int_0^{\infty}\left(\varphi(x,y)\Big|_{x=-\infty}^{x=\infty}\right)dy$$
$$= -\int_0^{\infty} 0\, dy = 0.$$
So $\dfrac{\partial}{\partial x}H(y) = \mathbf{0}$. Hence $\dfrac{\partial u}{\partial x} + 3yu = e^{-3yx}(-3y)H(y) + 3y\cdot e^{-3yx}H(y) = \mathbf{0}$.
Moreover, $u(0, y) = e^{-0}H(y) = 1\cdot H(y) = H(y)$.

Solution to Exercise 6.22, page 252

First we note that for all $\varphi \in \mathcal{D}(\mathbb{R})$, we have
$$\left\langle x \cdot \operatorname{pv}\frac{1}{x}, \varphi \right\rangle = \left\langle \operatorname{pv}\frac{1}{x}, x\varphi \right\rangle = \lim_{\epsilon \searrow 0} \int_{|x|>\epsilon} \frac{x\varphi(x)}{x} dx = \lim_{\epsilon \searrow 0} \int_{|x|>\epsilon} \varphi(x) dx$$
$$= \int_{\mathbb{R}} \varphi(x) dx = \langle \mathbf{1}, \varphi \rangle,$$
where $\mathbf{1}$ is the constant function $\mathbb{R} \ni x \mapsto 1$.

Suppose on contrary, it is possible to define an associative and commutative product such that for $\alpha \in C^\infty(\mathbb{R})$ and $T \in \mathcal{D}'(\mathbb{R})$, it agrees with Definition 6.6, page 249. Then
$$(\delta \cdot x) \cdot \operatorname{pv}\frac{1}{x} = (x \cdot \delta) \cdot \operatorname{pv}\frac{1}{x} \quad \text{(commutativity)}$$
$$= (\mathbf{0}) \cdot \operatorname{pv}\frac{1}{x} \quad \text{(Definition 6.6)}$$
$$= \mathbf{0} \quad \text{(Definition 6.6)},$$

whereas
$$\delta \cdot \left(x \cdot \operatorname{pv}\frac{1}{x} \right) = \delta \cdot \mathbf{1} \quad \text{(Definition 6.6)}$$
$$= \mathbf{1} \cdot \delta \quad \text{(commutativity)}$$
$$= \delta \quad \text{(Definition 6.6)},$$

and so $(\delta \cdot x) \cdot \operatorname{pv}\frac{1}{x} \neq \delta \cdot \left(x \cdot \operatorname{pv}\frac{1}{x} \right)$, violating associativity.

Solution to Exercise 6.23, page 252

(1) Let $\left(\dfrac{d}{dx} - \lambda \right) T = \mathbf{0}$. Then we have
$$(e^{-\lambda x} T)' = -\lambda e^{-\lambda x} T + e^{-\lambda x} T' = e^{-\lambda x}(-\lambda T + T') = e^{-\lambda x}\left(\frac{d}{dx} - \lambda \right) T = \mathbf{0}.$$
From Exercise 6.9, page 241, there exists a $c \in \mathbb{C}$ such that $e^{-\lambda x} T = c$, that is, $T = c e^{\lambda x}$.

(2) Since $f \in C^\infty$, there exists an $F \in C^\infty$ such that $\left(\dfrac{d}{dx} - \lambda \right) F = f$.

(In fact, an explicit expression for one such F (for which $F(0) = 0$), is given by $F(x) = e^{\lambda x} \int_0^x e^{-\lambda \xi} f(\xi) d\xi$. This can be checked by differentiation using the Product Rule and the Fundamental Theorem of Calculus.)

Hence we obtain $\left(\dfrac{d}{dx} - \lambda \right)(T - F) = f - f = \mathbf{0}.$

From part (1), $T - F = c e^{\lambda x}$ for some $c \in \mathbb{C}$. Hence $T = F + c e^{\lambda x} \in C^\infty$.

(3) Let $D = \sum_{k=0}^{n} a_k \left(\dfrac{d}{dx}\right)^k$, with $a_n \neq 0$.

Then $D = P\left(\dfrac{d}{dx}\right)$, where $P(\xi) = \sum_{k=0}^{n} a_k \xi^k = a_n(\xi - \lambda_1) \cdots (\xi - \lambda_n)$.

So $P(\xi) = (\xi - \lambda)Q(\xi)$, where $\lambda = \lambda_n$, and a suitable polynomial Q.

Correspondingly, with $D_1 := Q\left(\dfrac{d}{dx}\right)$, $D = \left(\dfrac{d}{dx} - \lambda\right) D_1$.

We'll use induction (on the order n of D) to prove
$$\left(DT = f,\ f \in C^\infty\right) \;\Rightarrow\; \left(T = T_F,\ F \in C^\infty\right).$$

This is true for $n = 1$, from part (2) above.

Suppose that the claim is true for all differential operators of order n.

Let D have order $n+1$, and write $D = \left(\dfrac{d}{dx} - \lambda\right) D_1$ where D_1 is order n.

If $DT = f \in C^\infty$, then $\left(\dfrac{d}{dx} - \lambda\right) D_1 T = f$, and so $D_1 T = T_g$ for some $g \in C^\infty$.

But by the induction hypothesis, it now follows that $T = T_F$, with $F \in C^\infty$.

(4) If E is also a fundamental solution, then $DE = \delta$.
But also $DE_* = \delta$, and so $D(E - E_*) = \mathbf{0}$.
Thus $E - E_* = F$, where F is a classical solution of the homogeneous equation $DF = \mathbf{0}$. So $E = E_* + F$.

Conversely, if F is a classical solution of the homogeneous equation $DF = \mathbf{0}$, then $E := E_* + F$ is a fundamental solution of D too: indeed, we have that $DE = DE_* + DF = \delta + \mathbf{0} = \delta$.

So we conclude that: $E \in \mathcal{D}'(\mathbb{R})$ satisfies $DE = \delta$ if and only if
$$E = E_* + F,\text{ where } F \in C^\infty \text{ is such that } DF = \mathbf{0}.$$

	Any two fundamental solutions
Summary:	of a linear, constant coefficient, ordinary differential operator differ by a classical solution to the homogeneous equation.

Solution to Exercise 6.24, page 252

If $T = c\delta$, where $c \in \mathbb{C}$, then clearly $xT = x(c\delta) = \mathbf{0}$.

Now suppose that $T \in \mathcal{D}'(\mathbb{R})$ is such that $xT = \mathbf{0}$.
This means that for all $\varphi \in \mathcal{D}(\mathbb{R})$, we have $0 = \langle xT, \varphi \rangle = \langle T, x\varphi \rangle$.
Hence $\{x\varphi : \varphi \in \mathcal{D}(\mathbb{R})\} \subset \ker T$. We will now identify the set on the left-hand side as $\ker \delta = \{\psi \in \mathcal{D}(\mathbb{R}) : \psi(0) = 0\}$, and then use part (1) of Exercise 6.9, page 241.
First, let us note that if $\psi = x\varphi$, where $\varphi \in \mathcal{D}(\mathbb{R})$, then $\psi \in \mathcal{D}(\mathbb{R})$, and moreover, $\psi(0) = 0\varphi(0) = 0$. So we have $\{x\varphi : \varphi \in \mathcal{D}(\mathbb{R})\} \subset \{\psi \in \mathcal{D}(\mathbb{R}) : \psi(0) = 0\}$.
Next, let us show the reverse inclusion. Let $\psi \in \mathcal{D}(\mathbb{R})$ be such that $\psi(0) = 0$.
We have, by the Fundamental Theorem of Calculus:
$$\psi(x) = \psi(0) + \int_0^x \psi'(\xi)d\xi = 0 + x\int_0^1 \psi'(tx)dt = x\int_0^1 \psi'(tx)dt.$$

Set $\varphi(x) := \int_0^1 \psi'(tx)dt$. Then $\psi(x) = x\varphi(x)$.

By differentiating under the integral sign we see that $\varphi \in C^\infty$.

If ψ is zero outside $[-a, a]$ for some $a > 0$, then as $\varphi(x) = \dfrac{\psi(x)}{x}$ ($x \neq 0$), it follows that φ also vanishes outside $[-a, a]$. Thus $\varphi \in \mathcal{D}(\mathbb{R})$.

So we have $\{\psi \in \mathcal{D}(\mathbb{R}) : \psi(0) = 0\} \subset \{x\varphi : \varphi \in \mathcal{D}(\mathbb{R})\}$ as well.

Thus $\ker \delta = \{x\varphi : \varphi \in \mathcal{D}(\mathbb{R})\} \subset \ker T$, and by part (1) of Exercise 6.9, page 241 there exists a $c \in \mathbb{C}$ such that $T = c\delta$.

Solution to Exercise 6.25, page 254

First, we prove by induction that $\dfrac{d^n}{dx^n}e^{-x^2} = p_n(x)e^{-x^2}$, where p_n is a polynomial.

This is indeed true for $n = 0$ and $n = 1$: $e^{-x^2} = 1 \cdot e^{-x^2}$ and $\dfrac{d}{dx}e^{-x^2} = -2xe^{-x^2}$.

If it is true for some n, then

$$\dfrac{d^{n+1}}{dx^{n+1}}e^{-x^2} = \dfrac{d}{dx}\left(\dfrac{d^n}{dx^n}e^{-x^2}\right) = \dfrac{d}{dx}p_n(x)e^{-x^2}$$

$$= p_n(x) \cdot (-2xe^{-x^2}) + p'_n(x) \cdot e^{-x^2} = (p'_n(x) - 2xp_n(x))e^{-x^2}$$

$$= p_{n+1}(x)e^{-x^2},$$

where $p_{n+1}(x) := p'_n(x) - 2xp_n(x)$ is a polynomial.

This finishes the proof of our claim.

Now to show $e^{-x^2} \in \mathcal{S}(\mathbb{R})$, it is enough to show that $\sup_{x \in \mathbb{R}} |x^\ell e^{-x^2}| < \infty$ for all nonnegative integers ℓ. For $\ell = 0$, this is clear since $|e^{-x^2}| \leq 1$ for all $x \in \mathbb{R}$.

We have $e^{x^2} \geq \dfrac{(x^2)^\ell}{\ell!} = \dfrac{|x|^\ell}{\ell!}|x|^\ell$, and so for $|x| \geq 1$, $|x^\ell e^{-x^2}| = |x|^\ell e^{-x^2} \leq \dfrac{\ell!}{|x|^\ell} \leq \ell!$.

Since $x^\ell e^{-x^2}$ is a continuous function, there is an $M > 0$ such that $|x^\ell e^{-x^2}| \leq M$ for $x \in [-1, 1]$. Consequently, $\sup_{x \in \mathbb{R}} |x^\ell e^{-x^2}| \leq \max\{M, \ell!\} < \infty$. So $e^{-x^2} \in \mathcal{S}(\mathbb{R})$.

Solution to Exercise 6.26, page 254

Since $\varphi_n \xrightarrow{\mathcal{D}} \mathbf{0}$, we know that there exists an $a > 0$ such that all the φ_n vanish outside $[-a, a]$, and moreover, φ_n and all its derivatives converge uniformly to $\mathbf{0}$ on $[-a, a]$. So for any nonnegative integers m, k, we have that

$$\sup_{x \in \mathbb{R}} |x^k \varphi_n^{(m)}(x)| = \sup_{|x| \leq a} |x|^k |\varphi_n^{(m)}(x)| \leq a^k \sup_{|x| \leq a} |\varphi_n^{(m)}(x)| \xrightarrow{n \to \infty} a^k \cdot 0 = 0.$$

So $\varphi_n \xrightarrow{\mathcal{S}} \mathbf{0}$.

Solution to Exercise 6.27, page 255

We have for $\varphi \in \mathcal{S}(\mathbb{R})$ that

$$|\langle T_f, \varphi \rangle| = \left| \int_{\mathbb{R}} f(x)\varphi(x)dx \right|$$

$$\leq \|f\|_\infty \int_{-\infty}^{\infty} |\varphi(x)|dx = \|f\|_\infty \int_{-\infty}^{\infty} \frac{1}{1+x^2} \left(|\varphi(x)| + x^2|\varphi(x)| \right) dx$$

$$\leq \|f\|_\infty \cdot \left(\int_{-\infty}^{\infty} \frac{1}{1+x^2} dx \right) \cdot \left(\|\varphi\|_\infty + \|x^2\varphi\|_\infty \right).$$

From here it follows that if $(\varphi_n)_{n\in\mathbb{N}}$ is a sequence in $\mathcal{S}(\mathbb{R})$ such that $\varphi_n \xrightarrow{S} 0$ as $n \to \infty$, then $\langle T_f, \varphi_n \rangle \to 0$. Thus $T_f \in \mathcal{S}'(\mathbb{R})$.

Solution to Exercise 6.28, page 256

For $\varphi \in \mathcal{S}(\mathbb{R})$, we have that

$$\langle \widehat{T_f}, \varphi \rangle = \langle T_f, \widehat{\varphi} \rangle = \int_{\mathbb{R}} f(x)\widehat{\varphi}(x)dx$$

$$= \int_{-\infty}^{\infty} f(x) \left(\int_{-\infty}^{\infty} \varphi(\xi) e^{-2\pi i x \xi} d\xi \right) dx = \int_{-\infty}^{\infty} \int_{-\infty}^{\infty} f(x)\varphi(\xi) e^{-2\pi i x \xi} d\xi dx$$

$$= \int_{-\infty}^{\infty} \varphi(\xi) \left(\int_{-\infty}^{\infty} f(x) e^{-2\pi i x \xi} dx \right) d\xi$$

$$= \int_{-\infty}^{\infty} \varphi(\xi) \widehat{f}(\xi) d\xi = \langle T_{\widehat{f}}, \varphi \rangle.$$

Note that in the last step, we have used the fact that the Fourier transform of an $L^1(\mathbb{R})$ function is bounded on $\mathbb{R}$, and hence it defines a tempered distribution.

The Lebesgue integral

In this appendix, we give a summary of the Lebesgue integral in one dimension, to give the reader some feeling for some of the examples we have treated in the book. This is of course no substitute for a thorough exposition to the subject, which can be found for example [Smith (1983)] or [Apostol (1974)]. We will exclusively work in one dimension, that is in $\mathbb{R}$, although one can more generally work in $\mathbb{R}^d$ in an analogous manner. Also, we will just work with real-valued functions, instead of complex-valued functions. Again, by decomposing a complex-valued function into its real and imaginary parts, one can carry over the definitions and results to this more general setting.

Measurable sets

The length of an interval $I \subset \mathbb{R}$ is defined to be

$$\lambda(I) := \begin{cases} b - a & \text{if } I = [a,b],\ (a,b),\ (a,b],\ [a,b) \text{ and } -\infty < a \leqslant b < \infty, \\ \infty & \text{if } I \text{ is unbounded.} \end{cases}$$

We shall now associate a "measure" to more general subsets of $\mathbb{R}$ following a method originally due to Henri Lebesgue (1875–1941). The more general sets which possess a measure will be called measurable sets, and we will denote the measure of a measurable set $A \subset \mathbb{R}$ by $\lambda(A)$. The associated integral, which we will define in the next section is called the Lebesgue integral.

Step 1: Compact sets. Let $K \subset \mathbb{R}$ be a compact set, that is, closed and bounded. Let K be covered by intervals $I_1, \cdots, I_n \subset \mathbb{R}$, $n \in \mathbb{N}$. Then we

expect $\lambda(K)$ to satisfy
$$\lambda(K) \leq \sum_{k=1}^{n} \lambda(I_k).$$
This should hold for every such cover of K, and we expect the right-hand side above to be close to the left-hand side when the "overlap" of the covering intervals becomes smaller. This motivates the following definition. We define
$$\lambda(K) := \inf \left\{ \sum_k \lambda(I_k) : K \subset \bigcup_k I_k \right\},$$
where the infimum is taken over all covers of K by a finite number of intervals I_k's. We note that if $K = [a,b]$, then our definition above delivers $\lambda(K) = b - a$, which is indeed the length of the interval $[a,b]$. Also, we note that $\lambda(K) < \infty$ for compact K.

Step 2: Open sets. The measure of an open set $U \subset \mathbb{R}$ is defined by
$$\lambda(U) := \sup\{\lambda(K) : K \subset U, \ K \text{ compact}\}.$$
For open sets U, $0 \leq \lambda(U) \leq \infty$. Also, if $U = (a,b)$, then $\lambda(U) = b - a$ for finite a, b, and is ∞ if $a = \infty$ or $b = \infty$.

Step 3: Bounded measurable sets. Let $A \subset \mathbb{R}$ be a bounded set. Consider all compact sets $K \subset A$ and all open sets $U \supset A$. Then we have $\lambda(K) \subset \lambda(U)$. Thus
$$\sup_{\substack{K \text{ compact,} \\ K \subset A}} \lambda(K) \leq \inf_{\substack{U \text{ open} \\ A \subset U}} \lambda(U).$$
We say that the bounded set A is measurable if there is equality above, and define its measure $\lambda(A)$ to be the common value, that is,
$$\lambda(A) := \sup_{\substack{K \text{ compact,} \\ K \subset A}} \lambda(K) = \inf_{\substack{U \text{ open} \\ A \subset U}} \lambda(U).$$
If A is compact, then this definition coincides with the ones from Step 1. Also, if A is open and bounded, then this definition coincides with the one from Step 2. It can be shown (invoking Zorn's Lemma) that there exist bounded subsets $A \subset \mathbb{R}$ that are not measurable; see [Apostol (1974), Exercise 10.36, page 304].

Step 4: Measurable sets. Let $A \subset \mathbb{R}$ be an arbitrary set. We call A *measurable* if for every compact set $K \subset \mathbb{R}$, the bounded set $A \cap K$ is measurable, and we define the *measure* $\lambda(A)$ *of* A by
$$\lambda(A) := \sup_{K \text{ compact}} \lambda(A \cap K).$$

If A is bounded, then this definition coincides with the one from Step 3.

This is how the (Lebesgue) measure $\lambda(A)$ is defined for (Lebesgue) measurable subsets A of $\mathbb{R}$. The following result can be shown.

Theorem 7.1. (Properties of measurable sets).

(1) *If A is measurable, then $\mathbb{R}\setminus A$ is also measurable.*
(2) *Let A be measurable and $x \in \mathbb{R}$. Set*
$$x + A := \{x + a : a \in A\}, \text{ and}$$
$$xA := \{xa : a \in A\}.$$
Then $x + A$ and xA are measurable, and
$$\lambda(x + A) = \lambda(A), \text{ and}$$
$$\lambda(xA) = |x|\lambda(A).$$

(3) *If A_1, A_2 are measurable and $A_1 \subset A_2$, then $\lambda(A_1) \leqslant \lambda(A_2)$.*

Now suppose that $(A_n)_{n \in \mathbb{N}}$ is a sequence of measurable sets.

(4) $\bigcup_{n \in \mathbb{N}} A_n$ *is measurable, and* $\lambda\left(\bigcup_{n \in \mathbb{N}} A_n\right) \leqslant \sum_{n=1}^{\infty} \lambda(A_n)$.

If $A_i \cap A_j = \emptyset$ whenever $i \neq j$, then $\lambda\left(\bigcup_{n \in \mathbb{N}} A_n\right) = \sum_{n=1}^{\infty} \lambda(A_n)$.

If $A_1 \subset A_2 \subset A_3 \subset \cdots$, then $\lambda\left(\bigcup_{n \in \mathbb{N}} A_n\right) = \sup_{n \in \mathbb{N}} \lambda(A_n)$.

(5) $\bigcap_{n \in \mathbb{N}} A_n$ *is measurable.*

Sets of measure 0. Sets of measure zero play an important role in measure theory (for example, they underly the notions of "almost everywhere" and "for almost all", as we shall see). Examples of sets A with $\lambda(A) = 0$ are:

(1) $A = \{a\}$, a singleton, because A is then an interval in $\mathbb{R}$, with
$$\lambda(A) = a - a = 0.$$

(2) $A = \{a_1, a_2, a_3, \cdots\} = \bigcup_{n \in \mathbb{N}} \{a_n\}$, a countable set. Then
$$\lambda(A) = \sum_{n=1}^{\infty} \lambda(\{a_n\}) = \sum_{n=1}^{\infty} 0 = 0.$$

We remark that there are uncountable sets with Lebesgue measure 0, for example, the standard Cantor set, recalled below.

Example 7.1. (Cantor set revisited). Recall the Cantor set C from Example 1.25 on page 47. Let us show that C is uncountable.

We will prove that there is a one-to-one correspondence between points of C and the points of $[0, 1]$. First note that any point x in C is associated with a sequence of letters "L" or "R" as follows. Let $x \in C$. Then for any n, $x \in F_n$, and when the middle thirds of each subinterval in F_n is removed, x is present either in the left part or the right part of the subinterval, and the nth term in the sequence of letters is L or R accordingly. For example, the points

$$0 \equiv \text{L,L,L,L,L,L}, \cdots,$$

$$1 \equiv \text{R,R,R,R,R,R}, \cdots,$$

$$\frac{1}{3} \equiv \text{L,R,R,R,R,R}, \cdots,$$

$$\frac{2}{9} \equiv \text{L,R,L,L,L,L}, \cdots,$$

$$\frac{20}{27} \equiv \text{R,L,R,L,L,L}, \cdots.$$

But points in $[0, 1]$ are also in one-to-one correspondence with such sequences. Indeed,

$$[0,1] = \left[0, \tfrac{1}{2}\right] \cup \left(\tfrac{1}{2}, 1\right]$$
$$= \left[0, \tfrac{1}{4}\right] \cup \left(\tfrac{1}{4}, \tfrac{1}{2}\right] \cup \left(\tfrac{1}{2}, \tfrac{3}{4}\right] \cup \left(\tfrac{3}{4}, 1\right]$$
$$= \left[0, \tfrac{1}{8}\right] \cup \left(\tfrac{1}{8}, \tfrac{1}{4}\right] \cup \left(\tfrac{1}{4}, \tfrac{3}{8}\right] \cup \left(\tfrac{3}{8}, \tfrac{1}{2}\right]$$
$$\cup \left(\tfrac{1}{2}, \tfrac{5}{8}\right] \cup \left(\tfrac{5}{8}, \tfrac{3}{4}\right] \cup \left(\tfrac{3}{4}, \tfrac{7}{8}\right] \cup \left(\tfrac{7}{8}, 1\right]$$
$$\cdots.$$

If $x \in [0, 1]$, then for each n, we can look at the nth equality, and see if x falls in the left or the right part of the new subintervals created when the each subinterval on the right-hand side of the nth equality is divided into two parts, and this gives the $(n + 1)$st term of the sequence of Ls and Rs associated with x: for example,

$$0 \equiv L,L,L,L,L,L, \cdots,$$

$$1 \equiv R,R,R,R,R,R, \cdots,$$

$$\frac{1}{2} \equiv L,R,R,R,R,R, \cdots.$$

As $[0,1]$ is uncountable, it follows that so is C.

In the construction of C, since the sum of the lengths of the intervals removed is

$$\frac{1}{3} + 2\frac{1}{3^2} + 4\frac{1}{3^3} + \cdots = 1,$$

(factor out $\frac{1}{3}$ and sum the resulting geometric series), the measure of F is $1 - 1 = 0$. So this is an example of an uncountable set with measure 0. ◇

Any subset of a measurable set of measure 0 is also measurable with measure 0. We say that two functions $\mathbf{x}_1, \mathbf{x}_2 : A \to \mathbb{R}$ defined on a measurable set A are equal *almost everywhere* if there exists a measurable set N with $\lambda(N) = 0$ such that $\mathbf{x}_1(t) = \mathbf{x}_2(t)$ for all $t \in A \backslash N$. Sometimes then we also say that $\mathbf{x}_1(t) = \mathbf{x}_2(t)$ for *almost all* $t \in A$.

Measurable functions

Let A be a measurable subset of $\mathbb{R}$. A function $\mathbf{x} : A \to \mathbb{R} \bigcup \{-\infty, +\infty\}$ is called *measurable* if $\mathbf{x}$ has any of the following equivalent properties:

(M1) For all $y \in \mathbb{R}$, $\{t \in A : \mathbf{x}(t) < y\}$.
(M2) For all $y \in \mathbb{R}$, $\{t \in A : \mathbf{x}(t) \leq y\}$.
(M3) For all $y \in \mathbb{R}$, $\{t \in A : \mathbf{x}(t) > y\}$.
(M4) For all $y \in \mathbb{R}$, $\{t \in A : \mathbf{x}(t) \geq y\}$.

Practically *all* functions are measurable, and they are abundant:

(1) All continuous functions are measurable.
(2) All functions that are continuous outside a set of measure 0.
    For example

$$\mathbf{x}(t) = \begin{cases} 1 & \text{if } t = \dfrac{1}{n\pi}, \ n \in \mathbb{Z} \backslash \{0\}, \text{ or } t = 0, \\ \dfrac{1}{\sin(1/t)} & \text{otherwise}, \end{cases}$$

is measurable.
Such functions are called *continuous almost everywhere*.

(3) All monotone functions are measurable.
(4) If A is a measurable set, then its indicator function $\mathbf{1}_A$, given by

$$\mathbf{1}_A(t) = \begin{cases} 1 \text{ if } t \in A, \\ 0 \text{ if } t \notin A, \end{cases}$$

is a measurable function. Indeed, we have

$$\{t \in \mathbb{R} : \mathbf{1}_A(t) \geq y\} = \begin{cases} \mathbb{R} \text{ if } y \leq 0, \\ A \text{ if } 0 < y \leq 1, \\ \varnothing \text{ if } y > 1. \end{cases}$$

(5) The sum, product and (if well-defined) the quotient of measurable functions are all measurable.
(6) If $\mathbf{x}$ is measurable, then so is $|\mathbf{x}|$.
Hence[1] if $\mathbf{x}_1, \mathbf{x}_2$ are measurable, then $\max\{\mathbf{x}_1, \mathbf{x}_2\}$ and $\min\{\mathbf{x}_1, \mathbf{x}_2\}$ are also measurable.
(7) If $(\mathbf{x}_n)_{n \in \mathbb{N}}$ is a sequence of measurable functions, which converges pointwise to a function $\mathbf{x}$, then $\mathbf{x}$ is measurable.

The integral of measurable functions

While defining the *Riemann* integral, we consider upper and lower sums corresponding to a partition $P = \{a = t_0, t_1, \cdots, t_{n-1}, t_n = b\}$ of the domain $[a, b]$ of the function $\mathbf{x}$, for example the lower sum

$$\underline{S}(\mathbf{x}, P) = \sum_{k=0}^{n-1} \left(\inf_{t \in [t_k, t_{k+1}]} \mathbf{x}(t) \right) (t_{k+1} - t_k).$$

The above is really the Riemann integral of a *step function*, which assumes finitely many values, and is constant on intervals.

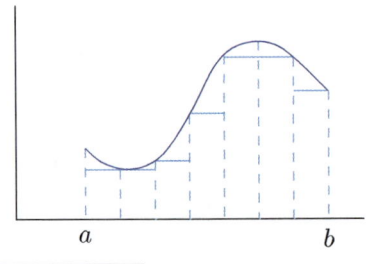

---
[1] For real a, b, $\max\{a, b\} = (a + b + |a - b|)/2$, and $\min\{a, b\} = a + b - \max\{a, b\}$.

While defining the *Lebesgue* integral, we shall consider *simple* functions. A simple function assumes finitely many values (just as before, with step functions), but now is constant (more generally than the case of step functions) on *measurable* sets (instead of mere intervals).

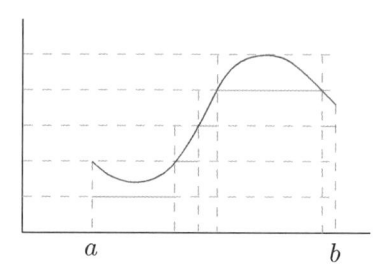

Roughly speaking, such simple functions arise from a partition of the *range* (rather than a partition of the *domain* for the step functions considered when defining the Riemann integral).

Every step function is a simple function (since every interval is measurable), but not every simple function is a step function (because not every measurable set is an interval).

Now let A be a measurable set, and let $\mathbf{s} : A \to \mathbb{R}$ be a simple function. This means that $\mathbf{s}$ assumes finitely many values, which we arrange in creasing order:
$$-\infty < y_1 < y_2 < \cdots < y_n < \infty,$$
and let $A_k = \{t \in A : \mathbf{s}(t) = y_k\}$, $1 \leqslant k \leqslant n$. Thus we may write
$$\mathbf{s} = y_1 \cdot \mathbf{1}_{A_1} + \cdots + y_n \cdot \mathbf{1}_{A_n}. \tag{7.1}$$
If $\mathbf{s}(t) \geqslant 0$ for all $t \in A$, then $y_1 \geqslant 0$, and in this case, we define[2]
$$\int_A \mathbf{s}(t)dt := y_1 \cdot \lambda(A_1) + \cdots + y_n \lambda(A_n).$$
The right-hand side is either a nonnegative real number or ∞ (if one of the A_k's has infinite measure).

The collection of all nonnegative simple functions on A is denoted by $S_+(A)$. For each $\mathbf{s} \in S_+(A)$, we have defined
$$\int_A \mathbf{s}(t)dt.$$

---

[2]We will write $\int_A$ instead of $\int_{A}_{\text{Lebesgue}}$.

If A is a set of measure 0, then for all $\mathbf{s} \in S_+(A)$,
$$\int_A \mathbf{s}(t)dt = 0.$$
Indeed, since every subset of a set of measure 0 is also a measurable set of measure 0, it follows, with the notation from (7.1), that $\lambda(A_k) = 0$ for all $1 \leq k \leq n$. The claim follows by the definition of the integral.

Now, let $\mathbf{x} : A \to \mathbb{R} \bigcup \{-\infty, \infty\}$ be a measurable function, and suppose that $\mathbf{x}(t) \geq 0$ for all $t \in A$. Then we define
$$\int_A \mathbf{s}(t)dt = \sup_{\substack{\mathbf{s} \in S_+(A) \\ \mathbf{s} \leq x}} \int_A \mathbf{s}(t)dt.$$
The right-hand side is either a nonnegative real number or $+\infty$. In this sense, we can say that for nonnegative measurable functions, their Lebesgue integral always exists, but this is not the case with Riemann integrals.

Example 7.2. Let $A = [0,1]$, and let $\mathbf{x} : [0,1] \to \mathbb{R}$ be defined by
$$\mathbf{x}(t) = \begin{cases} 1 & \text{if } t \text{ is irrational,} \\ 0 & \text{if } t \text{ is rational.} \end{cases}$$
The sets
$$A_0 = \{t \in [0,1] : \mathbf{x}(t) = 0\} \text{ and}$$
$$A_1 = \{t \in [0,1] : \mathbf{x}(t) = 1\}$$
are measurable. Since A_0 is countable, $\lambda(A_0) = 0$. On the other hand,
$$\lambda(A_1) = \lambda(A \backslash A_0) = \lambda(A) - \lambda(A_0) = 1 - 0 = 1.$$
Since $\mathbf{x} = \mathbf{1}_{A_1}$ is a simple function,
$$\int_A \mathbf{x}(t)dt = 1 \cdot \lambda(A_1) = 1.$$
But $\mathbf{x}$ is not Riemann integrable, as shown below. For every partition
$$P = \{0 = t_0, t_1, \cdots, t_{n-1}, t_n = 1\},$$
we have that each subinterval $[t_k, t_{k+1}]$ contains both rational as well as irrational points, and this observation shows that
$$\overline{S}(\mathbf{x}, P) = \sum_{k=0}^{n-1} \Big(\sup_{t \in [t_k, t_{k+1}]} \mathbf{x}(t)\Big)(t_{k+1} - t_k) = \sum_{k=0}^{n-1} 1 \cdot (t_{k+1} - t_k) = 1 - 0 = 1,$$
$$\underline{S}(\mathbf{x}, P) = \sum_{k=0}^{n-1} \Big(\inf_{t \in [t_k, t_{k+1}]} \mathbf{x}(t)\Big)(t_{k+1} - t_k) = \sum_{k=0}^{n-1} 0 \cdot (t_{k+1} - t_k) = 0.$$
Thus $\overline{S}(\mathbf{x}, P) - \underline{S}(\mathbf{x}, P) = 1$ for *all* partitions P, and so $\mathbf{x}$ is not Riemann integrable. ◇

Let A be a measurable set. Suppose that all the functions appearing in the list below are defined on A, take values in $[0, \infty) \cup \{+\infty\}$, and are measurable. Then we have:

(1) $\int_A \big(\mathbf{x}_1(t) + \mathbf{x}_2(t)\big) dt = \int_A \mathbf{x}_1(t) dt + \int_A \mathbf{x}_2(t) dt.$

(2) For $\alpha \geq 0$, $\int_A \alpha \mathbf{x}(t) dt = \alpha \int_A \mathbf{x}(t) dt.$

(3) If for all $t \in A$, $\mathbf{x}_1(t) \leq \mathbf{x}_2(t)$, then $\int_A \mathbf{x}_1(t) dt \leq \int_A \mathbf{x}_2(t) dt.$

(4) (Monotone Convergence Theorem). If $0 \leq \mathbf{x}_1(t) \leq \mathbf{x}_2(t) \leq \cdots$, and

$$\mathbf{x}(t) := \lim_{n \to \infty} \mathbf{x}_n(t).$$

Then $\int_A \mathbf{x}(t) dt = \lim_{n \to \infty} \int_A \mathbf{x}_n(t) dt.$

(5) If $\lambda(A) = 0$, then $\int_A \mathbf{x}(t) dt = 0.$

(6) If $\int_A \mathbf{x}(t) dt < \infty$, then there exists a set N of measure zero such that $\mathbf{x}(t) < \infty$ for all $t \in A \setminus N$.

Now let $\mathbf{x} : A \to \mathbb{R} \cup \{-\infty, +\infty\}$ be a measurable function defined on the measurable set A. Note that $\mathbf{x}$ is no longer assumed to be nonnegative. We can, nevertheless, write $\mathbf{x}$ as a *difference*, $\mathbf{x} = \mathbf{x}_+ - \mathbf{x}_-$, of the two nonnegative (and measurable) functions $\mathbf{x}_+ := \max\{\mathbf{x}, \mathbf{0}\}$ and $\mathbf{x}_- := \max\{-\mathbf{x}, \mathbf{0}\} = -\min\{\mathbf{x}, \mathbf{0}\}$. See the following picture.

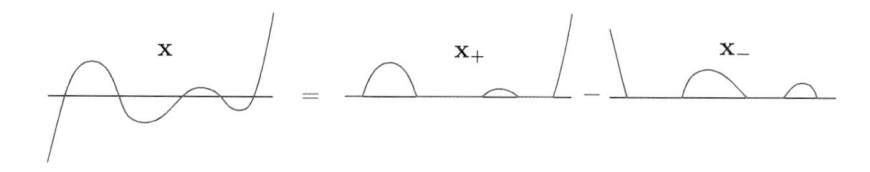

We say that $\mathbf{x}$ is *(absolutely) integrable* on A if

$$\int_A |\mathbf{x}(t)| dt < \infty, \qquad (7.2)$$

and, in this case, we define

$$\int_A \mathbf{x}(t) dt := \int_A \mathbf{x}_+(t) dt - \int_A \mathbf{x}_-(t) dt.$$

We note that since $0 \leqslant \mathbf{x}_\pm(t) \leqslant |\mathbf{x}(t)|$, and thanks to assumption (7.2), it follows from item (3) from the list of properties of the integrals of nonnegative measurable functions given on page 367, that
$$\int_A \mathbf{x}_+(t)dt, \int_A \mathbf{x}_-(t)dt < \infty,$$
and so their difference, $\int_A \mathbf{x}(t)dt$, is finite too.

We'll denote the set of all absolutely integrable functions on A by $\mathcal{L}^1(A)$. For $\mathbf{x}_1, \mathbf{x}_2, \mathbf{x} \in \mathcal{L}^1(A)$ and $\alpha \in \mathbb{R}$, we have the following:

(1) $\mathbf{x}_1 + \mathbf{x}_2 \in \mathcal{L}^1(A)$ and $\int_A (\mathbf{x}_1 + \mathbf{x}_2)(t)dt = \int_A \mathbf{x}_1(t)dt + \int_A \mathbf{x}_2(t)dt$.

(2) $\alpha \cdot \mathbf{x} \in \mathcal{L}^1(A)$ and $\int_A (\alpha \cdot \mathbf{x})(t)dt = \alpha \int_A \mathbf{x}(t)dt$.

(3) $|\mathbf{x}| \in \mathcal{L}^1(A)$ and $\int_A |\mathbf{x}(t)|dt \leqslant \left|\int_A \mathbf{x}(t)dt\right|$.

(4) If $\int_A |\mathbf{x}(t)|dt = 0$, then there exists a set $N \subset A$ of measure 0 such that $\mathbf{x}(t) = 0$ for all $t \in A\backslash N$.

(5) Let $A = B\bigcup C$, where B, C are measurable too and $B\bigcap C = \emptyset$. Then $\mathbf{x} \in \mathcal{L}^1(B)$, $\mathbf{x} \in \mathcal{L}^1(C)$ and $\int_A \mathbf{x}(t)dt = \int_B \mathbf{x}(t)dt + \int_C \mathbf{x}(t)dt$.

(6) If $\mathbf{y} : A \to \mathbb{R}\bigcup\{-\infty, +\infty\}$ is measurable and
$$|\mathbf{y}(t)| \leqslant \mathbf{x}(t) \text{ for almost all } t \in A,$$
then $\mathbf{y} \in \mathcal{L}^1(A)$ and $\left|\int_A \mathbf{y}(t)dt\right| \leqslant \int_A |\mathbf{y}(t)|dt \leqslant \int_A \mathbf{x}(t)dt$.

The parts (1), (2) assert that $\mathcal{L}^1(A)$ is a real vector space, and the integral
$$\mathbf{x} \mapsto \int_A \mathbf{x}(t)dt : \mathcal{L}^1(A) \to \mathbb{R}$$
is a linear transformation (or a linear functional, since the co-domain is the field of scalars $\mathbb{R}$).

We also remark that in part (4), under the given hypothesis, we cannot in general conclude that $\mathbf{x} \equiv 0$ on *all* of A. Indeed,
$$\int_0^1 \mathbf{1}_{\mathbb{Q}\cap[0,1]}(t)dt = \lambda(\mathbb{Q} \cap [0,1]) = 0,$$
as $\mathbb{Q}$ is countable, however the integrand is not identically zero: for example, its value at $1/\sqrt{2}$ is 1. On the other hand, if in (4), we are also given that $\mathbf{x}$ is continuous, then we can safely conclude that $\mathbf{x} \equiv 0$ on A.

The Dominated Convergence Theorem. The Dominated Convergence Theorem says that if all the terms $\mathbf{x}_n$ in a sequence of functions has an $\mathcal{L}^1$-majorant, then assuming that their pointwise limit $\mathbf{x}$ exists almost everywhere, is also an element of $\mathcal{L}^1(A)$.

(Dominated Convergence Theorem). Let A be measurable. Let $(\mathbf{x}_n)_{n \in \mathbb{N}}$ be a sequence in $\mathcal{L}^1(A)$, and $\mathbf{x}: A \to \mathbb{R}$, be such that
$$\lim_{n\to\infty} \mathbf{x}_n(t) = \mathbf{x}(t) \text{ for almost all } t \in A.$$
Suppose that $\mathbf{y} \in \mathcal{L}^1(A)$ is such that for all $n \in \mathbb{N}$,
$$|\mathbf{x}_n(t)| \leqslant \mathbf{y}(t) \text{ for almost all } t \in [0,1].$$
Then $\mathbf{x} \in \mathcal{L}^1(A)$, and $\lim_{n\to\infty} \int_A \mathbf{x}_n(t) dt = \int_A \lim_{n\to\infty} \mathbf{x}_n(t) dt = \int_A \mathbf{x}(t) dt.$

We remark that the hypothesis of the existence of an $\mathcal{L}^1$ majorant is essential, as demonstrated by the following two examples.

Example 7.3. (Lacking an $\mathcal{L}^1$ majorant). Let $A = \mathbb{R}$.
(1) Let $\mathbf{x}_n = \mathbf{1}_{[-n,n]}$, $\mathbf{x} = 1$. Then $\mathbf{x}_n \in \mathcal{L}^1(\mathbb{R})$, $(\mathbf{x}_n)_{n\in\mathbb{N}}$ converges pointwise everywhere on $\mathbb{R}$ to $\mathbf{x}$, but $\mathbf{x} \notin \mathcal{L}^1(\mathbb{R})$.

(2) Let $\mathbf{x}_n = \mathbf{1}_{[n,n+1]}$, $\mathbf{x} = 0$. Then $\mathbf{x}_n \in \mathcal{L}^1(\mathbb{R})$, $(\mathbf{x}_n)_{n\in\mathbb{N}}$ converges pointwise everywhere on $\mathbb{R}$ to $\mathbf{x}$, but
$$\int_\mathbb{R} \mathbf{x}(t) dt = 0 \neq 1 = \lim_{n\to\infty} \int_\mathbb{R} \mathbf{x}_n(t) dt. \qquad \diamond$$

Link with the Riemann integral. Let $\mathbf{x} \in C[a,b]$. Then $\mathbf{x} \in \mathcal{L}^1[a,b]$ and
$$\int_a^b \mathbf{x}(t) dt = \int_{a\ \mathrm{Riemann}}^b \mathbf{x}(t) dt.$$
That $\mathbf{x} \in \mathcal{L}^1[a,b]$ follows from the fact that $\mathbf{x}$ is measurable (since it is continuous), and it is bounded (Extreme Value Theorem).

The L^p spaces. We will just consider the case $p = 1$, and with $A = [0,1]$. More general situations can be handled in an analogous manner. Consider on $\mathcal{L}^1[0,1]$ the candidate for the norm
$$\|\mathbf{x}\|_1 := \int_0^1 |\mathbf{x}(t)| dt, \quad \mathbf{x} \in \mathcal{L}^1[0,1].$$
This map $\|\cdot\|_1$ fails to be a norm because functions that are almost everywhere 0 (e.g. $\mathbf{1}_{\mathbb{Q}\cap[0,1]}$) have zero norm. Hence we should essentially

"consider such functions to be also the zero vector in the vector space $\mathcal{L}^1[0,1]$". This intuitive remark can be made rigorous by considering the following relation on $\mathcal{L}^1[0,1]$. We say that

$$\mathbf{x} \sim \mathbf{y}$$

if there exists a set[3] $N \subset [0,1]$ of measure 0, such that

$$\mathbf{x}(t) = \mathbf{y}(t) \text{ for all } t \in [0,1] \backslash N.$$

It can be seen that $\sim$ is an equivalence relation on $\mathcal{L}^1[0,1]$, that is,

(ER1) (*Reflexivity*). $\mathbf{x} \sim \mathbf{x}$ for all $\mathbf{x} \in \mathcal{L}^1[0,1]$.
(ER2) (*Symmetry*). If $\mathbf{x}, \mathbf{y} \in \mathcal{L}^1[0,1]$ and $\mathbf{x} \sim \mathbf{y}$, then $\mathbf{y} \sim \mathbf{x}$.
(ER3) (*Transitivity*). If $\mathbf{x}, \mathbf{y}, \mathbf{z} \in \mathcal{L}^1[0,1]$, $\mathbf{x} \sim \mathbf{y}$ and $\mathbf{y} \sim \mathbf{z}$, then $\mathbf{x} \sim \mathbf{z}$.

Let $[\mathbf{x}]$ denote the equivalence class of $\mathbf{x}$:

$$[\mathbf{x}] = \{\mathbf{y} \in \mathcal{L}^1[0,1] : \mathbf{x} \sim \mathbf{y}\}.$$

Thus $[\mathbf{x}]$ is the collection of all elements of $\mathcal{L}^1[0,1]$ that are almost everywhere equal to $\mathbf{x}$ on $[0,1]$. Define

$$L^1[0,1] := \{[\mathbf{x}] : \mathbf{x} \in \mathcal{L}^1[0,1]\}.$$

Then we can endow a vector space structure on $L^1[0,1]$ by setting

$$[\mathbf{x}] + [\mathbf{y}] = [\mathbf{x} + \mathbf{y}],$$
$$\alpha \cdot [\mathbf{x}] = [\alpha \cdot \mathbf{x}],$$

for $[\mathbf{x}], [\mathbf{y}] \in \mathcal{L}^1[0,1]$ and $\alpha \in \mathbb{R}$. It can also be seen that the above operations $+, \cdot$ are well-defined, that is they do not depend on the chosen representatives $\mathbf{x}, \mathbf{y} \in \mathcal{L}^1[0,1]$ for the equivalence classes $[\mathbf{x}], [\mathbf{y}]$, respectively.

We now define the map $\|\cdot\|_1 : L^1[0,1] \to \mathbb{R}$ by

$$\|[\mathbf{x}]\|_1 := \int_0^1 \mathbf{x}(t) dt, \quad [\mathbf{x}] \in L^1[0,1].$$

Then it can be checked that $\|\cdot\|_1$ defines a norm on $L^1[0,1]$. In particular, now if $\|[\mathbf{x}]\|_1 = 0$, then it follows that $\mathbf{x}(t) = 0$ for almost all $t \in [0,1]$, and so $[\mathbf{x}] = [\mathbf{0}]$, that is, $[\mathbf{x}]$ is the zero vector from the vector space $L^1[0,1]$, as desired.

We recall that we had mentioned that $L^1[0,1]$ is complete, and in this sense the space $L^1[0,1]$ was "better" than $C[0,1]$. We supply a sketch of the proof below. Suppose that $([\mathbf{x}_n])_{n \in \mathbb{N}}$ is a Cauchy sequence in $L^1[0,1]$.

---
[3]Depending in general on $\mathbf{x}$ and $\mathbf{y}$.

In order to prove its convergence, it is enough to show the convergence of a subsequence. Hence we may assume (by passing to a subsequence if necessary) that
$$\|[\mathbf{x}_{n+1}] - [\mathbf{x}_n]\|_1 < \frac{1}{2^n}, \quad n \in \mathbb{N}.$$
Let $\mathbf{x}_0 := \mathbf{0}$, and set
$$\mathbf{y}_n(t) := \sum_{k=0}^{n} |\mathbf{x}_{k+1}(t) - \mathbf{x}_k(t)|,$$
$$\mathbf{y}(t) := \sum_{k=0}^{\infty} |\mathbf{x}_{k+1}(t) - \mathbf{x}_k(t)|.$$
By the Triangle Inequality, we have
$$\|[\mathbf{y}_n]\|_1 = \int_0^1 |\mathbf{y}_n(t)| dt \leq \sum_{k=0}^{n} \|[\mathbf{x}_{k+1}] - [\mathbf{x}_k]\|_1 \leq \|[\mathbf{x}_1] - [\mathbf{x}_0]\|_1 + \sum_{k=1}^{n} \frac{1}{2^k}.$$
By the Monotone Convergence Theorem,
$$\|[\mathbf{y}]\|_1 = \int_0^1 |\mathbf{y}(t)| dt = \lim_{n \to \infty} \int_0^1 |\mathbf{y}_n(t)| dt \leq \|[\mathbf{x}_1]\|_1 + 1 < \infty.$$
Hence the function $\mathbf{y}$ is finite almost everywhere on $[0, 1]$. So the series
$$\sum_{k=0}^{\infty} (\mathbf{x}_{k+1}(t) - \mathbf{x}_k(t))$$
is absolutely convergent for almost all $t \in [0, 1]$. For these t's, we set
$$\mathbf{x}(t) = \sum_{k=0}^{\infty} (\mathbf{x}_{k+1}(t) - \mathbf{x}_k(t)).$$
But $\mathbf{x}_n(t) = \sum_{k=0}^{n-1} (\mathbf{x}_{k+1}(t) - \mathbf{x}_k(t))$ for all $t \in [0, 1]$, and so
$$\lim_{n \to \infty} \mathbf{x}_n(t) = \mathbf{x}(t).$$
Furthermore, $|\mathbf{x}_n(t)| \leq \sum_{k=0}^{n-1} |\mathbf{x}_{k+1}(t) - \mathbf{x}_k(t)| \leq \mathbf{y}(t)$ for almost all $t \in [0, 1]$.
By the Dominated Convergence Theorem,
$$\int_0^1 |\mathbf{x}(t)| dt = \lim_{n \to \infty} \int_0^1 |\mathbf{x}_n(t)| dt \leq \int_0^1 \mathbf{y}(t) dt < \infty.$$
Hence $[\mathbf{x}] \in L^1[0, 1]$. Also, note that $[|\mathbf{x}|] + [\mathbf{y}] \in L^1[0, 1]$, and furthermore $|\mathbf{x} - \mathbf{x}_n| \leq |\mathbf{x}| + \mathbf{y}$ for all n. The Dominated Convergence Theorem again gives
$$\lim_{n \to \infty} \|[\mathbf{x}_n] - [\mathbf{x}]\|_1 = \lim_{n \to \infty} \int_0^1 |\mathbf{x}_n(t) - \mathbf{x}(t)| dt = 0,$$
showing that $([\mathbf{x}_n])_{n \in \mathbb{N}}$ converges to $[\mathbf{x}]$ in $(L^1[0, 1], \|\cdot\|_1)$.
Consequently, $(L^1[0, 1], \|\cdot\|_1)$ is complete.

Notes

The path to defining the Lebesgue measure we have adopted, stems from the notes on Integration Theory by Erik Thomas, University of Groningen [Thomas (1998)], [Dijksma (1997)]. This is equivalent to alternative standard definitions of the Lebesgue measure used in the literature, for example, the Caratheodory approach, giving the same set of measurable sets as well as the measure. The author is grateful to Raymond Mortini, University of Metz, for pointing this equivalence out.

The proof of the completeness of $L^1[0,1]$ is based on [Limaye (1996), Theorem 4.6, page 51].

Bibliography

Apostol, T. (1974). *Mathematical Analysis, Second Edition* (Narosa Publishing House).
Bremermann, H. (1965). *Distributions, Complex Variables, and Fourier Transforms* (Addison-Wesley).
Bryant, V. (1990). *Yet Another Introduction to Analysis* (Cambridge University Press).
Day, M.M. (1973). *Normed Linear Spaces, Third Edition* (Springer-Verlag).
Dijksma, A. (1997). *Diktaat bij Wiskund 4, 5 en 6 Natuurkunde en Sterrenkunde* (Rijksuniversiteit Groningen, Vakgroep Wiskunde).
Hirsch, M. and Smale, S. (1974). *Differential Equations, Dynamical Systems, and Linear Algebra* (Academic Press).
Hörmander, L. (1990). *The Analysis of Linear Partial Differential Operators I, Second Edition* (Springer).
Kreyszig, E. (1978). *Introductory Functional Analysis with Applications* (Wiley).
Limaye, B. (1996). *Functional Analysis, Second Edition* (New Age International Limited Publishers).
Luenberger, D. (1969). *Optimization by Vector Space Methods* (John Wiley).
Neuenschwander, D. (2011). *Emmy Noether's Wonderful Theorem* (Johns Hopkins University Press).
Pedersen, G. (1989). *Analysis Now.* (Springer).
Pinchover, Y. and Rubinstein, J. (2005). *An Introduction to Partial Differential Equations* (Cambridge University Press).
Rudin, W. (1976). *Principles of Mathematical Analysis, Third Edition* (McGraw-Hill).
Rudin, W. (1991). *Functional Analysis, Third Edition* (McGraw-Hill).
Singh, D. (2006). The Spectrum in a Banach Algebra. *American Mathematical Monthly* **113**, 8, pp. 756–758.
Sasane, A. (2015). *The How and Why of One Variable Calculus* (Wiley).
Sasane, A. (2016). *Optimization in Function Spaces* (Dover).
Schwartz, L. (1966). *Mathematics for the Physical Sciences* (Addison-Wesley).
Smith, K. (1983). *Primer of Modern Analysis* (Springer).
Steen, L.A. and Seebach, J.A. (1995). *Counterexamples in Topology, Second Edi-*

tion (Dover).
Taylor, A. and Lay, D. (1980). *Introduction to Functional Analysis, Second Edition* (John Wiley).
Thomas, E.G.F. (1996). *Distributietheorie* (Rijksuniversiteit Groningen).
Thomas, E.G.F. (1997). *Lineaire Analyse* (Rijksuniversiteit Groningen).
Thomas, E.G.F. (1998). *Integraalrekening* (Rijksuniversiteit Groningen).

Index

$B(x,r)$, 18
$C(X,Y)$, 67
$CL(X)$, 67
$CL(X,Y)$, 67
$C[a,b]$, 3
$C^1[a,b]$, 4, 44
$C^n[a,b]$, 17
$D^{\mathbf{k}}$, 231
$H(x)$, 234
$H^n(a,b)$, 242
$K(X,Y)$, 211
$L(X)$, 57
$L(X,Y)$, 57
$L^1_{\text{loc}}(\mathbb{R}^d)$, 233
$L^\infty[a,b]$, 7, 14
$L^p(\Omega)$, 14
$L^p[a,b]$, 5, 14
$O(1,1)$, 50
$O(2)$, 50
$[\cdot,\cdot]$, 84
$\mathcal{D}'(\mathbb{R}^d)$, 233
$\mathcal{D}(\mathbb{R}^d)$, 230
$\mathcal{L}^1(A)$, 368
$\mathcal{S}'(\mathbb{R})$, 255
$\mathcal{S}(\mathbb{R})$, 254
ℓ^∞, 4
ℓ^p, 4
$\mathbb{S}^2$, 21
$\mathbb{S}^{d-1}$, 50
$\{\cdot,\cdot\}$, 152
c, 5
c_{00}, 5, 21, 43

c_0, 5
p-adic norm, 17

absolute time, 141
absolutely convergent series, 37
absolutely integrable function, 367
action, 146
adjoint operator, 108, 190
algebra, 83
almost all, 363
almost everywhere, 363
Appollonius identity, 162
approximate spectrum, 103
arc length, 61
Archimedean property, 21
averaging operator, 77
Axiom of Choice, 113

Baire Lemma, 93
Banach algebra, 83
Banach limit, 116
Banach space, 27
Banach, Stefan, 27
Banach-Steinhauss Theorem, 97
Bernstein polynomials, 22
Bernstein, Sergei, 22
Besicovitch almost periodic function, 188
Bessel's inequality, 172
bidual of a normed space, 110
Bolzano-Weierstrass Theorem, 30
bounded above, 112

bounded linear operator, 73
bounded variation, 105

Cantor set, 47
Cauchy sequence, 26
Cauchy-Schwarz inequality, 157
Cayley transform, 199
Cesáro summation, 77
chain, 112
chain rule, 125
classical mechanics, 145
closed ball, 20
Closed Graph Theorem, 96
closed interval, 20
closed set, 20, 43
closed unit ball, 16
closure (of a set), 43
coarser topology, 79
commutant, 252
commutator, 84, 153
compact operator, 103, 210
compact set, 45, 51
complementary observables, 205
complete normed space, 27
completion of an inner product space, 163
composition of maps, 64
composition of operators, 82
conservation of angular momentum, 143
continuity at a point, 60
continuous map, 60
continuously differentiable function, 4
convergence in $\mathcal{D}$, 232
convergence in $\mathcal{S}$, 254
convergent sequence, 24
convex function, 128
convex set, 16, 128
convolution, 257
Convolution Theorem, 256

d'Alembert's formula, 56
delta distribution, 235
delta function, 229
dense set, 21, 43
derivative, 119

diagonal operator, 103, 214
diagonalizable matrix, 91
differentiability at a point, 119
differentiable, 119
diffusion equation, 90
dipole, 238
Dirac delta distribution, 235
Dirac distribution, 103
Dirac, Paul, 229, 251
Dirichlet problem, 86
discrete topology, 79
distance, 8
distribution, 233
distributional derivative, 237
dual operator, 106
dual space, 79, 104
dual space of a normed space, 104

eigenvalue, 98
eigenvector, 98
energy, 148
Enflo, Per, 77
epigraph, 129
equivalence class, 163
equivalence relation, 163
equivalent norms, 33
essential supremum norm, 14
Euclidean norm, 9
Euclidean plane, 141
Euler-Lagrange equation, 134
Extreme Value Theorem, 10

factorial of a multi-index, 249
finite rank operator, 211
Fréchet derivative, 119
Fredholm integral equation, 87
functional analysis, vii
functional, bounded linear, 104
fundamental solution, 241, 242, 251, 252, 257
Fundamental Theorem of ODEs, 39

Galerkin approximation, 219
Galerkin method, 217
Galilean spacetime, 141
Gateux derivative, 122

Gelfand-Beurling formula, 102
generalised Fourier series, 185
Gram-Schmidt orthonormalisation, 167
graph, 96
Green's function, 257

Hölder's inequality, 14, 15
Hahn-Banach Lemma, 111
Hahn-Banach Theorem, 108
Hamel basis, 94, 115, 183
Hamiltonian, 150
Hamiltonian equations, 151
Hamiltonian mechanics, 150
harmonic oscillator, 172
Heaviside function, 234
Heaviside, Oliver, 234
Heisenberg's uncertainty principle, 84, 204
helicoid, 144
Hermite functions, 172
Hermitian operator, 195
Hilbert cube, 48
Hilbert space, 160
Hilbert-Schmidt norm, 76, 163, 196
Hilbert-Schmidt operator, 196, 215
hyperbolic rotation, 50
hypoelliptic differential operator, 252

ideal of an algebra, 216
identity operator, 83
image of a set, 62
index, 20
index set, 20
induced norm, by an inner product, 157
induced norm, in a subspace, 15
initial value problem, 39, 89
inner product, 156
inner product space, 156
integrable function, 367
integral equation, 87
integral operator, 70
invariant subspace, 77
invariant subspace problem, 77
inverse image of a set, 62

invertible operator, 84
isometric isomorphism, 185
isomorphic normed spaces, 96
isomorphism, 96
Isoperimetric Theorem, 187

Jacobi identity, 153
Jordan canonical form, 92
jump rule, 239

Kepler's second law, 143
kernel of an integral operator, 71
kinetic energy, 146

Lagrangian, 146
Lagrangian density, 143
Laplace equation, 171, 241, 242, 257
Laplacian, 242
law of conservation of energy, 148
law of conservation of momentum, 148
least squares approximation problem, 174
least squares regression, 176
Lebesgue integral, 6, 165
left shift operator, 54, 68, 102
Legendre polynomial, 168, 184
Leibniz rule, 249
Leibniz's formula, 169
Lie algebra, 152
limit point of a set, 43
linear isometric embedding, 110
linear transformation, 54
Lipschitz condition, 39
Lipschitz function, 42
locally integrable function, 233

matrix mechanics, 200
maximal element, 113
measurable function, 363
measurable set, 360
measure, 360
metric space, 8
mid-point convexity, 94
minimising sequence, 178
minimum surface area, 144

Minkowski spacetime, 141
module, 250
momentum, 148
multi-index notation, 231
multiplication, of a distribution by a smooth function, 249
multiplicative identity element, 83

Neumann series, 86
Neumann Series Theorem, 86
Neumann, Carl, 86
Newton's second law, 145
Newtonian mechanics, 145
nilpotent matrix, 92
Noether's Theorem, 148
norm, 8
normed algebra, 83
normed space, 8
numerical analysis, 217

observables, 84
ODE, 39
open ball, 17
open cover, 51
open interval, 18
Open Mapping Theorem, 95
open operator, 93
open set, 18
open set in a topological space, 23
open unit ball, 16
operator norm, 74
order of a multi-index, 249
ordinary differential equation, 39
orthogonal, 161, 166
orthogonal complement, 172
orthogonal projection of a vector, 180
orthonormal basis, 183

parallelogram law, 160
Parseval's identity, 186
partial order on a set, 112
partially ordered set, 112
partition of an interval, 105
perturbation theory, 237
phase plane, 151
phase plane trajectory, 151

phase portrait, 151
Poisson bracket, 152
Poissonian mechanics, 152
polarisation formula, 161
position operator, 103, 251
positive operator, 226
potential energy, 145
principle of stationary action, 145
product of normed spaces, 44
projection approximation, 220
proper time, 142
Pythagoras's Theorem, 161

quantum mechanics, 84, 103, 153, 171, 172, 200, 229, 237, 251

reflexive space, 110
regular distribution, 234
relation, 163
representation theorems, 104
resolvent, 99
resolvent identity, 102
Riesz Representation Theorem, 104, 189
Riesz, Frigyes, 106, 189
Riesz, Marcel, 189
right shift operator, 54, 68
Rodrigues's formula, 169

scalar multiplication, 3
Schauder basis, 76
Schrödinger equation, 172
Schwartz space of test functions, 254
seaview property, 28
self-adjoint operator, 195
separable normed space, 21
separable space, 77
simple function, 365
skew-adjoint operator, 195
Sloan approximation, 220
smaller topology, 79
Sobolev spaces, 242
special relativity, 142
Spectral Mapping Theorem, 103
spectral radius, 102

Spectral Theorem for Compact
    Operators, 103
spectrum, 99
sphere, 50
square root operator, 226
state, 151
stationary for a functional,
    curve/function/solution, 134
step function, 364
strong operator topology, 80
stronger topology, 79
subcover, 51
subspace, 4
supremum norm, 10
symmetric set, 94

taxicab norm, 10, 19
tempered distribution, 254
test function, 230
topological space, 23
topology, 23
total variation, 105
triangle inequality, 8, 9, 14
trivial topology, 79
twin paradox, 142

unbounded operator, 103
Uniform Boundedness Principle, 96
uniform operator topology, 80
unit sphere, 21
unitary operator, 195
upper bound, 112

vector addition, 3
vector space, 3

wave equation, 56, 144
wave mechanics, 200
weak operator topology, 81
weaker topology, 79
Weierstrass's Approximation
    Theorem, 22
Weierstrass's Theorem, 66

zero linear transformation, 57
zero vector, 3
Zorn's Lemma, 112, 183